UNDERSTANDING TELECOMMUNICATIONS

Systems, Networks and Applications

Vol. 1

Ming-Chwan Chow

Retired District Manager
AT&T Education Center

Distinguished Member of Technical Staff
Lucent Technologies, Inc.

Andan Publisher
New Jersey

Publisher's Cataloging in Publication Data

Chow, Ming-Chwan
    Understanding Telecommunications: Systems, Networks and Applications/
Ming-Chwan Chow. - - 1st ed.
    v. cm.
    Includes bibliographical references and index.
    LCCN: 99-97227.
    ISBN: 0-9650448-3-1 (Vol. 1)
    ISBN: 0-9650448-8-2 (Vol. 2)

    1. Telecommunications systems.   I. Title.

TK5101.C46 2000   621.382   QBI99-1700

Printed in the United States of America

First edition (2000) by:

Andan Publisher
4 Aufra Place
Holmdel, New Jersey 07733
732-946-4155 (Voice)
732-946-4717 (Fax)

Library of Congress Catalog Card Number: 99-97227

ISBN: 0-9650448-3-1

To my wife, Joann, our children
Anne and Daniel, and their spouses
Robert and Martha, and
the memory of my parents

Books published by Andan Publisher:

- Ming-Chwan Chow, "*Understanding SONET/SDH: Standards and Applications*," 1ˢᵗ ed., 1996, ISBN 0-9650448-2-3

- Ming-Chwan Chow, "*Understanding Wireless: Digital Mobile, Cellular and PCS*," 1ˢᵗ ed., 1998, ISBN 0-9650448-5-8

- Ming-Chwan Chow, "*Understanding Telecommunications: Systems, Networks and Applications (Vol. 1)*," 1ˢᵗ ed., 2000, ISBN 0-9650448-3-1

- Ming-Chwan Chow, "*Understanding Telecommunications: Systems, Networks and Applications (Vol. 2)*," 1ˢᵗ ed., 2000, ISBN 0-9650448-8-2

Contents

Volume 1

CHAPTER 2   Network, Services and Call Connection

CHAPTER 3   Access: Traditional and Broadband - HDSL, ADSL, VDSL & Wireless

CHAPTER 4   Signaling Concepts and Applications

CHAPTER 5   Switching Principles and Applications

CHAPTER 6   Transmission Systems and Applications

CHAPTER 7   Optical Fiber Communications

Answers to Review Questions

References

Abbreviations and Acronyms

Index

Volume 2 (A separate book; brief outline shown as follows)

CHAPTER 10   Network Architectures: LANs and WANs

CHAPTER 11   TCP/IP Voice & Fax over IP (VoIP & FoIP)

CHAPTER 12   Network Management and TMN

CHAPTER 13   Error Control Technologies

CHAPTER 14   Timing and Synchronization

(*)   Further description beyond this level

Preface

Telecommunications services have evolved from point-to-point connections to global interconnections; and the required communications equipment are no longer just the essential telephones (voice terminals) and simple manual switches, but the software controlled integrated network nodes besides the intelligent customer premises equipment. Engineers were the only group of people had to get involve with telecommunications system design, deployment and maintenance. In modern society, even it is essential for the end users to learn about telecommunications to a certain degree so that they can conduct their daily business effectively.

Telecommunications networks have evolved from all analog, to mixtures of analog and digital, finally to modern digital dominated global networks. Telecommunications applications were originally designed for point-to-point services. Now, multicast, broadcast and global interconnection technologies have brought the world closer. The traffic carried by telecommunications networks is no longer just voice (speech) signals, it has been statistically studied/concluded that half of telecommunications traffic is data, including Internet traffic. The research/development of telecommunications networks must thereby be multimedia and ISDN.

An end-to-end communication link can be viewed as the connection of two networks: the access and the backbone (backhaul) networks. In the US, the backbone networks are no longer plesiochronous digital hierarchy (PDH) dominating networks. The networks have been implemented using SONET standards, such as OC-48 (Optical Carrier signal-48) with a data rate of 2.44832 Gbps (a voice capacity of 32,256 channels). The networks of the Internet have also been implemented using SOENT OC-12 signals. For many other countries, SDH STM-16 and STM-4 are used instead of OC-14 and OC-12. The (speed) **bottleneck** of an end-to-end connection is "access networks". Traditional analog (metallic) loops associated with local exchange office are designed to carry one voice signal. Since digital computers was invented in the early 1960s, various modem techniques have been developed for carrying digital data over these analog loops. However, due to the rapid advance of Internet services, broadband access demand is continuously increasing, the present modem technique can no longer meet the increasing speed needs. Therefore, besides intelligent signaling network, high throughput digital switches, and high-speed backbone networks, it is essential to develop broadband access networks. Presently, integrated digital loop carrier systems provided business customers to access the backbone network via OC-3 or OC-12 in North America, and STM-1 and STM-4 in other countries. Cable modems and Hybrid Fiber Coax (HFC) have been deployed for generic applications. In addition, service providers with well-established metallic copper loop plants are seeking for compatible alternate techniques to provide subscribers broadband access capability. Among various techniques, High-speed Digital Subscriber Line

(HDSL), Asymmetrical Digital Subscriber Line (ADSL) and Very high-speed Digital Subscriber Line (VDSL) have potential to be adopted as broadband access standards. HDSL systems may eventually be able to operate on 99% of the copper twisted-pairs that conform to the Carrier Serving Area (CSA) guidelines in the US.

"Understanding Telecommunications: Systems, Networks, and Applications" is a two-volume book. The content of this two-volume is listed as follows:

Volume I: ISBN 0-09650448-3-1:

Volume II: ISBN 0-09650448-8-2:

Through the years, I have taught several telecommunications courses including: digital transmission, optical fiber, wireless, SONET, SDH and ATM courses [at universities, Bell laboratories, AT&T and Lucent Technologies in the USA, and abroad]. The materials within this book comes from my teaching notes. Without valuable insights and feedback from students who have taken my courses, "Understanding Telecommunications" would not be possible. I would like to give my sincere thanks to all my students and to my colleagues at Lucent Technologies,. Special thanks go to Mr. Carl Mason, who has not only carefully reviewed, and edited the book, but also provided valuable information on several topics. A applause to my wife, as usual, she provides her support in the preparation of this book. She skillfully transformed my hand-written notes and graphs into this final form. My thanks also go to the following people who have contributed their time in review of this book: Ms. Anne Chow, Dr. Daniel Chow, Mr. Robert Moore, Ms. Martha Taylor, and Mr. Mike Seidel.

Ming-Chwan Chow

CHAPTER 1

Introduction:
Basic Definitions and Terminologies

Chapter Objectives

Upon the completion of this chapter, you should be able to:

- Describe the signals (amplitude, frequency, wavelength, phase, and bandwidth) used to carry speech, data and video. Distinguish between analog and digital signals, and various types of digital signals.

- Discuss the signaling used for call-associated functions, simplified call processing, dial tone, numbering plans, ringing tones, audible ringing tones, busy tones, audible busy tones, and Common-Channel Signaling (CCS) concept.

- Describe the principles of switching and cross-connection, various switching applications (local, exchange, and gateway), and customer switches.

- Describe transmission facilities (twisted-pair wires, coaxial cables, waveguides, airway, and fibers), transmission system bandwidth, and transmission equipment [e.g., Pulse Code Modulation (PCM), Integrated Digital Loop Carrier (IDLC), Business Remote Terminal (BRT), modems, multiplexers, Add-Drop Multiplexers (ADMs), Digital Cross-connect Systems (DCSs), regenerators, echo cancelers, speech encoders, channel encoders, scramblers, and line encoders].

- Define network architectures [e.g., Public Switched Telephone Network (PSTN), public data network, private data network, Integrated Service Digital Network (ISDN)], and network services (e.g., customer premises services, local exchange services, exchange services, and international services)

- Describe telecommunications standards, and standard organizations (e.g., ITU-T, ITU-R, ISO, ANSI, ECSA, EIA, IEEE, CEPT, ETSI, ECMA, etc.)

1.1 INTRODUCTION

Telecommunications services have evolved from local connection to global inter-connectivity, from voice-only transport to ISDN (Integrated Service Digital Network), from wired (wireline, landline) to wired plus wireless networks, from analog to digital, ..., etc. The terminology used for the telecommunications is also increasing with the explosive growth of this industry.

This chapter will introduce (with brief descriptions) the most common forms of telecommunications terminology. Each of these topics are described in detail throughout subsequent chapters of this book. The terminologies covered include:

- Signal characteristics:

    * Signal amplitude
    * Signal frequency
    * Signal phase
    * Signal bandwidth or rate
    * Frequency and wavelength
    * Single tone and multi-tone signals
    * Binary, ternary and multi-level signals
    * A special analog signal: speech signal
    * Human ear sensitivity

- Units and Measurements

- Signaling (control signals)

    * Human ear sensitivity
    * In-band and out-of-band signaling
    * A simplified call processing
    * Dial tone
    * Numbering plan
    * Dual Tone Multi-Frequency (DTMF; touch tone)
    * Ringing tone and audible ringing tone
    * Busy tone and audible busy tone
    * Common Channel Signaling (CCS)

- Switching

    * Switching versus cross-connection
    * Blocking versus non-blocking
    * Information transfer modes
    * Local versus toll switches
    * Customer switches

- Transmission

    * Transmission facilities
    * System bandwidth
    * Transmission equipment
    * Transmission media

- Networks
  - * Network nodes
  - * Network elements
  - * Network types
  - * Network management
  - * Network services
  - * Network standards

1.2 SIGNALS

Signals carry information to be transported from one location to another: this includes "point to point" or "point to multi-points" (known as broadcasting or multi-casting: see Figure 1-1) connections. Any signal possesses four characteristics as listed as follows:

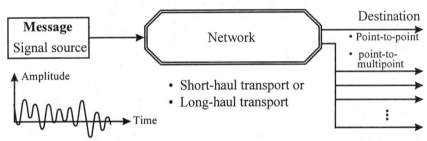

Figure 1-1   Signal Source, Transport, and Destination.

1. **Amplitude (magnitude)**: Amplitude represents the strength of a signal. The typical parameters used to represent signal are:

   ⊗ Voltage: It has the unit of *v* (volts), *mv* (milli-volts), *μv* (micro-volts), ...
   ⊗ Current: It has the unit of *a* (amperes), *ma*, *μa*, ...
   ⊗ Power: It has the unit of *w (watts)*, *mw*, ..., and dBm (described in appendix 1-2).

2. **Frequency**: Frequency represents the rate of change of a signal. The measurement unit used is *Hz* (Hertz; named for a scientist) or *cps* (cycles per second). Signals can have only one frequency (known as single-tone signals), dual frequencies, or multiple frequencies. Most practical signals, such as speech, possesses multiple frequencies.

3. **Phase**: Phase represents the relative position of a signal with respect to a reference signal. Degrees (°) is often used to measure a signal's phase.

4. **Bandwidth/Rates**: A signal carrying a message can either be an analog signal or digital signal (described later in this chapter). An analog signal has a certain signal bandwidth, which is expressed as a unit of *Hz* (as in frequency). The bandwidth of any signal to be transported from its source to its destination must have its bandwidth defined. A digital signal theoretically has an infinite bandwidth. Instead of defining

the bandwidth for digital signal transport, the signal speed (rate) is typically specified. Signal rates are usually expressed as bits per second (bps or b/s), kilo-bits per second (kbps or kb/ps), or Mega-bits per second (Mbps or Mb/s), and so on.

1.2.1 Units and Measurements

For telecommunications applications, the units used in system analysis and design (such as Hz, ft, m, and second) may be too small for some applications, and too large for others. Several units are introduced for expressing those units (Table 1-1).

Table 1-1   Common Units.

Expression	Time (seconds)	Frequency (Hertz)	Distance (meters/ft)	Prefer name
10^{12}	*Ts*	*THz*	*Tm (or Tft)*	*Tera*
10^9	*Gs*	*GHz*	*Gm (or Gft)*	*Giga*
10^6	*Ms*	*MHz*	*Mm (or Mft)*	*Mega*
10^3	*ks*	*kHz*	*km (or kft)*	*kilo*
1 (or 10^0)	*s*	*Hz*	*m(or ft)*	-
10^{-3}	*ms*	*	*mm (or mft)*	*milli*
10^{-6}	*μs*	*	*μm (or μft)*	*micro*
10^{-9}	*ns*	*	*nm (or nft)*	*nano*
10^{-12}	*ps*	*	*pm (or pft)*	*pico*

* Not practical.

Example 1-1: Assume a communication satellite, known as a geostationary satellite, has an altitude of about 22 *miles* from the earth's surface. Give examples of how this distance can be expressed.

This distance (22 miles) can be expressed in other units: about 35 kilometer (*km*), 35,405 meters (*m*), 3,540,500 centimeters (*cm*), 116,160 feet (*ft*), or 1,393,920 inches (*in*). For practical use either miles of kilometer is appropriate. (Note: *1 mile = 1.609344 km, and 1 mile = 5,280 ft*). In detail:

$$22 \times 1.609344 = 35.405568 \text{ km}$$
$$22 \times 5280 = 116,160 \text{ ft}$$

Example 1-2: Describe the telecommunications time unit that is known as a "frame".

A commonly-used time unit in modern telecommunications is known as **frame**. It has an interval of **125 *μs***. One frame of 125 *μs* is equal to a time interval of 1/8000 second. That is, there are **8000 frames per second**, for a system based on 125 *μs* frame interval. For example, a DS1 signal in North American digital networks carries twenty-four 8-bit channels, and an E1 signal (worldwide) carries thirty-two 8-bit channels (Chapter 6).

1.2.2 Signal Amplitude (Power)

The amplitude of a typical signal varies over the time varies [Figure 1-2(A)]. For some signals, the amplitude may remain constant for long periods of time [Figure 1-2(B)].

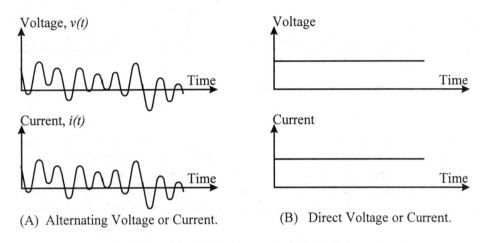

(A) Alternating Voltage or Current.　　　　(B)　Direct Voltage or Current.

Figure 1-2　　Signal Amplitude: ac or dc signal.

- **Alternating current/voltage**: When a signal current (or voltage) varies as the time varies, it is an alternating current (ac/AC; commonly called ac current even though this is somewhat redundant.), or alternating voltage (commonly called ac voltage even though the letter "c" represents current, which is conflicting). AC signals are found in more practical telecommunications systems than the Direct Current (DC) signals.

- **Direct current/voltage**: When a signal current (or voltage) remains constant as the time varies, it is a direct current (dc; or dc current) or direct voltage (or dc voltage). This is not a "stand-alone" signal, but is actually a component of a "practical signal" (this concept is discussed later in this chapter). That is, a dc voltage alone is typically not used to carry information in telecommunications services.

One derived parameter used to express signal amplitude is called the root mean square (rms) value of the signal voltage or current. The rms value is defined as follows:

$$V_{rms} = \sqrt{\frac{1}{T}\int_0^T v^2(t)dt} \qquad \text{and} \qquad I_{rms} = \sqrt{\frac{1}{T}\int_0^T i^2(t)dt} \qquad (1\text{-}1)$$

As indicated in Eq.(1-1), a square **root function** is used in this definition. The integration from time t = 0 to t = T, with a factor of 1/T represents the average or **mean** value of the signal. The **square** of the voltage (*v*) or the current (*i*) represents the signal power. Therefore, the voltage or current represented by Eq. (1-1) is called the root mean square (rms).

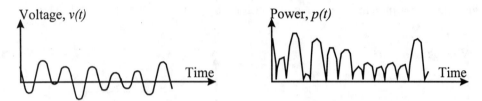

Figure 1-3   Signal Voltage (Amplitude) and Signal Power.

It is often important to know the absolute amplitude of a signal carrying a message, rather than the "sign" (positive or negative) of the signal. Another term, the "signal **power**" serves this purpose. The unit used to represent the signal power is w (watt), with appropriate prefixes (see Table 1-1); mw, μm, etc. Another frequently-used unit, dBm is used more often to represent a signal power (dBm will be described further in Appendix 1-2 of this chapter). The signal power is related to its voltage or current as given by Eq.(1-2). (Note that the symbol "$\propto$" represents "directly proportional to").

$$P \propto V^2 \qquad or \qquad P \propto I^2 \qquad\qquad (1\text{-}2)$$

From the definitions given in Eq.(1-2), it can be seen that the value of signal power is always "positive". For example, if a signal passing through a device, with a resistance of 1 ohm (Ω) has an amplitude of 2 v, the signal is said to have a power of 4 w ($P = V^2R = 2^2 \times 1$). Similarly, if the amplitude is -2 v, the signal power is still 4 w. (Note: a simplified definition of resistance is given in Appendix 1-1).

1.2.3 Signal Frequency/Wavelength

The frequency of a practical telecommunications signal [e.g., the signal in Figure 1-2(A)] is not easy to determine, unless it is a sinusoidal wave. The frequency is always defined for a sinusoidal signal (Figure 1-4), and is expressed by the following equation:

$$v(t) \text{ or } i(t) = A(t) \sin(\omega t + \theta) = A(t) \sin(2\pi f t + \theta) \qquad (1\text{-}3)$$

where

$A(t)$ = the signal amplitude (described earlier)

ω (f) = frequency (more precisely, f is called the frequency of the sinusoidal wave and ω is called the radian frequency of the sinusoidal wave)

θ = the phase (described next).

Definition 1-1:   A signal with a sinusoidal waveform (as shown in Figure 1-4) is called a *single tone* signal because it possesses only one (single) frequency.

Example 1-3:   Referring to Figure 1-4, the frequency of a sinusoidal signal can be determined (calculated) for the following parameter values:

- If t_1 = 1 sec
- If t_1 = 1 ms
- If t_2 = 1 sec
- If t_2 = 1 μs
- If t_3 = 1 ps

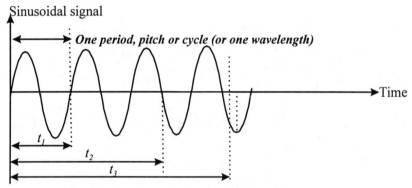

Figure 1-4   A sinusoidal Signal with Four Cycles.

For the case that $t_1 = 1$ sec, the time interval covers one cycle (or one period) of the single tone signal. The signal changing rate is one cycle per second. By definition, this signal has a frequency of 1 *cps* or 1 *Hz*.

For the case that $t_1 = 1$ ms (= 10^{-3} second), the signal will have a changing rate of 1,000 changes per second. The signal has a frequency of 1 kHz. The signal frequency can also be obtained using the following relationship:

$$f = \frac{1 \, cycle}{1 \, ms} = \frac{1}{1 \times 10^{-3}} \; Hz = 1 \, kHz$$

For the case that $t_2 = 1$ sec, this time interval covers 2½ cycles. The signal changing rate is 2 ½ cycles. Therefore, the signal has a frequency of 2.5 Hz or 2.5 cps.

For the case that $t_2 = 1$ μs, the signal frequency can be calculated as follows:

$$f = \frac{2.5 \, cycles}{1 \, \mu s} = \frac{2.5}{1 \times 10^{-6}} \; Hz = 2.5 \, MHz$$

For the case that $t_3 = 1$ ps, the time interval covers 3.625 cycles. The frequency can be obtained as:

$$f = \frac{3.625 \; cycles}{1 \; ps} = \frac{3.625}{1 \times 10^{-12}} \; Hz = 3,625 \; GHz$$

The single tone signals $[i(t) = A(t) \cdot sin \; (\omega t + \theta) = A(t) \cdot sin \; (2\pi f t + \theta)]$ have many applications in the telecommunications field. For example, the signal used in a carrier system is often a single tone signal (Note: carrier signals and carrier systems are presented in a separate chapter in this book). However, a single tone signal alone can not carry messages effectively. A signal carrying a message has a typical waveform as shown in Figure 1-2(A) or 1-3. These signals are not single-tone signals.

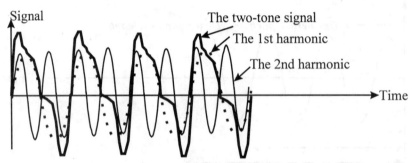

Figure 1-5   A Two-tone Signal (The Bold Line Waveform).

Definition 1-2: A double tone, triple tone, or a multi-tone signal has two, three, or more frequencies, respectively as follows.

- A single tone signal: $i(t) = A(t) \; sin \; \omega t$ (1-4)

- A double tone signal: $i(t) = A_1(t) \; sin \; \omega_1 t + A_2(t) \; sin \; \omega_2 t$ (1-5)

- A multi-tone signal: $i(t) = A_0 + \sum_{i=1}^{N} A_i \; sin \; i\omega t + \sum_{i=1}^{N} B_i \; cos \; i\omega t$ (1-6)

In a double tone signal, Eq. (1-5), the amplitudes of the two tones may or may not be equal [i.e., $A_1(t) \neq A_2(t)$]. Similarly, the frequencies of these two tones may or may not having a harmonic relationship, $\omega_2 = 2\omega_1$. The signal indicated by a bold line in Figure 1-5 is a two tone signal. In this example it is assumed that the two-tone signal consist of two harmonic components with $\omega_2 = 2\omega_1$.

The expression shown in Eq.(1-6) is known as the Fourier series of a signal. A typical signal in a telecommunications network and has a waveform as shown in Figure 1-2(A) or Figure 1-3 is a multi-tone signal, which generally consists of:

- A dc component, with a zero frequency of A_0.

- Two first harmonic (i.e., **fundamental** harmonic) components, with a frequency of ω, expressed as $A_1 \sin \omega t$ & $B_1 \cos \omega t$.

- Two 2nd harmonic components, with a frequency of 2ω, expressed as $A_2 \sin 2\omega t$ & $B_2 \cos 2\omega t$.

It must be emphasized that any signal, even a very complex speech signal, can be represented by a Fourier series as in Eq.(1-6). The signal can be displayed in the frequency domain (Figure 1-6) instead of using the time domain graph [as shown in Figure 1-2(A) or Figure 1-3]. That is, a signal possesses a dc component, a 1st-order harmonic, a 2nd-order harmonic, a 3rd-order harmonic, etc.

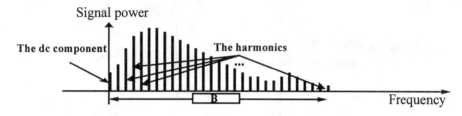

Figure 1-6   A Typical (Multi-tone) Signal.

In Figure 1-4, the wavelength of a sinusoidal signal is shown. Although the wavelength of this sinusoidal signal is appears to be the same as the period of the signal, the wavelength is not identical to the frequency of a signal. Instead, the relation between the signal frequency (f) and the signal wavelength (λ) is given by the following equation:

$$\lambda f = c \tag{1-7}$$

where c is the speed of light in a vacuum, which has a value of approximately 3×10^8 m/s. Figure 1-7 lists some typical telecommunications applications and their corresponding

In Figure 1-7, it is obvious that the higher the frequency of a signal, the shorter the wavelength. It should be understood that *as technology advances, telecommunications system frequencies becomes higher.*

- For radio broadcasting, the first technology developed was Amplitude Modulation (AM; described in Chapter 6) utilizing the frequency band around (560 kHz ~ 1600 kHz), followed by Frequency Modulation (FM; described in Chapter 6) utilizing the frequency band around (88 MHz ~ 118 MHz). For modern radio broadcasting, FM stations are generally more popular than AM stations (i.e., because of sound quality).

- For television broadcasting, Very High Frequency (VHF;TV channels No2.2 through 13) was developed before Ultra High Frequency (UHF).

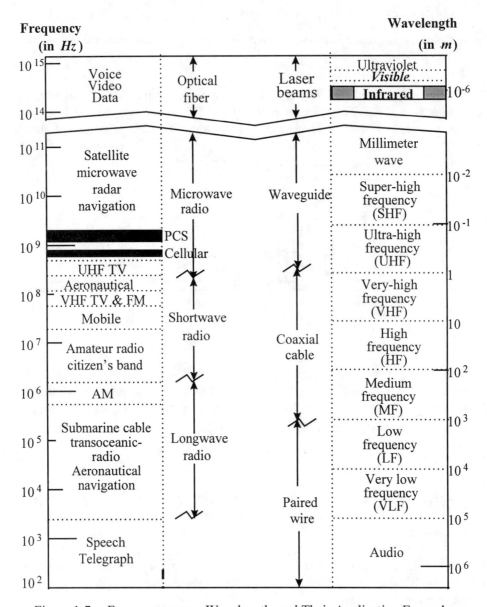

Figure 1-7   Frequency versus Wavelength, and Their Application Examples.

- For wireless services, Mobile Phone Service/Advanced Mobile Service (MPS/AMPS) was developed before cellular and PCS technology. The cellular communications network uses the frequency spectrum of 800~900 MHz. As the technology improved, wireless service evolved into the Personal Communication Service (PCS), which occupies the frequency spectrum of 1850~1990 MHz.

- The Wireless In-building Networking (WIN) service uses a variety of frequencies. One frequency used by WIN is Infra-Red (IR) technology, which is in the range of 10^{14} Hz. The same frequency spectrum is used for optical fiber communications systems.

- For optical fiber communication systems, "wavelength" instead of "frequency" is used to designate various technologies. These technologies are listed as follows, and will be discussed in Chapter 7.

  * The 820 *nm* (0.82 *μm*) technology
  * The 1310 *nm* (1.31 *μm*) technology
  * The 1550 *nm* (1.55 *μm*) technology

These three technologies are known as infrared because the wavelength of the red light is approximately 700 *nm* (0.7 *μm*).

1.2.4 Signal Bandwidth

As previously mentioned, a typical signal carrying a message is a multi-tone signal, which has the power spectrum shown in Figure 1-6. The signal power spectrum is a common method used to represent the power (or energy) distribution with respect to the frequency. For example, 98% of the power in a typical speech signal is distributed from 200 to 3,000 Hz, but 90% of the power is concentrated in the range between 200 Hz and 1,000 Hz.

Many signals have its spectrums extending into a very high frequency ranges. For practical system design and analysis, the frequency range is typically limited to a well-defined value. This frequency range is called the signal bandwidth. Although speech signals have 98% of their power confined to 3,000 Hz, the highest speech frequency has actually reached 10,000 Hz. For system design purpose, the speech signal is assumed to have a signal bandwidth of 4,000 Hz. That is, speech signals are assumed to have very limited amount of energy beyond 4,000 Hz.

In contrast, a digital signal [Figure 1-11(B), (C), and (D)] theoretically has infinite signal bandwidth. As shown in Figure 1-8, the spectrum extends indefinitely. A typical digital signal has power spectrum (see Figure 1-8), which consists one main lobe and infinite number of side lobes. The frequency range occupied by the main lobe is twice the frequency range occupied by each side lobe, and the energy contained in the main lobe is about 80 to 95% of the total signal energy. Therefore, in a practical digital system, the required bandwidth typically covers the main lobe only. The details of the digital signal power spectrum waveform (known as the sinc or sampling function) and the transmission quality degradation due to bandwidth limitation (which is typically negligible), are described in a later chapter of this book.

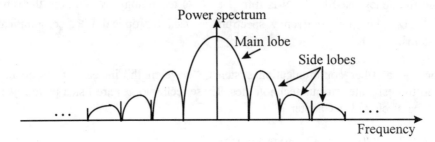

Figure 1-8   Digital Signal Power Spectrum.

1.2.5 Signal Speed (Rate)

If a digital signal is transported from one point to another, the signal speed (not the bandwidth) must be specified. For example, the digital signal used in North America to carry 24 digitized voice channels has a speed of 1.544 Mbps. Similarly, the digital signal used in other parts of the world to carry 30 digitized voice channels has a speed of 2.048 Mbps. Additional details associated with the speed of digital signals is discussed later in this chapter.

1.2.6 Signal Phase

The phase of a signal is normally insignificant for voice signal transport. However, the phase of a data signal is very important. A detailed discussion of a signal phase is given in this section by referring to Figure 1-9.

In Figure 1-9, the signal indicated by a "solid line" is chosen to be the reference, which has a phase of $0°$ by definition. Hence, this signal can be represented ass shown in Eq.(1-8):

$$x(t) = A \bullet \sin(\omega t + 0°) = A \bullet \sin \omega t \tag{1-8}$$

At $t = 0$, this signal, $x(t)$ with an amplitude of 0 (volt or ampere) starts at the point "A" [refer to points "A" of both Figure 1-9(A) and 1-9(B)]. As time progresses, the signal travels from point "A" to "C", and then to "B", as the signal completes one cycle (i.e., one period). At point "B", the cycle will start all over again (i.e., a new cycle starts from point "A"). As the signal travels from "A" to "B", it transverses an angle of "360°". Referring to the "dotted" line in Figure 1-9(A), the signal $y(t)$ travels from "A" to "D" , it has an angle of 90° (= ¼ × 360°) since the distance between "A" and "D" is equal to ¼ of the distance from "A" to "B". Similarly, point "D" is 90° "ahead of" point "C". Therefore, by definition, the "dotted" signal, $y(t)$ has a signal phase of +90° with respect to the reference signal, $x(t)$. That is, signal $y(t)$ can be represented as shown in Eq.(1-9):

$$y(t) = A \bullet \sin(\omega t + 90°) \tag{1-9}$$

where the reference signal x(t) has a phase of 0°.

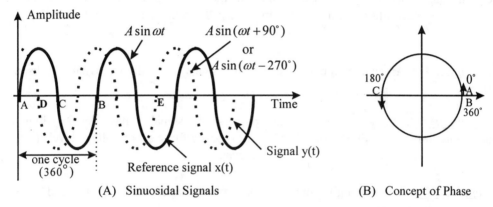

(A) Sinuosidal Signals   (B) Concept of Phase

Figure 1-9   The Definition of a Signal Phase.

By changing reference points, the relationship between $x(t)$ and $y(t)$ can be viewed differently. For example, point "E" of signal $y(t)$ is "270°" (= 360° − 90°) "behind" point "B" of signal $x(t)$. Therefore, $y(t)$ has a signal phase of −270° with respect to $x(t)$, and can re-written $y(t)$ as shown in Eq.(1-10):

$$y(t) = A \bullet \sin(\omega t - 270°) \tag{1-10}$$

Example 1-4:   Use a graph to illustrate the condition when two signals are 180° "out of phase".

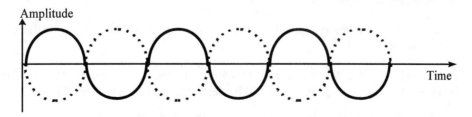

Figure 1-10   Two Signals that are 180° out of Phase.

Figure 2-10 illustrates to signals that are 180° out of phase. The relationship between these signals are shown as follows: when one signal is at its "positive" peak the other signal will be at its "negative" peak, and vice versa. It should be understood that when defining the phase of a signal, the reference signal must have the ***same frequency***.

1.2.7 Signal Types

A signal of any form (electrical, radio, or light) can be classified as either an analog or a digital signal. There are different ways to define these two types of signals. Definition 1-3 is a "textbook definition" while Definition 1-4 is a "common language definition".

Definition 1-3: An analog signal is a continuous signal, but a digital signal is a discrete signal.

Definition 1-4: An analog signal may possess any signal voltage level (within an allowable limit), but a digital signal can only possess a finite set of voltage levels (e.g., 2 voltage levels, 3 voltage levels, four voltage levels, etc.).

Definition 1-5: A binary digital signal can have two voltage levels (e.g., +5v and −5v; 0v and 5v; , +3v and −3v, etc.).

Definition 1-6: A ternary signal can have three voltage levels (e.g., +5v, 0v, and −5v).

Definition 1-7: A multi-level signal can have (2N+1) voltage levels, e.g., +Nv, +(N−1)v, +(N−2)v, ..., +1v, 0v, −1v, ..., −(N−2)v, −(N−1)v, and −Nv). A multi-level digital signal may have 2N voltage levels.

Example 1-5: Referring to Figure 1-11, identify (name) each signal.

- Figure 1-11(A) is an analog signal since it possesses infinite many voltage levels.

- Figure 1-11(B) is a binary signal, which is also called a unipolar signal (this signal will be described further in Chapter 6).

- Figure 1-11(C) is a binary signal known as a polar signal (this signal will be described further in Chapter 6).

- Figure 1-11(D) is a ternary signal, known as a bipolar signal or Alternate Mark Inversion (AMI) signal (this signal will be described further in Chapter 6).

- Figure 1-11(E) is a 10-level digital signal (additional details: see Example 1-6).

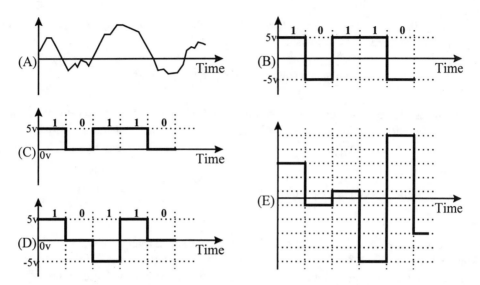

Figure 1-11   Various Example Signals.

Example 1-6: Describe the multi-level digital signal shown in Figure 1-12.

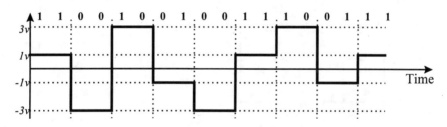

Figure 1-12   A Four-level Digital Signal.

This multi-level digital signal has four voltage level. Since $4 = 2^2$, every voltage level will carry **2** digits.(i.e., *–3v* represents "00", *–1v* represents "01", *1v* represents "11" while *3v* represents "10").

Example 1-7: Explain why a multi-level digital signal can carry more information (i.e., a multi-level signal has a higher signal rate). Figure 1-13(A) shows an 8-level signal.

- For an 8-level signal, there are 3 bits (because of $8 = 2^3$) of information carried for each one voltage level. For example, a "7v" interval that is 1 ms long may be assigned to carry "000"; a "5v" interval of 1 ms long carries "001"; a "–5v" interval of 1 ms carries "101"; ... , and a "−7v" interval of 1 ms carries "100". The speed (rate) of this signal can be derived as follows:

$$\text{Signal speed (rate)} = 3 \text{ bits per } 1 \text{ ms} = 3 \text{ kbps}$$

- In contrast, for the 2-level signal [see Figure 1-13(B)] each 1 ms interval carries only one information bit. For example, a "3v" interval of 1ms long carries a "1", while a "-3v" interval of 1ms long carries a "0". The speed of this signal is derived as follows:

$$\text{Signal speed (rate)} = 1 \text{ bit per } 1 \text{ ms} = \frac{1 \, bit}{1 \, ms} = \frac{1 \, bit}{1 \times 10^{-3} \, s} = 1 \text{ kbps}$$

It can be seen that the 8-level signal has a higher speed than a 2-level signal (that is, 3 kbps > 1 kbps). Since $8 = 2^3$, the speed of an 8-level signal is three (3) times the speed of a 2-level signal. For an M-level ($M = 2^n$) signal, the speed will be increased by a factor of $\boldsymbol{n}$, with respect to a 2-level signal.

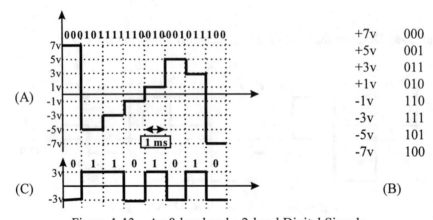

Figure 1-13   An 8-level and a 2-level Digital Signals.

This chapter defines the characteristics of an analog signal and a digital signal, which has various forms. Later in this book, each of these signal types will be further divided into several subgroups. For example,

(1)   An analog signal can be an Amplitude Modulated (AM) signal, a Frequency Modulated (FM) signal, an Amplitude Shift Keyed (ASK) signal, a Frequency Shift Keyed (FSK) signal, etc.

(2)   A digital signal can be a Pulse Code Modulated (PCM) signal, a Differential PCM (DPCM) signal, a Bipolar 3 zero Substitution (B3ZS) signal, a Bipolar 8 zero Substitution (B8ZS) signal, a High Density Bipolar 3 (HDB3) signal, etc.

1.2.8 Speech Signal: a Special Analog Signal

What is a sound? From the scientific viewpoint, a sound has two distinct meanings: (1) to a psychologist it means a sensation; and (2) to a physicist it means an atmospheric disturbance or a stimulus whereby a sensation is produced in the human ear. A sound is a wave produced by some vibrating body such as a bell, tuning fork, the human vocal chords, or similar objects capable of producing rapid "to-and-fro" vibratory motion of air particles.

For a speech signal to be transmitted over a long distance, it is obviously impractical to transport it in its original acoustic (sound) form. During the late 1800s speech signals were converted into electrical signal for long distance telephony transport. Later the electrical signals were converted into radio signals for wireless transport. Today, many long-distance high speed optical fiber systems transport speech as light signals. However, acoustic speech is first transformed into electrical format before it is converted into an optical signal.

After the speech (sound) has been converted from its acoustic form into an electrical signal, these waveforms typically possess frequencies from several tens of hertz up to 10 kHz. However, high-quality sound reproduction systems often cover the band from 20 Hz to about 40 kHz. Even though the human ear does not respond to frequencies above 15 kHz (the human ear sensitivity is shown in Figure 1-15), the higher frequencies (i.e., > 15 kHz) contribute to an overall sensory appreciation of sound.

Note that the speech signal's energy peaks between 800 and 1000 Hz, and that about 98% of the energy lies below 3 kHz. The actual speech signal power spectrum curve varies with age, sex, and country (i.e., the language being spoken), so the curve in Figure 1-14 should be interpreted as an average approximation (i.e., there are documented cases of human vocal sounds reaching 10,000 Hz).

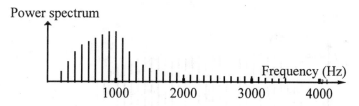

Figure 1-14   A Typical Speech Signal Spectral Density.

The speech spectrum can be represented either by a continuous curve from 200 Hz to 4000 Hz, or by a "discrete spectral curve". Figure 1-14 is a discrete spectral representation of a speech signal. This graph is often used to indicate that a speech signal contains many different frequency components, known as harmonics. Theoretically, if a transmission system could carry a infinite number of harmonics, the speech quality of the restored signal at the receiver (after long-distance transmission) would be perfect (i.e., without distortion and having negligible noise). However, systems have finite bandwidth (i.e., only a finite numbers of harmonics can be carried) and also contain channel noise, hence distortion in the restored signal is unavoidable.

For design and analysis of a voice network, a speech signal is normally assumed to have a bandwidth of 4000 Hz (4 kHz). That is, **a voice signal is said to have a signal bandwidth of 4 kHz**. It should be understood that not all human voices have a spectrum as shown in Figure 1-14. However, approximately 90% of the speech energy for a typical person is concentrated around 1,000 Hz and below. For telecommunications applications, the frequency range from 200 Hz to 3200 Hz (sometimes, 3400 Hz) is considered the voice-(frequency) band. Any signal within this range is called an "in-band signal". All other frequencies are referred to as an "out-of-band signal".

In modern communications networks, the equipment, terminals, switching machines, and transmission facilities are almost all digital. Therefore, it is necessary to convert 4 kHz analog voice into a digital signal so that it can be transported over digital networks. Voice digitization is the process used for converting voice (speech) into a digital format. The major functions in voice telecommunications networking are voice digitization, bit compression (including speech encoding), digital signal transport, switching, and management.

Speech communications include the "end-to-end" process of transmission and reception of a speech signal. The transmission starts with the vocal track acoustic signal, and reception ends at the human ears. The human ear senses various frequency tones with different degrees of sensitivity. For example, humans can distinguish a 1,000 Hz tone much easier than other tones. Figure 1-15 shows ear sensitivities to different tones. Due to the shape of this curve, it is often referred to as the C-curve (e.g., C-message curve, or C-message weights), and is commonly applied to any studies associated with human hearing.

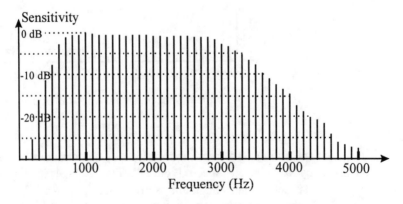

Figure 1-15   Ear Sensitivity in Terms of Frequency.

Referring to Figure 1-15, note that the ear sensitivity around 1,000 Hz is 0 dB. The sensitivity at 220 Hz is about −24 dB ($\equiv 1/2^8 = 1/256$, where the number "8" comes from 24 dB ÷ 3 dB = 8). This indicates that the human ears can detect a 1,000 Hz tone at a power level of 1 watt, with the same degree of sensitivity as a 200 Hz tone with a power level of 256 watts. In other words, the human ears are much more sensitive to a 1,000 Hz tone than a 220 Hz tone. Similarly, the sensitivity at 4,600 Hz is also about −24 dB. Therefore, a 220 Hz tone at any power has the same level of perception by the human ear as a 4,600 Hz tone with equivalent power. Appendix 1-2, at the end of this chapter, provides tutorial information describing the concept and use of decibels (dBs) in telecommunications applications.

Some additional data points derived from Figure 1-15 are: (400Hz, −11.4 dB), (800 Hz, −1.5 dB), (1,200 Hz, −0.2), (1,500 Hz, −1.0dB), (1,800 Hz, −1.3 dB), (2,000 Hz, −1.3 dB), (2,500 Hz, −1.4 dB), (3,000 Hz, −2.5 dB), (3,500 Hz, −7.6 dB), (4,000 Hz, −14.5 dB), (4,500 Hz, −21.5 dB), and (5,000 Hz, −28.5 dB). Beyond 5,000 Hz, the attenuation continues to increase at a rate of not less than 12 dB/octave until the attenuation reaches a value of −60 dB.

"The C-message curve is important in designing a handset for any telephony service. The receiver (i.e., ear phone or earpiece) of the handset must use the sensitivity factors (represented in Figure 1-15) as the frequency weights to compensate for the ear's sensitivities to various frequencies."

1.3 SIGNALING

To send a information (e.g., voice/data) from its source to its destination, many types of (control) signals must be processed before, during, and/or after the connection between the source and the destination is established. These controls are referred to as "*signaling*". Figure 1-16 is a conceptual block diagram of an end-to-end connection. The actual transmission equipment is not shown in Figure 1-16. It should be noted that for a long-haul connection, to complete the connection two types of networks must be involved:

- **Transport network**: The transport network is the portion of the network used to carry the signal from its source to its destination. This part of network consists of:

  * Switching machines
  * Transmission facilities
  * Transmission equipment

- **Signaling network**: The signaling network is used for call setup, trunk testing, call release, and other functions associated with call processing. The components used in the signaling network are essentially the same as the transport network. It also consists of switching machines, transmission terminals, and transmission facilities.

However, these components may not be as high speed as those used in the transport network.

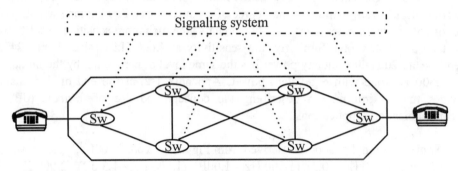

Figure 1-16   Transport and Signaling Networks.

The technology of signaling has evolved considerably over a long period. For example, the first generation of signaling used in-band signaling technology, which has since evolved into out-of-band signaling. This evolution was one of many important technological development in the modern telecommunications industry.

1.3.1 In-band Signaling/Out-of-band Signaling

In the early stages of the communications era, signaling for call setup was carried by signals occupying the frequency range between 200 Hz ~ 3,200 Hz. This technology is known as in-band signaling. For example, when a data signal is carried over a voice-grade circuit, the data signal is preceded by a 2,200 Hz tone for a 400 ms interval. This in band signal is called a "pilot tone", and is used to announce the arrival of the data signal.

As signaling technology matured it moved out of the 200 Hz ~ 3,200 Hz range. Therefore, this technology is referred to as out-of-band signaling (Figure 1-17).

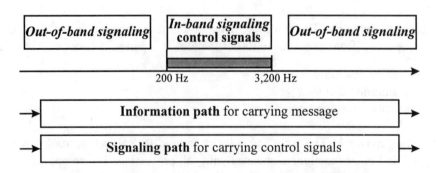

Figure 1-17   In-band, Out-band and Common-channel Signaling.

Many modern networks use signaling technology that requires a separate signaling path (other than the information path) for carrying control signals required for call processing. This approach is known as Common Channel Signaling (CCS described in Chapter 4) in conjunction with Signaling System 7 (SS7) protocol.

1.3.2 A Simplified Call Process

A telephone call involving two switching offices (Figure 1-18) is used to illustrate the signaling concept. The steps involved in a typical call are described below and correspond to the circled numbers in Figure 1-18.

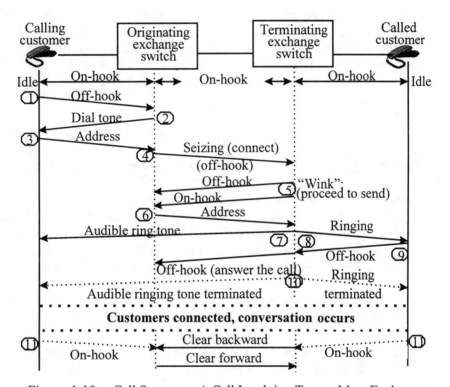

Figure 1-18   Call Sequence: A Call Involving Two or More Exchanges.

1.   A request for service is initiated when the caller lifts the telephone handset off the cradle or switchhook (loop closed – initiates an "off-hook" seizure signal: see Figure 1-19).

2.   After the current sensor circuit at the originating exchange switching office detects the "off-hook" signal, the switching machine sends a dial tone to the caller. Now, the calling party may begin dialing the called party's telephone number.

3. The address (normally numerical digits of the dialed phone number) of the called customer are transmitted to the first switching machine when the caller (the originating party) dials the called number (either dial pulses, dual tone multi-frequency or touch-tone pulses; discussed later in this chapter).

4. The originating exchange seizes the inter-exchange circuit, which is typically a digital carrier system such as a "T1" (for other countries other than USA, the equivalent trunk circuit between two switches is the E1 or the PS-1 digital carrier system) by sending an "off-hook" (seizing or connect) signal to the terminating exchange. The reception of the "off-hook" signal will initiate attachment of an incoming register for the first switching machine.

5. The terminating exchange (in this case, the second switch) sends an "off-hook" signal followed by an "on-hook" signal (the "on-hook" and "off-hook" signals together is called a "*wink*" signal) indicating that a register has been attached and is ready to receive the address information. This "wink" function is needed between any two adjacent switches if more than two switches are used in an end-to-end connection.

6. After the originating switch has received the "wink" signal, its register transmits the address (i.e., dialed number) using either dial pulses or Dual-Tone Multi-Frequency (DTMF) signals "down stream to the next equipment entity (e.g., exchange).

7. If the called customer's loop is not busy, the terminating exchange (i.e., the last switch that connects to the called party) will alert the called customer by sending a ringing signal. At the same time the terminating switch will send an audible alert tone (via the all the upstream switches) to the calling party indicating that the call set-up process is in progress, according to the conditions described in step 8.

8. Different information is provided to the originating customer by the terminating exchange according to various called party or network conditions:

    - *If the called customer is not busy*, the terminating exchange will return an audible ringing tone to the caller.

    - *If the called customer is busy*, the terminating exchange will send an audible busy tone to the caller (not shown in Figure 1-18).

    - *If the call cannot be completed through either exchange*, the exchange involved will send an audible congestion or busy tone to the caller (also known as "fast busy" tone).

9. Assuming the called customer indicates acceptance of the incoming call by lifting the telephone handset (loop closed – initiates an "off-hook" answer signal), the answer signal is conveyed to the originating exchange as an "off-hook" signal.

10. The terminating exchange recognizes the customer's answer (off-hook) signal and removes the ringing signal and the audible ringing tone. The customers at both ends are now connected, and the conversation may begin. When the conversation has been terminated, step 11 will take place immediately. The termination originated by either party is known as the "call release".

11. The exchanges recognize the customer's terminate (hang-up) signals, and return to "on-hook" states on for both loops and the inter-exchange circuits by transmitting "clear forward" and "clear back" signals. This completes the call sequence.

1.3.3 Dial Tone

When a telephone is not in-use, it is referred to as being "on-hook". *In-band signaling using 3,825 Hz (2,600 Hz in North America) "tone-on" (on-hook), and "tone-off" (off-hook) line signaling sequences are used to convey supervisory signals.*

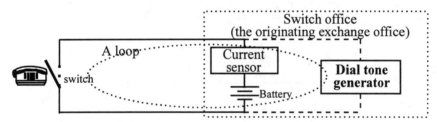

Figure 1-19   A Dial Tone Generator.

As shown in Figure 1-19, when the phone is on-hook, the switch is open. Therefore no electrical current is flowing through the telephone set. If the subscriber decides to place a call, he/she picks up the phone (or touches a special button on the phone) to request service, thereby entering the off-hook state. The switch is closed and a loop is established. The battery in the switch office (known as the originating office) generates a current that is detected by a current sensor. This sensor triggers a dial tone generator, which is also located in the switching office. The serving (originating) end exchange office generates a dial tone, and sends this dial tone to the caller indicating that he/she may begin dialing. ***"A dial tone is a continuous signal that typically has a frequency of 350 Hz mixed with a frequency of 440 Hz."*** Note that dial tone is classified as a form of in-band signaling.

1.3.4 Address of a Call: Telephone Number (Numbering Plan)

After a subscriber receives the dial tone, he/she can start to dial the number. The telephone number is transported by using tones or dial pulses, and is another form of in-band signaling.

1.3.4.1 ITU-T Numbering Plan

The world is divided into nine zones (*1 through 9*) as shown in Table 1-2 which indicates the principal areas covered by each zone number. Note that the number "0" is not assigned in the world zone numbering plan.

A numbering plan provides the structure for a pre-determined set of logical patterns for interpreting digits (i.e., it addresses the digits dialed by the subscriber). The number is used by the network to route the call to its destination, and also for billing computation by telecommunications service providers.

Table 1-2   Assignment of Country Codes

Zone	Principal areas covered
1	Canada & United States
2	Africa
3 & 4	Europe
5	Mexico &South/Central America
6	South Pacific
7	USSR (old)
8	North Pacific
9	Far/Middle East

0: is spare

1.3.4.2 E.163 Numbering Plan

The original numbering plan for the global telephone network is defined in ITU-T Rec. E.163 (International Telecommunications Union - Telecommunications Standardization Sector; formally CCITT, Recommendation E.163).

Each country or zone (Table 1-2) is assigned a Country Code (CC), which can consists of 1, 2 or 3 digits. The second field of a global telephone number is called the National Significant Number (NSN) as shown in Figure 1-20. The NSN is divided into two sets: (1) the trunk code, and (2) the subscriber number. The maximum total length is 12 digits (1 + 11, 2 + 10, or 3 + 9) for the CC and SNS, respectively. For example, if a country decides to use two (2) digits for the country code, then the maximum NSN length is 10 digits.

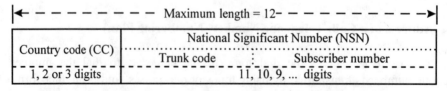

Figure 1-20   ITU-T Rec. E.163 Numbering Plan.

1.3.4.3 E.164 Numbering Plan

The E.163 numbering plan was used until the early 1980s, before the numbering plan for ISDN was proposed. ITU-T Rec. E.165 details the arrangements for implementing the E.164 numbering plan and specifies the date of December 31, 1996 for bringing E.164 into effect. E.164 applies the same principles of the E.163 recommendations, however, E.164 increases the total length from 12 digits to 15 digits. Also, the trunk code is changed to network destination code, which includes the trunk code and a code used to identify special networks (see Figure 1-21).

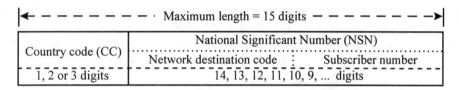

Figure 1-21   ITU-T E.164 Numbering Plan.

1.3.4.4 Numbering Plan for Mobile Subscriber

ITU-T Rec. E.212 specifies the identification plan for mobile subscribers, and Rec.E.213 specifies the Mobile Subscriber Roaming Number (MSRN). The number consists of the country code in which the mobile subscriber is registered. The country code is followed by the national significant mobile number, which consists of a Network Destination Code (NDC) and a Subscriber Number (SN) as shown in Figure 1-22. The mobile identification code is actually limited to 10-digit number made up of the country code, the network destination code, and the mobile subscriber number.

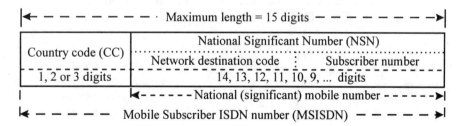

Figure 1-22   Mobile Subscriber Identification Number.

The following recommendations are proposed in ITU-T E.213:

- The numbering plan should apply standard telephone charging and accounting principles.

- Each administration should be able to develop its own numbering plan.

- The plan should allow the cellular system to change the international mobile roaming identity number without changing the telephone number assigned to the mobile unit.

- Roaming, without constraints, should be possible.

1.3.4.5 North American Numbering Plan (NANP)

The general format for domestic applications of the numbering plan for North American networks is shown in Figure 1-23. Note that "N" and "X' represent possible numerical values in the respective positions indicated in Figure 1-23.

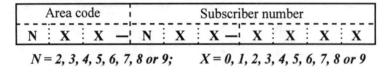

$N = 2, 3, 4, 5, 6, 7, 8$ or 9; $X = 0, 1, 2, 3, 4, 5, 6, 7, 8$ or 9

Figure 1-23   North American Numbering Plan Format.

1.3.5 Dual Tone Multi-Frequency or Touch Tone

There are 16 positions (buttons) on a "touch tone" phone. This dialing arrangement is called a Dual Tone Multi-Frequency (DTMF) scheme. DTMF is also an in-band signaling technology. When a number "1" is dialed, a two-tone signal (697 Hz and 1,209 Hz) is sent; likewise, if an "8" is dialed, another two-tone (852 Hz and 1,336 Hz) signal is sent, etc. At the present time, there are four additional positions reserved (i.e., the 4th column) for possible future use. The DTMF frequencies assignments are given as follows:

1	ABC 2	DEF 3		697 Hz
GHI 4	JKL 5	MNO 6		770 Hz
PRS 7	TUV 8	WXY 9		852 Hz
*	0	#		941 Hz
1,209 Hz	1,336 Hz	1,477 Hz	1,633 Hz	

Row No. 1:   697 Hz
Row No. 2:   770 Hz
Row No. 3:   852 Hz
Row No. 4:   941 Hz

Column No. 1:   1,209 Hz
Column No. 2:   1,336 Hz
Column No. 3 :   1,477 Hz
Column No. 4:   1,633 Hz

Figure 1-24   Dual Tone Multi-frequency (DTMF).

1.3.6 Ringing Tone and Audible Ringing Tone

If the called customer's loop is not busy, the exchange alerts the called customer by sending a ringing signal. This ringing signal is implemented by equipping the phone set with a ringing circuit (as shown in Figure 1-25).

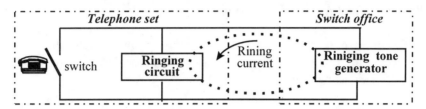

Figure 1-25   The Ringing Circuit of a Telephone Set.

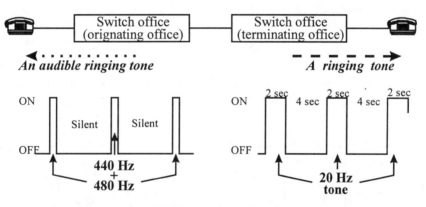

Figure 1-26   Ringing and Audible Ringing Tones.

Even when the telephone set is "on-hook" (i.e., the voice signal loop between the user and the central office switch is open as shown Figure 1-25), the ringing circuit inside the telephone set is still available to receive the alerting signal from the exchange office, via the ringing signal loop as shown by the dotted line in Figure 1-25.

In the central office there is a ring signal generator which produces the ringing (alerting) signals. These alerting signals include the ringing signal delivered to the called party, and the audible ringing signal delivered to the calling party. These two alerting signals (Figure 1-26) have some similarities and some differences, which are briefly described as follows:

•   The ringing signal is produced at the central office by a ring signal generator, and has a frequency of 20 Hz. This signal is generated with a 2-second "ON" and 4-second "OFF" so that the customer (the called party) can be alerted.

- The audible ringing tone has a frequency of 440 Hz mixed with 480 Hz. This composed signal is a slow period tone, in which tone period is shorter than the silent period, with a cadence similar to that of the ringing signal. This signal is used to inform the calling party that the ringing activity is in progress.

1.3.7 Busy Tone and Audible Busy Tone

There are two types of busy tones: one is called the **busy tone**, and the other is called the (network) **congestion or reorder tone**. Both the busy tone and the network congestion tone are in-band signaling (see Figure 1-27).

- Busy tone: *If the called customer is busy*, the terminating exchange will send an audible busy tone to the caller.

- Congestion or reorder tone: *If the call cannot be completed through either exchange*, the exchange involved will send an audible congestion or busy tone to the caller.

The same frequency is applied to both types of busy tone. The frequency is typically 480 Hz mixed with 620 Hz. There is one difference between them. Busy tone has **60 interruptions per minute,** while the reorder tone has **120 interruptions per minute**, and is commonly called a "fast busy" signal.

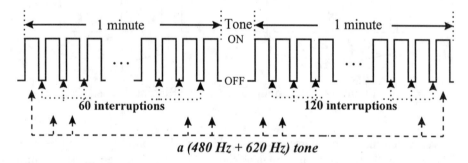

Figure 1-27   Busy Tone and Network Congestion Tone.

1.3.8 Common Channel Signaling

As previously shown in Figure 1-16, a Common Channel Signaling (CCS) system is a separate network that has the network architecture as the transport network (i.e., consists of transmission equipment, transmission facilities, and switching machines). The common channel signaling system has the building blocks shown in Figure 1-28. There are many duplicated Signaling Transfer Points (STPs), connected by duplicated transmission facilities known as "A", "B" and "C" links. A STP can also be connected to Operator Position System (OPS) and Service Control Node (SCN). A functional description of these building blocks is provided in Chapter 4.

In modern long-distance transmission networks, common-channel signaling technology has replaced the older per-trunk signaling. In the per-trunk signaling approach, the signaling activity takes place before transmission of the actual user message because both the signaling and messages share the same trunk (i.e., physical connection). Therefore, the per-trunk signaling is not as effective as common channel signaling technology.

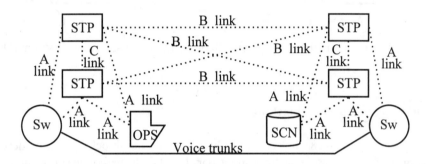

Figure 1-28   Common Channel Signaling System.

1.4 SWITCHING

Selecting the transmission path to route a signal being transported is the key function of a switch in a telecommunications network. Various switching definitions and terminologies are used to describe switching systems. They are: switching versus cross-connection; blocking versus non-blocking; information transfer modes; local versus trunk (toll); and public versus (customer) private switches.

1.4.1 Switching versus Cross-connection

Conceptually there is no difference between a switch and a Digital Cross-connect System (DCS) from an information moving aspect. Both are used to connect a signal from one input port to any specific output port (Figure 1-29). For example, the signal from input port No. 1 may be connected to output port No. 1 at one instance (i.e., $t = t_1$). Later (i.e., $t = t_2$), the same signal from input port No. 1 may be "switched" to output port No. 2. This function can be implemented using either by a switch or a DCS. The fundamental difference is that the connection paths in a DCS usually remain unchanged for a relatively long-period of time (e.g., days, weeks, or months) compared to "real time switch" connections (e.g., typically minutes or hours).

The functional description of a digital switch (e.g., 4ESS and 5ESS) is given in Chapter 5 and the functional description of a DCS is given in Chapter 6.

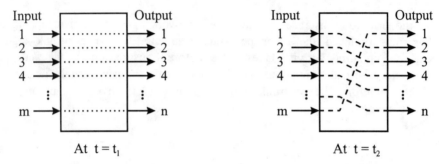

Figure 1-29   Functional View of a Switch or a Digital Cross-connect System (DCS).

1.4.2 Blocking versus Non-blocking

By referring to Figure 1-28, the concept of blocking and non-blocking will be described. Assume at the instance a signal from input port No. 1 is to be connected to output port No. 2, there are two situations that can happen to the connection:

(1)   The signal can be connected successfully. That is, the connection is "non-blocking". Typically, a digital cross-connect system is designed to be a "non-blocking" system.

(2)   The signal can not be connected successfully. That is, the connection is blocked because there is no transmission path available (e.g., all time slots have been occupied by other signals). Typically, a digital switch is designed to have a specific blocking probability, but it is relatively small (e.g., 0.1%).

The blocking of a digital switch is expressed as 1%-blocking, 0.1%-blocking, etc. With a 0.1% blocking probability, statistically the switching system is expected to complete 999 call attempts (and to fail 1 call attempt) out of a total of 1000 call attempts since

$$P_e = \frac{Failed\ call\ attempt(s)}{Completed\ call\ attempts\ +\ failed\ call\ attempt(s)} = \frac{1}{999 + 1} = 0.1\% \qquad (1\text{-}11)$$

Generally, the blocking probability requirement for a "landline" network serving a traditional telecommunications customer is much smaller (e.g., 0.1%) than the blocking probability requirement of a cellular communications network serving mobile customers (typically 1% in USA; but, much higher in some developing countries). A higher blocking probability is sometimes permitted to allow an existing switching system to serve more customers.

1.4.3 Information Transfer Modes

The ITU-T defines a **transfer mode** as "*a technique used in telecommunications networks covering aspects related to transmission, multiplexing, and switching*." Three transfer modes have been developed and implemented for moving information (voice, data, video) from their sources to distant destinations:

(1)   The circuit transfer mode: This mode is typically applied for traditional telephone services. Figure 1-30(A) illustrates a circuit-switched connection (indicated by bold trunks "*a*", "*b*", and "*c*" between two adjacent Network Elements (NEs). These trunks are dedicated for the information transfer between users A and B, and are not to be shared by any other users.

(2)   The message transfer mode: This mode is used for telegram and e-mail services.

(3)   The Packet transfer mode: This mode primarily used for data signal transport since being deployed in the 1970s. However, new "fast packet "switching technology (developed in the late 1980s) is suitable for ISDN [Specifically, Broadband ISDN (B-ISDN)] and multimedia applications. Figure 1-30(B) illustrates two packet-switched connections (one by bold trunks, and the other by dotted trunks). The packets transfer between users C and D are carried over trunks marked by "*a*", "*b*", and "*c*". The packets transfer between users E and F are carried over trunks marked by "*a*", "*b*", "c", and "*d*". It is clear that, unlike in the case of circuit-switched connection, trunks "*a*", "*b*", and "*c*" are **not** dedicated to any specific users. Instead, they are shared by two pairs of users. In practical systems, these trunks can be shared by more than two pairs of users.

It is clear that bandwidth utilization is much more effective in packet-switched networks than in circuit-switched networks.

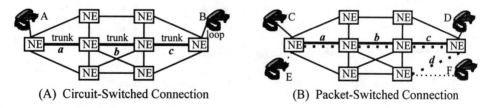

(A) Circuit-Switched Connection   (B) Packet-Switched Connection

Figure 1-30   Circuit-Switched versus Packet-Switched Connections.

Modern telecommunications networking utilizes digital technologies, and conventionally divides "time" into frames. A frame typically has a time interval of 125 µs (i.e., 10^{-6} second; described in Chapter 6). Furthermore, a frame is divided into time slots (timeslots, or channels). There are several methods used to assign these timeslots to customers as shown in Figure 1-31. Two commonly used methods are Synchronous Transfer Mode (STM) and Asynchronous Transfer Mode (ATM):

- Synchronous Transfer Mode (STM): Referring to Figure 1-31, one frame (125 μs interval) is divided into "N" timeslots. Note that two frames are shown in this figure. In the STM mode, timeslot No. 1 of **every** frame is always assigned to the same user (user No. 1) repeatedly. Likewise, timeslot No. 2 of **every** frame is always assigned to user No. 2 repeatedly; ...; and, timeslot No. N of **every** frame is always assigned to user No. N repeatedly. The multiplexing technique used for this mode is called Time Division Multiplexing (TDM), which is further described in Chapter 6. This technology (STM) has been adopted by circuit switching networks to handle voice signal transport.

- Asynchronous Transfer Mode (ATM): Referring to the lower portion of Figure 1-31, timeslots Nos. 1, 2 and 3 are assigned to user No. 2 (bandwidth on demand), timeslot No. 4 is assigned to user No. 1, ..., and timeslot No. N is assigned to user No. N for frame No. 1. During frame No. 2 timeslot No. 1 is assigned to user No. N again; timeslot No. 2 and No. 3 are both assigned to user No. 3; timeslot No. 4 is assigned to user No. 5, ..., timeslot No. N is assigned user No. N. Based on this example, it is clear that timeslot No. k ($k = 1, N$) of various frames may or may not be assigned to the same user. The multiplexing technique used for ATM is referred to as statistical multiplexing (i.e., a modified TDM).

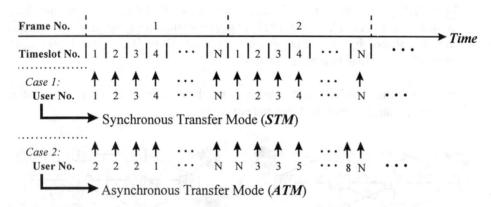

Figure 1-31   Synchronous and Asynchronous Transfer Modes.

Besides bandwidth utilization effectiveness, a comparison (from data rate, connection duration, resource allocation, multiplexing technique, delay, error and processing viewpoint) between STM and ATM is given in Table 1-3. It should be understood that the terminology "ATM" used here is a *generic* term. This is not the term

"ATM" generally used in the telecommunications industry, which implies a particular standard (e.g., 53-otectet packet-size ATM; discussed in Chapter 5).

Table 1-3   Circuit Switching versus Packet Switching.

	Circuit switching (STM)	**Packet switching (ATM)**
Date rate	*Constant*	*Variable (burst)*
Connection duration	*May be long*	*Typically short*
Resource allocation	*Allocated at call setup time*	*Allocated as needed*
Multiplexing	*TDM*	*Statistical multiplexing*
Delay	*Constant*	*Variable with delay jitter*
Error	*Relatively high*	*Typically low*
Processing	*Simple*	*Complex*

1.4.4 Local versus Toll Switch

Figure 1-32 shows an end-to-end connection of a telecommunications network, which consists of at least two local exchanges (switches) and many toll (trunk) exchanges (switches). In the local exchange office, besides the local switches, there are several types of transmission terminals such as Pulse Code Modulation (PCM) terminals for converting analog voice signal into digital signal, and low-to-medium speed digital Add-Drop Multiplexer (ADM) for multiplexing/demultiplexing digital signals. At the trunk exchange office, the typical transmission terminals are medium-to-high speed add-drop multiplexer and Digital Cross-connect Systems (DCSs). Between any two exchanges, there are various types of transmission media (twisted-pair wire, coaxial cable, radio, and optical fibers) used to carry the digital bit stream.

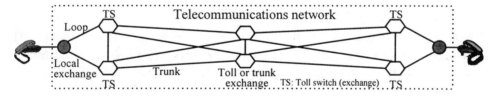

Figure 1-32   Local and Trunk Exchange Switches.

- Local exchange: A local exchange switch is used to connect the end user to the telecommunications network. The facility connecting the end user and the local exchange is known as "loop". The majority of end users are Plain Old Telephone Service (POTS) customers with traditional analog loops. Some specific loops are qualified to serve as the facilities for high-speed access applications: Digital Subscriber Line (DSL), High-speed DSL (HDSL), Asymmetrical DSL (ADSL), and

Very high-speed DSL (VDLS) services (described in Chapter 3). A typical local exchange switch is 5ESS (described in Chapter 5).

- Tandem exchange: A tandem exchange configuration is used to provide a transit function for connecting two or more local exchange switches. It connects "trunks to trunks" within a local geographical area such as a metropolitan service area. The tandem exchange arrangement also provides greater traffic efficiency and routing diversity. Recently, tandem exchange switch applications have been extended to trunk exchange areas.

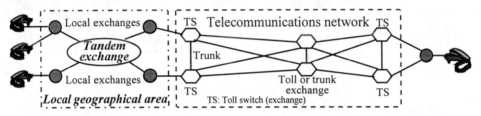

Figure 1-33   Tandem, Local and Toll Exchanges.

- Trunk (toll) exchange: Trunk exchange switches are typically used to connect long-distance calls. 5ESS or 4ESS switches (discussed in Chapter 5) can serve as trunk exchanges. Trunk exchanges are used to connect local exchanges in different geographical areas. Normally, no subscribers are directly connected to a trunk switch. That is, trunks carrying many multiplexed subscribers are the inputs and outputs of a trunk switch. Hence, the name "trunk switch" reflects this input-output arrangement. In North America, calls passing through trunk exchanges usually involve "toll" charges. Therefore, a trunk exchange is commonly called a "toll exchange" or "trunk switch".

Definition 1-8: Trunk is the facility connecting two exchange offices (switches). The trunk connecting a local exchange to a toll exchange is called toll-connecting trunk or access trunk (access trunk group). The trunk connecting two toll exchanges is called toll trunk (inter-toll trunk).

Definition 1-9: From traffic engineering and network management viewpoint, "trunk" has another definition. Trunk can be used to express voice capacity of a specific transmission facility. For example, a T1 digital carrier system (described in Chapter 6) has a capacity of 24 voice trunks since a T1 system can transport 24 voice users simultaneously.

- Gateway exchange: Gateway switches are used to connect international calls. A gateway exchange is a specialized type of trunk exchange that provides access to other countries via international networks. It performs all the trunk exchange functions along with some special functions required to connect networks having different operating parameters (e.g., signaling format, transmission speed, billing, etc.). Operator services are provided to assist in completing collect, calling card, person-to-person, "charge-to-third-number" calls, etc. Figure 1-34 illustrates an international connection utilizing three gateway exchanges (A, B and C).

Definition 1-10: The originating and terminating gateway exchanges are known as "terminating" gateway (e.g., A and C) and the others are called "transit" gateways.

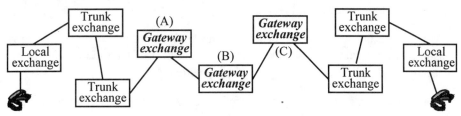

Figure 1-34   An International Connection via Gateway Exchanges.

1.4.5 Public versus Private (Customer) Switch

Public switches are typically local or toll switches that are operated and maintained by service providers. These services, provided by the public switches, are considered a public utility, and are usually regulated by government agencies. In North America the Public Switched Telephone Network (PSTN) has an overall objective of providing reliable universal services. A customer switch is tailored to meet specific needs associated with flexible services that include both voice and data communications. Customer switching can be provided by either: (1) special systems that are owned or leased by the business customer and located on the customer's premises, or (2) special services offered by a service provider via the local exchange.

The basic functions of a customer switch are to transport intra-location calls, accept incoming calls, send outgoing calls, provide system management, and enable office automation (further described in Chapter 5).

Common customer switching types are listed as follows:

- Key Telephone System (KTS)
- Private Branch eXchange (PBX)
- Hybrid system
- Centrex service
- Specialized switching system: (1) Automatic call distributor and (2) Telephone answering systems

Review Questions I for Chapter 1:

(1) Any signal possesses the following four characteristics: _____ (or _____), _____ , _____ , and _____ / _____ .

(2) (True, False) The operating frequency range for optical fiber communication using infrared technology is around (190 ~ 500) THz.

(3) (True, False) A voice or data signal always contains multiple frequencies.

(4) The signals used as control signaling, such as dial tone, are typical _____ -tone, or _____ -tone, but not multi-tone.

(5) For telecommunications network design and analysis purposes, a speech signal is assumed to have (signal) bandwidth of _____ kHz. For landline applications, the speech signal is digitized into a _____ kbps bitstream.

(6) (True, False)   For simplicity, one can make the following definition: an analog signal can possess any voltage values (within system limits), and a digital signal usually possesses 2 or 3 voltage values (e.g., unipolar, polar, and bipolar formats).

(7) To indicate human ear sensitivity, the ___ -message curve is often used. The human ear is most sensitive to a tone having a frequency around _____ Hz.

(8) Modern telecommunications adopt signaling scheme, known as _____ _____ signaling, using separate network to transport call-related control signaling. Two common used schemes are _____ and _____ .

(9) Each zone in the world is assigned a Country Code (CC) for telecommunications: zone 1 is for _____ , zone 2 is for _____ , zone 3/4 is for _____ , zone 5 is for _____ , zone 6 is for _____ , zone 7 is for _____ , zone 8 is for _____ , zone 9 is for _____ , and zone 0 is _____ .

(10) The international numbering plan allows a telephone number to have a maximum length of _____ digits.

(11) In a Dual Tone Multi-Frequency (DTMF) phone, each key is represented by _____ frequencies. For example, the number "0" is transmitted by a tone consisting of _____ , and _____ Hz.

(12) There are two types of busy tones: known as _____ tone and (network) _____ or _____ tone.

(13) (True, False)   Typically, a DCS is designed to perform non-blocking cross-connection (routing) function, and a switching machine is designed to perform a similar routing function but with a small blocking probability.

1.5 TRANSMISSION

Transmission is considered to be an essential part of the "telecommunications pyramid" (Figure 1-35). The pyramid is completed by the addition of "switching", "signaling", and "network management".

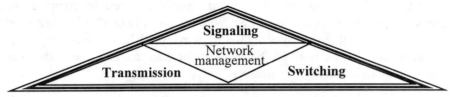

Figure 1-34   The Telecommunications Pyramid.

Signaling network performs call processing functions. Switches are responsible for routing traffic from its source to the intended destination. Transmission consists of the physical components (e.g., transmission equipment and facilities) needed an end-to-end communications link. Network management includes Operations, Administration, Maintenance and Provisioning (OAM&P).

1.5.1 Transmission Facilities

Media used as transmission facilities are described as follows:

- Wire facilities: Two types of wire facilities have been used for communications -

  * Twisted pair wire is used in two areas: (1) traditional analog subscriber loop for carrying one voice or voice-grade data channel, and (2) short-haul (≤ 200 miles) digital carrier systems for carrying tens of voice or voice-grade data channels. For example, the T1 digital carrier system used in North America carries 24 voice channels, while the E1 digital carrier system used worldwide, has 32 voice channel capacity (described in Chapter 6). Note that the cooper twisted pair wires are often referred to as metallic loops. The wires must be twisted (with equal distance between twists) so that the crosstalk noises between adjacent pairs of wire can be minimized.

  * Coaxial cables are used for cable television (TV) and long-haul digital carrier systems. For example, a long-haul carrier system used in North America is known as the T3 digital carrier system (Chapter 6) and is used to transport 672 voice channels.

- Waveguides: A waveguide facility is a hollow metal tube (cross-section is typically either circular or rectangular in shape) that carries radio waves. It can be used for long-haul transport and has a capacity slightly higher than coaxial cables, but lower than optical fibers. The long-haul waveguide applications are not as popular as coaxial cables or optical fibers. In North America, waveguides are generally used for short-haul radio applications.

- Airway (airlink): The atmosphere is a natural transmission facility. It is used for many applications: digital radio, satellite systems, mobile and cellular communications, Personal Communication Service (PCS), Wireless Local Loop [WLL, fixed wireless, airloop, Wireless Subscriber System (WSS)], wireless PBX, wireless computers, and Wireless In-building Network (WIN). More applications are expected in the future as the telecommunications industry evolves.

- Optical fibers: Fiber was introduced to communications networks in the early 1980s. Presently, more than 90% of the long distance traffic in North America is carried by optical fibers. It is expected that the same trend will follow throughout the world. Two characteristics have been the main contributors to wide deployment of fiber in modern communications networking. First, fiber links have a much lower Bit Error Rate (Ratio, BER), is typically three to five orders of magnitude better than other media types. Second, fiber links have high capacity. For example, an OC-192 (Chapter 6) optical link can carry 129,024 voice channels simultaneously for North American applications. In contrast, a twisted-pair T1 digital carrier system can carry 24 voice channels. With the development of Dense Wavelength Division Multiplexing (DWDM; Chapter 7), a single fiber link's capacity can be increased by a factor of 8, 16, 32, 80, 128, or even higher in the near future. For example, a system using 40-WDM carrying OC-192 signals has a voice capacity of 5,160,960 channels. This system has been deployed in North America, and known as WaveStar™.

1.5.2 System Bandwidth

One of the most important characteristics of a transmission system is "system bandwidth". This parameter determines the system's capacity or transmission speed, and is illustrated in Figure 1-35. Several bandwidth definitions have been used in the communication industry. However, a common reference is the 3-dB bandwidth model.

The 3-dB bandwidth model is based on a system having a frequency response (sometimes, called system loss characteristic) as shown in Figure 1-35. For example, if an input signal is applied to the system with a power of 10 dBm at 100 Hz, the received signal power is measured to be 3.0 dBm. By definition, this system has a loss of 7.0 dB at 100 Hz (dB and dBm are defined in Appendix 1-2 at the end of this chapter), which contributes a point (100 Hz, 7 dB loss) in the frequency response curve. If the frequency of the input signal is increased to 200 Hz, but the input power remains to be 10 dBm, the received signal power is increased to 3.5 dBm, with a corresponding system loss at 200 Hz of 6.5 dB, which establishes another point of the system frequency response curve. This procedure is continued until the frequency response curve is plotted for the operational range of the system (see Figure 1-35). This frequency response curve is different for different systems. However, they have a similar shape.

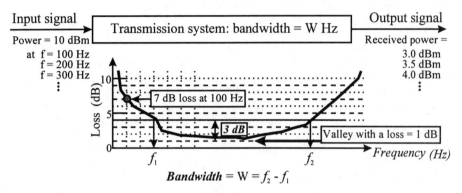

Figure 1-35   System (3-dB) Bandwidth Definition.

Once the frequency response curve is complete, the "valley loss value" can be identified. In this example, the "valley" has a loss of 1 dB. A horizontal line that is 3 dB higher than the valley is drawn across the frequency response curve. This line intersects the curve at two points, whose frequencies are f_1 and f_2. Thus, the system 3-dB bandwidth is defined as:

$$W = f_1 - f_2 \quad \text{(Hz)} \tag{1-12}$$

In summary, the system 3-dB bandwidth is "***The frequency range, $f_1 - f_2$, between the two 3-dB points of the system frequency response curve.***"

Example 1-8: Using the 3-dB bandwidth concept, derive the bandwidth and capacity for a system that has a frequency response curve as shown in Figure 1-36.

Assuming an analog carrier system has the frequency response shown in Figure 1-36, the system bandwidth and capacity for voice applications is determined as follows:

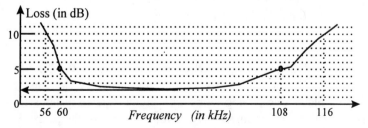

Figure 1-36   An Example of System Frequency Response.

From the frequency response curve, it can be seen that the lowest system loss is about 2 dB (i.e., the valley value). Therefore, the two 3-dB points are established by a horizontal line passing through the 5-dB point on the y-axis. This line intersects the curve at the 60 and 108 kHz points. The system bandwidth is derived as:

$$\text{System bandwidth} = 108 - 60 = 48 \text{ kHz}$$

If the system is used for voice applications, the capacity is 12 voice channels [= (48 kHz)/(4 kHz)], assuming the voice signal has a bandwidth of 4 kHz. Such a signal (carrying 12 speech signals) is known as a "***group***" signal. Its digital counterpart, the DS1 signal is called a "***digroup***" signal since it carries 24 speech signals simultaneously.

1.5.3 Transmission Equipment

In addition to transmission facilities and switches, an end-to-end communication link requires many types of transmission equipment. The following is a list of the most common transmission equipment:

- Pulse Code Modulation (PCM) analog to digital converter: Speech is an analog signal that must be converted into a digital bitstream for transmission over digital networks. The functions performed by a PCM transmitter [Figure 1-37(A)] are: (1) sampling (the analog voice), (2) quantizing the sampled signals into specific quantization levels, (3) coding each quantized sample into an 8-bit codeword, (4) multiplexing many digitized voice bitstreams into a single bitstream, and (5) inserting control/call-related signals into the bitstream that is transmitted over a digital network. Practical commercial PCM systems use 8 bits to carry a voice sample, and achieve acceptable speech quality at the receiving end (discussed in Chapter 6).

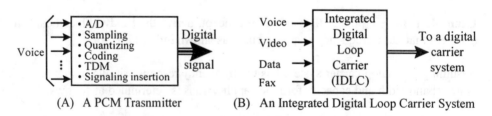

(A) A PCM Trasnmitter   (B) An Integrated Digital Loop Carrier System

Figure 1-37   A PCM Transmitter and An IDLC.

- Integrated Digital Loop Carrier (IDLC) systems: The development of IDLC is a natural evolution from Subscriber Loop Carrier (SLC) systems. SLC was first designed to replace long analog loops with digital carrier systems to improve voice transmission quality. The objective was for subscribers far away from the central office to have the same service quality as subscribers located close to the central office. That is, SLC was originally designed for voice applications. In contrast, IDLC can be applied for voice, data, video, and fax signals [Figure 1-37(B)].

- Business Remote Terminal (BRT): A SLC or IDLC can be modified for business use. A BRT (Figure 1-38) is often located on a customer's premises, and is connected to the network via a central office. For large business customers, the input to a BRT may be DS1 digital signals with a speed of 1.544 Mbps (Chapter 6), and the connecting facilities can be twisted-pair wires, coaxial cables, or optical fibers.

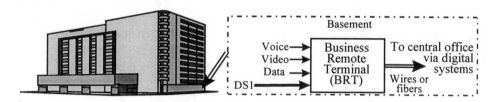

Figure 1-38   A Business Remote Terminal.

- MODEM (MOdulator/DEModulator): A modulator function is required to convert computer data into a special signal format that is suitable for transmission over a traditional analog loop. A demodulator performs the inverse function. Various modulation methods (e.g., M-FSK, M-PSK and M-QAM; Figure 1-39) are described in details in Chapter 6.

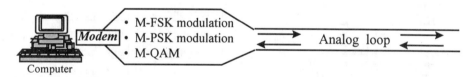

Figure 1-39   A Computer Modem (Modulator/Demodulator).

- Multiplexer (mux): A multiplexer is a commonly used network component in modern high speed digital networks. Its major function is to multiplex several low-speed digital signals into a single high-speed signal for long distance transport. It should be understood that a multiplexer is always designed to have a demultiplexer so that the equipment can be used for bidirectional communications. A demultiplexer performs the inverse functions of a multiplexer.

- Add-Drop Multiplexer (ADM): An ADM is a modern digital multiplexer that not only multiplexes digital signals, but also performs the "add and drop" function for lower-speed tributaries. An ADM is considered to be a "back-to-back" multiplexer/demultiplexer configuration. An modern ADM can add/drop electrical or optical signals of specific data rates. An ADM can also be configured as a digital regenerator. In modern digital networks, ADMs have become important "user-to-network" interfacing equipment (Figure 1-40). Note that in Figures 1-40 (B) and (C), the lines with different sizes/shapes are used to indicate different digital facilities with different data rates.

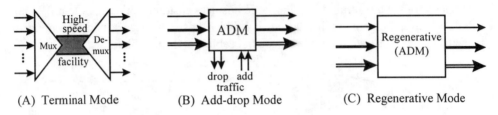

(A) Terminal Mode   (B) Add-drop Mode   (C) Regenerative Mode

Figure 1-40   Three Operation Modes of an ADM.

• Digital Cross-connect System (DCS): DCSs have been widely used in digital networks since the early 1980s (a functional description of a DCS is given in Chapter 6). In addition to cross connection [Figure 1-41(A)], a DCS can be used to consolidate traffic from several partially-filled digital carrier facilities to optimize the system bandwidth utilization [Figure 1-38(B)]. For example, trunks No.1 and No.2 are partially filled, by using a DCS, the traffic on these two trunks can be cross-connected to trunk No.5. Thus, the outgoing trunk No.6 can be used to carry additional traffic instead of carrying the partially-filled traffic from trunk No. 2.

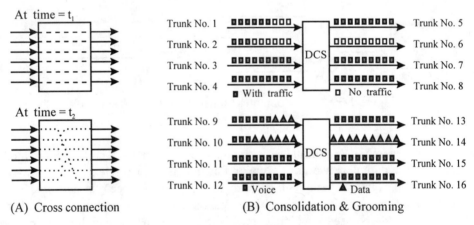

(A) Cross connection   (B) Consolidation & Grooming

Figure 1-41   DCS: Function and Applications.

A DCS can be used to groom traffic so that facilities of different types can be easily routed to its application. For example, data signals often require facilities with higher quality than voice signals, and it is economical to separate voice signals from data signals (into different facilities) by using a DCS. In Figure 1-41(B), both trunks

No. 9 and No.10 carry a mixture of voice and data signals. Without using a DCS, both trunks must be provisioned for data quality, which requires a more expensive facility and equipment. In contrast, if a DCS is used, voice traffic from both trunks can be cross-connected to trunk No.13 while data traffic from both trunks can be cross-connected to trunk No.14. Now, only one trunk (No.13) instead of two trunks must be provisioned for data quality. Thus, the communications cost can be reduced.

Another important application of a DCS is network restoration, as illustrated in Figure 1-42. Three DCSs are used in the network, the normal traffic route between DCS No.2 and the destination, has experienced a facility failure (or equipment malfunction), and the affected traffic has been re-routed from DCS No.2 to DCS No.3 to reach the destination.

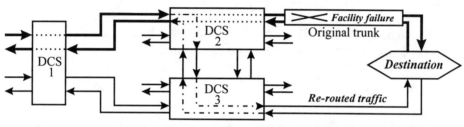

Figure 1-42   Network Restoration.

- Regenerator: In a long distance communications link, regenerators must be deployed in appropriate locations to regenerate the digital bitstream for further transmission. Three major functions performed by a regenerator are:

(1) Pre-amplification (re-amplification) and equalization
(2) Re-timing
(3) Re-generating

Further description on regenerators is given in Chapter 6. In an optical fiber link, optical amplifiers may also be applied to reduce the number of regenerators required, and thus reduce the overall system costs.

- Echo canceller: In a voice network, there are times that a strong reflected signal returns back to the speaker as an echo that degrades the communications quality. Two factors contribute to "echo":

(1) Long signal propagation delay
(2) Poor impedance mismatch along the transmission path.

An echo canceller is used to reduce the degree of echo annoyance (Figure 1-43). An echo cancellor is a device that first predicts the echo ($Echo_{predicted}$) based on the signal characteristics. The echo canceller then subtract the predicted echo from the actual echo ($Echo_{actual}$) that returns via the echo path. The "final echo" ($Echo_{final}$) returned to the talker has been sufficiently attenuated (i.e., cancelled) so that it is no longer objectionable to the talker.

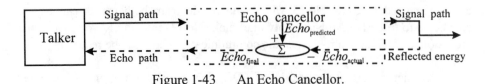

Figure 1-43   An Echo Cancellor.

Example 1-9: Illustrate the effect of using an echo canceller with a "30-dB cancellation" capability. Figure 1-44 illustrates the most common network configuration (a two-wire/4-wire hybrid) where echo is generated.

A cancellation capability of 30 dB can be considered to be a "loss of 30 dB". Therefore, an echo cancellor with a 30-dB cancellation implies that:

$$Echo_{final} = (10^{(-30)/10}) \times Echo_{actual} = 0.1\% \text{ of } (Echo_{actual})$$

Based on the attenuation of the echo signal ($Echo_{actual}$), it can be concluded that this is an excellent echo cancellor.

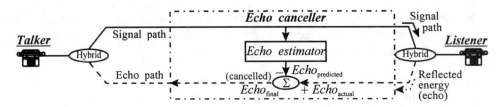

Figure 1-44   Echo Canceller used in Hybrid Network Configuration.

A hybrid is a circuit (using a 3-port transformer) to convert a 2-wire facility into a 4-wire facility (for long distance transport), and vice versa. The needs for using "hybrids" are explained next. Long-distance transport requires signal regeneration, that utilizes amplifiers for rebooting signal power. Amplifiers can only strengthen signal power for one direction. Thus, a 4-wire facility must be used for long distance transport. Due to impedance mismatch at the far end hybrid (the one near the listener), portion of "talker's energy" reflects via the echo path and returns to the talker as an annoying echo.

- Speech encoder: There are systems (especially wireless) in which the speed of a voice signal generated by a conventional analog-to-digital converter is too high for practical applications. For example, an 8-bit companding PCM terminal converts a 4-kHz voice signal into a 64 kbps signal, an Adaptive Differential PCM (ADPCM) terminal can produce a 24 kbps signal, and an Adaptive Delta Modulation (ADM) terminal can generate a digitized voice signal rate of 16 kbps, that are all too high for wireless applications. Therefore, a speech encoder can be used to compress the digitized voice signals to 8 kbps or a lower speed (Figure 1-45).

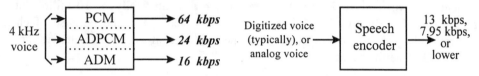

Figure 1-45   PCM, ADPCM, ADM and Speech Encoder.

- Channel control encoder: When a digital bit is transmitted over a channel, the bit may be erroneously restored (as a logical "1" instead of a logical "0", or vice versa). The term "Bit Error Rate (BER)" is a measurement of bitstream-transport performance (accuracy). To improve this BER performance, a channel control code (e.g., CRC, Hamming, Reed-Solomon, BCH, etc.) is used (Figure 1-46). The channel control code can be used to: (1) detect errors only (if errors are found, re-transmission is often requested), or (2) detect and correct channel errors. Chapter 13 is devoted to the description of channel error control codes.

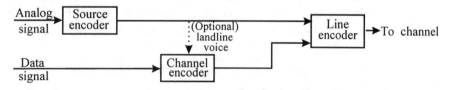

Figure 1-46   Three Coding Application Areas.

Figure 1-46 illustrates three different coding application areas: (1) source encoder for converting an analog signal into digital bitstream (e.g., PCM terminals), (2) line encoder for performing zero suppression for system timing synchronization (further description in Chapter 14), and (3) channel encoder for channel error control. For landline applications transporting speech signals, no channel encoders are used. In contrast, channel encoders are always used to improve speech quality.

Example 1-10: Illustrate the difference between a system (system "A") with a BER of 2.5×10^{-3}, and a system (system "B") with a BER of 2.5×10^{-9}.

The two BERs can be expressed as follows:

$$\text{(System A)} \quad \text{BER} = 2.5 \times 10^{-3} = \frac{2.5}{10^3} = \frac{25,000,000}{10,000,000,000} \tag{1-13}$$

and,   (system B) $\text{BER} = 2.5 \times 10^{-9} = \dfrac{2.5}{10^9} = \dfrac{25}{10,000,000,000}$ (1-14)

From Eqs.(1-13) and (1-14), it can be seen that: (1) system "A" will receive 25,000,000 bits that are "in error" (statistically), if 10 billion bits are transmitted; and (2) system "B" will receive 25 bits that are "in error" (statistically) if 10 billion bits are transmitted. Clearly, system "B" is superior to system "A". The purpose of channel control code is to improve system "A's" performance, if needed, so that it can have system "B's" performance.

- Scrambler: In a digital link, the receiver often relies on the incoming bitstream to derive timing (clock) synchronization to restore the digital bitstream. However, the receiver does not always have the same (identical) clock source that was used at the transmitter. Therefore, zero suppression, using a scrambler, is a common method used by a digital receiver to synchronize its clock with the transmitter's clock. The scrambler eliminates "strings of consecutive zeroes" to insure an adequate number of data transitions (i.e., "0" to "1" and "1" to "0") are available to derive accurate timing so the incoming data bitstream is properly restored.

- Line encoder: As shown in Figure 1-46, a digital system typically uses a line encoder at the transmitter to implement zero suppression. If a line encoder is not applied, a scrambler can be used to achieve the same goal. Table 1-4 summarizes the most commonly used zero suppression techniques (discussed in Chapter 6) used in modern telecommunications networks (e.g., the μ-law, the A-law, SONET/SDH and ATM).

Table 1-4   Zero Suppression (Substitution) Techniques.

Signal	Line code	Signal	Line code
DS1	B8ZS	E1	HDB3
DS2	B6ZS	E2	HDB3
DS3	B3ZS	E3	HDB3
FDDI	4B/5B	E4	CMI
Ethernet	Manchester	Ethernet	Manchester
SONET/SDH: Scrambler with $g(x) = 1 + x^6 + x^7$			
ATM: Scrambler with $g(x) = 1 + x^{43}$			

1.6 NETWORK

A telecommunications network consists of many interconnected network nodes. In addition to the information network (the top graph of Figure 1-47) that carries subscriber traffic, a separate signaling network (the bottom graph of Figure 1-47) is often used for call associated functions.

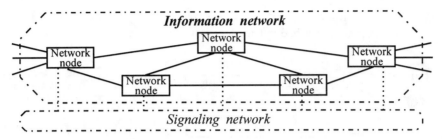

Figure 1-47   Simplified Communications Network.

In a Public Telephone Switched Network (PSTN), the information (transport) network nodes are local exchanges, trunk exchanges, and gateway exchanges (Figure 1-48). For wireless services, the network nodes can be Mobile Switching Centers (MSCs) and Base Stations (BSs). Within an exchange office, there are switching machines, transmission equipment, and transmission terminals [e.g., Pulse Code Modulation (PCM) equipment, Integrated Digital Loop Carrier (IDLC) systems, multiplexers, Add-Drop Multiplexers (ADMs), Digital Cross-connect Systems (DCSs), and echo cancelers]. Connecting the exchanges, MSCs, and BSs are transmission facilities (e.g., twisted pair wires, coaxial cables, airlink, waveguides, and optical fibers).

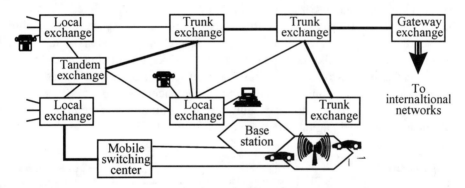

Figure 1-48   Typical Network Nodes of an Information Network.

The network nodes shown in Figure 1-48 are classified as follows:

- Local exchange network node (Figure 1-49): The function of a local exchange is to allow subscribers to access the telecommunications network. The interfaces of a local exchange office are voice signals, data signals (including Internet signals), fax signals, PBX trunks, and low-to-medium speed digital facility connections to other network nodes. A local exchange network node typically contains local exchange switching machines (e.g., 1AESS, or 5ESS), pulse code modulation terminals (e.g., D4 channel banks), central office terminals, Add-Drop Multiplexers (ADMs), multiplexers/demultiplexers, and Digital Cross-connect Systems (DCSs). These network elements are described in Chapters 5 and 6.

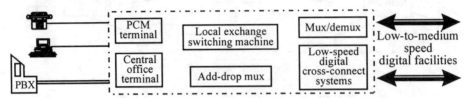

Figure 1-49   Interfaces and Network Elements of a Local Exchange.

- Tandem exchange network node: The function of a tandem exchange node is to interconnect several local exchanges so that the traffic flow between switches can be handled efficiently. In addition, the tandem exchange interfaces operate at higher speeds with increased traffic capacity, therefore the number of facility connections are reduced. Tandem exchanges typically contain tandem switching equipment and medium to high speed facility interfaces. Generally, voice signals, data signals, fax signals, PBX trunks are not connected directly to a tandem exchange node.

- Trunk exchange network node (Figure 1-50): The function of a trunk exchange node is to provide interconnection for long distance calls. Because these calls usually involve additional charges, trunk exchanges are also commonly called toll switches. The interfaces of a trunk exchange office are low-to-medium speed digital facilities, and medium-to-high speed digital facilities. Generally, voice signals, data signals (including Internet signals), fax signals, and PBX trunks are not connected directly to a trunk exchange network node. Within a trunk exchange network node, the key network elements are trunk exchange switching machines (e.g., 5ESS, or 4ESS), central office terminals, add-drop multiplexers, multiplexers/demultiplexers, and medium-to-high speed Digital Cross-connect Systems (DCSs).

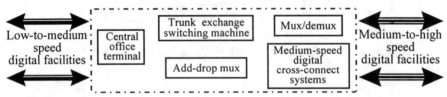

Figure 1-50   Interfaces and Network Elements of a Trunk Exchange.

- Gateway exchange network node: The function of a gateway exchange node is to provide access to international networks. A gateway exchange is similar to a trunk exchange. However, because the telecommunications systems are different globally, it is necessary for a gateway exchange to perform protocol conversion (signal rates, control signaling, transmission speed, billing, etc.) in addition to handling switching and transport of international traffic.

- Mobile Switching Center (MSC): The function of a MSC is to handle traffic switching for digital wireless networks. A MSC is similar to trunk exchange node, however, it serves as an interface between the base stations and the telecommunications landline network.

- Base Station (BS) wireless network node (Figure 1-51): Wireless technologies have shifted from analog to digital. Thus, a digital base station (in a cell site) interfaces with a MSC via digital facilities (e.g., T1 or E1 digital facilities; discussed in Chapter 6). It also interfaces with mobile (handset) stations via radio waves (i.e., airlink).

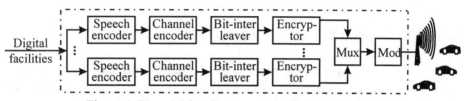

Figure 1-51   Major Components of a Base Station.

The BS node components are a demultiplexer (not shown in Figure 1-51) that is used to decompose a Time Division Multiplexed (TDM) signal (carried by digital facilities) into individual signals. The separate signals are fed to a speech encoder (in some systems, the speech encoder has been implemented within a MSC) to reduce the bits required for carrying a voice signal. A channel encoder is used to detect/correct bit errors, and a bit interleaver is used to break up bursty errors into random errors. An encryptor is always used in digital wireless system for security applications. Finally, all the individual signals are multiplexed together and modulated for transmission via airway. It should be understood that a corresponding process (in reverse order) is used to receive signals originated by the mobile stations.

1.6.1 Network Types

From a traffic viewpoint, networks can be classified as public and private, or voice and data networks. A public network is typically used for voice signal transport while a private voice network is known as Private Line Network (PLN, dedicated, or nailed-up network).

For data applications, a public network can either be a Circuit-Switched Public Data Network (CSPDN), or a Packet-Switched Public Data Network (PSPDN). Local Area Networks (LANs) are typically used as private line networks for data applications (see Figure 1-52). The Integrated Service Digital Network (ISDN) has been implemented for carrying both voice and data traffic over the same network for either public or private domain applications. Each of these network types (PSTN, PDN, PSPDN, ISDN) is described separately in the following sections.

	Public network		*Private network*
Voice transport	Public Switched Telephone Network (PSTN)	Integrated Service Digital Network (ISDN)	Private Line Network (PLN)
Data transport	Circuit-/Packet-Switched Public Data Network (C/PSPDN)		PLN or Local Area Network (LAN)

Figure 1-52   Network Types.

1.6.1.1 Public Switched Telephone Network (PSTN)

Figures 1-16 and 1-48 illustrate typical Public-Switched Telephone Network (PSTN) connections. Figure 1-53 shows PSTN applications: voice, data, video, PBX, Business Remote Terminal (BRT), etc. The transmission (physical) facility connecting a voice customer to a local exchange is known as a local loop (Figure 1-19 illustrates the function of a loop) or a line. For business applications, voice services are typically provided by using a Private Branch Exchange (PBX) connected to the local exchange via a PBX trunk (tie trunk). In addition to PBX, BRTs are often used to connect business traffic to public telecommunications network (i.e., a PSTN).

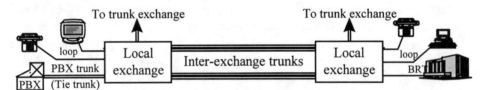

Figure 1-53   Public-Switched Telephone Network Applications.

1.6.1.2 Public Data Network (PDN)

A Public Data Network (PDN) can be either (1) a Circuit-Switched Public Data Network (CSPDN), or (2) a Packet-Switched Public Data Network (PSPDN) as shown in Figure 1-54. The operation of a CSPDN is similar to voice transport. Once a circuit is "set-up", the customer "owns the bandwidth" until the call is completed. This type of data transport is called voice grade or voiceband data. A PSPDN accepts data packets from the customer as they become ready for transport. This type of data transport is called statistical multiplexing (modified time division multiplexing, asynchronous transfer mode, or bandwidth on demand). Several generations of packet-switching technologies have been implemented since the 1970s. The first standardized packet switching technology was the ITU-T X.25 protocol. Frame relay packet switching technology is known as the second generation packet technology, and applies variable packet size. Third generation (3G) packet switching includes ATM protocol. Due to the fixed and small size of ATM packet, ATM technology is also called cell switching technology (further description of ATM technology is given in Chapter 5).

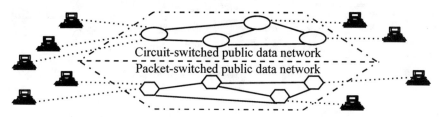

Figure 1-54   CSPDN and PSPDN High-level Architecture.

1.6.1.3 Private (Packet-Switched) Data Network

Private data networks are typically implemented by using a Local Area Network (LAN) that allows several independent devices to communicate directly with each other within a moderately sized geographic area. Data traffic carried over a physical communications channel operating at moderate data rates. Figure 1-55 illustrates three Local Area Network architectures that are typically used for private data communications. Data transport uses packet-switching protocol that can be ITU-T X.25 packet-switching for low data rates, frame relay packet-switching for moderate data rates, and ATM cell-switching for moderate to high data rates.

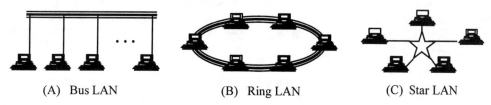

(A) Bus LAN   (B) Ring LAN   (C) Star LAN

Figure 1-55   Various LAN Architectures.

1.6.1.4 Integrated Service Digital Network (ISDN)

Figure 1-56 illustrates a conceptual Integrated Service Digital Network (ISDN) application with a 2B+D basic rate services. It has several unique attributes:

(1) Provides end-to-end digital connectivity
(2) Provides access and service connectivity
(3) Provides customer control and service features
(4) Providing upward compatibility of network features
(5) Only a limited set of standardized user-network interfaces are defined.

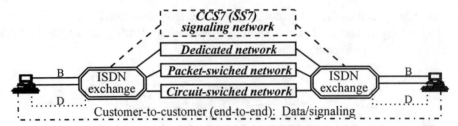

Figure 1-56   ISDN Conceptual View.

Table 1-5   ISDN Rates.

Channel	1.544 Mbps hierarchy	2.048 Mbps hierarchy
B	64 kbps	
H_0	384 kbps	
H_1	1.536 Mbps (H_{11})	1.92 Mbps (H_{12})
H_3	43 Mbps (H_{32})	32.768 Mbps (H_{31})
H_4	132.032 ~ 138.24 Mbps	

Table 1-5 lists all the presently defined ISDN rates. A information-**B**earing ("**B**") channel has a rate of 64 kbps, the same rate as a digitized voice signal (DS0 channel). A "H_0" channel is equivalent to six **B**-channels, and has a rate of 384 kbps. A "H_{11}" channel is equivalent to 24 B (**DS0**) channels and has a rate of 1.536 Mbps. A "H_{12}" channel is equivalent to 30 B (**E0**) channels and has a rate of 1.92 Mbps. A "H_{31}" channel has a rate of 32.768 Mbps. A "H_{32}" channel has a rate of approximately 43 Mbps. A "H_4" channel has a rate of 132.032~138.24 Mbps. Several ISDN rate structures have been defined as follows:

- Basic rate ISDN access having a rate of 144 kbps, and a structure of "2B + D16".

- Primary rate ISDN access having a rate of 1.544 Mbps, and a structure of "23B + D64", "3H_0 + D64", "H_{11}", etc.

- Primary rate ISDN access having a rate of 2.048 Mbps, and a structure of "30B + D64", "5H_0 + D64", "H_{12} + D64", etc.

- Broadband-ISDN (B-ISDN), is yet to be standardized (e.g., "H_4").

1.6.2 Network Services

Telecommunications services have evolved from simple Plain Old Telephone Service (POTS) to Internet services, which may dominate modern telecommunications for some time. Telecommunications services can be divided into three general areas:

1. Customer premises services: Currently, there are six service categories that involve customer premises equipment.

   - Voice and data services
   - Computer services
   - Internet services
   - Customer switching services
   - Local Area Network services
   - Supplementary services:

     ◆ Call waiting
     ◆ Call forwarding
     ◆ Abbreviated dialing
     ◆ Automatic redial
     ◆ Selective call rejection or acceptance
     ◆ Do not disturb
     ◆ Malicious call tracing

2. Local exchange services: Currently, there are seven service categories that involve local exchange equipment.

   - Residential services (e.g., basic and customer calling services)
   - Business services (e.g., Centrex and ISDN)
   - Access services (connection to long-distance telephone networks)
   - Public telephone services (includes coin phone services)
   - Directory services
   - Information services
   - Mobile (cellular) communication services

3. Inter-exchange carrier services: Currently, there are three service categories that involve inter-exchange(toll) equipment.

   - Switched network services

     ◆ Long-distance services
     ◆ Wide Area Telecommunications Service (WATS)
     ◆ 800/Advanced 800 services
     ◆ Switched digital services

   - Non-switched network services

     ◆ Analog private line services
     ◆ Accunet® T1.5, T45, etc.
     ◆ Skynet® services

   - Private switched network services

     ◆ Electronic Tandem Network (ETN) services
     ◆ Customer Network Option (CNO) services
     ◆ Virtual Private Network (VPN) services
     ◆ Customized private services
     ◆ Enhanced Private Switched Communications Services (EPSCS)

1.6.2.1 Network Services Example: Centrex Services

Centrex Services are typically provided by a local exchange switch. Figure 1-57 shows the network configuration of a Centrex service (typically provided by 1ESS or 1AESS switches). A Centrex services are similar to PBX services used by business customers. However, Centrex services are provided from a local exchange instead of customer premises equipment. That is, except for station sets and attendant positions, there is no equipment located in the customer premises. Subscribers are connected directly to the local exchange office that serves several Centrex customers.

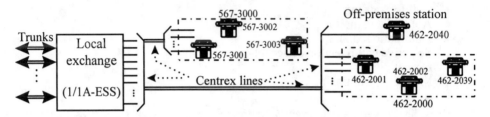

Figure 1-57   Centrex Services: Switching, Centrex Lines, and Stations

Originally Centrex was designed for large business customers to provide services such as call holding, call transfer, and abbreviated dialing. Today, Centrex services are available for small business and residential customers. The features of Centrex services are provided by 1ESS/1AESS switches, and may be selected from the following list:

- Customer calling
- Direct inward dialing
- Abbreviated dialing for often-called numbers
- Transfer of incoming and outgoing calls
- Call forwarding
- Extension-to-extension dialing
- Automatic identified outward dialing
- Restricted stations
- Single-digit dialing
- Three-way calling
- Consulting hold
- Do not disturb (slumber service)
- Centralized attendant service
- Special night service options
- Tie trunk service

1.6.2.2 Network Services Example: Automatic Message Accounting (AMA)

Automatic Message Accounting (AMA) service is usually provided by a 4ESS switch, as shown in Figure 1-58. Automatic Message Accounting (AMA) is augmented by an

Attached Processor System (APS) for disk storage and updates. A 4ESS switch can terminate up to 72 Centralized AMA (CAMA) operator positions, and can terminate a maximum of 8160 CAMA trunks. The major AMA components are described as follows:

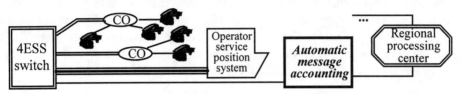

Figure 1-58   Automatic Message Accounting (AMA).

- Automatic Message Accounting Standard Entry (AMASE): AMASE enables billing data to be "teleprocessed" to a regional processing center (Figure 1-58). Billing information is collected and sent to the regional processing center via an APS (typically a 3B20D computer).

- Automatic Message Accounting Recording (AMAR): AMAR provides the capability of keeping records for various types of calls entering the switch. Call records may be used for billing, access charging, and recovery of transport costs from other carriers.

- Automatic Message Accounting (AMA): AMA produces records for incoming calls using equal access signaling, WATS, 800, teleconference records, call terminated due to call denial feature, test calls originating outside the inter-exchange network, international calls, and Software Defined Network (SDN; one of the many 4ESS switch capabilities, using the architecture shown in Figure 1-53).

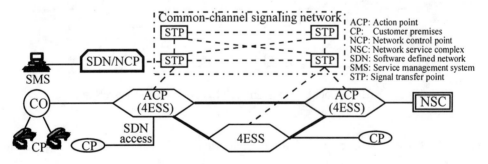

Figure 1-59   Software Defined Network (SDN).

Example 1-12: Describe the functional architecture of an Operator Service Position System (OSPS).

OSPS is an integrated option of a 5ESS switch and has a block diagram as shown in Figure 1-60. In addition to 5ESS switch hardware, an OSPS requires an Operator

Position Controller (OPC). OSPS is an application of ISDN within a 5ESS switch, that uses the ISDN access protocol to support the integrated voice and data capability required for operator interaction. The OPC allows the operators' equipment to be located remotely from the host 5ESS switch by using a digital facility (e.g., T1 digital carrier systems). Three primary OSPS functions are: (1)Traffic assistance: Traffic assistance can be achieved by operators equipped with Intelligent Communications Workstations (ICWs), that are designed with a call data display, a customer keyboard, and headsets, (2) Directory inquiry, and (3) Automatic Call Distribution (ACD).

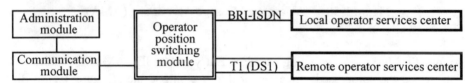

Figure 1-60   Operator Service Position System.

1.6.2.3 Network Services Example: Local Area Signaling Service (LASS)

Figure 1-61 shows the Local Area Signaling Service (LASS) architecture. LASS provides the following functions: (1) Record line history of the last call, including both incoming and outgoing, (2) Screen list (see Example 1-13) of numbers for distinction alert, and selective call forward, (3) Retrieval of calling line information (e.g., calling party information, and privacy indicator for multiple directory numbers), and (4) Retrieval of distant line information (e.g., feature compatibility, busy, and idle status).

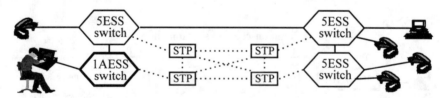

Figure 1-61   Local Area Signaling Services (LASS) Architecture.

Example 1-13:   Describe the features supported by Local Area Signaling Services (LASS).

The LASS provides screen list editing, which includes the following features:

- Selective Call Forwarding (SCF): A subscriber can pre-select which calls will be forwarded, based on the identity of the calling party. The customer creates a SCF screening list that contains the directory numbers of all callers that he/she desires to have forwarded. Calls from the parties on the list will be forwarded, but all others will receive standard terminating treatment.

- Selective Call Acceptance (SCA): A subscriber can pre-select which calls will be accepted, based on the identity of the calling party. The customer creates a SCA screening list that contains the directory numbers of callers whose calls should be accepted, all other calls are blocked or terminated.

- Selective Call Rejection (SCR): A subscriber can pre-select which calls will be rejected. The customer creates a SCR screening list that contains the directory numbers of all callers whose calls should be rejected. The rejected calls are routed to a rejection announcement, and the receiving party is not alerted for those calls.

- Selective Distinct Alerting (SDA): A customer can specify a list of calling directory numbers for which special alerting will be given: (1) When the customer is idle and receives a call from one of the number on the list, the station set will ring with a distinctive ringing (alerting) tone, or (2) If the customer, who has subscribed to call waiting, is busy when a call from a number on the list is received, a distinctive message waiting tone will notify the customer of the call.

1.7 TELECOMMUNICATIONS STANDARDS & ORGANIZATIONS

Standardization of parameters and features for telecommunications is critical in a telecommunications industry that is extremely diverse, rapidly growing, and complex. The need for standards as a catalyst for efficient development has been recognized globally by scientists and engineers working in the telecommunications field. The establishment, evaluation, and implementation of standards has become an important daily task of many telecommunications engineers.

Definition 1-11 (Cited from the U.S. National Telecommunications and Information Administration): A standard is a prescribed set of rules, conditions, or requirements concerning (1) definition of terms, (2) classification of components, (3) specification of materials, performance, or operations, (4) delineation of procedures, or (5) measurement of quantity and quality in describing materials, products, systems, services, or practices.

1.7.1 Types of Telecommunications Standards

There are several problems associated with arriving at standards. For example, even when standards are clearly written for all to follow, individual interpretations of meaning can still occur. Some interpretations are inadvertent, others are deliberately used to provide an individual corporation with a perceived advantage. The quest for a "standard" is not always determined by a simple criteria, such as the best or the most advanced technology that is available. Instead, a "standard" may be specifically engineered (i.e., selected) to optimize the vested interests of a few companies that have strategic influence in standard committees. A telecommunications standard is typically adopted for: (1) compatibility, or (2) interconnection of multi-vendor equipment.

A standard can be one of the following four types (some standards may satisfy the criteria for more than one type):

(1) De facto standards: A de facto standard is normally accepted based on the ubiquity of a product or service that adheres to the standard. For example, the MS-DOS operating system is a de facto standard. Likewise, the DS1 digital signal protocol originated as a de facto standard, and has evolved into semi-international standard (especially for North American digital networking).

(2) Proprietary standards: Theses "standards" are identified with a particular product or vendor, and may or may not be generally accepted by the industry. Common examples of telecommunications proprietary standards include:

- UNIX® operating system
- MS-DOS operating system
- OS/2 operating system
- PC windows
- BX.25 AT&T packet switch protocol
- Datakit® II Universal Receiver Protocol (URP)
- Digital Communications Protocol (DCP) ISDN-like Basic-Rate Interface to System75/85 PBXs (customer switches), and Merlin II key telephone systems.

(3) National standards: These standards are endorsed within a nation, but are not necessarily adopted by the international community. Common examples include:

- The ISDN Primary Rate Interface (PRI) for 1.544 Mbps standard has been adopted in North America, and the ISDN PRI for the 2.048 Mbps standard has been generally adopted worldwide.
- Common Channel Signaling System 7 (CCS7) has been adopted in North America, while Signaling System 7 (SS7) has been generally adopted worldwide.
- The US dialing plan is used in the US, but has not been adopted worldwide.

(4) International standards: These standards are endorsed by an industry worldwide. Several common international standards are listed as follows:

- Basic Rate Interface ISDN(BRI-ISDN)
- Signaling System 7 (SS7) signaling protocol
- Open System Interface (OSI)
- Synchronous Optical NETwork/Synchronous Digital Hierarchy (SONET/SDH)
- Asynchronous Transfer Mode (ATM)
- International dialing plan

1.7.2 Standards Development Process

The formal standard development process can be divided into four phases: (1) conceptualization, (2) discussion, (3) exposition, and (4) implementation. Figure 1-62 illustrates how a standard is typically developed within a corporate entity.

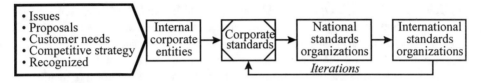

Figure 1-62   Typical Standard Development Process.

1.7.3 Standard Organizations

Primary telecommunications national standards organizations in the USA include: American National Standards Institute (ANSI), Accredited Standards Committee X3 (ASC X3), Accredited Standards Committee T1 (ASC T1), Exchange Carrier Standards Association (ECSA), Electronic Industries Association (EIA), Institute of Electrical and Electronic Engineer Association (IEEE), and National Bureau of Standards (NBS).

Prominent telecommunications international standards organizations include: International Telecommunications Union (ITU) [ITU consists of several sectors: ITU - Telecommunications Standardization Sector (ITU-T), ITU-Radio Communications Standardization Sector (ITU-R), etc.], Conference of European Posts and Telecommunications (CEPT) Administrations, International Electro-technical Commission (IEC), and International Standards Organization (ISO). A brief description of these organizations and their functions are given in the following sections.

1.7.3.1 International Telecommunications Union (ITU)

International Telecommunications Union is a specialized agency of the United Nations composed of representations of 160 countries. ITU is responsible for the regulation and planning of worldwide telecommunications. ITU assumed the present organization structure in 1993. The three ITU organizations are described as follows:

•   ITU-Telecommunications Standardization Sector (ITU-T): ITU-T was formerly the International Telegraph and Telephone Consultative Committee (CCITT) of the International Telecommunications Union. ITU-T is chartered to study and issue recommendations for technical, operating and tariff questions relating to telegraphy and telephone. An ITU-T "recommendation" is equivalent to a "standard" adopted by other standard organizations. ITU-T's approval process is a four-year cycle (e.g., 1988, 1992, 1996, 2000, …) of recommendations and discussion.

Address: ITU-T
           Place des Nations   Telephone: 41-22-995111
           CH-1211, Geneva 20, Switzerland   Telex: 421000 UITCH

The following five categories of ITU-T membership are recognized by ITU:

* Administration members: The principal member is the ITU-T member country's government. For example, the US Department of State, and the Department of Communications in Canada are ITU-T administration members.

* Recognized Private Operating Agencies (RPOA): The RPOA represents private or government corporations that provide telecommunications services, actively participate in all aspects of ITU-T work, and serve as Advisors to the administration in their countries. Examples are AT&T (USA), GTE (USA), British Telecom (UK), Telecommunications Carriers Association (Canada), etc.

* Scientific and Industrial Organization (SIO): The SIO consists of private corporations that are concerned with problem solving, designing, and/or manufacturing telecommunications equipment. Active members include: Lucent, Alcatel, Omnicom, Northern Telecom, Tellabs, etc.

* International organizations: ISO and ECMA, for example, have an interest in telecommunications standards and participate/contribute to the ITU-T work as observers.

* Specialized treaty agencies: Any large user of telecommunications services can be considered a specialized treaty agency having a direct interest in ITU-T work (e.g., World Meteorological Organization).

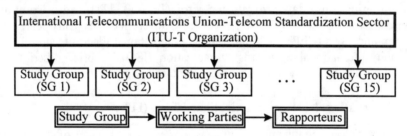

Figure 1-63   ITU-T Organization: 15 Study Groups.

Figure 1-63 illustrates the ITU-T organization, that currently consists of 15 Study Groups (SGs; see Appendix 1-3 at the end of this chapter). Within each Study Group, the actual work is performed by Working Parties (WPs) that study a particular set of recommendations. Within a Working Party, there are rapporteurs who study and report on specific points as required. Study Groups and Working Parties meet annually. At the meetings, the written contributions submitted by the members are reviewed and draft recommendations are produced for the Plenary Assembly. The rapporteurs work individually or with a group of collaborators by correspondence or in separate meetings.

Example 1-14: Describe the process used by ITU-T to establish Synchronous Digital Hierarchy (SDH) standards.

The worldwide Synchronous Digital Hierarchy (SDH) for new physical layer protocol has been recommended by ITU-T, whose Rec. G.707 describes the standards for SDH. Four ITU-T Study Groups (i.e., SG15, SG4, SG11 and SG13) are actively involved in the development of SDH standards as shown in Figure 1-64.

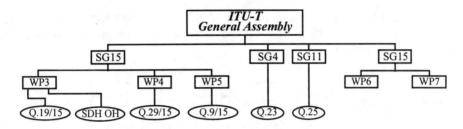

Figure 1-64   ITU-T Organizational Structure Supporting Synchronous Digital Hierarchy (SDH) Standards Generation.

ITU-T Study Group 15 (SG15) is involved with standardization of transmission systems and equipment. SG4 sets the recommendations for SDH network maintenance. SG11 deals with standard for SDH switch signaling, while SG13 is focussed on SDH network aspects (e.g., network architecture and configuration).

- ITU-Radio Communication Standardization Sector (ITU-R): The ITU-R was formerly the International Radio Consultative Committee (CCIR), formed in 1927. ITU-R is charted to study technical and operating questions in radio-communications, and to issue recommendations. The present ITU-R structure was adopted in 1948. ITU-R activities can be divided into as two main lines: (1) the technical aspects of radio spectrum use, and (2) the performance criteria and system characteristics for compatible internetworking. ITU-R currently has 11 study groups: (1) SG1 for spectrum utilization and monitoring, (2) SG2 for space research and radioastronomy, (3) SG3 for fixed services at frequencies below 30 MHz, (4) SG4 for fixed satellite service, (5) SG5 for propagation in non-ionized media, (6) SG6 for ionospherical propagation, (7) SG7 for standard frequency time signals, (8) SG8 for mobile services, (9) SG9 for fixed services using radio relay systems, (10) SG10 for broadcasting services (sound), and (11) SG11 for broadcasting services (television).

      Address: ITU-R
                  Place des Nations
                  CH-1211, Geneva 20, Switzerland   Telephone: 41-22-995111

- International Organization for Standardization (ISO): ISO was formed in 1947. ISO is charted to promote the development of standardization and related activities in the world with a view to facilitate international exchange of products and services, and to develop cooperation in the sphere of intellectual, scientific, technological, and economic activity. ISO is not a United Nations organization. It is a private organization, and ISO members are national standards bodies. The Open System Interface (OSI) reference model was initiated and developed in 1977, and adopted by ITU-T for establishing a universally accepted structured approach for communication between open systems.

- The American National Standards Institute (ANSI): ANSI was formed in 1918. ANSI represents USA as an ISO member and a voluntary, non-profit/non-government organization (see Figure 1-65).

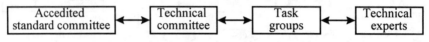

Figure 1-65   ANSI's Organizational Structure.

The standards approved by ANSI have been supported, since its establishment, by trade organizations, professional societies, and manufacturing companies. ANSI members are representatives from more than 200 professional societies, and over 1,000 US corporations. ANSI members coordinate the development of voluntary national standards in the United States, approves American national standards, and maintain interfaces with US government agencies.

Address:	American National Standards Institute	Telephone:	212-642-4900
	1430 Broadway	Telex:	424296 ANSI UI
	New York, NY 10018	Fax:	212-302-1286

Example 1-15: Describe the ANSI process for administering SONET standards.

The T1 (T stands for telecommunications, and 1 for the first ANSI activity) committee provides a public forum for developing interconnection standards. T1 has several subcommittees (committees): (1) T1E1 for network interface, (2) T1M1 for operations technology, (3) T1Q1 for performance, (4) T1S1 for service, architecture, and signaling, (5) T1X1 for digital hierarchy, and (6) T1Y1 for specialized subjects. Figure 1-66 illustrates the ANSI administrative structure.

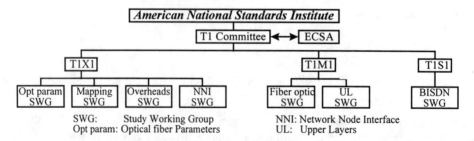

Figure 1-66   ANSI Organizational Structure Supporting Synchronous Optical
NETwork (SONET) Standards.

- Exchange Carriers Standard Association (ECSA): ECSA was formed at the time of the Bell System Divestiture (i.e., ECSA was created in response to an FCC request for public comments regarding the post-divestiture standards process). ECSA members include exchange carriers, inter-exchange carriers, telecommunications resellers, and manufacturers. ECSA is the sponsor of standards committee T1.

- Electronic Industries Association (EIA): EIA was formed in 1924, and has developed more than 400 standards and publications (mostly hardware oriented). EIA members include more than 4,000 industry and government agencies. EIA has contributed to both ITU-T and ISO. A recent EIA standard is RS-232-C.

- Institute of Electrical and Electronics Engineers (IEEE): The IEEE was formed in 1884, and is chartered to advance the theory and practice of electrical engineering; electronics, radio, allied branches of engineering, or related arts and services. For example, IEEE continuously contributes to the creation of standards in the following areas: data communication systems, transmission systems, telecommunications switching, and computer communications. IEEE also performs standards coordinating and liaison activities. IEEE includes 17 technical committees that are accredited by ANSI, since IEEE standards are routinely submitted for adoption by ANSI. Two recent IEEE standards are IEEE.802 standard for local are networks, and IEEE.828 standard for software configuration management.

- Conference of European Posts and Telecommunications (CEPT): CEPT was formed in 1959, and charted to improve postal and telecommunications relations between European countries to form a homogeneous, coherent, and efficient unit on a continental scale. CEPT activities are now part of ETSI

- European Telecommunications Standards Institute (ETSI): ETSI was formed in 1988 to replace CEPT as a regional standards body. Its charter is to implement European policies for separation of operational and regulatory functions.

- European Computer Manufactures Association (ECMA): ECMA was formed in 1961, and currently includes 32 technical committees. ECMA is involved in the following areas: (1) Media and hardware (including product safety and acoustics), (2) Coding of software, (3) Document architecture and interchange, (4) Communication, network, and system level interconnections, and (5) Database and portable common tool environment.

- Other standards organization: Various US Government agencies, Asia Pacific Telecommunity (APT), Japan Standards Association (JSA), Bell Communications Research (Bellcore, now Telcordia), etc.

Review Questions II for Chapter 1:

(14) The telecommunications pyramid requires four parts: transmission, _____, _____, and _____.

(15) (True, False) Transmission media include twisted-pair wires, coaxial cables, waveguides, airway, and optical fibers. But, in the US, waveguides are not used as long-haul transmission facilities.

(16) Among transmission media, presently one single fiber can carry up to _____ voice channels simultaneously.

(17) Several definitions have been developed to describe system bandwidth. The most common used one is the _____-bandwidth, which is the frequency range between the two _____ points of the system _____.

(18) List the common used transmission equipment: _____ terminals, _____ systems, _____, _____, _____, _____, _____, _____, etc.

(19) List the most important applications of a DCS: _____ of traffic, _____ of traffic, providing _____, _____ and _____.

(20) A echo canceller with a 50-dB cancellation capability will reduced an echo to _____ % of its power level.

(21) The primary functions performed by a cellular network, besides routing the information, include: _____, _____, _____, _____, _____, _____, etc.

(22) Zero suppression technique is used for timing synchronization in a digital link. Two methods have been used for zero suppression: _____ (e.g., in the US, DS3 uses _____, DS2 uses _____, and DS1 uses _____), and _____. (e.g., SONET/SDH uses _____).

(23) Gateway exchanges perform, in addition to routing information, _____, _____, and _____.

(24) For data transport, there are two public networks that can be used. They are _____-switched _____, and _____-switched _____.

(25) Both common used ISDN interfaces are: _____ and _____ (i.e., _____ISDN and _____-ISDN).

(26) A service, provided by local exchange, that provides business customers similar services as PBX is called the _____ service.

Appendix 1-1

Resistance, Inductance, and Capacitance

This appendix provides a brief review of circuit characteristics: resistance, inductance, and capacitance. Their relationships to voltage, current, and power are also presented.

A1-1   RESISTOR/RESISTANCE

Resistance is the friction (i.e., the opposition component) that acts upon a force in any form. For telecommunications, resistance directly determines the current, the voltage or the power delivered to the signal destination (Figure A1-1). It is important to understand the relationship between circuit voltage, current, and power. For the network shown in Figure A1-1, the signal source is assumed to deliver a voltage of V_s volts, which varies with time, and represents the signal strength. Assume the network resistance is R_s (commonly referred to as the source resistance), and the load resistance is R_L. The current delivered to the load, I_L, the voltage cross the load resistor (V_L), and the power delivered to the receiver (P_L) are given, by the following equations:

$$I_L = \frac{V_s}{R_s + R_L} \tag{A1-1}$$

$$V_L = I_L \times R_L \tag{A1-2}$$

$$P_L = I_L^2 \times R_L \tag{A1-3}$$

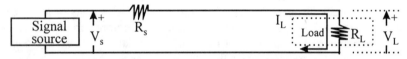

Figure A1-1   Source, Load and Resistance.

The relationship stated in Eq. (A1-2) and Eq.(A1-1) is known as Ohm's law. From Eq. (A1-1) it can be seen that the current flowing through the load (the receiver) (R_L) will decrease as the overall resistance (contributed by the network, the source resistance, and/or the load resistance) increases.

Eq.(A1-3) indicates that the power delivered to the load is proportional to the square of the load current multiplied by the value of the load resistance.

The basic unit used to express electrical resistance is the ohm (Ω; other units are kΩ, MΩ, etc.). The basic unit used to express electrical current is the ampere (A; other units are mA, μA, etc.). The basic unit used to express electrical voltage is the volt (V; other units are kV, mV, μV, etc.). The basic unit used to express electrical power is the watt (W; other units are kW, mW, MW, etc.).

A1-2   INDUCTOR/INDUCTANCE

An inductor is an electrical network element that stores magnetic energy (generated by a magnetic field) and has a basic unit of the henry (h; other units are: *mh*, *µh*, etc.). They are components found in network circuits, and are fundamental elements of a transformer.

The inductance (L) describes the property of an electrical coil (i.e., inductor) that contributes to the network impedance as a reactive component (described later in this appendix). The impedance ($Z_L = jX_L$) due to an inductor is related to f (the network frequency) as shown in Eq.(A1-4).

$$X_L = \omega L = 2\pi f L \qquad (in\ \Omega;\ if\ f\ in\ Hz,\ and\ L\ in\ henry) \qquad (A1\text{-}4)$$

The parameter "ω" is called the radian frequency of the circuit, and has a unit in radians. Note that "ω" is related to the frequency (f) as $\omega = 2\pi\ f$.

A1-3 CAPACITOR/CAPACITANCE

A capacitor (C) is one of the three major passive components (i.e., resistor, inductor, and inductor) in telecommunications network circuits. Its function is different from an inductor. That is, it stores electrical energy, rather than magnetic energy. The capacitor has a basic unit of the Farad (f; other units are: *mf, µf, pf,* etc.) .

Capacitance is a property (of a capacitor) that contributes to the over network impedance as a reactive component. The impedance ($Z_C = 1/jX_C$) due to a capacitor is related to f the network frequency by the following equation:

$$X_C = \frac{1}{\omega C} = \frac{1}{2\pi f C} \qquad (in\ \Omega;\ if\ f\ in\ Hz;\ and\ C\ in\ farad) \qquad (A1\text{-}5)$$

A1-4 IMPEDANCE

Eqs.(A1-1), (A1-2) and (A1-3) can be used exclusively if the circuit contains only resistors. However, a practical circuit never contains just resistance, but consists of all three passive components (i.e., resistors, capacitors, and inductors). Therefore, an electrical circuit is typically represented as an impedance that has both the resistive and reactive components. The resistive component is contributed by the circuit resistance, and the reactive component is contributed by the circuit capacitance and inductance. Therefore, the parameters, R_s and R_L, in Eqs.(A1-1), (A1-2) and (A1-3) must be replaced by source and load impedances: Zs, and ZL. For example, the voltage that appears across the load impedance (ZL) is given by Eq.(A1-6), instead of Eq.(A1-2). Note that the load is a combination of the resistance "R", inductance "L", and capacitance "C" as shown in Figure A1-2.

$$V_L = I_L \times Z_L \qquad (A1\text{-}6)$$

$$Z_L = R + j\omega L - j\frac{1}{\omega C} = R + j\left(2\pi f L - \frac{1}{2\pi f C}\right) \qquad (A1\text{-}7)$$

In Eq.(A1-7) the load impedance (Z_L; of the receiver) contains the parameter "j" (an imaginary number), indicating the impedance value is a complex number (i.e., not a real number).

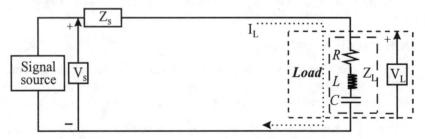

Figure A1-2   The Load Impedance (Z_L).

Appendix 1-2

dB, dBm, TLP, dBm0, dBrn, dBrnC and dBrnC0

Two units, dB (decibel) and dBm (dB with respect to 1 mW), are used throughout this book. These two units are essential for communication systems study, analysis, and design. Their definitions and applications are discussed in this appendix. For advanced network design and analysis, engineers are also required to understand the following units: Transmission Level Point (TLP), dBrn (dB with respect to reference noise), dBrnC (dB with respect to reference and using the C-message weight) and dBrnC0 (dB with respect to reference, using the C-message weight, and measured at 0TLP point).

A2-1   dB (DECIBEL)

This unit is often mistakenly used in the telecommunications industry. The unit dB is appropriate for the following four application fields:

   (1)   Gain (power, voltage or current; most frequently power)
   (2)   Loss (power, voltage or current; most frequently power)
   (3)   Signal-to-noise ratio (S/N)
   (4)   Echo cancellation

This section provides the mathematical definitions for the first three applications, and an example is used to describe echo cancellation.

- Gain/Loss:

   The mathematical definition of a (power) gain or loss is given Eq.(A2-1):

$$\text{Gain (in dB)} = \text{Loss (in dB)} = 10 \times \log \frac{p_1}{p_2} \qquad \text{(A2-1)}$$

The inverse of dB is expressed as:

$$gain\ (unitless) = 10^{\frac{Gain\ in\ dB}{10}} \qquad \text{(A2-2)}$$

Or,
$$loss\ (unitless) = 10^{\frac{Loss\ in\ dB}{10}} \qquad \text{(A2-3)}$$

- S/N Ratio:

   The Signal-to-Noise ratio (S/N) of a communication system is often expressed in dB [shown in Eq.(A2-4)] rather than a unitless ratio [shown in Eq.(A2-5)]:

$$\frac{S}{N}\ (in\ dB) = 10 \times \log \frac{s}{n} = 10 \times \log \frac{Signal\ power}{Noise\ power} \qquad \text{(A2-4)}$$

$$\frac{s}{n} \ (unitless) = \frac{Signal\ power}{Noise\ power} = 10^{\frac{S/N\ dB}{10}} \tag{A2-5}$$

Example A2-1: Assuming an echo canceller has a cancellation of 30-dB, express the resultant (final) echo in terms of the original echo before cancellation was performed.

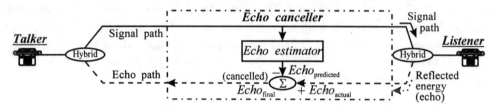

Figure A2-1   Application of an Echo Cancellor.

The processor of an echo cancellor (i.e., echo estimator) generates a signal that emulates the predicted echo (Echo$_{predicted}$; E$_p$) that is expected to be returned to the talker. The actual echo (Echo$_{actual}$; E$_a$) generated by the impedance mismatch of the return path. This actual echo is used to cancel the predicted echo so that the final echo (Echo$_{final}$; E$_f$) heard by the talker can be reduced in power.

With 30-dB cancellation, the following relationship can be used to obtain the final echo (E$_f$) in terms of the actual echo (E$_a$):

$$30\ dB = 10 \times \log \frac{E_a}{E_f} \quad or \quad \frac{E_a}{E_f} = 10^{\frac{30\ dB}{10}} \tag{A2-6}$$

From which the following expression can be obtained:

$$\text{Echo}_{final} = [1 / (10^{30/10})] \times \text{Echo}_{actual} = 0.001 \times \text{Echo}_{actual} = 0.1\% \text{ of Echo}_{actual}$$

The result of using an echo canceller reduces the final echo power to approximately 0.1% of the original echo strength. An echo cancellor with this amount of echo cancellation is considered to have excellent performance because the "talker" will hear very little echo. In practical applications echo cancellation typically ranges (20~ 50 dB).

A2-2   POWER UNIT: dBm

In a communication system, two different types of units have used to express the signal power strength:

(1)   Traditional units: Watt (w, mw, kw, μw, etc.)
(2)   Special unit: dBm (dB with respect to 1 mw of power)

The unit dBm (dB with respect to the 1 mw reference power level) is used to express the signal power. It can also be expressed by conventional unit such as watts (mw, μw, etc.) When signal power is given in dBm, the entire system can be expressed in dBm and dB (several examples are given later in this appendix). Power (P) in dBm is defined by the following equation:

$$P \text{ (in dBm)} = 10 \times \log p \text{ (in mw)} / (1 \text{ mw}) \qquad \text{(A2-7)}$$

$$p \text{ (in mw)} = 10^{\frac{P \text{ (in dBm)}}{10}} \qquad \text{(A2-8)}$$

A similar unit, dBμ (dB with respect to the 1 μw reference power level) can be used for a system that transmits/receives a weak signal. Eqs.(A2-7) and (A2-8) are modified to:

$$P \text{ (in dB}\mu\text{)} = 10 \times \log p \text{ (in } \mu w) / (1 \ \mu w) \qquad \text{(A2-9)}$$

$$p \text{ (in } \mu w) = 10^{\frac{P \text{ (in dB}\mu\text{)}}{10}} \qquad \text{(A2-10)}$$

A2-3 SEVERAL USEFUL SHORT-CUT FORMULARS FOR dB AND dBm

The following four relations between the units dB and dBm are very important in the analysis and design of telecommunications networks.

(A dB)	$\pm$	(B dB)	=	(A $\pm$ B) dB	(A2-11)
(A dBm)	$\pm$	(B dB)	=	(A $\pm$ B) dBm	(A2-12)
(A dBm)	$-$	(B dBm)	=	(A $-$ B) dB	(A2-13)
(A dBm)	"+"	(B dBm)	=	? dBm	(A2-14)

Eq.(A2-14) is often incorrectly stated as (A dBm + B dBm) = (A + B) dBm. For example, 10 dBm + 20 dBm ≠ 30 dBm. The accurate expression of Eq.(A2-14) is given as follows:

$$A \ dBm \ "+" \ B \ dBm = 10 \times \log (10^{4/10} + 10^{B/10}) \ dBm \qquad \text{(A2-15)}$$

In general, the sum of n powers in dBm, ($P_T = \Sigma P_i$ with i = 1, n), P_T is given as follows:

$$P_T = 10 \times \log \sum_{i=1}^{n} 10^{\frac{P_i \text{ (dBm)}}{10}} \quad \text{(dBm)} \qquad \text{(A2-16)}$$

The mathematical definitions given in Eqs. (A2-7) throughout (A2-16) are the basic formulas required for telecommunications network design and analysis. However, a couple of short-cut approximations are useful when dealing with practical system applications. The approximations are given in Table A2-1.

Table A2-1   Several Useful Approximations (Power only).

3 dB (3.01 dB)	gain ➔ x 2	7 dB (6.99 dB)	gain ➔ x 5
	loss ➔ x (1/2)		loss ➔ x (1/5)

The approximations in Table A2-1 imply that

(1)   If a signal has a power gain of 3 dB (more precisely 3.01 dB), the signal's power is doubled (i.e., the power has been increased by 2 times).

(2)   If a signal experiences a power loss of 3 dB, the signal's power is reduced to ½ of its original power.

(3)   If the signal has a power gain of 7 dB (more precisely 6.99 dB), the signal's power is increased by 5 times.

(4)   If the signal has a power loss of 7 dB, the signal's power is to 1/5 of its original power.

Example A2-2: Express the relationship between the signal power and the noise power, assuming the signal-to-noise (S/N) ratio is 18 dB and 21 dB, respectively.

A S/N = 18 dB implies that the received signal power is stronger than the noise power. Thus, the concept of gain (i.e., gain makes power stronger) is applied. Applying Eq. (A2-11) to decompose 18 dB as (3 dB + 3 dB + 3 dB + 3 dB + 3 dB + 3 dB) implies a gain of $2 \times 2 \times 2 \times 2 \times 2 \times 2 = \underline{\textbf{\textit{64}}}$ (recall each gain of 3 dB implies a power doubling). Therefore, it can be interpreted that:

$$S/N = 18 \text{ dB} \quad \blacktriangleright \quad \text{Signal power} = \underline{64} \times \text{noise power.}$$

and,

$$S/N = 21 \text{ dB} \quad \blacktriangleright \quad \text{Signal power} = \underline{128} \times \text{noise power.}$$

It should be understood that the above calculations are approximation, but they are accurate enough for practical applications. By using Eq.(A2-5), the exact answer can be derived as follows:

$$\frac{s}{n} = 10^{18/10} = 10^{1.8} = 63 \quad \blacktriangleright \quad \text{Signal power} = \underline{63} \times \text{noise power.}$$

and,

$$\frac{s}{n} = 10^{21/10} = 10^{2.1} = 125.9 \quad \rightarrow \quad \text{Signal power} = \underline{126} \times \text{noise power.}$$

Clearly, the exact answers can be derived using mathematical formulas, but, the approximation approach described here is acceptable in most cases.

Example A2-3: If a system has a transmitted power is 100 w, facility losses/gains are shown in Figure A2-2, and a receiver sensitivity of −50 dBm (from the power budget viewpoint), determine whether the system has suitable regenerators (Gains shown in Figure A2-2 represent regenerators used along the transmission path).

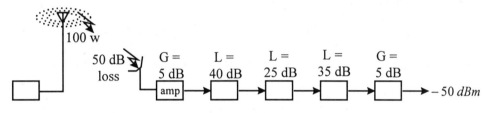

Figure A2-2   Cascading Facility Losses and Gains.

First, by applying Eq.(A2-7), the transmitted power of 100 w is converted into units of dBm. That is, 50 dBm [= 10 log (100,000 in mw)]. This step is required because:

> *"In analyzing/designing a system from a power budget viewpoint, the dB and dBm units should be used"*.

Next, by applying the power budget for the end-to-end connection, the received power can be obtained as follows:

$$P_R = 50 \text{ dBm} + 5 \text{ dB} - 40 \text{ dB} - 25 \text{ dB} - 35 \text{ dB} + 5 \text{ dB} = -40 \text{ dBm} \quad \text{(A2-17)}$$

From Eq.(A2-17) it can be seen that − 40 dBm is stronger than the receiver sensitivity (− 50 dBm). Therefore, the system in Figure A2-2 equipped with two regenerators (a gain of 5 dB) is acceptable from the power budget view. In Eq.(A2-17) (i.e., the power budget formula), either one of the following two conventions can be used as long as the convention is consistent throughout the entire design/analysis calculation:

> (1)   Use "+" for gains, and use "−" for losses
>
> (2)   Use "−" for gains, and use "+" for losses

Example A2-4: Explain the meaning of a signal having a signal power of 11 dBm using the approximation method, and then check the answer by using the formula.

First, by applying the approximation listed in Table A2-1, the power of 11 dBm can be written as follows:

11 dBm = + 11 dBm = 11 dB stronger than 1 mw   ⇔   "+" implies stronger

$$= (7 + 7 - 3) \text{ dB stronger than 1 mw}$$

$$= (7 \text{ dB gain})(7 \text{ dB gain})(3 \text{ dB loss}) \text{ with respect to 1 mw}$$

$$= 5 \times 5 \times (1/2) \times (1 \text{ mw}) = 12.5 \text{ mw}$$

Second, by applying the formula given in Eq.(A2-8), the (exact) signal's power of 11 dBm can be derived as:

$$p = 10^{(11 \text{ dBm})/10} = 10^{1.1} = 12.59 \text{ mw}$$

It can be seen that 12.5 mw ≈ 12.59 mw ⇔ *Table A2-1 approximation is useful*.

Example A2-5: Explain the meaning of a sensitivity of −14 dBm.

First, −14 dBm can be viewed as 14 dB below (i.e., weaker than) 1 mw power. If a signal becomes weaker, a loss must have occurred in the system. A 14 dB (= 7 + 7) loss implies that signal strength is (1/5) × (1/5) [from Table A2-1: 7 dB → (1/5)] of the original (reference) signal power, which in this case is 1 mw. Therefore, a signal power of −14 dBm is equivalent to:

$$\text{Power of } -14 \text{ dBm} = (1/5) \times (1/5) \times 1 \text{ mw} = 40 \text{ } \mu\text{w}$$

Therefore, a receiver with a sensitivity of −14 dBm must receive a minimum signal power of 40 μw to function properly.

Therefore, it can be seen that *a −14 dBm power level does not imply the signal has a "negative power"*. Instead, it implies that the signal power is 14 dB weaker than a 1 mw reference power level. Another common mistake is assuming a *0 dBm* signal has a power of 0 w. It actually has a power of *1 mw*, since the signal power is neither stronger nor weaker than 1 mw. Another power level, used quite often is 10 dBm. A *10 dBm* signal is equivalent to *10 mw* power.

Eq.(A2-18) is a short-cut for calculating the receiving power for a system with a loss of L dB, and a transmit power of p_T:

$$p_R = p_T \times 10^{-\frac{L \text{ } dB}{10}} \tag{A2-18}$$

Example A2-6:   Determine the received power if a system has net loss of 60 dB, and the transmit power is 100 w.

By applying Eq. (A2-18), the received power can be derived as follows:

$$p_R = p_T \times 10^{-\frac{L\,dB}{10}} = (100\ w) \times 10^{-\frac{60}{10}} = 10\ mw$$

By using the approximation in Table A2-1, 60 dB = (3 + 7) + (3 + 7) + (3 + 7) + (3 + 7) + (3 + 7) + (3 + 7). = (3 + 7) × 6 ⇔ (1/2 × 1/5)^6. Thus, the same result can be obtained:

$$p_R = (100\ w) \times (1/2 \times 1/5)^6 = 10\ mw$$

Example A2-7: There conditions when the received signal power is stronger than the maximum power that a receiver can handle. Assuming the system described in Example A2-3 only operates properly with a received power in the range of (−50 dBm ~ −45 dBm), determine what action should be taken.

The concept is rather simple. The received signal has a power of −40 dBm (as shown in Example A2-3), which is 5 dB [≡ −40 dBm − (−45 dBm)] "*hotter*" than the maximum power that the receiver can handle. Therefore, at the receiver, an "*attenuator*" with a loss in the range of (5 dB to 10 dB) must be applied since:

$$-40\ dBm - 5\ dB\ \ = -45\ dBm$$
$$-40\ dBm - 10\ dB = -50\ dBm$$

which will bring the signal within the receiver's operating range, (−50 dBm ~ −45 dBm).

Example A2-8: Determine the total power (in dBm) if the two individual signals are (1) 10 dBm and 10 dBm; (2) 10 dBm and 20 dBm; and (3) 10 dBm and 50 dBm.

(1)   For two signals of 10 dBm and 10 dBm, apply Eq.(A2-15) or (A2-16) to obtain:

$$P_T = 10\ dBm + 10\ dBm = 10 \times \log(10^{10/10} + 10^{10/10}) = 13\ dBm$$

(2)   For two signals of 10 dBm and 20 dBm, apply Eq.(A2-15) or (A2-16) to obtain:

$$P_T = 10\ dBm + 20\ dBm = 10 \times \log(10^{10/10} + 10^{20/10}) = 20.4\ dBm$$

(3)   For two signals of 10 dBm and 50 dBm, apply Eq.(A2-15) or (A2-16) to obtain:

$$P_T = 10\ dBm + 50\ dBm = 10 \times \log(10^{10/10} + 10^{50/10}) = 50\ dBm$$

This example confirms the following approximations can be used for practical applications:

- A dBm "+" A dBm = (A + 3) dBm
- A dBm "+" B dBm ≈ (B + a number < 3) dBm, if A < B
- A dBm "+" B dBm ≈ B dBm, if A << B

Example A2-9: Assume a coupler, a device that combines two signals into one signal [e.g., a coupler is used in Wavelength Division Multiplexing (WDM) systems], has an insertion loss of 2 dB, determine its output signal (Figure A2-3).

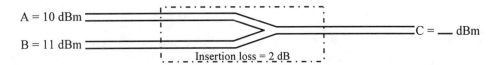

Figure A2-3 Coupler with Two Inputs.

By applying Eq.(A2-15), the sum of signals A and B (10 dBm and 11 dBm) can be derived as follows:

$$P_T = 10 \text{ dBm} + 11 \text{ dBm} = 10 \times \log (10^{10/10} + 10^{11/10}) = 13.54 \text{ dBm}$$

Thus,

Signal C has a power of 11.54 (= 13.54 − 2) dBm.

A2-4   TRANSMISSION LEVEL POINT (TLP)

The power levels of various points in a communications network are usually different. For network design/analysis, and controlling network performance requirements, each node (point) in a network is identified by its Transmission Level Point (TLP) [e.g., 6 (dB) TLP, or 6TLP, −3TLP, etc.]. In a network, the reference point is defined to be at the "*0 dB TLP*". The reference point can be an actual node or a "virtual node" (i.e., a theoretical point in the network).

Definition A2-1: The dBm0 represents the signal power level in dBm when the signal power is measured at network reference point.

$$x \textbf{ dBm0} \rightarrow$$

The signal power level = x **dBm** if the power is measured at **0 TLP**   (A2-19)

Example A2-10:   Figure A2-4 shows a seven-node network. Assuming Node E is the network reference point (0 dB TLP; simply 0 TLP), the transmission direction and the facility (regenerator included) loss or gain are as shown (Figure A2-4), and the signal level at Node E is −3 dBm. Determine the TLP and the signal level (dBm and dBm0) for each node in the network. That is, verify the data in Table A2-2.

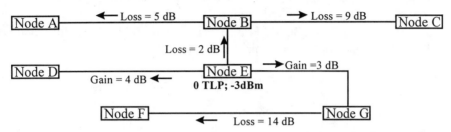

Figure A2-4   A Seven Node Network Example.

Table A2-2   Signal Levels and TLPs.

Level	A	B	C	D	E	F	G
TLP	−7	−2	−11	4	**0**	−11	3
dBm	−10	−5	−14	1	*−3*	−14	0
dBm0	−3	−3	−3	−3	*−3*	−3	−3

Since Node E is the network reference point, by definition it is at the 0 (dB) TLP. The signal power measured at "E" is assumed to be −3 dBm (about 0.5 mw). For node E (the reference point), by the definition given in Eq.(A2-19), its dBm0 value is equal to its dBm value. That is, at node "E", the signal power level is −3 dBm or −3 dBm0.

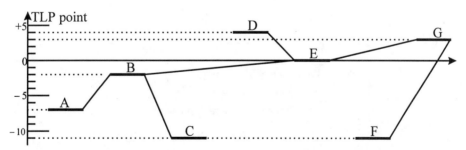

Figure A2-5   Various TLPs of the Network Shown in Figure A2-4.

There is a gain from "E" to "D", and the signal at "D" is 4 dB stronger than the signal level at "E". Therefore, the node "D" is at "4 TLP". As the signal travels from "E" to "D", the signal will gain a power of 4 dB (−3 dBm + 4 dB = 1 dBm). The facility connecting "E" and "G" has a gain of 3 dB, "G" is at +3 TLP, and has a power of 0 dBm

(= −3 dBm + 3 dB). From "G" to "F" there is a loss of 14 dB, thus, from "E" to "F" the net loss is 11 dB. Therefore, "F" is at −11 TLP, and has a power of −14 dBm [= −3 dBm + (−11 dB)]. By repeating the same procedure, the "TLP" and the "dBm" values listed in Table A2-2 can be derived for the other network nodes.

By definition, "***The dBm0 at any point in the network should be identical.***" In this example, Node "E" is at −3 dBm0. Therefore, the "dBm0" row in Table A2-2 lists the values at each node as "−3 dBm0". From the results in Table A2-2, the relationship among TLP, dBm and dBm0 for any point is given as follows.

$$X \text{ dBm} - Y \text{ TLP} = Z \text{ dBm0} \qquad (A2\text{-}20)$$

A2-5 REFERENCE NOISE

A noise signal with a power level of 10^{-12} w (or −90 dBm) is defined as a "***reference noise***". In analyzing a communication system's performance, the noise level is often expressed in units defined with respect to the reference noise level. The unit is "dBrn" (i.e., dB with respect to the reference noise).

Example A2-11: Express a noise signal with a power level of 5×10^{-9} watts in dBrn.

Anytime this type of information is to be derived, it is recommended that a vertical line be drawn with three points as shown in Figure A2-6.

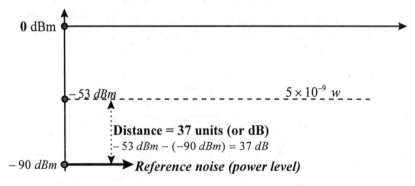

Figure A2-6   Reference Noise and dBrn.

First, convert the noise power of 5×10^{-9} watts into −53 dBm by applying Eq.(A2-7) [i.e., $10 \times \log (5 \times 10^{-9}/10^{-3}) = -53$ dBm]. The result is plotted in Figure A2-6, from which it can be seen that the power level point of −53 dBm is 37 units (dB) above the reference noise level. By definition, the noise signal has a power level of 37 dBrn (+37dBrn, which is +37 dB above the reference noise) .

A2-6 C-MESSAGE CURVE (C-MESSAGE WEIGHT)

Communication system study, analysis, and design associated with voice functions requires consideration of the human ear sensitivity factor (known as the C-message curve; shown in Figure 1-15). This topic has been presented in Section 1.2.8. Two important units tie the communication system performance with human reactions: dBrnC and dBrnC0. In these units, C refers to as C-message. That is:

> "A" dBrnC $\Rightarrow$ noise power: "A" dB above the reference noise
> taking "C weight" into consideration

and,

> "B" dBrnC0 $\Rightarrow$ noise power: "B" dB above the reference noise
> taking "C weight" into consideration
> if the noise is measured at 0 TLP

Example A2-12: Express a 200 Hz noise with a noise power of 5×10^{-9} watts in dBrnC.

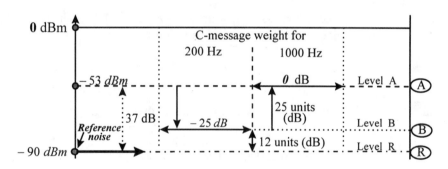

Figure A2-4 Relationship between dBm, dBrn and dBrnC.

As it has been calculated (in Example A2-11) that a power of 5×10^{-9} watts is equivalent to a power of −53 dBm as shown (level A) in Figure A2-4. That is, a noise signal with a power of 5×10^{-9} watts is equivalent to a power of 37 dBrn (from level "R" to level "A"). From Figure 1-15, the C-message weight for a 200 Hz tone is −25 dB. This implies that a 200 Hz noise has less effect to human ear than a 1000 Hz noise by an amount of 25 dB. This is shown as level "A" for a 1000 Hz noise, and level "B" for a 200 Hz noise (see Figure A2-4). It can be seen that the difference between level "B" and level "R" is 12 dB. Therefore, the 200 Hz noise with a noise power of 5×10^{-9} watts is known to have a noise power of **_12 dBrnC_** (from level "R" to level "B". That is, 12 <u>dB</u> <u>rn</u> <u>C</u> is 12 <u>dB</u> stronger than the <u>reference noise</u> when the <u>C</u>-message weighting is used. It can be seen that 12 dBrnC = 37 dBrn + (−25) dB of C-message weight. From this example, the definition of the unit dBrnC can be derived as follows:

$$\boxed{\textbf{X} \text{ dBrnC} = \textbf{Y} \text{ dBrn} + \text{``C-message'' weight}} \qquad \text{(A2-21)}$$

Example A2-13:   If the 200 Hz noise (with a power of 5×10^{-9} watts) in Example A2-11 is observed at a node which is at -7 TLP [e.g., Node "A" in Figure A2-4 (Table A2-2)], express its power in unit of dBrnC0.

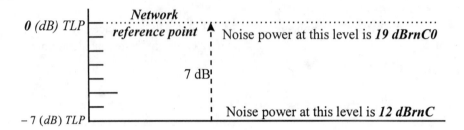

Figure A2-5   Relationship between dBrnC and dBrnC0.

As shown in Figure A2-5, the 200 Hz noise with a power of 5×10^{-9} watts is equivalent to 12 dBrnC. It is assumed this noise is measured at -7 (dB) TLP as indicated in Figure A2-5. This 200 Hz noise can then be expressed as follows:

$$12 - (-7) = 19 \text{ dBrnC0}$$

In summary, the definition of dBrnC0 is given by the following:

$$\boxed{\textbf{X} \text{ dBrnC0} = \textbf{Y} \text{ dBrnC} - (\textbf{Z} \text{ TLP})} \qquad \text{(A2-22)}$$

Appendix 1-3

ITU-T Study Groups and Recommendations

This appendix is a brief summary of the functions of each ITU-T Study Group and ITU-T series of Recommendations. A summary of ITU-T recommendation examples is also provided for illustration purpose.

Table A1-3.1   ITU-T Study Groups/Functions.

Study Group	Functions
1	(Telecommunications) service definition
2	Network operation
3	Tariff and accounting principles
4	Network maintenance
5	Protection against electromagnetic environmental effects
6	Outside plant
7	Data networks and open system communications
8	Terminal for telematic services
9	Television and sound transmission
10	Languages for telecommunications applications
11	Switching and signaling
12	End-to-end transmission performance of networks and terminals
13	General network aspects
14	Modems and transmission techniques for data, telegraph, and telematic services
15	Transmission systems and equipment

Table A1-3.2   ITU-T Recommendations

Rec. series	Scope of Recommendation
A	Organization of the work of the ITU-T
B	Means of expression (definitions, symbols, classification)
C	General telecommunication statistics
D	General tariff principles
E	Overall network operation, telephone service, service operations, and human factors
F	Non-telephone telecommunication service
G	Transmission systems and media, digital systems, and networks
H	Audiovisual and multimedia systems
I	Integrated Services Digital Network (ISDN)
J	Transmission of television, sound programme, and other multimedia signals
K	Protection against interference

Table A1-3.2   ITU-T Recommendations (continued).

Rec. series	Scope of Recommendation
L	Construction, installation and protection of cable, and other elements of outside plant
M	TMN and network maintenance: international transmission systems, telephone circuits, telegraphy, facsimile, and leased circuits
N	Maintenance: international sound programme, and television transmission circuits
O	Specification of measuring equipment
P	Telephone transmission quality, telephone installations and local line networks
Q	Telephone switching and signaling
R	Telegraph transmission
S	Telegraph services terminal equipment
T	Terminal equipment for telematic services
U	Telegraph switching
V	Data communication over the telephone network
X	Data communication networks and open system
Y	Global information infrastructure
Z	Languages and general software aspects for telecommunication systems

Table A1-3.3   ITU-T Recommendation Examples.

Rec.	Scope of recommendation
G.701	Vocabulary of digital transmission and multiplexing, and Pulse Code Modulation (PCM) terms
G.702	Digital hierarchy bit rates
G.704	Synchronous frame structures used at 1544, 6312, 2048, 8448 and 44736 kbps hierarchical levels
G.707	Network node interface for the Synchronous Digital Hierarchy (SDH)
G.726	40, 32, 24, 16 kbps Adaptive Differential Pulse Code Modulation (ADPCM)
G.729	Coding of speech at 8 kbps using Conjugate-Structure Algebraic-Code-Excited Linear-Prediction (CS-ACELP)
G.753	Third order digital multiplex equipment operating at 34.368 Mbps, and using positive/zero/negative justification
G.754	Fourth order digital multiplex equipment operating at 139.264 Mbps, and using positive/zero/negative justification
G.784	SDH management
G.832	Transport of SDH elements on PDH networks
G.842	Interworking of SDH network protection architectures

CHAPTER 2

Networks, Services, and Call Connections

Chapter Objectives

Upon the completion of this chapter, you should be able to:

- Describe telecommunications service platforms, various classes of service (e.g., messaging, retrieval, distribution, etc.), and service examples.

- Describe (1) an inter-office (inter-toll) connection (including call origination, originating register, dial tone, translation, routing, path selection, outpulsing, automatic message accounting, trunk seizing, alerting, audible ring, ringing, etc.); (2) a local (intra-office) call; (3) examples of completed connections in a typical long distance voice call; (4) an international call carried via gateway exchanges; (5) a private network connection; and, (6) a wireless connection.

- Describe an Internet connection: including TCP/IP capabilities/functions, routers, gateways, IP connection for Voice over IP (VoIP), dial-up Internet connections, and dedicated link to IP Service Provider (ISP).

- Describe ATM services: including switched ATM connections, ATM virtual path connections, ATM virtual channel connections, ATM permanent virtual connections, ATM-based LAN connections, and ATM-based LAN emulation.

- Describe an SONET/SDH connection: SONET paths, lines, and sections; SONET access, junction and core layer applications; SONET intra-office and inter-office connections, IP over ATM, ATM over SONET/SDH, IP over SONET/SDH, SONET/SDH over DWDM, and IP over DWDM.

2.1 INTRODUCTION

The telecommunications industry originated from a simple point-to-point connections and switched networks, that only provided local services. Over time, it evolved into regional, national, and finally a worldwide network. The physical connection had to be manually made by telephone operators for many decades. The telecommunications network gradually evolved to Direct Distance Dial (DDD), and then International DDD (IDDD).

The "network" was originally designed and implemented to only carry speech (voice) signals. Today, the network can carry voice, data, video, etc. over the same facilities. Telecommunications trends will prepare modern networks to provide multimedia capabilities, and will integrate information movement with network management. The present trends in telecommunication networks are:

- Migration from analog to digital technology.

- Transition from metallic cooper wiring to coaxial cables; from long wavelength radio to short wavelength; very short wavelength, microwave, and eventually infrared; and greater utilization satellite links, and optical fiber as transmission media.

- Conversion from "hardware-driven" to "software-driven" technologies.

- Evolution from "closed network" to "open network" [e.g., Open System Interface (OSI), Transmission Control/Internet Protocol (TCP/IP), frame relay, Fiber Distributed Data Interface (FDDI), Asynchronous Transfer Mode (ATM), Synchronous Optical NETwork (SONET), Synchronous Digital Hierarchy (SDH), and Dense Wavelength Division Multiplexing (DWDM)].

- Upgrading from "voice only" to "integrated voice, data, image, and video" services.

The primary factors that influence telecommunications network evolution are classified into: (1) technical, and (2) non-technical categories. For example:

(1) Technical factors that influence telecommunication networking technologies are:

    * Transmission technologies: The transmission technologies that play important roles in telecommunications network evolution are:

       ♦ Digital technologies: There are numerous advantages gained by implementing networks using digital technologies instead of analog technologies. These advantages include: (1) easier multiplexing for integrating voice, data, video, etc., (2) nearly noise immune (when properly designed) which yields excellent service quality, and (3) enhanced security through encryption and/or scrambling techniques.

       ♦ Optical technologies: This technology supports increased system bandwidth (capacity) and offers excellent service quality.

♦ Wireless technologies: Wireless applications allow subscriber mobility, in addition to offering quick deployment in various geographic/political regions.

♦ Internet services: This service provides broad, quick, and constantly-changing capabilities for a wide variety of end users.

♦ Broadband access technologies: xDSL [e.g., High-speed Digital Subscriber Line (HDSL), Asymmetrical DSL (ADSL), and Very high-speed DSL (VDSL)], cable modems, Hybrid Fiber Coax (HFC), wireless access [Wireless Local Loop (WLL)] and satellite broadband access technologies. All provide the high speed infrastructure needed to deliver enhanced services demanded by end-users.

♦ Optical networking and DWDM: In addition to super high system capacity, they break the "speed bottleneck" of electronic technologies, thereby allowing future enhancements of telecommunications services.

* Switching technologies: Interconnection of end users is supported by low-speed switching techniques (e.g., step-by-step and cross-bar switching) to "high-speed and high throughput" electronics switching systems (ESS). Conversion from analog switching to totally digital switching offers "ease of integration" with digital transmission equipment.

* Signaling technologies: Low-speed in-band signaling is used for performing call-related functions, and dedicated high-speed Common-Channel Signaling (CCS) speeds up call processing with improved service reliability. CCS has since evolved into the Intelligent Network (IN), and the Advanced IN (AIN) to achieve seamless call-related and network management functions.

* Microelectronics: With advances in technology, system size has been reduced drastically. Very Large Scale Integrated (VLSI) technologies are constantly reducing power and size, while increasing speed. Fast processing time and less expensive buffer memory devices will continue to advance.

* Optoelectronics: This technology will continuously improve device speed (faster processing times) and reduce device size.

* Software: Telecommunications networking is evolving from being hardware-driven to software-driven. Continuously improving software technologies will be critical for implementing modern telecommunications networks.

* Operations, Administration, Maintenance and Provisioning (OAM&P): These activities have evolved from a "fire-fighting" approach to routine OAM&P. Automated Operations Systems (OSs) are the only way to manage complex of telecommunications networks. Two prominent protocols are ATM and SONET/SDH.

(2) Non-technical factors: Four significantly influential factors are:

* Deregulation and competition: In the United States of America, the most famous deregulation was the 1984 Bell System's divestiture, in accordance with the FCC's

Modified Final Judgement (MFJ), and 1996 Telecommunications Bill. Many countries has begun to deregulate their telecommunications industry. "Deregulation" always triggers the creation of many independent telecommunications service providers, and equipment vendors, thus encouraging product/service competition. High-speed and excellent quality systems/services, offered by responsive vendors, will dominate future telecommunications development.

* Global standards for telecommunications technologies: The benefits of having standards are countless. Some examples are: improved product quality, technical compatibility, ease of integration, lower development costs, lower manufacturing costs, lower maintenance costs, reduced vendor risk.

* Economic factors: Continuously searching for better and least-expensive services has triggered many new technologies. For example, IP, multimedia, ISDN, broadband access, and wireless services will dominate modern telecommunications industry.

* Customer expectations: Residential applications, customers are seeking the least expensive and most user-friendly access communications methods available (e.g., wireless, broad access to Internet services, etc.). Business customers tend to have the same needs, with additional emphasis on high speed and reliability. For new subscribers, "fast deployment" is critical.

2.2 TELECOMMUNICATIONS SERVICES

"Intelligence" and "complexity" in telecommunications are continuously expanding in networks all over the world. As a result, multiple services are being offered to users in a "global marketplace". The "task" of converting "intelligence" to "pragmatic solutions" for meeting telecommunications needs in both global, and national markets must be treated with care to insure this "world society" is properly served. Therefore, understanding behavior, needs, and social customs are essential for becoming a major player in global telecommunications marketplace.

Telecommunications services is a vital part of daily operations in all societies. Often it is difficult for service providers and end users to make long-term decisions because new networking technologies are on the horizon. The growing complexity of designing, provisioning, and managing a seamless worldwide network is demanding increasing amounts of resources. For example, the integration of existing circuit-switched and packet-switched networks is a technically excellent solutions that requires extensive coordination.

2.2.1 Telecommunications Service Platforms

Telecommunications services are based on "platforms" that interface with the transmission media - the "information superhighway". These platforms process the information, and route it to appropriate destinations (e.g., between switching points and/or users' access points).

The features and capabilities of service platforms are continuously evolving, some examples are described as follows:

- Loop plant: It is used to transport voice, animation, video, computer data, libraries, government records, and other information via common carrier networks to proper destinations, with "fast" response time and high reliability. The metallic loop plant has dominated the North American infrastructure for many years, but is now competing with wireless and fiber optic technology.

- Wireless service promises "fast", "anywhere", and "anytime" access capability.

- Optical fibers offers "excellent" signal quality and "broadband" access/transport.

- Internet computer access is used to send/retrieve, display, print, or forward information to multiple destinations, and has become a worldwide service.

- Switches: They are used to address, translate, and route information to its appropriate destination.

- Common channel signaling: It reliably transports administrative control information between all points in telecommunications networks.

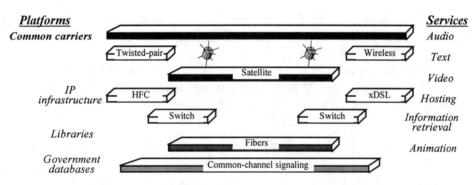

Figure 2-1   Telecommunications Platforms and Services.

As shown in Figure 2-1, the modern architecture for the "information superhighway" involves both optical fiber and wireless services to complete seamless interconnection. The service switching points located close to the customer's topology incorporate creation of services and requests which are subsequently delivered to "backbone" networks.

Example 2-1: Describe the basic elements in the call flow sequence that are used to ensure service viability for telecommunications networks.

Figure 2-2 illustrates the routing of a business (or residential) call, that typically involves two sets of customer premises equipment, several switching machines, transmission equipment, and transmission facilities. The service feature/definition, order, processing, provisioning, billing, maintenance, etc., associated with a call are also shown in Figure 2-2.

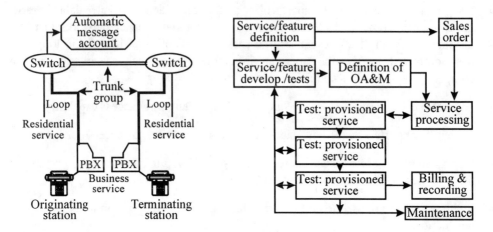

Figure 2-2   Service Order, Tests, and Connection.

Service features are defined when a service order is initiated. The following major features associated with a new service must be defined as follows:

- Numbering: "What number is dialed by the customer?" "What number is used for routing through the network?" "What numbers are recorded on the Automatic Message Account (AMA) tape?" "What numbers are outpulsed at the terminating end?" "What numbers are used for billing the call?"

- Trunking: "What specific trunks are required for this service (e.g., access, inter-toll, international)?" "What specific trunk characteristics are required for this service (e.g., type of signaling, trunk hunt sequence, billing number, call screening)?"

- Routing: "How is the call routed?" "What digits are used for routing?" "How many digits need to be translated?" "Is a database used?" "If used, how is the database accessed?" "Is any call screening required?"

- Screening: "What calls are allowed?" "What calls are denied?" "What calls need to be recorded?"

- Recording: "How is the call recorded?" "What triggers the recording of AMA?" "How are the basic elements recorded?"

- Billing: "What steps are needed to ensure the call is properly billed?" "Is a unique service ID required so that the AMA record for the service can be uniquely identified?" "What tariff rates need to be applied?" "How is the AMA data and customer data correlated to produce a bill?" "Is the call origination/termination clearly identified by AMA information?" "How is the customer's billing information and the recorded AMA call details combined to produce/mail a bill?"

2.2.2 Classification of Telecommunications Services

Communications facilities in a network (e.g., analog or digital, wired, wireless, coaxial cable, or fibers) are made available to the user in the form of defined telecommunications services. The term "*services*" includes all the telecommunications methods and facilities provided to users by the telecommunications carrier (e.g., local or interexchange) for communication over public and private networks. Examples are: telephony, teletext, telefax, videotext, and data transmission services. The services are characterized by their technical, operational, and administrative attributes.

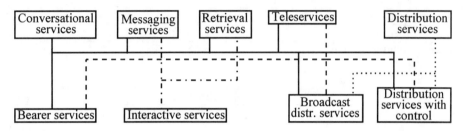

Figure 2-3   Classification of Telecommunications Services and Interrelations.

- Teleservices: These services support "user to user", or "user to host" communications. Examples are: telephone, teletext, telfax, and videotext.

- Conversation services: In general, these services provide the means for bidirectional dialog communication, and real-time (not store-and-forward) end-to-end information transfer from "user to use", or "user to host" (e.g., data processing).

- Messaging services: They provide user to user communication between individual users via storage devices with store-and-forward, mailbox and/or message handling (e.g., information editing, processing and conversion) functions. Examples are message handling services and mail services for audio information, text, data, graphics and high resolution images.

- Retrieval services: These services allow users to access data stored in information centers for public reach. The information is sent on demand basis only.

- Distribution services: These services can either be provided with or without individual user controls. They allow the distribution of information from a central source to an unlimited number of authorized receivers connected to the network. Distribution services without individual user controls are also known as "broadcast" services.

2.2.2.1 Service Example: Fax Services/Standard Fax Groups.

Facsimile (fax) service was originally developed to transmit photographic and documentary information over standard telephone lines. However, it can now be implemented over all transmission media. Several fax standards have been developed/used for several years. Group 1 terminals are now almost obsolete. Most modern fax services are carried by machines of Groups 2 and 3 (Table 2-1). Group 4, which has high-speed capabilities that can reproduce an *A4*-size (*210 mm* by *297 mm*) document in about 2 seconds, has been developed for digital network applications.

Table 2-1   Standard Fax Groups.

	Group 2	*Group 3*	*Group 4*
Transmission time required (A4 document)	3min	1 min	2-4 sec
Modulation technology	AM/PM-VSB	DPSK	Digital[1]
Carrier frequency	2,100 Hz	1,800 Hz	64 kHz
Vertical resolution (1/mm)	3.85 or 5.3	3.85 or 7.7	7.7 or 15
Horizontal resolution (pel/mm)	5.3	8	16
Image signal	Analog	Digital	Digital
Handshake signal	Audio tones	300 bps FSK (ITU-T V21)	ISDN compatible[2]

(1) 64 kbps   (2) ITU-T V29/33

The information in a document that is being transmitted usually represented by dark markings on a light background. The document is segmented into elemental areas small enough to resolve the finest detail needed. The document is scanned sequentially by a light beam, in a pattern consisting of a series of very narrow horizontal strips. The magnitude of the reflected light from each picture element (pel or pixel) is then used to generate an electrical signal. The resolutions are 20 pels/mm^2 ($\approx$ 3.85 1/mm $\times$ 5.3 pel/mm) or 28 pels/mm^2 ($\approx$ 5.3 $\times$ 5.3) for Group 2, 30 pels/mm^2 ($\approx$ 3.85 $\times$ 8) or 62 pels/mm^2 ($\approx$ 7.7 $\times$ 8) for Group 3, and 123 pels/mm^2 ($\approx$ 7.7 $\times$ 16) or 240 pels/mm^2 ($\approx$ 15 $\times$ 16) for Group 4 terminals. An A4 page of 210 mm by 297 mm contains about 1.9 $\times$ 10^6 pixels if a Group 3 terminal is used.

2.2.2.2 Service Example: Video Conference

A video server is a special network element that enables video traffic to be integrated with present and next generation LANs. There are several video conference arrangements used in the field. Some typical examples are described as follows:

(1) Fixed configurations utilize a specially configured room equipped with audio and video hardware.

(2) A more flexible arrangement is a rollabout unit that can be brought into traditional conference rooms and offices.

(3) Another option is traditional video conference and multiport control units used in conjunction with desktop video-conferencing systems (Figure 2-4) and video servers.

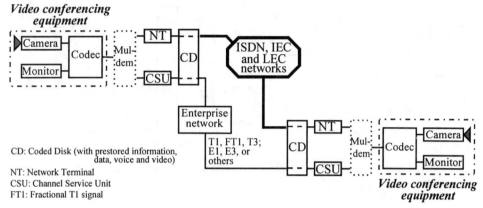

CD: Coded Disk (with prestored information, data, voice and video)
NT: Network Terminal
CSU: Channel Service Unit
FT1: Fractional T1 signal

Figure 2-4   Video-Conferencing System.

Typically more than two sites participate in a video conference meeting. Therefore, a Multipoint Control Unit (MCU) is a device used to link three or more video-conferencing locations. The MCU provides (one MCU can support 4 to 16 sites) interactive video and audio bridging. Cascaded MCUs can be used for very large video conferences. In desktop video conferencing, the multipoint functionality can be provided by on-premises video servers, MCUs, or ATM switches.

In 1990, the ITU-T introduced Rec. H.320 that applies compression technology to increase the video resolution by changing the speed from 56 kbps to 2.048 Mbps. The following are major ITU-T Recommendations associated with video conferencing:

- ITU-T Rec. H.200   Audio only and still graphics with audio and full-motion video conferencing
- ITU-T Rec. H.221   Definitions for audio/video compression and framing for transmission
- ITU-T Rec. H.242   Definition of communication protocols
- ITU-T Rec. H.261   Video coding algorithms
- ITU-T Rec. H.320   Definition of control and indication data

Video servers utilize standard components in conjunction with proprietary video compression hardware to implement video-conferencing systems (see Figure 2-4).

Example 2-2: Describe the Accunet Switched Digital Service (SDS; Figure 2-5) used for video conferencing application.

Accunet Switched Digital Services (SDS) enables business customers to increase efficiency and quality (see Figure 2-5). SDS are applied in all areas of data communications. With Accunet SDS, business customers experience transmission performance that has same level of reliability as dedicated private line services. SDS capabilities include:

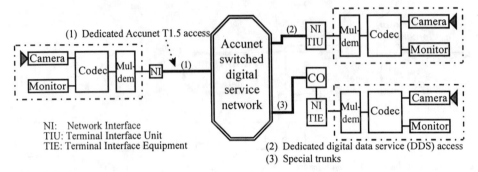

Figure 2-5   Accunet SDS for Video Conferencing Application.

- Send/receive high-speed or high volumes of data during certain time intervals known as "windows".

- Provide bandwidth on demand, and usage based on price (i.e., applications that are active for a limited duration).

- Transmit to multiple end-point destinations (i.e., serial or non-simultaneous communications between an originating point and several end points).

- Establish networking flexibility to accommodate specific traffic patterns with daily or seasonal variations that may change drastically.

Some Accunet Switched Digital Service (SDS) applications are:

- Video conferencing (see Figure 2-5)
- Bulk data transfer
- High-speed fax
- Digital broadcast teleconferencing
- Image viewing and transfer
- PC-to-PC file transfer
- Voice and data encryption
- Peak time overflow
- LAN-to-LAN interconnections

Accunet SDS are a family of high-speed "*dial-up*" digital data services that offer a full duplex digital transmission interface at a speed of 56 kbps/64 kbps (DS0A, DS0B, or DS0), 384 kbps (six contiguous DS0 channels), 1.536 Mbps (24 contiguous DS0 channels), or other bit streams provided by Customer Premises Equipment (CPE). SDS is a true "dial-up" service, it doesn't require customer to make pre-call arrangements to reserve or reconfigure interoffice facilities. SDS is a true "bandwidth on-demand" service.

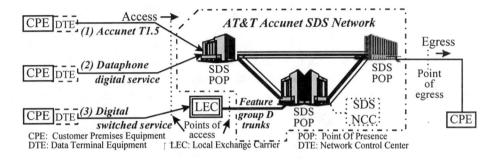

Figure 2-6   Accunet SDS Network Architecture and Access Arrangements.

Figure 2-6 shows the Accunet Switched Digital Services (SDS) network architecture and access arrangements. SDS can be accessed via (1) Accunet T1.5 service, (2) Dataphone® digital service, or (3) Digital Switched Access (DSA). Several local exchange carriers (LECs) provide connection to AT&T's Accunet SDS network via a "1 + xxx + yyy + zzzz" dialing scheme, the same as for voice long distance service [called Digital Switched Access (DSA) as shown in Figure 2-6]. Unlike dedicated access services, DSA does not terminate directly at a SDS Point Of Presence (POP) in the Accunet SDS network. Instead, DSA lines from the customer premises terminate in the LEC central office, which are connected to the SDS network via "Feature group D" trunks.

Example 2-3: Describe the Accunet Spectrum Digital Service (ASDS) which uses the Accunet SDS network.

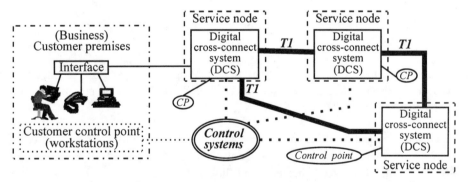

Figure 2-7   Accunet Spectrum Digital Service (ASDS).

Due to the broad range of Accunet capabilities, several names have been associated with Accunet services: Accunet 56 services, Accunet 64 services, Accunet T1.5 services,

Accunet Bandwidth Management (ABM) services, and Accunet Spectrum Digital Services (ASDS; see Figure 2-7). In addition to the features mentioned earlier, Accunet services allow reconfiguration of interoffice channels, bandwidth on demand, network alarm notification, automatic routing to diverse locations, and automatic restoration of failed channels.

When ASDS customers are provided with time slots at the T1 access facility, each service node (gateway node, including international digital access nodes) is accessed via a Digital Cross-connect System (DCS; Figure 2-7). These cross-connect systems are interconnected within the network via 1.544 Mbps digital stream [Clear-Channel Capability (CCC) over 24 Extended SuperFrame (ESF) channels using B8ZS zero-substitution line code (Chapter 6), plus Cyclic Redundancy Check (CRC: Chapter 13)] error detection/correction. ASDS has been implemented over fiber trunks using in-service monitoring interface units to achieve CRC functionality. The DCSs are connected to control systems for reliability monitoring, dialing circuit codes, and protection switching.

Example 2-4: Connecting sequence for Accunet SDS video conferencing services.

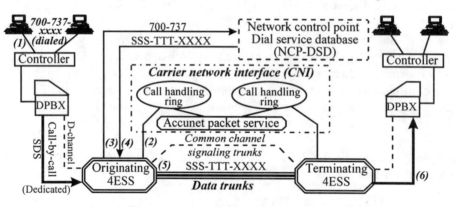

Figure 2-8   Call Sequence for Accunet SDS Video Conferencing.

The following are typical Accunet SDS dialing plans: (1) For Accunet 56 kbps services dial "700-560-XXXX" or "700-561-XXXX", or (2) For Accunet 64 kbps, 384 kbps, or 1.536 Mbps dial "700-737-XXXX".

- For direct connection: The dialed-up number (e.g., 700-560-XXXX) received from the LEC is translated in a Network Control Point (NCP) to "SSS-TTT-XXXX" for 4ESS identification. A dedicated trunk/line is "marked" (i.e., allocated) after the dialed number is routed through the LEC. The "originating 4ESS" routes the number to the "terminating 4ESS", which then routes the data connection to the customer with all the correct digits and outpulsing (see Figure 2-8). The call set-up functionality is routed through the common-channel signaling network, while the data information bit stream is routed via the data trunk.

- For 56-kbps Digital Broadcast Capability (DBC): The customer dials the number (e.g., 700-560-XXXX). The call is sent to a NCP-Dial Service Database (NCP-DSD). The 4ESS derives the service identity, and the NCP returns an Action Point Number (APN; SSS-TTT-XXXX). The originating switch routes the call to the terminating switch, which then routes the call to the Multiple Access System (MAS) to establish the conference.

- For Accunet switched digital services (e.g., video conferencing): As shown in Figure 2-8, the call sequence is summarized in the following five steps (the numbers below corresponds to the numbers in Figure 2-8):

  (1) The customer dials 700-737-YYYY over ISDN.

  (2) The digits received via the D-channel are interpreted by the Carrier Network Interface (CNI).

  (3) The dialed (called) number is translated in the Network Control Point (NCP).

  (4) The NCP-DSD (NCP Dial Service Database) returns "SSS-TTT-XXXX" for use as the Action Point Number (APN).

  (5) The originating switch translates the APN and routes it to the terminating switch.

  (6) The terminating switch then routes the number (i.e., connection) to the appropriate customer.

2.2.2.3 Service Example: Alliance 1000/2000

The Alliance 1000 is a "*voice teleconferencing*" service that supports up to 15 conferees (including the originator), without operator assistance. The service utilizes a Type I bridge located in one of several serving offices [Figure 2-9(A)]. The originating station must be a "touch-tone" device, however conferees can have either "touch-tone" or "rotary" (i.e., dial pulse) devices. Dialing "0-700-456-1000" routes the call to its nearest bridge. Examples of Type I bridges are: "Reno: 0-700-456-1001; Chicago: 0-700-456-1002; White Plains: 0-700-456-1003, and Dallas: 0-700-456-1004".

Customer assistance is provided by the serving office for "credit" or "service problems". The system is "timed out" when the originator disconnects from the call. Figure 2-9(A) shows a simple arrangement in the 4ESS environment, however a similar arrangement is possible for local exchange switching offices.

The Alliance 2000 is a "*data teleconferencing*" service that supports up to 15 PCs (including the originator). It supports non-simultaneous two-way transmission over a switched network. This service utilizes a Type II bridge located in one of several serving offices [Figure 2-9(B)]. The originating station must be a "touch-tone" device, however the conferees can have either "touch-tone" or "rotary" devices. Dialing "0-700-456-2000" routes the call to its nearest bridge. Examples of Type II bridge are: "Reno: 0-700-456-2001; Chicago: 0-700-456-2002; White Plains: 0-700-456-2003, and Dallas: 0-700-456-2004".

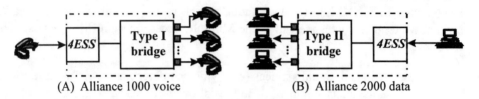

(A) Alliance 1000 voice   (B) Alliance 2000 data

Figure 2-9   Alliance 1000/2000 Teleconferencing Services.

Simultaneous conferencing requires separate Alliance bridges. Local switch bridging arrangements operate using identical methods. Alliance dedicated teleconferencing provides switched network teleconferencing, operating in conjunction with a business's dedicated private switched network. This service accommodates audio, audio-graphics, freeze-frame video, and other forms of teleconference media (e.g., display terminal and computer communications). It provides audio quality comparable to a two-point long-distance call. Broadcast bridge configurations can increase the number of "*add-on*" listeners to 24, 32, 40, 48, 56, or 240 ports.

2.2.2.4 Service Example: 800-type Service

The 800/888 basic, enhanced, and advanced services are "one-way incoming services" designed for customers who receive large volumes of calls from diverse points within the United States plus Puerto Rico, the U.S. Virgin Islands, etc. Advanced 800 services typically have the following example features:

- Time and data manager: This feature allows calls to be routed to various regional offices based on the time of day and the day of the week.

- Single number service: This service enables identification of a company/business by using just one 800 number nationwide for both interstate and intrastate customer calls.

- Area code routing: This feature divides incoming calls into geographical areas and routes those calls to the appropriate locations or offices.

- Courtesy response: This feature allows the caller to hear a message indicating his/her call will not be completed when the answering locations are closed or inactive.

Two other 800 service groups are inbound 800 service-international, and outbound 800 service-international. The inbound service allows a business in the U.S. to purchase a service so that many countries can call the U.S. on a toll-free basis. The caller dials an international access code (unique for each country), plus a toll-free access code and a seven-digit number [e.g., international access code (AC) + 05 + 7 digits]. The local switching office in a country selects a trunk for International Transit Exchange (ITE) and outpulses the toll-free access code plus 7-digit to the ITE. The ITE queries a database where the "AC + 7 digits" is converted to the format of "SSTTT-XXXX", and a billing record is made. The ITE then selects a trunk connected to an International Switching Center (ISC) in the U.S. The ISC receives the number and converts "SSS-TTT-XXXX" into an "NPA + YYY-ZZZZ" format using a Direct Signaling Database (DSD in the NCP). Finally, the ISC routes the call over the switched network for call completion.

The 800 outbound service-international allows a business outside the United States to purchase an international 800 service that allows callers from the U.S. to terminate "toll free" calls in their countries. An equivalent process (adopted by the inbound 800 service) is used so that the inbound service is able to complete the call.

2.2.2.5 Service Example: Multimedia Service

ITU-T Recommendations identify network capabilities needed to support multimedia services. Multimedia services are defined as those involving at least two different types of information. Multimedia services are implemented in ISDN environments that incorporate both narrowband and broadband services. Services include all types (e.g., interactive services, distribution services, etc.) and different configurations (e.g., "*symmetric*" and "*asymmetric*" services).

Definition 2-1 Symmetric and asymmetric services: These services can be defined from the service types or data rate viewpoint. If both directions of transmission carry the same types of signal, it is a symmetric service [Figure 2-10(A)]. Similarly, if both upstream and downstream signals are transported at the same data rate, it is considered symmetric configuration. In contrast, if different signal types or rates are handled [Figure 210(B)], it is considered an asymmetric service.

(A)   Symmetric Services (Configuration)

(B)   Asymmetric Service (Configuration)

Figure 2-10   Symmetric versus Asymmetric Services (Configurations).

Example 2-5: Describe an ATM-based network supporting multimedia services. (Note that ATM network connections are discussed further in a later section of this chapter).

ATM protocol enables high-speed transport and switching of multiple user data bit streams using packet (cell) format. ATM can support different speeds, effectively multiplex signals carrying different traffic types (e.g., voice, data and video), and provides several classes of service to meet different Quality of Service (QoS) requirements.

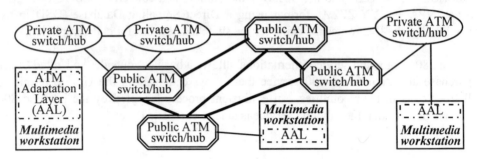

Figure 2-11   ATM Network for Multimedia Applications.

ATM cells can be transmitted between two remote multimedia workstations, that are connected via either a private or a public ATM switch (hub). The public ATM switches (hubs) are parts of a Public Switched Telephone Network (PSTN). ATM transport over physical medium is based on Virtual Channels (VCs) or Virtual Paths (VPs) (described in Chapter 5). The multimedia workstations convert user information data bit streams into ATM cells (packets) using ATM Adaptation Layer (AAL) protocol conversion (see Figure 2-11). The ATM cells are delivered to an ATM switch/hub (private or public) for routing and transported via transmission facilities (e.g., T1 or T3 digital carrier or SONET/SDH OC-3c, OC-12c, etc.) to remote destinations.

From a system perspective, multimedia services include the following functions:

- Capture: These functions deal with collecting and transforming external signals associated with multimedia (typically, analog voice, data, sound and video) into a form that can be utilized by a computer system (e.g., PC). Numerous capture devices are commercially available: microphones, Musical Instrument Digital Interface (MIDI) devices, video cameras, fax devices, Computer-Aided Topo-graphical (CAT) devices, scanners, optical character recognition systems, etc.

- Storage: These functions use appropriate hardware (e.g., disk storage) to retain multimedia information in a form so that a user can quickly access it. Examples are: videotapes, video disks, hard disks, and optical storage systems.

- Presentation: These functions involve the delivery of information to the user at the logical level via the personal machine interface. The information is physically delivered by a display device (e.g., video monitor) and/o audio speakers.

- Retrieval: These functions involve sophisticated database access, navigation software, and communication links.

- Transfer: These functions refer to the preparation of multimedia messages for delivery to remote recipients.

- Group sessioning (connectivity): These functions refer to the logical connection of two or more parties, allowing them to participate in multimedia sessions with minimum complexity, protocol conversion, and multi-step or multi-command sequences. The term "group-to-group conferencing" is also used to describe "group sessioning".

2.2.2.6 Service Example: Software Defined Network (SDN)

Software Defined Network (SDN) services provide many unique features, and can serve as an alternative to private network offerings. SDN makes use of the public switched network [e.g., AT&T Switched Network (ASN)] to offer "premises-to-premises" voice and data transport, along with a set of customer controlled call management and monitoring features. By using SDN access arrangements, users can place "on-net calls", "off-net calls", and calls to/from private networks (see Figure 2-12). Several access capabilities are available for SDN access to public switched networks (e.g., ASN):

- Direct analog access with E&M supervision
- Direct digital access
- Direct digital access for SDN via PRI-ISDN (Primary Rate Interface-ISDN)
- Switched access via an end offices

SDN can provide 56 kbps "premises-to-premises" data transport via other network services such as Switched Digital Service (SDS). SDS is a separate and unique 56/64 kbps network, compatible with B8ZS facilities (described in Chapter 6) and "per call controlled" echo cancellers. A variety of call management features are available for SDN users:

- Calls originating from SDN stations can be screened for allowable destinations (e.g., "on-net stations", "off-net stations", etc.).

- SDN provides Direct Inward Dialing (DID) for PBX applications.

- SDN supports voice and voiceband data at speeds up to 19.2 kbps.

- SDN caller dialed authorization codes (entered via a Touch-Tone keypad) can be used to override restrictions placed on the originating SDN station.

- SDN basic service can adopt standard and special ASN tones and announcements.

- Interconnection between SDN and private networks is available.

- SDN dial capabilities include standard 7-digit dialing, and International Long Distance Translation (ILD-T).

- Network remote access can provide toll-free access to SDN and non-SDN stations.

- Location dependent blocking allows SDN customers (e.g., managers) to define a list of numbers that ***cannot*** be called from a given SDN location.

- 56 kbps end-to-end service-integrated access can be provided between access stations connected to the SDN.

- The "off-net overflow" feature allows automatic overflow from all dedicated SDN direct egress or switched egress lines to be processed by "off-net" egress routing functions.

- The "forced on-net" feature recognizes and routes 10-digit calls destined for "on-net locations".

- SDN can prompt a caller to enter an authorization code when the requested call type is "off-net", international, or "international off-net".

- "Express connect" is the automatic connection service feature that allows an SDN customer to connect two pre-designated SDN stations together.

Since SDN is a virtual private network that shares public switching machines and transmission facilities while using special software and databases to control call-related processing, it can easily be extended for global applications [i.e., become a Global SDN (GSDN)], involving several gateway exchanges.

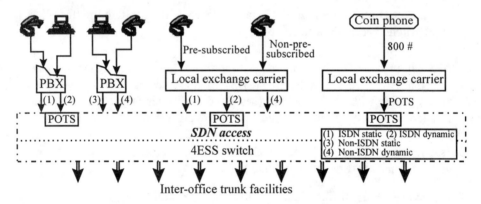

Figure 2-12   SDN Access Arrangements.

2.2.2.7 Service Example: MultiQuest Service

MultiQuest 900 is an information service, that was introduced by AT&T to provide 900-inward calling to customers with direct egress from a network service node. The service is targeted for high-volume inward calls that terminate at the sponsor's location. It is a "value-added" service for subscribers to offer callers (clients) various operational features. Sponsors can choose the premium charges they wish to apply under a flexible billing agreement. MultiQuest callers can dial a 900 number, connect to the CPE multiplexing equipment via T1.5 egress facilities, and communicate with live attendants, voice messaging equipment, or computer databases to obtain "value-added services". This interactive capability distinguishes MultiQuest 900 service from "dial-it 900" services. The MultiQuest primary service features are:

- Call detail report: The billing analysis feature provides the customer (at no extra charge) a list of individual calls with complete call details.

- Executive summary report: A monthly report for the prior billing period (at no extra charge) provides the customer's summary usage information.

- Flexible billing: This feature allows a sponsor to charge any rate that they desire on a "per-900 number" basis, subject to the service provider's parameters.

- Electronic funds: This feature allows a sponsor to transfer premiums (e.g., payments) directly into its account via an electronic transfer procedure.

The call management features for MultiQuest service are identical to Advanced 800 services. The MultiQuest services are:

- Technical, software, and service support (for an additional fee)
- On-line service for home PC users (for an additional fee)
- Information retrieval/database service for business applications
- Nationwide doctor/lawyer/dentist (or others) service access
- Help for homeowners/hobbyist/do-it-yourself (or other) activities

2.2.2.8 Service Example: Advanced Intelligent Network

Figure 2-13 shows the Advanced Intelligent Network (AIN) architecture. The AIN Service Switching Point (SSP) function allows a switching system to identify calls associated with AIN services. When the SSP detects that conditions for AIN services are met, it will initiate a dialogue with the AIN Service Control Point (SCP) in which the information for the requested service resides.

A SSP can communicate with one or more SCPs by means of Signaling Transfer Points (STPs) within the Common Channel Signaling system (CCS: described in Chapter 4). Likewise, a SCP can communicate with multiple SSPs. It should be understood that STPs are treated as part of the existing CCS network, hence no new procedural requirements are imposed on STPs when AIN services are added.

When an SSP detects that AIN service control is needed, it sends a CCS7 (Signaling System 7; SS7) message containing service-related information (e.g., calling and called parties' identities, and call processing data) to the appropriate SCP. The SCP uses service control logic and subscriber information to return a message to the SSP, requesting it to perform additional processing associated with the call or customer service request. The Service Management System (SMS) is one of several Operations Systems (OSs) that can be used in the AIN architecture. The OSs, together with capabilities provided by SSPs and SCPs, can support the functions necessary for provisioning, maintaining, and administering AIN services. AIN provides the following functions:

- Hot line: AIN can automatically complete a call to a pre-designated telephone number when the calling party's terminal equipment is "off-hook" and dials that number.

- Inter-dialing: This function enables private-line service subscribers to dial other selected station users with a 1- or 2-digit code instead of 7 or more digits.

- Toll usage control: This function allows subscribers to control their toll usage by implementing restrictions that are tailored to specific user needs.

- Computer access restriction: This function allows subscribers to have a list of phone numbers and Personal Identification Numbers (PINs) that are used to identify authorized and unauthorized callers.

- Personal Communications Services (PCS): With PCS, a user can have a Universal Personal Telecommunications (UPT) number that is used for both wireless and wireline access to the network.

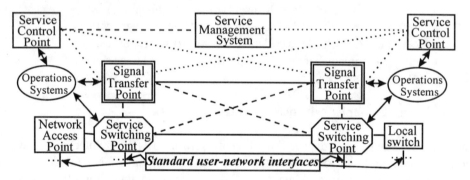

Figure 2-13   Advanced Intelligent Network Architecture.

2.2.2.9 Service Example: AT&T Skynet

Skynet service provides high-speed private line services between North America and foreign countries via satellite links owned and operated by the INTELSAT business service. Many companies use this service for high-quality and cost-effective communications for transporting voice, data, and video signals. Skynet operates in the 4, 6, 12, 14, 17, or 27 GHz band. For "earth station to earth station" service, a 99% error-free-seconds objective measured over 24 consecutive hours (equivalent to a bit error rate of 10^{-7}) with the application of error correct detection/correction codes, in wet weather conditions, is the typical criteria for acceptable performance. The performance can be much better if the weather is clear. Skynet transmits/receives on a usage-sensitive basis at 64 kbps, 1.544 Mbps, 2.048 Mbps, or higher rates. Transmission costs typically increase with decreasing distances between earth stations. Skynet becomes most cost effective if the earth stations are at least 500 miles apart.

2.3 VOICE-CHANNEL CONNECTION

This section describes an inter-office (inter-toll) call connection, followed by an intra-office call connection, and finally an international call connection.

2.3.1 An Inter-office (Inter-Toll) Connection

Figure 2-14 shows a typical interoffice call involving two local exchange switches. Customer A, whose number is 747-3333 (747 is known as the user's exchange number), is assumed to be the subscriber that originated the call. The local exchange (i.e., exchange A) serving customer A is called the originating office (exchange). Customer A is connected to the local exchange via a subscriber loop. In this call example, called party (whose number is 789-5555) is served by another local exchange (i.e., exchange B), which is known as the terminating office (exchange).

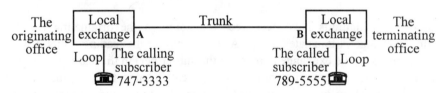

Figure 2-14   An Interoffice Call Involving Two Local Exchanges.

There are many call processing functions associated with an inter-office call. The main functions listed here are illustrated in this interoffice call description and are discussed further in Chapters 4 and 5.

- Call origination
- Digit collection
- Digit analysis/translation
- Routing
- Alerting
- Answering
- Disconnecting

2.3.1.1 Call Origination, Originating Register and Dial Tone

When subscriber A lifts the telephone handset (if the user has a speaker phone, he/she touches the "speaker" key), an off-hook signaling is delivered to the local exchange switch A (Figure 2-15). The control equipment at local exchange A detects the change from on-hook to off-hook status, and interprets this change as **a request for service**. Simultaneously, if an originating register (i.e., a touch-tone receiver or a dial pulse receiver) is available, the **originating register** is reserved and dial tone is connected to subscriber A. After customer dials the first digit of the called party's number, the **dial tone** is disconnected. All the digits dialed by subscriber A are received and stored in the originating register.

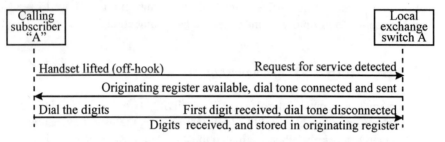

Figure 2-15   Call Origination, Originating Register and Dial Tone.

2.3.1.2 Translation, Routing, Path Selection, Outpulsing Register, and AMA

After all the dialed digits have been received/stored in the originating register, the originating exchange A translates them. After examining the leading digits (usually the first digits, known as the exchange number of the subscriber) in this case exchange A will determine the call is an interoffice (or inter-switch) call. That is, the call must be routed to another exchange. Therefore, this call must be connected to a outgoing trunk of exchange A. Routing information, which is stored in the system to indicate the appropriate paths (trunk groups) for the routing the outgoing signal, must be determined.

An Automatic Message Accounting (AMA) register is requested for all billable calls. At this time, the exchange control equipment transfers the call information from the originating register into the outpulsing register. This releases the originating register of the exchange so it can be used to process another call.

Now, the control equipment begins to scan the outgoing trunks in order to find an idle trunk between exchange A and exchange B. The outgoing trunks may all be busy. If this is the case, a reorder tone (known as a fast busy tone; described in Chapter 1) will be sent to the calling party, and the call is terminated as incomplete. These steps are summarized as follows:

- All digits received (by exchange A) are stored in the originating register (at exchange A).

- Translation area consulted (by Exchange A).

- Leading digits are analyzed (by Exchange A), and recognized as an interoffice call.

- Routing information determined (by Exchange A).

- AMA register is requested (by Exchange A) for all billable calls.

- Transfer (by Exchange A) all the call information from the originating register into the outpulsing register.

- The originating register is released (by Exchange A) for other calls.

- Outgoing trunks are scanned (by Exchange A) to find an idle trunk for connection.

- If an idle trunk can not be found, a reorder tone is sent (by Exchange A) to the calling party, and the call is terminated as incomplete.

2.3.1.3 Trunk Seizing, Ready Signal, Outpulsing and Seizing Incoming Register

After exchange A has scanned the outgoing trunks and found an idle trunk available for connecting exchange A to exchange B, the idle trunk will be seized. A two-way trunk can be seized by the switching system at either end to originate a call, but a one-way trunk can only be seized from one end (see Figure 2-16).

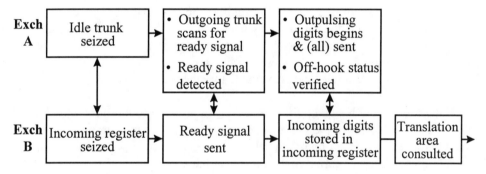

Figure 2-16   Trunk Seizing and Outpulsing.

The customer's line is connected to the outgoing trunk through a path within the switching system. The identity of the trunk, the number of digits to be transmitted, and additional information necessary for call setup are recorded in an outpulsing register. In exchange B, an incoming register of the switch is seized. Exchange B responds to the incoming call by signaling its readiness to receive address information from exchange A.

Exchange A periodically scans for the "ready" signal generated by exchange B. After the "ready" signal has been detected by exchange A, outpulsing of digits begins at exchange A. Before the last digit is sent, the control equipment checks to see that the calling customer's line is still off-hook. If the calling customer has hung up (abandoned the call), the control equipment will terminate the call-processing sequence and release the associated equipment and circuits. Assuming the calling customer is still off-hook, the last digit is outpulsed from exchange A, and the outpulsing register is released. At this point, the digits are stored in the incoming register of exchange B.

2.3.1.4 Alerting, Ringing, Audible Ring and Connection Complete

After all the outpulsing digits from exchange A are stored in the incoming register of exchange B (i.e., the terminating exchange), the translation area is consulted. The incoming trunk from exchange A is connected through the switching network to the called party's line. A ringing register is seized, the incoming register is released from this call, and the called party's line is checked for busy/idle status. That is, a ringing signal is sent to the line (loop). Simultaneously, an audible ring tone is sent back to the calling party.

The control equipment at exchange B scans the called party's line status for an answer indication (off-hook). If it is detected, the exchange terminates the ringing signal, and returns an answer supervision signal to exchange A (i.e., the originating exchange). The answer supervision signal is used to record answers and/or connect time for billable calls.

Assume the called party answers the phone, and conversation begins. During the conversation, the originating exchange A monitors the outgoing trunk (used for this call) for a disconnect signal. If the calling party hangs up first, the connection is released, and a disconnect supervision message is sent to the terminating exchange. After the terminating exchange detects this message and sends back an "on-hook" supervision message, the trunk is then idle, and is available for other calls.

If the called party hangs-up first, there are two possible conditions that will result in the trunk being released. In the first case, if the called party has hung-up, but the system has not yet detected the "on-hook" signal from the calling party, the system will release the trunk after a time-out period of 10-11 sec. In the second case, if the called party has hung-up, and the calling party also hangs-up within 10-11 sec, the trunk will be released.

If the call is billable (i.e., the call is not covered by a fixed monthly charge or a flat rate) completion of the call is detected and recorded at exchange A (the originating exchange) for accounting purposes. When the call is first initiated, the control equipment in exchange A determines whether or not the call is billable. In the USA, this is done by examining the routing information associated with the first three digits (i.e., the exchange number) of the seven-digit customer address. If the call is indeed billable, then a register is requested from an automatic message accounting system to store the information that must be recorded for the call. The information typically recorded for billing purposes are: the caller's telephone number, the called party's telephone number, the time the called party answered, and the time the connection was release. Data for billable calls is forwarded from the originating exchange to a data-processing/accounting center which periodically computes the customers' charges. The charges appears in the customers' monthly statement. If the Common Channel Signaling system 7 (SS7) is used, a similar process occurs between the originating and the terminating exchanges. However, the information is communication by means of signaling messages, which are transmitted over a separate signaling network.

2.3.1.5 Access Feature Groups

To provide standard interconnections between telecommunications carriers (e.g., local exchange and inter-exchange carriers), different feature categories have been defined for network access. Four common categories are described as follows:

(1)   Feature group A (FGA): Feature Group A defines the voice transmission path (with a bandwidth of 3 kHz; approximately 200 to 3200 Hz) between the First Point Of Switching (FPOS) within a Local Access and Transport Area (LATA) and the Point Of Termination (POT) (see Figure 2-17).

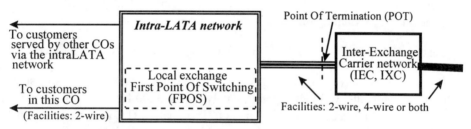

Figure 2-17   Feature Group A Service Configuration.

FGA is used to support existing Foreign Exchanges (FXs) and similar services. The FPOS provides a line-side termination that is assigned a 7-digit local telephone number, or a unique number (e.g., 800 number) that identifies the service provider. The FGA services from an inter-exchange carrier (IEC or IXC) within a LATA are switched from line-side terminations to end users. Miscellaneous originating calls from the LATA are provided access to FGA services via the intraLATA network; as determined by the 7-digit telephone number assigned to the line termination. The interface group designations apply only to the configuration at the POT (Point Of Termination). The connection at the local exchange switch "line-side termination" is typically a 2-wire metallic facility. The facility between the FPOS and POT can be 2-wire, 4-wire, or both.

(2)   Feature Group B (FGB): Feature Group B also defines the voice transmission path (with a bandwidth of 3 kHz; approximately 200 to 3200 Hz) between the FPOS and POT within a LATA (see Figure 2-18), however FGB terminates on the trunk side of a Stored Program Control (SPC) element which is the FPOS. Service may be provided directly to an SPC end-office, or via an SPC Access Tandem (AT). The AT may also serve customers calls terminating within in the same switch. Connections between the AT and end-office switches in the LATA are carried by the LEC intraLATA network, and may be routed through an additional Local Tandem (LT) switch.

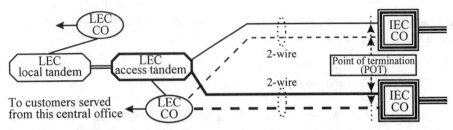

Figure 2-18   Feature Group B Service Configuration.

FGB services used for calls originating from an Inter-Exchange Carrier (IEC) are switched from the trunk side termination to end users in the LATA. End users originating calls from the LATA are provided access to FGB via uniform access codes "950-0XXX" or "950-1XXX". Connections are made directly to the FGB service of the IEC in the end office, or via the intraLATA network to the Local Tandem (LT) switching office.

(3)   Feature Group C (FGC): FGC services have been replaced by FGD services.

(4)   Feature Group D (FGD): Feature Group D also defines the voice transmission path (with a bandwidth of 3 kHz; approximately 200 to 3200 Hz) between an end-office (serving the end users) and the Point Of Termination (POT) (Figure 2-19). FGD provides any IEC with the capability of achieving Switched Access Services (SAS) transmission. LATA access is provided through trunk-side switching at an SPC end office or Access Tandem (AT) switch. When routed via an AT, the transmission parameters of each segment are more stringent than those available for other feature groups or via direct FGD access.

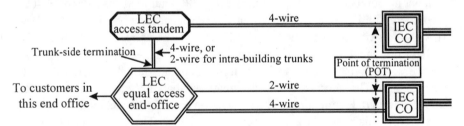

Figure 2-19 Feature Group D Service Configuration.

2.3.2 An Inter-toll Connection Example (NYC to San Francisco)

The call processing, switching, and transmission path for a direct-dial call from New York City to San Francisco, California is illustrated in this section. There are numerous ways to establish a communication path between NYC and SF. The connection illustrated in this section is an example of using an inter-exchange carrier. The connection is divided into three segments: (1) access, (2) backbone, and (3) egress.

2.3.2.1 The Access Connection

In New York the caller's touch-tone phone is connected to the New York Local Exchange Carrier (LEC) central office by a twisted-pair local loop. The connection is typically a metallic copper wire providing capacity for one voice channel (Figure 2-20).

There are many types of equipment housed in the New York central office, including a local exchange switch (e.g., an electronic No. 1 ESS switch). In an electronic switching system the logical steps involved in making a call connection are stored in software. A high-speed processor controls call processing functions, and allows users to share equipment. This concept is called "***Stored Program Control*** (SPC)".

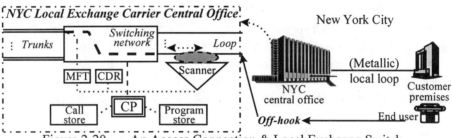

Figure 2-20   An Access Connection & Local Exchange Switch.

To initiate a call, the customer (end user) lifts the handset from the customer equipment (e.g., a telephone set). A scanner at the LEC switch detects the off-hook status and establishes a message path from the loop (customer end) to the Customer Digit Receiver (CDR). The CDR provides a dial tone to alert the calling party that the LEC switch is ready (active). As every digit is dialed, two tones (Chapter 1) are received at the CDR via the loop. The CDR then passes the address (phone number) to the Central Processor (CP), which transfers the address to the Call Store (CS). The CS is a memory "scratch pad" that temporarily stores the address for the duration of the call.

Next, (under the control of software) the Call Processor (CP) analyzes the called address, consults the routing/translation table stored in the Program Store (PS), and seizes an outgoing trunk. The CP establishes a connection to the Multi-Frequency Transmitter (MFT), and an exclusive physical connection for voice transmission is reserved between the customer line (loop) and the outgoing trunk. Finally, the MFT transmits ("***outpulses***") the called address via the reserved trunk.

2.3.2.2 Inter-exchange Trunk (Backbone) Connections

This example is assumed to be an inter-toll call, therefore the signal must be routed to an Inter-Exchange Carrier (IEC) office via an access trunk. In this case, the access trunk connecting the NYC Local Exchange Carrier (LEC) office to the NYC IEC office is typically a T1 digital carrier system (described in Chapter 6) as shown in Figure 2-21.

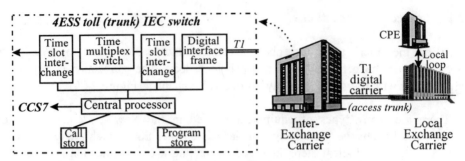

Figure 2-21   Inter-exchange Office and Access Trunk.

T1 is a facility that only transmits digital signals. Therefore, the voice and address signals from the MFT at LEC office must be converted (digitized) into digital bit stream by using Pulse Code Modulation (PCM) techniques (Chapter 6). After the address and voice signals have been converted into bitstream, they can be transmitted over the T1 carrier. When transmission is complete, both the call store and the MFT in the LEC office (Figure 2-20) will be released for another call setup.

The digital bitstream is transported by the T1 carrier to the IEC office, which typically contains a 4ESS toll switch (Chapter 5). The 4ESS has the capacity to handle 650,000 busy-hour calls, and takes over routing the call through the IEC (backbone) network. The 4ESS handles the two signals (the address and the voice signal) differently:

- Address signal: The Digital Interface Frame (DIF) at the IEC office receives the address information from the LEC office, and forwards it to the IEC-office's Central Processor (CP). For speed and efficiency purposes, address information is not sent over the voice trunk group. It is carried by a separate network called the Common-Channel Signaling (CCS) network (Chapter 4). In the New York region, a mated Signaling Transfer Point (STP) acts as the switching device that routes the address information to a terminating STP in the designated destination region (in this case, the terminating STP is in San Francisco). This function is performed by the 4ESS switch in New York. First, its central processor analyzes the address information and assigns a timeslot associated with the outgoing voice trunk that terminates on a 4ESS in San Francisco. In addition, a continuity test signal is attached. The address information is sent in packet form from the STP in New York to the STP in San Francisco. After the address information is received by the STP associated with the 4ESS inter-exchange office in San Francisco, the 4ESS sends a test signal via the voice path back to the 4ESS in New York. In a fraction of a second, this 3,000 mile trunk is tested and verified to be ready for voice signal transport.

- Voice signal: The 4ESS switching network is a time division device that consists of four major components: (1) Digital Interface Frame (DIF), (2) input Time Slot Inter-changer (TSI), (3) Time Multiplex Switch (TMS), and (4) output TSI. As previously described, the DIF handles signaling. The input TSI stores the incoming PCM signal for a particular call. The high-speed TMS is a space coordinating device that provides a path connecting to the output TSI. The trunk associated with the output TSI in this example is a "long-haul" lightwave system connecting New York: to Philadelphia, Chicago, Denver and Salt Lake City. Between Salt Lake City and SF the signal is transmitted by microwave radio.

Figure 2-22 shows a typical lightwave connection between two IEC offices (e.g., from NYC to Philadelphia; Philadelphia to Chicago; etc.). In this example, a FT-G lightwave system in New York accepts DS3 signals from a digital multiplexer (DDM-1000), and other transmission facilities. The FT-G transmitter can multiplex 36 or 72 DS3 signals into a single bitstream with a data rate of 1.7 Gbps (a capacity of 24,192 voice or voice-equivalent channels) or 3.4 Gbps (a capacity of 48,384 voice or voice-equivalent channels).

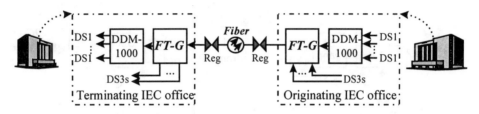

Figure 2-22   Lightwave Systems Connecting Two IEC Offices.

The multiplexed 1.7 Gbps [or 3.4 Gbps using the 2-wavelength Wavelength-Division Multiplexing (WDM) technique] digital bitstream is converted into light pulses using a high-speed laser diode. The light pulses signal, with a wavelength 1310 *nm*, is carried by single-mode optical fibers (described in Chapter 7) to the terminating IEC office. The lightwave signal degrades (attenuates) as it propagates along the optical fibers, therefore regenerators spaced between 25 and 75 miles must be installed along the transmission path.

At the terminating IEC office, the lightwave pulses are converted back into electrical signal with a speed of 1.7 Gbps (or 3.4 Gbps). The conversion is performed by a device known as PIN photodiode [a semiconductor device using Positive (P), Intrinsic (I) and Negative (N) semiconductor material types]. For longer spans, an APD photodiode is used instead of a PIN (described in Chapter 7; APD stands for Avalanche Photo-Diode and has a much better sensitivity, than a PIN receiver, to detect a weaker lightwave signal being transmitted over a longer span).

The electrical bitstream is demultiplexed into DS3 or DS1 signals for add/drop functions. Signals that need to be transmitted to the next IEC office are multiplexed with added signals. This procedure is the same as the one used at the originating IEC office. That is, an IEC office contains "back-to-back" equipment as shown in Figure 2-23. The same procedure (including multiplexing/demultiplexing, add/drop, and conversion between electrical and lightwave signals) is applied at every IEC office along the transmission path until the signal reaches Salt Lake City. At Salt Lake City the electrical signal is converted into a microwave signal for transmission over a digital radio system.

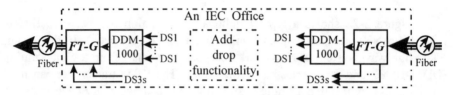

Figure 2-23   A Back-to-Back FT-G Transmission Equipment.

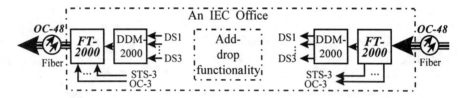

Figure 2-24   Lightwave System: A Newer FT-G Equivalent.

It should be noted that the transmission systems described in Figures 2-22 and 2-23 have been replaced in many areas within the United States by newer SONET (Chapter 6) equipment (i.e., DDM-1000 replaced by DDM-2000, and FT-G replaced by FT-2000) as shown in Figure 2-24. The differences between the new lightwave system and existing FT-G system are: (1) the low-speed port handles DS1, DS1C, DS2, DS3, STS-3 and/or OC-3 signals instead of DS1, DS1C and DS2 signals; (2) the high-speed interface is OC-48 with a speed of 2.48832 Gbps instead of 1.7 or 3.4 Gbps; and (3) the wavelength transmitted can be either 1310 or 1550 *nm*, depending on the transmission path length. Lower SONET data rate interfaces (e.g. OC-12) are supported.

Between Salt Lake City and San Francisco a microwave digital radio system is used as the transport system. The microwave radio regenerators are spaced 25-30 miles apart, and *"line of sight"* is required. The major building blocks of a radio transmitter are shown in Figure 2-25. The digital signal is modified by adding redundant parity-check digits [channel encoder; such as CRC, Reed-Solomon; BCH, or convolutional codes (described in Chapter 13 of this book)] for error control (detection/correction) purposes. Several signals are then multiplexed together using either Frequency Division Multiple Access (FDMA), Time Division Multiple Access (TDMA), or Code Division Multiple Access (CDMA) technology (described in Chapters 6 and 8) for sharing the microwave channel bandwidth. M-ary Phase Shift Keying (M-PSK; e.g., 64-PSK) or M-ary Quadratural Amplitude Modulation (M-

QAM; e.g., 128-QAM or 256-QAM) modulation is applied for effective transmission. Radio-frequency modulation is then used to raise the signal spectrum from base-band spectrum to the microwave frequency range (2, 4, 8, 11, or 18 GHz) for long-distance transmission. An power amplifier is typically required to gain sufficient power before the radio signal is transmitted, via "line of sight", to a distant destination.

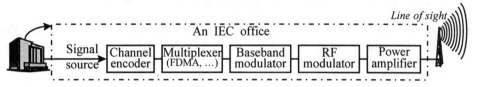

Figure 2-25   Major Radio System Transmitter Components.

2.3.2.3 The Egress (Access) Connection

The microwave radio signal terminates at a 4ESS switch (consisting of DIF, TSIs, TMS, etc.; see Figure 2-21) in the San Francisco IEC office as shown in Figure 2-26.

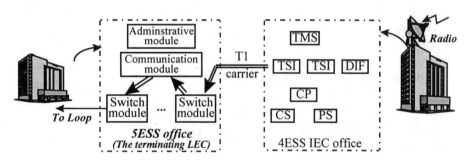

Figure 2-26   A Terminating Local Exchange Office.

The microwave radio signal, after being demodulated, demultiplexed, and decoded (using the inverse functionality of Figure 2-25) into an electric signal, is processed by the 4ESS trunk exchange switch and routed to the "addressed" central office (the terminating Local Exchange Carrier office) via a digital carrier system. The T1 digital carrier system is used in this example. The LEC office in this case is equipped with a 5ESS (Chapter 5) electronic switch, whose major building blocks are an Administrative Module (AM), a Communication Module (CM), and several Switch Modules (SMs).

The SM at the input port terminates the incoming T1 signal and routes it through the CM to the outgoing ("addressed") SM. The addressed loop is tested for availability by the outgoing SM. If the line is not busy, an alert audio signal is sent to the calling party. If the called party answers the phone, communication can be started.

This connection example illustrates the complexity of an inter-toll voice connection. It is obvious that many different equipment and transmission facilities are required:

- Switching machines: 1AESS, 4ESS and 5ESS systems (described in Chapter 5) are used as local exchange and toll exchange switches (gateway exchange is not used in this connection, 4ESS and 5ESS can serve as gateway exchanges).

- Signaling network: This separate network handles call setup functions. The Common Channel Signaling network 7 (CCS7) is used in North America, Signaling System 7 (SS7) is used worldwide.

- Transmission media: Twisted-pair metallic wires, coaxial cables, microwave radio, satellite radio, and optical fibers transport signals in different forms (e.g., electrical current or voltage, microwave radio wave, and infrared light).

- Transmission terminals (equipment): They perform various functions (Chapter 6).
  - Pulse Code Modulation (PCM) terminal
  - Digital Add-Drop Multiplexer (ADM)
  - Digital Cross-connect System (DCS)
  - Integrated Digital Loop Carrier (IDLC) system
  - Regenerators
  - Echo cancelers

2.3.3 Local (Intra-LATA, or Intra-Office) Call Connection

When a call does not require connection via a trunk (toll) exchange switch, it is an intra-office or an intra-LATA call (which may or may not be a "free" local call) as indicated by "circled 1" and "circled 2" notations, respectively, in Figure 2-27.

For an intra-office call, the local exchange switch (e.g., 1AESS) serves as both the originating and terminating office. The local exchange office contains many types of equipment: switching machines, transmission terminals, and operations systems. As in the case of a toll call, the customer picks up the handset (or pushes a "phone" button on the customer premises equipment) to request service. The off-hook signal is detected by the scanner at the switch machine (Figure 2-20). A path between the Customer Digit Receiver (CDR) and the calling line is then established. The CDR provides a dial tone signal produced by the "dial tone generator" to the calling customer indicating that he/she can begin dialing. As every digit of telephone number is dialed, two tones (with two distinct frequencies) are received by the CDR. It then passes this address information (i.e., the telephone number) to the Central Processor (CP). The CP passes the address information to the call store which temporarily holds this information for the duration of the call.

The CP, under the direction of software control, analyzes at the called address and consults the routing/translation table stored in the "program store". It then seizes an outgoing line (not a trunk since the call is an intra-office call). The CP establishes a connection between the calling loop and the outgoing line via the aid of the Multi-Frequency Transmitter (MFT). The MFT "outpulses" the address to the outgoing line, and the status of the line is tested. If the line is available, a ringing signal is sent to the called party, and another audible alerting ringing tone is sent back to the calling party. Communication will start if the called party answers the call.

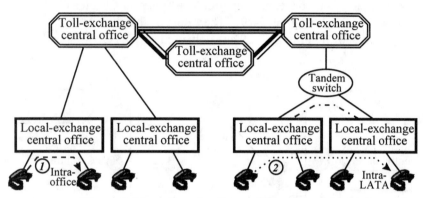

Figure 2-27   Local or Intra-office Connection.

A "non-toll" call can also be an intra-LATA call that involves two local exchange offices. Sometimes, a tandem-switch office may be involved in the connection as shown on the bottom right of Figure 2-27. The transmission facilities connecting two local exchange offices (or between a local exchange office and a tandem office) are typically T1 digital carrier systems. Both the local exchange switch and the tandem switch can be an electronic 1AESS or a 5ESS switching system. The call procedure and call connection are similar to the "access" and the "egress" offices previously described for inter-exchange connection.

2.3.4 International Call Connection

There are several unique characteristics of an international call connection. First it should be understood that the world is divided into several zones (Table 1-2 in Chapter 1). Therefore, when making an international call, a telephone number having 15 digits (Figures 1-20, 1-21, and 1-22 in Chapter 1) may have to be dialed. Second, international gateway exchange offices (Figure 2-28) are required for this type of connection.

There are two types of transport networks used for the international connections: (1) the undersea optical fiber lightwave systems, and (2) the geostationary satellite network (Figure 2-28).

In an international connection, both the local exchange and trunk exchange perform the exact same functions as previously described in the inter-trunk connection example.

However, the gateway exchange performs unique functions. First, it must convert from one protocol to another. For example, digital PCM bitstreams in the U.S. are implemented using the μ-law companding algorithm, while the A-law algorithm is used in other areas of the world. Hence, for international connections, the conversion between the A-law and the μ-law must be implemented. In addition, the U.S. digital bitstream adopts a 24-timeslot format (within a 125-μs frame) while A-law adopts a 32-timeslot format. A conversion (the conversion algorithms is given in Chapter 6) between these two formats is essential (Figure 2-29). The "dotted" blocks, one on each gateway exchange office as shown in Figure 2-29, are only used in SDH environments.

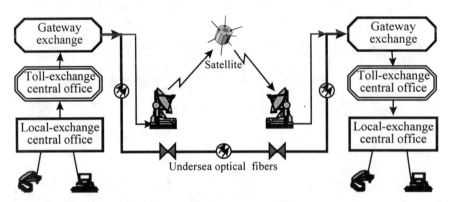

Figure 2-28   An International Connection via Gateway Exchanges.

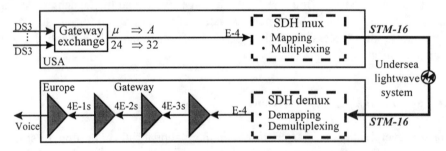

Figure 2-29   The International Gateway Interconnection.

Since the 1960s, μ-law signals (e.g., DS1 and DS3) have been converted into A-law companded signal (Chapter 6), and 24-timeslot format replaced by a 32-timeslot format for international applications. The converted A-law signals transported over undersea optical fiber system can be 2.048 Mbps E1, 8.448 Mbps E2, 34.368 Mbps E3, or 139.264 Mbps E4 signals. However, E4 signals have been the most popular transport signals used for connecting two international gateway offices. For example, TAT-8 and TAT-9 (Trans-Atlantic Transport system-8 and -9) are undersea lightwave systems that connect the US and Europe, and TPC-3 (Trans-Pacific Cable system-3) is a undersea lightwave system that connect the US and Asia. These systems have data rates of 280 Mbps or 560 Mbps (2 × E4, 4 × E4 speed), respectively.

In an SDH (Chapter 6) environment (Figure 2-29), sixteen (16) E4 signals are multiplexed to form a SDH signal, known as STM-16 (OC-48; a SONET equivalent), which has a speed of 2.48832 Gbps. With additional advanced techniques, new undersea lightwave systems can increase its speed from 2.48832 to about 20 (19.90656 Gbps). There are several undersea lightwave systems deployed worldwide, for example: TAT-12, TAT-13, TPC-5, and SL-2000 (Submarine Lightwave system-2000).

After arriving Europe, a STM-16 is demultiplexed into sixteen E4 signals. An E4 signal is demultiplexed to four E3s; an E3 is demultiplexed into four E2s; an E2 is demultiplexed into four E1s; and finally, an E1 is demultiplexed into 32 time slots. In practical applications, the 32 time-slot capacity is used to carry 30, instead of 32, voice users while two remaining time slots are reserved for network management functions.

2.3.5 A Private Network Connection

There are two types of private networks: (1) private switched, and (2) private data networks. The private switched network has same architecture as public switched networks with some slight modifications.

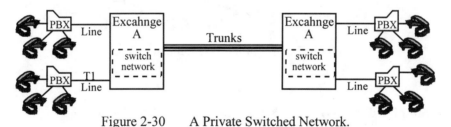

Figure 2-30   A Private Switched Network.

The access network is the portion of a private network that is different from the public switched network. In most applications, the T1 digital carrier system is used as the private access line (Figure 2-30). Private Branch eXchange (PBX) are the switching machines located on the customer premises. In addition to performing circuit switching functions (e.g., call setup, establishing connections, and routing calls from the originating terminal to the addressed station), the PBX imposes call authorization for a pre-selected group of callers. That is, the PBX grants special call privileges (e.g., toll or international calls) to some callers.

Local Area Network (LAN) is typically implemented for private data (data and ISDN described in Chapter 9) communications within a moderately sized geographic area. The network architecture and applications of LANs are described in Chapter 10.

2.3.6 A Wireless (Cellular) Connection

The principles of wireless (cellular) networks are described in Chapter 8. This section illustrates two wireless call connection scenarios: (1) a mobile to land call (Figure 2-31), and (2) a land to mobile call (Figure 2-32). In addition, the handoff procedure involving just one Mobile Switching Center (MSC).

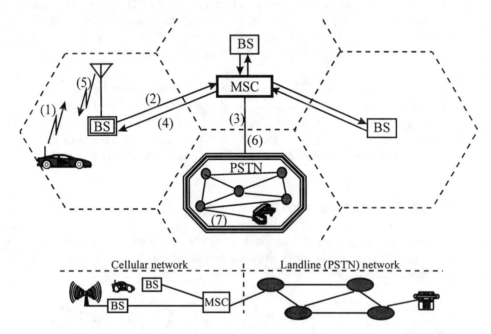

Figure 2-31   Call Flow: Mobile to Land.

2.3.6.1 Simplified (Generic) Call Flow Scenario: Mobile to Land

The flow of a call made by a mobile unit that is intended for a land-line subscriber is shown in Figure 2-31. The cellular network providing wireless services can share the same geographic areas with the Public Switched Telephone Network (PSTN) as shown in the top graph of Figure 2-31. However, the cellular network and PSTN are two totally separated networks from network architecture viewpoint, as shown in the bottom graph of Figure 2-31. Note that the numbered steps for this "mobile to land" call correspond to the numbers inside the ( ) in Figure 2-31.

(1)   The mobile unit subscriber dials the directory number of the land-line (called) party and presses the "send" button on the handset.

(2)   The Base Station (BS) recognizes the call request originated by the mobile unit, measures the signal strength, and checks the mobile identity. If the signal level is

strong enough and the identity is verified, the BS will refer the call attempt to the Mobile Switching Center (MSC) that serves this particular BS.

(3)   The MSC sets up the landline portion of the connection to the Public Switched Telephone Network (PSTN).

(4)   The MSC then informs the BS that the connection to the called party is being established. This allows the BS to perform step (5).

(5)   The BS prepares (selects) a Radio Frequency (RF/rf) channel, and then orders the mobile unit to automatically tune to this particular channel.

(6)   After the connection between the mobile unit and the MSC is established, the MSC routes the call to the PSTN thereby completing the connection between the mobile unit and the PSTN.

(7)   The PSTN routes the call to the land-line (called) party. The call setup is complete when the landline subscriber "answers the phone", and communication between the mobile party and the landline party can start.

2.3.6.2 Simplified (Generic) Call Flow Scenario: Land to Mobile

The mobile unit's power is "**ON**" whenever the mobile subscriber wants to make a call or **is willing to** receive a call. As long as the mobile unit's power is "ON", it is scanning all surrounding cell sites looking for one with a strong paging channel signal. When a signal with sufficient strength is located, the mobile unit "listens" to that paging channel. Note that the numbered steps for the mobile to land call sequence corresponds to the numbers inside the ( ) in Figure 2-32.

(1)   Assume that the land-line subscriber originates the call, and he/she dials the directory number of a mobile unit. The switching office that serves the land-line customer recognizes the directory number as being associated with a mobile switching center. That is, the landline switch recognizes this is a mobile call.

(2)   The PSTN routes the call to the MSC (i.e., the called party's number is sent to MSC).

(3)   The MSC recognizes the call is intended for one of its mobile units, and orders **all** of its cell sites (base stations) to page the mobile subscriber. Since the mobile unit could be anywhere in the serving area, all Base Stations (BSs) connected to this MSC are involved in the paging process. In the event that the mobile is not located in the serving area, the mobile unit will have already registered in another serving area. In this case, the "visiting MSC" will conduct the paging process for its serving area.

(4)   All the cell sites attempt to page the mobile unit.

(5)   The mobile unit responds to the page from the cell site with the strongest signal.

(6)   This cell site relays the page-response to the MSC.

(7)   The MSC informs the BS that it has recognized the mobile unit's page response, which allows the BS to perform step (8).

(8)   The base station selects a RF channel, orders the mobile unit to tune to this particular RF channel, and alerts the mobile subscriber by sending a ringing signal to the mobile unit.

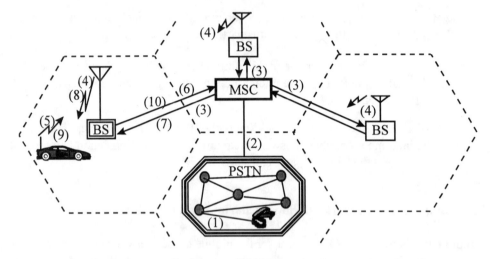

Figure 2-32   Call Flow: Land to Mobile.

(8)   The mobile phone rings, and the mobile subscriber answers the call. The mobile unit then sends a message informing the base station that the mobile subscriber has answered the call.

(9)   The base station alerts the MSC that the mobile unit has answered the call. The MSC removes the (alerting) ringing signal being sent to the calling party, and completes the connection between the PSTN and the cell site. Communications between the land-line (calling) party and the mobile subscriber can now begin.

2.3.6.3 The Handoff (One MSC)

The handoff process is described for the case when only one MSC is involved. In this example, it is assumed that the serving base station resides in cell site *A*, and the mobile unit is moving away from cell site *A* towards cell site *B* (see Figure 2-33). The handoff involves the MSC that serves the base stations of cell sites *A*, *B*, and *C*. Note that the numbered steps for this "mobile to land call" correspond to the numbers inside the ( ) in Figure 2-33. The discussion of a handoff involving two MSCs is beyond the scope of this book.

(1)   When the BS in cell site *A* detects that the signal power of the mobile unit is getting weaker, the BS serving the mobile unit initiates the handoff process.

(2)   The serving BS requests **all** the ***neighboring*** cell sites (including cell site *C*) via the MSC to measure the signal strength of the particular mobile unit.

(3)   All neighboring cell sites measure the signal strength of the particular mobile unit. Recall that weak signal power was initially detected by the BS in cell site *A*.

(4)   All the cell sites report signal strength measurements for the particular mobile unit to the base station at cell site *A*, via the MSC.

(5)   The base station in cell site *A* selects the "best cell site" (i.e., the greatest signal strength), to serve the mobile subscriber. Cell site *A* then requests a handoff be initiated by the MSC.

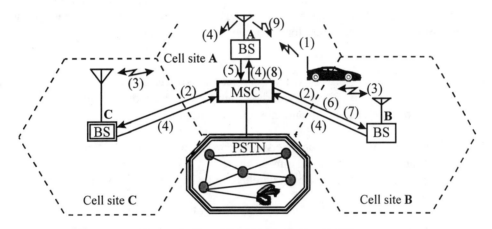

Figure 2-33   The Handoff: One MSC.

(6)   In this example, the MSC informs the new base station in cell site *B* to prepare for accepting a handoff.

(7)   The MSC also orders the new base station in cell site B to select a radio channel for the mobile.

(8)   The MSC then informs the base station of the original cell site (i.e., cell site A) to send a final message to the mobile unit.

(9)   The original base station (in cell site A) orders the mobile unit to tune to the newly selected radio channel of the base station in cell site B. Finally, the MSC switches the call from cell site *A* to cell site *B*, thereby completing the handoff procedure

Review Questions I for Chapter 2:

(1) Telecommunications industry originated from a simple point-to-point connections, and switched networks that only provided for local services. Now, it has evolved into _____, _____, and finally a worldwide network.

(2) Technical factors that influence telecommunications network evolution are: _____, _____, _____, _____, _____, _____, and _____.

(3) Currently, fax services are carried by so-called Groups __ and ___ standards. But, Group __ has high-speed capabilities; and can reproduce an ___-size (210 nm by 297 mm) document in about 2 seconds.

(4) Video conferencing equipment include _____, _____ and _____.

(5) (True, False) Basic, enhanced, and advanced 800/888 services are one-way incoming services designed for customer who receive large volumes of calls from diverse points.

(6) Software Defined Network (SDN) services make use of the _____ to offer "premises-to-premises" voice and data transport. It can provide a set of _____ controlled call management and monitoring features.

(7) A Personal Communications Service (PCS) with Advanced Intelligent Network (AIN) capabilities provides the user with a _____ number that can be used for both wireless and wireline access to the network.

(8) An inter-office (inter-toll) voice connection typically involves two _____ offices, and at least one _____ office. For an international connections, in addition to those offices, at least two _____ offices are required.

(9) The call processing functions in an inter-office or international call include: call origination, _____ collection, digit _____/_____, routing, _____, answering, and _____.

(10) In the US, several access Feature Groups (FGs) have been specified. Examples are: FG ___, FG ___, FG ___, FD ___, etc.

(11) (True, False) Prior to SONET networks, the commonly used optical links for inter-office connections are known as FT-G optical fiber systems with a rate of 1.7 Gbps, and can carry 24,192 simultaneous voice channels on a single fiber.

(12) (True, False) The gateway exchange offices for providing international calls in the PDH environment must perform protocol conversion functions.

2.4 AN INTERNET CONNECTION

The details of Transmission Control Protocol/Internet Protocol (TCP/IP) is given in Chapter 11. An overview, and several typical Internet connections are described here.

Definition 2-2: An "Internet" is a network connecting many different networks which are owned, operated and managed by individual service providers. An Internet is not a single entity, and there is no central policy-making/regulatory agency. All the networks are linked together by a common set of technical protocols known as "the TCP/IP protocol suite". This TCP/IP protocol suite makes it possible for Internet users connected to any network (within the "Internet") to communicate with or use services located on any other network. Internet technical specifications [called Requests for Comment (RFCs)], are developed by the Internet Engineering Task Force (IETF).

2.4.1 TCP/IP Protocol Overview

Like many other protocols, TCP/IP is packet switching protocol that segments data bitstreams into packets for switching/routing/transport. In addition to packets, the data bitstream can be formed into cells or frames that they can be delivered to their respective destinations. Methods used to deliver frames, packets, or cells can be different, depending upon the specific application.

Table 2-2   7-layer Protocols for Internet Applications.

OSI layer	Protocols		
7	• File Transfer Protocol (FTP) • Simple Mail Transfer Protocol (SMTP)		• Novell Mail Handling Service (NHS)
5/6	• Novell NetWare Core Protocol (NNCP) • Pretty good privacy		• RSA* data encryption (* Rivest, Shamir and Adleman)
3/4	• TCP/IP	• DECnet	• Appletalk
2	• Point-to-point protocol (PPP)		• Frame relay • ATM
1	• POTS • Private lines • xDSL		• SONET/SDH • ISDN

It is clear that TCP/IP is a vital part of the Internets worldwide interconnection. TCP is a "layer 4 end-to-end protocol" providing virtual connection reliability. IP is a "layer 3 protocol", providing addressing/routing for connectionless network services. As IP applications expand, the TCP/IP protocol suite will become inadequate. Therefore, TCP/IP must be incorporated with the 7-layer OSI reference model that has been applied in data communications since the late 1960s. Table 2-2 shows a set of protocol for each of the 7 layers that are suitable for the Internet applications: (1) Layer 1-POTS, private line, xDSL, ISDN and SONET/SDH; (2) Layer 2-PPP, frame relay, and ATM; (3) Layers3/4-TCP/IP, DECnet, and Appletalk; (4) Layers 5/6-NNCP, RSA data encryption, and pretty good privacy; and (5) Layer 7-FTP, NMHS, and SMTP.

TCP/IP can be viewed as a 5-layer protocol suite as illustrated in Figure 2-34: (1) Layer 1 - the physical layer [e.g., Plain Old Telephone Service (POTS), private lines, High-speed Digital Subscriber Line (HDSL), Asymmetrical DSL (ADSL), Very high-speed DSL (VDSL), Integrated Service Digital Network (ISDN), and Synchronous Optical NETwork/Synchronous Digital Hierarchy (SONET/SDH)]; (2) Layer 2-the data link layer [e.g., Asynchronous Transfer Mode (ATM), Frame relay, Point-to-Point Protocol (PPP), and Local Area Network (LAN)], (3) Layer 3-the network layer [e.g., Internet Protocol (IP), (4) layer 4-the transport layer [e.g., Transmission Control Protocol (TCP), and User Datagram Protocol (UDP)], and (5) layer 5-the session layer [e.g., Simple Network Management Protocol (SNMP), Real Time Protocol (RTP), Domain Name System (DNS), HyperText Transfer Protocol (HTTP), Simple Mail Transfer Protocol (SMTP), File Transfer Protocol (FTP)].

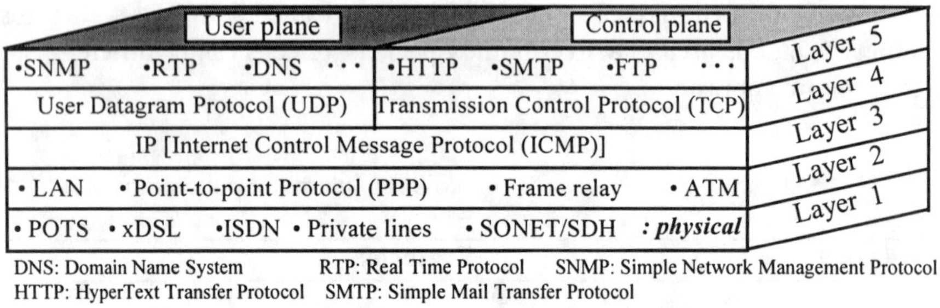

User plane			Control plane			
•SNMP	•RTP	•DNS ···	•HTTP	•SMTP	•FTP ···	Layer 5
User Datagram Protocol (UDP)			Transmission Control Protocol (TCP)			Layer 4
IP [Internet Control Message Protocol (ICMP)]						Layer 3
• LAN	• Point-to-point Protocol (PPP)		• Frame relay		• ATM	Layer 2
• POTS	•xDSL	•ISDN	• Private lines	• SONET/SDH	*: physical*	Layer 1

DNS: Domain Name System   RTP: Real Time Protocol   SNMP: Simple Network Management Protocol
HTTP: HyperText Transfer Protocol   SMTP: Simple Mail Transfer Protocol

Figure 2-34   TCP/IP 5-layer Protocol Suite.

2.4.1.1 TCP/IP Capabilities and Functions

TCP/IP has many capabilities and functions (Chapter 11) includes:

- Transmission Control Protocol (TCP):

  * TCP supports applications that require reliable virtual circuits and flow control.
  * TCP detects and adapts to network congestion conditions.
  * TCP is processed in the endpoints, not in network routers or switches.

- Internet Protocol (IP):

  * IP enables switching of data packets from the source to their proper destinations via routers.

* Routers/switches read the IP destination address, and deliver packets to the appropriate interfaces (toward destination).

* Each switch has a routing table that indicates the "best" output interface, depending upon the particular IP address.

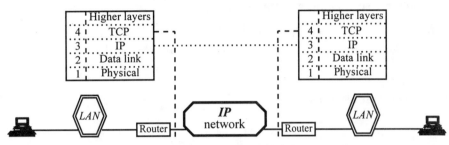

Figure 2-35   TCP/IP Capabilities/Functionality.

The method used to deliver frames, packets, or cells to their destinations can be different depending on the particular application. Two commonly used methods (described in Chapter 1) are: (1) connectionless [e.g., Internet, LAN, Switched Megabit Digital Service (SMDS), post service, etc.], and (2) connection oriented (X.25, frame relay, ATM, POTS). Note that X.25, frame relay, and ATM are connection-oriented *packet* switching technologies, while POTS is a connection-oriented *circuit* switching technology (Chapter 1).

2.4.1.2 Routers/Gateways

Since the internet connects many networks that can have different protocols, routers and gateways are devices that must be included n an Internet connection (Figure 2-35). A router performs the following functions:

* A router directs datagrams/packets to the appropriate interfaces that are connected to the final destination, via a complex maze of networks.

* A router uses network layer protocols (e.g., TCP/IP, Novell, DECnet, AppleTalk, etc.) to access to the network. Routers can handle single or multiple protocols.

* A router can re-format frames or packets for transport over different networks (including LANs).

* A router can segment (fragment) packets.

* A router can control access to other networks/resources.

* A router can ensure efficient use of redundant network paths.

A gateway function can be built into a router, or it can be implemented as a standalone device. It is designed to perform protocol conversion. Generally, gateways are not recommended because of the following reasons:

- The number of gateways required in an Internet increases exponentially as the number of protocols grow.

- Each gateway must be upgraded whenever a new protocol version is deployed.

- A gateway does not integrate e-mail directories with service, even though it enables e-mail users to communicate with each other.

It is recommended that all protocols migrate to a common format (e.g., SMTP over TCP/IP) rather than deploying gateways to perform multiple translations.

2.4.2 Internet Connections

The "Internet" has changed the way millions of people (worldwide) conduct their daily lives. There are various types of traffic carried on the Internet. A majority of the traffic is World Wide Web (WWW) and e-mail, flowing between users that have browsers and public web sites.

Definition 2-3 (World Wide Web/Web site): A website is a computer that contains pages of documents that are written in a standard format called the HyperText Markup Language (HTML). The documents can be downloaded and displayed by any browser (e.g., Netscape's Navigator and Microsoft's Internet Explorer that contains a HTML interpreter.

The Internet can also host bulletin board services (e.g., UseNet), "chat and sessions", and numerous financial/entertainment applications. File downloading via File Transfer Protocol (FTP), and a variety of other business functions (e.g., publishing general information, distributing promotional material, product descriptions, etc.) are available for commercial applications. Three Internet services are described as follows:

(1)   Information access: Services include WWW browsing, information search engines, and "point and click" access to multimedia.

(2)   UseNet newsgroup services: Services are made possible by posting bulletin boards.

(3)   Voice on IP and fax on IP: Needs are growing and quality is improving.

2.4.2.1 IP Connections for Voice over IP (VoIP)

Voice over IP (VoIP) and Fax over IP (FoIP) details are discussed in Chapter 11. Therefore, this section contains only high level descriptions of various IP connection: PC-IP-PC; PC-IP-PSTN; PSTN-IP-PSTN; and PBX-IP-PSTN.

- Computer-IP-computer (PC-IP-PC): This is the system architecture that was first implemented (Figure 2-36) to provide VoIP service.

In this configuration the computers (PCs) must be equipped with sound cards (full duplex for two-way conversation), microphones, speakers, handsets, and appropriate software. The advantages of this configuration are: (1) A flat charge for Internet access is applied for domestic or international calls [i.e., no additional fee is paid to Local Exchange Carrier (LEC) and/or International Exchange Carrier (IEC) companies]; (2) Several software packages are available (some are free of charge, such as Microsoft's NetMeeting); and (3) Video, audio, and data can be implemented over the same facilities.

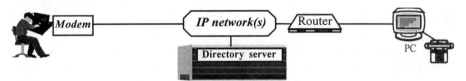

Figure 2-36   PC-IP-PC VoIP System Architecture.

There are several issues associated with this PC-IP-PC configuration: (1) voice service is not "PSTN quality" (i.e., voice quality is affected by compression technology, hardware, etc.); (2) Delay varies with network configuration (i.e., delays of 5 seconds are not uncommon); (3) Both parties (calling and called) must have special equipment and be connected simultaneously to the network to use this service.

Definition 2-4 (PSTN or trunk voice quality): Voice quality is typically measured subjectively by using a scale of 1 to 5. The measurement is known as Grade Of Service (GOS) or Mean Opinion Score (MOS). A score of "5" indicates "excellent"; "4" is "good"; "3" is "satisfactory"; "2" is "fair"; and "1" is "poor". A voice service with a MOS (GOS) rating of 4 or better is defined as "trunk quality" (i.e., equivalent to PSTN voice service).

- PC-IP-PSTN: Figure 2-37 shows the system architecture for providing VoIP service via the Public Switched Telephone Network (PSTN).

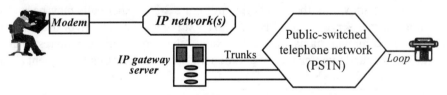

Figure 2-37   PC-IP-PSTN VoIP System Architecture.

The most significant aspect of this architecture is that the service can reach anyone served by the Public Switched Telephone Network (PSTN), which covers practically every corner in the US, and many other countries. In addition to its broad coverage, there is no need to pre-arrange the call connection. The software used for this configuration is provided by the carrier service provider.

Disadvantages (for PC-IP-PSTN) are: (1) per-minute charging is applied instead of a flat access fee; and (2) calls are generally pre-paid. Voice quality is still lower than trunk quality, and delays can cause degradation. In general, service quality is affected by computer hardware, software, and network traffic conditions.

- PSTN-IP-PSTN: The system architecture for VoIP service using a PSTN-IP-PSTN connection is shown in Figure 2-38. This architecture was first proposed by Inter-Exchange Carriers (IECs) to remain competitive in the communications industry. The goal for the IEC is to offer voice services at lower rates (competitive to the Internet service providers) for both domestic and international calls.

It is clear that rates must be much lower than the conventional switched calls. In addition, the service should not require any extra equipment (e.g., PCs) or special arrangements. However, the voice quality depends on equipment/software capabilities, and may not be equivalent to switched calls. Longer telephone numbers (i.e., greater than 15 digits) may also be required, and compatibility between IP carriers could be an issue.

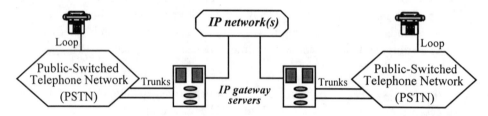

Figure 2-38   PSTN-IP-PSTN VoIP System Architecture.

- PBX-IP-PSTN: Figure 2-39 shows a system architecture for a business site using a hybrid service arrangement. Some calls may be routed by the PBX directly to the PSTN for switched "trunk quality" services, which have a voice quality with a Mean Opinion Score (MOS) of 4 (or better) out of 5 (the best defined MOS). This type of connections is shown by the "bold" line in Figure 2-39. Other calls may be routed to obtain a flat rate service via a PBX-IP-PSTN arrangement. The hybrid arrangement reduces telecommunications costs. Generally, this arrangement requires purchase, administration, and maintenance of additional Customer Premises Equipment (CPE).

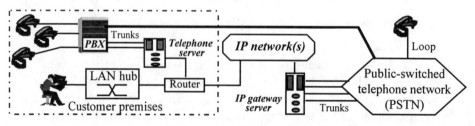

Figure 2-39   PBX-IP-PSTN VoIP System Architecture.

2.4.2.2 Dial-up Connections

IP connections for generic applications (services) are discussed in this section. Internet users share a "common 32-bit" addressing scheme known as "IP address". Access to the Internet is usually via (1) dial-up connections, or (2) a dedicated link to an Internet Service Provider (ISP). Since Internet services originated with dial-up connections, this connection is described in detail. At the customer premises, the home PC (desktop or laptop) must be equipped with a modem (modulator/demodulator, either an internal or external configuration). The customer must sign a contract with an IP service provider (ISP) to obtain a logon, and a dial-up number used to access the Internet.

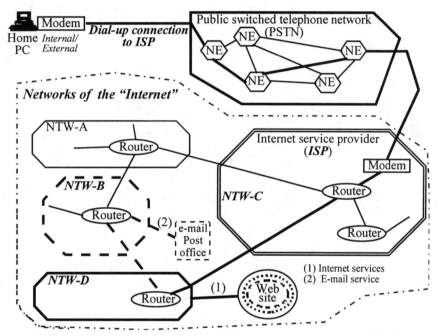

Figure 2-40   Connection between a Home PC and Internet Via PSTN.

The PC is connected directly to the Public Switched Telephone Network (PSTN). The PSTN contains local exchange and trunk exchange switches. The ISP dial-up may only involve a local exchange switch, or may require both local exchange and trunk exchange switches. Besides switches, transmission equipment and facilities are always needed to complete the IP connection.

An Internet is a network composed of different networks that are owned, operated and managed by individual service providers. Figure 2-40 shows an Internet with four networks (Network A, B, C, and D). Network C is assumed to be the ISP that serves this particular dial-up user. The PSTN connects the user to this network via a modem, and the connection is continued via several routers until the destination, either a Web site for IP services or a post office for e-mail service, is reached.

Note that some of the networks within an Internet may be owned by public or government organizations. These networks are equipped with various Web sites that post different "bulletin boards" or pages of public-domain documents, and are accessible via ISP connections. That is, IP or e-mail services may be provided by the network owned by the Internet Service Provider (ISP). In this example, the IP services are provided by Network D, the e-mail services are provided by Network B, and Network C serves as the ISP for this particular user (see Figure 2-40).

2.4.2.3 Dedicated Link to ISP

The end user in an office environment may access the Internet via a corporate network (e.g., Cornet) as shown in Figure 2-41. Figure 2-41 illustrates the interconnection between a corporate network (Cornet), the PSTN and the "Internet".

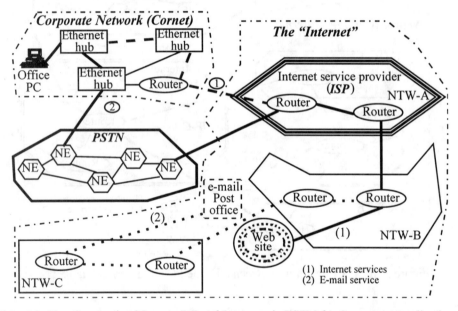

Figure 2-41   Connection between PC and Internet via PSTN for Corporate Applications.

The stations within the corporate network are typically connected using Ethernet LAN hubs, as shown in Figure 2-41. The routers in the corporate network are used for routing the IP traffic directly to the Internet Service Provider (ISP). In this example a "direct connection" between the end user and the ISP is shown by "dashed lines". As indicated, the end user can also be connected to the ISP via the PSTN.

Access to IP or e-mail services is the same as described in the previous section (Figure 2-40; for the connection between a Home PC and Internet). For example, the Web site is located in Network B, and e-mail service is provided by Network C.

Challenges facing IP services are: (1) developing a more robust version of IP protocol; (2) coordinating IP network management; (3) implementing network security features; (4) improving Quality of Service (QoS); (5) developing fast networking techniques (e.g., super high-speed routers, multi-protocol label switching, traffic flow techniques, etc.); and (6) utilizing broadband network technologies for Internet networking technique (e.g., frame relay, ATM, SONET, or SDH networks).

2.5 AN ATM CONNECTION

ATM technology is considered to be third generation packet-switching technology. That is, X.25 is considered the first generation, frame relay is the second generation, and ATM is the third generation of packet switching. The differences between circuit-switching and packet-switching are discussed in Chapter 1. The major differences between X.25, frame relay and ATM are briefly listed as follows:

- X.25: It is based on the concepts of data transport over voice-grade lines.

  * X.25 is a narrowband technology.
  * X.25 uses connection-oriented switching technology.
  * X.25 is designed for data transport only.
  * The maximum X.25 data rate is 56 kbps.

- Frame relay: This technology takes advantages of the improved transmission facilities provided for ISDN services.

  * Frame relay is a broadband technology.
  * Frame relay uses connection-oriented switching technology.
  * Frame relay is designed for data transport only.
  * The maximum frame relay data rate is 2 Mbps.

- ATM: This technology takes advantage of improved digital transmission facilities, and uses optical fiber for high-speed and excellent Quality of Services (QoS). ATM also utilizes faster processing techniques and high speed memory devices.

  * ATM is a broadband technology.
  * ATM uses connection-oriented switching technology.
  * ATM is designed for multimedia (ISDN) transport.
  * The maximum data rate is determined by the transmission facility capability. That is, ATM technology is rate-insensitive for services within the overall system capacity.

Note that additional details regarding X.25, frame relay, and ATM is provided on Chapter 5.

2.5.1 Overview of ATM Technology

ATM uses a small, fixed length packet size of 53 octets. Since it is a much smaller packet size compared to frame relay, which can contain up to 4096 octets, ATM is known as "*cell switching*" technology, which is considered to be a special "packet" technology.

Figure 2-42 shows the ATM cell organization. The 5-byte header can have two different formats: one for the use from user-to-network [Figure 2-42(A)], and another for network-to-network interfaces [Figure 2-42(B)]. The header consists of a Generic Flow Control field (GFC; 4 bits), Virtual Path Identifier (VPI, 8 or 12 bits), Virtual Channel Identifier (VCI, 16 bits), Payload Type Identifier (PTI, 3 bits), Cell Loss Priority (CLP, 1 bit), and Header Error Control byte (HEC, 8 bits). A detailed functional description of the 5 header bytes is provided in Chapter 5. The 48-byte payload field can be used to carry ATM user data, idle channel signals, unassigned channel signals, OA&M (Operations, Administration and Operations) signals, or system/network (control) signaling.

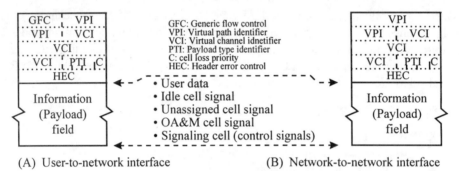

(A) User-to-network interface   (B) Network-to-network interface

Figure 2-42   ATM Cell Organization.

The concepts of ATM Virtual Path (VP) and Virtual Channel (VC) are briefly described in this section (see Figure 2-43). This technique allows channel bandwidth to be utilized efficiently for customers with different bandwidth needs. A physical (medium) channel (e.g., a SONET OC-3C channel with a bandwidth of 155.52 Mbps) can be divided into a maximum of 2^{12} Virtual Paths (VPs). Each VP can be further divided into a maximum 2^{16} Virtual Channels (VCs). It should be noted "*Each VP has a specific speed, but individual VCs can serve customer that require different speeds*."

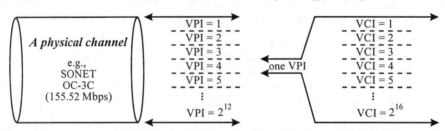

Figure 2-42   An ATM Physical Channel, Virtual Paths, and Virtual Channels.

Definition 2-5 : A Virtual Channel (VC) is a unidirectional communication link (capability) used for transporting ATM cells. A VC is basically an "end-to-end circuit" (entity) for ATM services. A Virtual Path (VP), that is a bundle of VCs, is also a unidirectional communication link (capability) for ATM cell transport. All the VCs within a VP must have the same endpoints (i.e., origin and destination).

2.5.2 ATM Services

A generic ATM connection is shown in Figure 2-44, which is also known as a "switched ATM connection". A switched ATM service requires signaling for call setup. A (VPI, VCI) pair uniquely identifies a virtual channel within a specific physical connection (link). The *(VPI, VCI) address is only locally significant*. That is, for switched ATM services, each ATM switch contains two pre-assigned (VPI, VCI) pairs that are used during the call setup process for routing purposes. *A virtual channel is identified by a series of concatenated (VPI, VCI) pairs*.

For example, in Figure 2-44 the user "**X**" is connected (via ATM switches A, C and D) to the destination "**Y**" through routing controlled by the ATM signaling network. As soon as the connection has been established, three routing tables are assigned for the associated ATM switches. Each routing (translation) table contains an input pair (VPI, VCI) and an output pair (VPI, VCI). The ATM cells from user "**X**" are originally contain in (VPI, VCI) = (99, 88), which is assigned by the network. When switch **B** receives a cell from user "**X**", it routes the cell via the **BC** ATM link to switch **C** [based on the routing table (99, 88) ⇒ (77, 66)]. Switch **C** [based on its routing table (77, 66) ⇒ (55, 44)] routes the cell via the **CD** link to switch **D**. Finally, switch **D** [based on its routing table (55, 44) ⇒ (65, 86)] routes the cell to destination "**Y**". Note that this is a connection-oriented switching operation.

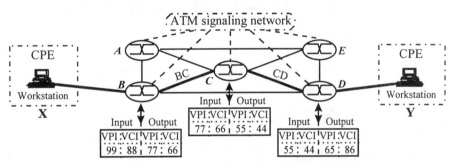

Figure 2-44   An ATM Connection for Switched ATM services.

From a provisioning viewpoint, ATM services can be grouped into two categories:

- Virtual Path Connection (VPC) for "switched" or "switched virtual" connections : A VPC is a concatenation of VP links (Figure 2-45) and segments that exists between the point where VCIs are assigned and the point where VCIs are translated/removed.

A VPC may exist between "user and user", "user and network", or "network and network". It is important to know that "*a VPC must meet the most demanding Quality of Service (QoS) of the VCCs within it*." The establishment and release of a VPC conforms to the following principles:

* VPCs can also be established/released by the network by using signaling procedures.

* VPC establishment/release is on a subscription basis, therefore signaling procedures are not necessary.

* VPC establishment/release functions may be controlled by the customer.

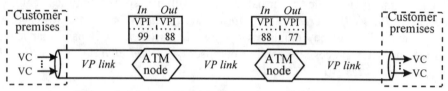

Figure 2-45   ATM Virtual Path Connection (VPC).

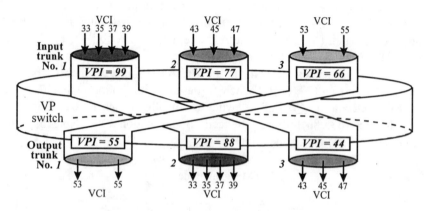

Figure 2-46   ATM VPC Switching.

Figure 2-46 shows an example of ATM VPC switching in an ATM network serving Virtual Path Connections (VPCs). The routing (translation) table contains only (VPI_{in}, VPI_{out}), **VCI values remain unchanged**. For simplicity, assume that VPI = 99 is input from trunk No. 1, VPI = 77 is input from trunk No. 2, and VPI = 66 is input from trunk No. 3 (in actual practice, all three VPIs can be input from the same trunk). After processing by the ATM VPC switch, VPI = 99 is output to trunk No. 2, VPI = 77 is output to trunk No. 3, and VPI = 66 is output to trunk No. 1 (in actual practice, they all can be output to the same trunk). The four VCs (VCI = 33, 35, 37 and 39) within VPI = 99 have been switched together to the output VPI = 88 (VPI = 88 corresponds to input VPI = 99).

* Virtual Channel Connection (VCC) for "permanent virtual" connections: A VCC is a concatenation of VC links that exist between two points where the ATM adaptation layer is accessed. Similar to VPC, VCC may exist between a "user and user", "user and network", or "network and network". The routing (translation) table for VCC switching requires both VPI and VCI information as shown in Figure 2-47.

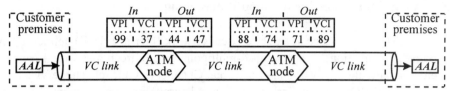

Figure 2-47   ATM Virtual Channel Connection (VCC).

The establishment and release of a VCC conforms to the following principles:

*   A VCC is established/released by using a meta-signaling procedure. However, semi-permanent/permanent VCCs are established during subscription (provisioning) time, therefore signaling procedures are not necessary.

*   A switched "end-to-end" VCC can be established/released by using a "user-to-network" signaling procedure.

*   A VCC within a VPC where two B-ISDN user-network interfaces already exists, can be established/released by employing "user-to-user" signaling procedures.

Unlike VPC switching, a VCC must switch both VPs and VCs as shown in Figure 2-48. That is, at each ATM switch node, a routing table (see Table 2-3) containing both (VPI_{in}, VPI_{out}) and (VCI_{in}, VCI_{out}) is required.

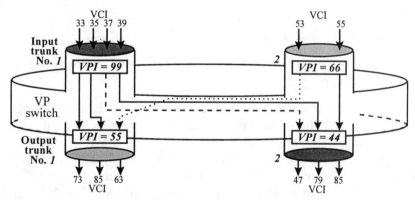

Figure 2-48   ATM VCC Switching.

Table 2-3   Typical Routing Table for a Virtual Channel Connection.

VPI_{in}	VCI_{in}	VPI_{out}	VCI_{out}
99	33	55	73
99	35	55	85
99	37	44	47
99	39	44	79
66	53	55	63
66	55	44	85

(Note that Table 2-3 corresponds to the example in Figure 2-48.)

2.5.3 ATM Connection Subscription/Provisioning

An ATM bearer service at a public User-Network Interface (UNI) defines a "point-to-point", "bidirectional", or "point-to-multipoint", "unidirectional virtual connection at either a Virtual Path (VP) and/or a Virtual Channel (VC) level. The service must meet the specified Quality of Service (QoS) and throughput requirement. For users that desire only VP service from the ATM network, the user is able to allocate individual VCs within the VP connection (provided none of the VCs required a higher QoS than the VP connection). Two provisioning options are described as follows:

- Permanent Virtual Connection (PVC): A PVC can be provisioned at the VP or VC level. Call establishment procedures are not needed because PVCs are provisioned manually. The routing (translation) table for a PVC may remain unchanged for weeks, months, or even years. Figure 2-49 shows an example of a typical PVC.

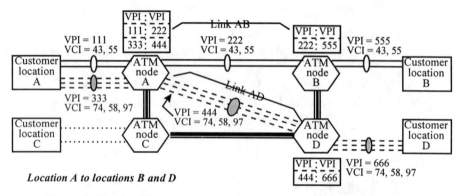

Location A to locations B and D

Figure 2-49   ATM Permanent Virtual Connection (Provisioning).

In Figure 2-49, the customer in location A is provisioned to have a Permanent Virtual Connection, with ($VPI_{in} = 111$, $VPI_{out} = 222$) and carrying two virtual channels (i.e., VCI = 43 and VCI = 55), to customer location B. This Virtual Path (VPI = 111) carrying two virtual connections (VCI = 43 and VCI = 55) is routed permanently to location B via ATM nodes A and B (typically ATM switches) over ATM link *AB*. When this VP arrives at ATM node B from node A, the old VPI = 222 is changed to VPI = 555 (see Figure 2-49). The routing table used at ATM node B is ($VPI_{in} = 222$, $VPI_{out} = 555$). ATM node B routes VP = 555, using its routing table, to customer location B. Note that VCI values (VCI = 43 & 55) remain unchanged from "end-to-end" for this Permanent Virtual Connection (VPC).

    Note that customer location A is also provisioned to have another Permanent Virtual Connection, with ($VPI_{in} = 333$, $VPI_{out} = 444$) and carrying three Virtual Channels (i.e., VCI = 74, VCI = 58, and VCI = 97) to customer location D. This Virtual Path (VPI = 333) carrying three Virtual Channels (VCI = 74, VCI = 58, and VCI = 97) is routed permanently to customer location D via ATM nodes A and D, and ATM link ***AD***. When this VP arrives ATM node D the VPI = 444 is changed to VPI = 666. The routing table used at ATM node D is (VPI$_{in}$ = 444, VPI$_{out}$ = 666). ATM node D routes this VPI = 666, using its routing table, to customer location D. Note that VCI values (VCI = 74, 58, and 97) remain unchanged from "end-to-end" for this Permanent Virtual Connection.

- Switched Virtual Connection (SVC): A SVC can only be provisioned at the VC level (i.e., not at the VP level). A SVC requires call establishment/release procedures (i.e., signaling). The call setup is dynamic (i.e., not permanent), therefore, the routing table exists only for the duration of a given call.

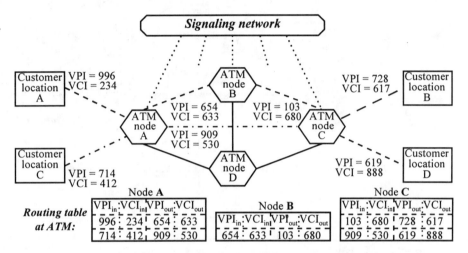

Figure 2-50   ATM Switched Virtual Connection (Provisioning).

In this example (Figure 2-50) the customer at location A, with (VPI, VCI) = (996, 234) (these values are assigned by the ATM switch network node A), has been setup to have a connection-oriented switched virtual channel from customer location A, to ATM network node A, network node B, network node C and finally the destination at customer location B. To understand the establishment of this ATM "end-to-end" connection, one needs to understand the usage of routing tables.

The routing tables associated with this VC are shown in Figure 2-50: (VPI$_{in}$, VCI$_{in}$) = (996, 234) to (VPI$_{out}$, VCI$_{out}$) = (654, 633) for node A, (VPI$_{in}$, VCI$_{in}$) = (654, 633) to (VPI$_{out}$, VCI$_{out}$) = (103, 680) for node B; and (VPI$_{in}$, VCI$_{in}$) = (103, 680) to (VPI$_{out}$, VCI$_{out}$) = (728, 617) for node C. The customer at location C, with (VPI, VCI) = (714, 412) (these values are assigned by the ATM switch node A), has also been setup to have a connection-oriented switched virtual channel from customer location C, to ATM node A, node C and finally the at customer location D. The routing tables associated with this VC are: (VPI$_{in}$, VCI$_{in}$) = (714, 412) to (VPI$_{out}$, VCI$_{out}$) = (909, 530) for node A; and (VPI$_{in}$, VCI$_{in}$) = (909, 530) to (VPI$_{out}$, VCI$_{out}$) = (619, 888) for node C (see Figure 2-50).

2.5.4 ATM Application Connections

From business perspective ATM has numerous applications. ATM may replace some traditional equipment, and add new capabilities required for modern telecommunications networking. Several examples are described in this section.

Example 2-6: Describe the differences between a ATM-based LAN and traditional leased-line LAN interconnections.

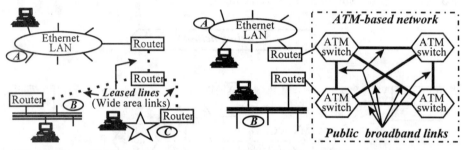

(A) Traditional LAN interconnection.　　(B) ATM-based LAN interconnection.

Figure 2-51　　LAN Interconnections: Traditional vs. ATM-based Technologies.

Figure 2-51(A) illustrates a LAN (Ethernet rings, bus or ring) interconnection that has been implemented by using routers and leased lines (e.g., 1.544 Mbps T1 lines in the U.S.A., or 2.048 Mbps E1 lines in other countries). In Figure 2-51(A), three LANs ("A": a ring LAN; "B": a bus LAN; and "C": a star LAN) are interconnected by a traditional scheme using leased lines. This architecture is very expensive, because the average monthly cost for one T1 line is about US$2,000 per month.

If an ATM-based network is used to interconnect LANs [in Figure 2-51(B) a ring Ethernet LAN, and a bus LAN), the public links are broadband, they can be shared by many users, and the costs are reduced considerably.

Example 2-7: Describe LAN emulation on an ATM switched Virtual Connection (see Figures 2-50 and 2-52).

Figure 2-52 shows the architecture for LAN emulation on an ATM switched Virtual Connection. An Address Resolution Protocol (ARP) is used to translate and map Medium Access Control (MAC) addresses and ATM identifiers (i.e., VPIs, VCIs). The LAN Emulation Server (LES) performs the address mapping function.

Assume a LAN end user (designated "A" in Figure 2-52) sends out a MAC frame to a router [designated "B"; called a LAN Emulation Client (LEC)] for routing to another

LEC router (designated "E") that serves the destination station (designated "Z"). The LEC router "B" assembles an ARP request, which includes the destination MAC address in the target physical address of the ARP packet. The ARP request is sent to LES "C" from LEC router "B", via the ATM switch and LEC router "D". The LES "C" retrieves the ATM ID (i.e., VCI value) of the destination LES router, and returns it to the requesting LEC (i.e., LEC router "B") in an ARP reply "response message" packet. The LEC router "B" then sets up an ATM switched Virtual Connection, using this information, to deliver a MAC frame to LEC router "E", and destination "Z". It should be noted that ***ATM network appear transparent to the end users*** in this application. That is, the service gives the illusion of being connected directly to the end user.

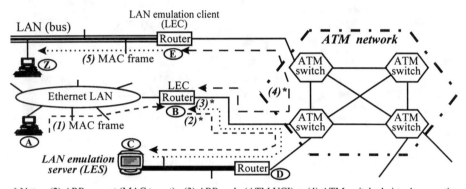

* Notes: (2) ARP request (MAC target); (3) ARP reply (ATM VCI); (4) ATM switched virtual connection

Figure 2-52   LAN Emulation on an ATM Switched Virtual Connection.

Example 2-8 (Virtual networking): Describe the ATM LAN and Virtual LANs Configuration.

Figure 2-53(A) shows a configuration consisting of Ethernet LANs (or other traditional LANs) that compel end users to share the communications medium. All the users connected to a traditional LAN must communicate at the same rate over the same medium. In a bus architecture, the users farther away from the server will experience longer delays than those close to the server. The delay difference in some cases could be large enough to cause unacceptable signal degradation.

ATM LANs delegate all control to a central ATM switch that provides each user with access to the full bandwidth of the network for as long as necessary. This arrangement does not require all users to communicate at the same rate or even over the same medium. The integration of LANs with ATM technology offers the benefits of virtual networking, which completely separates the physical and logical network infrastructure. As shown in Figure 2-53(B), ATM LANs can be "virtually" divided into multiple segments (there are three segments shown in this example) that can be organized along administrative boundaries to provide enhanced security and scalability.

Bridges/routers are used to implement connections between the partitioned virtual ATM segments [emulated LANs 1, 2 and 3 in Figure 2-53(B)]. Membership in an emulated LAN is characterized logically, rather than physically (as in traditional LANs). This capability offers increased flexibility in terms of terminal mobility and network management. A client may be a member of more than one emulated LAN, but still have only one physical connection to the network. Likewise, a client remains a member of an emulated LAN, even if they move from one physical location to another within the ATM network.

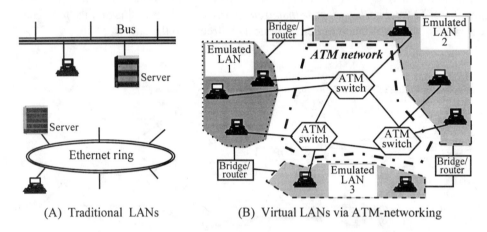

| (A) Traditional LANs | (B) Virtual LANs via ATM-networking |

Figure 2-53   Virtual Networking: Emulated LANs using ATM Technology.

Example 2-9: Describe the ATM network topology for an enterprise, access, and core networks (Figure 2-54).

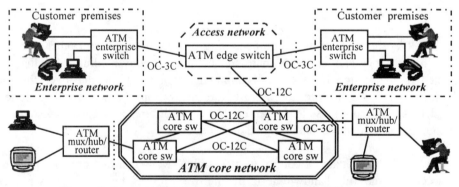

Figure 2-54   ATM Public and Private Network Topologies.

Presently, the primary applications for ATM networks are associated with business. A customer may own an "enterprise" ATM switch, that together with other ATM equipment (e.g., ATM workstations, PCs, voice terminals and video terminals) is located on the customer's premises. This enterprise switch also serves as an ATM that is used to combine customer signals (multimedia) into an ATM cell stream. An ATM "edge switch" is used to switch/multiplex different ATM cell streams coming from different ATM enterprise switches (shown on the top half of Figure 2-54). The output of the ATM edge switch is typically a much higher rate signal (e.g., OC-12C), which is routed to the ATM core network for transport to its final destination.

In addition to business applications, an ATM core network can also serve public customers as shown in the bottom half of Figure 2-54. ATM multiplexer, hub, or router can be used to connect customer ATM cell streams to the core network.

Example 2-10 (Circuit-switching emulation): Describe why ATM is the only packet technology that supports circuit emulation (Figure 2-55).

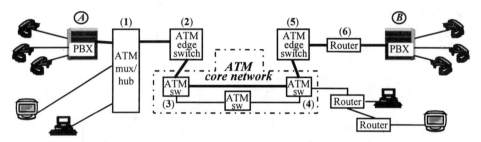

Figure 2-55   ATM Supports Circuit (Switched) Emulation.

A circuit-switched service (described in Chapter 1) establishes the "ownership" of an end-to-end connection during the entire call period. That is, the subscribers have the circuit dedicated for their exclusive use throughout the call. ATM technology allows users to have services allocated in a manner that is similar to circuit-switched lines (i.e., ATM technology supports circuit emulation).

The circuit emulation capability is implemented by using ATM technology as shown in Figure 2-55. The user served by PBX "A" and the user terminated by PBX "B" are assumed to have circuit-switched services. That is, they must have a connection having exactly the same characteristics as a circuit-switched line, even though the end-to-end connection (indicated by "bold" line segments along the transmission path) is composed of an ATM mux/hub [network element denoted as (1)], an ATM edge switch at each end [elements (2) and (5)], several ATM (core) switches [elements (3), (4), etc.] and one router [element (6)]. The signal from the station served by PBX "A" is segmented and formatted into ATM cells by the ATM mux/hub device. The cell stream is routed through the ATM network via ATM edge switches, core switches, and routers to destination. The ATM network is completely transparent to the two end users. Therefore, circuit-switched emulation has been effectively implemented.

2.6 A SONET/SDH CONNECTION

Synchronous Optical NETwork/Synchronous Digital Hierarchy (SONET/SDH) is an international physical layer protocol that was standardized in 1988, and since then high-speed backbone networking has adopted it as the "one and only standard". Detailed SONET/SDH protocol descriptions are given in Chapter 6. Applications for SONET/SDH have been extended from backbone networks to access networks, and finally to customer premises equipment. This section describes the common applications of SONET/SDH networks. As background for understanding the SONET/SDH connections, an overview of SONET/SDH protocol is also given in this section.

2.6.1 SONET/SDH Protocol Overview

An important features of SONET/SDH protocol is that specific data rates have been uniquely defined as listed in Table 2-4.

Table 2-4   SONET/SDH Signal Hierarchy.

SONET	SDH	Data rate (Mbps)	Capacity
STS-192	STM-64	9,953.28	192 STS-1s; 64 STM-1s
STS-48	STM-16	2,488.32	48 STS-1s; 16 STM-1s
STS-12	STM-4	622.08	12 STS-1s; 4 STM-1s
STS-3	*STM-1*	155.52	3 STS-1s; *1E4*
STS-1	STM-0	51.84	*1 DS3*
*	TU-3	48.96	1E3
VT6	TU-2	6.912	1DS2
VT3	*	3.456	1DS1C
VT2	TU-12	2.304	1E1
VT1.5	TU-11	1.728	1DS1

*: Invalid configuration

2.6.1.1 SONET Signal Hierarchy

SONET standards have been widely implemented in North America. There are five Synchronous Transport Signal (STS) data rates currently available: STS-N (N = 1, 3, 12, 48 and 192; additional higher data rates may soon become standard.). The STS-1 signal is known as the SONET basic rate (data rate of 51.84 Mbps) and is designed to carry one DS3 signal (the most popular long-haul digital signal in North America). Similarly, the basic rate for SDH is STM-1 (data rate of 155.52 Mbps) designed to carry one E4 signal or three STS-1 signals. For ease in multiplexing, both SONET and SDH signal hierarchy based on "modularity", as expressed by the following equation:

$$\text{Speed of STS-N/STM-N} = N \times (\text{speed of STS-1/STM-1}) \qquad (2\text{-}1)$$

For example, using Eq.(2-1) the speed of STS-12 is 622.08 Mbps (= 12 × 51.84 Mbps; twelve times the speed of STS-1). Similarly, The speed of STM-64 is 9,953.28 Mbps (= 64 × 155.52 Mbps; sixty-four times the speed of STM-1). Therefore, one STS-12 signal can carry 12 DS3 signals.

Definition 2-6 (*OC-N*): When an STS-N signal is carried by optical fiber links, it is called an OC-N signal (OC: optical carrier).

Definition 2-7 (*VT-n*): A SONET basic signal (STS-1) consists of many Virtual Tributaries (VTs). There are four sizes (types) of VT signals: VT6, VT3, VT2 and VT1.5. One STS-1 can contain seven VT6 signals, 14 VT3 signals, 21 VT2 signals, or 28 VT 1.5 signals. VT signals are referred to as "SONET logical signals".

A VT1.5 is designed to carry one 1.544 Mbps DS1 (or equivalent) signal. When 28 VT1.5 signals are multiplexed together, this forms a STS-1 signal. Several STS-1 signals can then be multiplexed with other STS-N signals to form a higher-rate SONET signal for long-haul transmission. A VT2 is designed to carry one 2.048 Mbps E1 (or equivalent) signal. When 21 VT2 signals are multiplexed together, this can also form a STS-1 signal. As previously indicated, several STS-1 signals can then be multiplexed with other STS-N signals to form a higher-rate SONET signal for long-haul transmission. The same principle can be applied to 14 DS1C (VT3) signals, or 7 DS2 (VT6) signals. Mixtures of DS1, E1, DS1C and/or DS2 signals can also be carried by SONET signals.

It should be understood that a VT1.5 signal has a data rate of 1.728 Mbps instead of 1.544 Mbps (data rate of DS1 signal) even that one VT1.5 is used to carry one DS1. The difference of 188 kbps (= 1.728 Mbps − 1.544 Mbps) data rate is designated as SONET overhead bits. They include: (1) VT pointer bytes designated to perform "new data", frequency justification, and concatenation indications; (2) VT path overhead bytes designated to perform error monitoring and reporting functions, and several other functions. Similarly, VT2 has a data rate of 256 kbps, VT3 has a data rate of 304 kbps, and VT6 has a data rate of 600 kbps designated for the following functions: new data flag, frequency justification and concatenation indication, error monitoring/reporting, etc.

2.6.1.2 SDH Signal Hierarchy

SDH standards have been implemented in many countries (other than North America), with very few exceptions. There are five Synchronous Transport Module (STM) data rates currently available: STM-N (N = 0, 1, 4, 16 and 64; additional higher data rates may soon become standard. Also, it should be understood that the STM-0 data rate has not been adopted globally, but only adopted by a very few countries). The signal STM-1 is known as SDH basic rate (155.52 Mbps), and is designed to carry one E4 signal (the most popular international long-haul digital signal). For ease of multiplexing, the SDH signal hierarchy also adopts "modularity". For example, the speed of STM-4 is 622.08 Mbps (= 4 × 155.52 Mbps; four times the speed of STM-1). SIMILARLY, The speed of STM-64 is 9,953.28 Mbps (= 64 × 155.52 Mbps; sixty-four times the speed of STM-1), and one STM-64 signal can carry 64 E4 signals.

Definition 2-8 (***TU-n***): A SDH basic signal (STM-1) consists of many Tributary Units (TU-ns). There are different sizes (types) of TU-n signals: TU-3, TU-2, TU-12 and TU-11. One STM-1 can contain three TU-3 signals, 21 TU-2 signals, 63 TU-12 signals, or 84 TU-11 signals. TU-n signals are referred to as "SDH logical signals".

A TU-11, which is equivalent to SONET's VT1.5 signal, is designed to carry one 1.544 Mbps DS1 (or equivalent) signal. When 84 TU-11 signals are multiplexed together, this forms a STM-1 signal. Several STM-1 signals can then be multiplexed with other STM-N signals to form a higher-rate SDH signal for long-haul transmission. A TU-12, which is equivalent to SONET's VT2 signal, is designed to carry one 2.048 Mbps E1 (or equivalent) signal. When 63 TU-12 signals are multiplexed together, this forms a STM-1 signal. As previously indicated, several STM-1 signals can then be multiplexed with other STM-N signals to form a higher-rate SDH signal for long-haul transmission. The same principle can be applied to 21 DS2 (TU-2) signals, or 3 E3 (TU-3) signals. Mixtures of DS1, E1, DS2, and/or E3 signals can also be carried by SDH signals.

Similar to SONET applications: (1) TU-11 has a data rate of 188 kbps; (2) TU-12 has a data rate of 256 kbps; (3) TU-2 has a data rate of 600 kbps; and (4) TU-3 has a data rate of 4.592 Mbps designated for the following functions: new data flag, frequency justification and concatenation indication, administration, operations, control, alarm surveillance, error monitoring/reporting, etc.

From Table 2-4, it can be seen that (1) SONET VT3 signals are not allowed in the SDH networks, and (2) SDH TU-3 signals are not allowed in the SONET networks. The rationale is provided as follows. A SONET VT3 signal is used to carry one DS1C (data rate of 3.152 Mbps), which is a specific digital signal used in a limited area in North America. On the contrast, DS1 signals have been widely adopted, even outside the US. Therefore, there is no need for SDH (an international standard) to handle VT3 data rates. Similarly, ITU-T E3 signals are not as popular as E1 signals internationally. That is, E3 signals regional signals scattering in many countries, and are not used to interconnect global networks. Since a SDH TU-3 signal is designated to carry one E3 signal, there is no need for SONET (North American standard) to handle TU-3 data rates.

2.6.2 SONET/SDH Generic End-to-end Connection

It is easier to understand the concepts of a generic SONET/SDH end-to-end connection by analyzing a simple SONET application (see Figure 2-56).

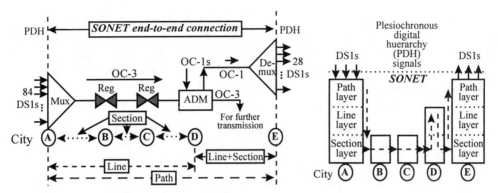

Figure 2-56   A SONET End-to-end Connection.

In Figure 2-56, 84 DS1 signals are multiplexed to form an OC-3 signal at City A. Because the OC-3 signal degrades as it travels on optical fiber, it is regenerated at Cities B and C so that the OC-3 signal can be transmitted farther. When the OC-3 signal arrives at City D, it is first regenerated, then an Add-Drop Multiplexer (ADM) is used to "*drop*" one OC-1 signal (one of the three OC-1 signals carried by the OC-3 signal), and "*add*" one OC-1 signal thereby forming a new OC-3 signal. This newly formed OC-3 signal is not relevant to the terms defined in this section, thus it is ignored in the subsequent discussion. The "*dropped*" OC-1 signal is demultiplexed into 28 DS1 signals for PDH (see Definition 2-9) network transport at City E, after being transmitted as an OC-1 signal over a SONET network for some distance. It should be mentioned that at City E, the OC-1 signal must first be regenerated before it is demultiplexed into 28 individual DS1 signals.

Definition 2-9 [***Plesiochronous Digital Hierarchy*** (PDH)]: Prior to the SONET/SDH digital hierarchical standards (Table 2-4), several digital hierarchies have been used by the telecommunications industry. Two widely standards are: (1) the μ-law digital hierarchy, and (2) the A-law digital hierarchy. Signals that follow the μ-law digital hierarchy are DS1, DS1C, DS2, DS3, DS3C, and DS4. The signals that follow the A-law digital hierarchy are E1, E2, E3 and E4 (Chapter 6). These digital signal hierarchies adopt completely different physical layer protocols, and thus are called PDH.

The term "*plesiochronous*" comes from a combination of Greek words: (1) "*plesio*" meaning "close", "near", or "almost the same"; and (2) "*synchronos*" meaning "happening at the same time", "occurring together", "simultaneous", or "having the same period between occurrences". In telecommunications, plesiochronous is used to describe any two signals that have the same "nominal rate", but do not originate from the same (identical) timing (clock) source. Hence, a Plesiochronous Digital Hierarchy (PDH) is a protocol that consists of signals that have relatively the same bit rates and characteristics, but they are not synchronized to a common timing source (timing and synchronization is described in Chapter 14 of this book). Generally, the physical layer protocols that do not adopt SONET/SDH standards are referred to as PDH protocols.

Figure 2-56 illustrates the three SONET terms: a SONET path, a SONET line, and a SONET section that are defined in Table 2-5. Note that SONET's "line" is equivalent to SDH's "multiplex section", and SONET's "section" is equivalent to SDH's "regenerator section". It is actually easier to relate the SDH names (terms) to the functional meaning of these terms (path, line or multiplex section, etc.).

Table 2-5   Terms Used to Define a SONET/SDH End-to-end Connection.

SONET	*SDH*	*Example* (City⇒City)	*Brief definition* (meaning)
Path	Path	A ⇒ E	Demarcation (PDH-to-SONET interface) to demarcation (SONET-to-PDH interface)*
Line	Multiplex section	A ⇒ D D ⇒ E	Link connecting two *adjacent* multiplexers or equivalent elements [e.g., Add-Drop Multiplexer (ADM) or Digital Cross-connect System (DCS)
Section	Regenerator section	A ⇒ B; B ⇒ C C ⇒ D; D ⇒ E	Link connecting two *adjacent* SONET network elements (e.g., regenerators)**

* Or, SDH-to-PDH, PDH-to-SDH interfaces   ** Or, SDH network elements

Every SONET/SDH signal (Table 2-4) contains additional bandwidth reserved for network Operations, Administration, Maintenance and Provisioning (OAM&P). These OAM&P functions are performed by (1) Path OverHead (POH) bytes, (2) Line OverHead (LOH), or Multiplex Section (MSOH) bytes, and (3) Section OverHead (SOH), or Regenerator Section OverHead (RSOH) bytes. The SONET/SDH network elements "*generate*, *modify*, and/or *interpret*" the POH, LOH (MSOH) and SOH (RSOH) bytes.

___

__Definition 2-10__ (**PTE, LTE/MSTE, and STE/RSTE**): The SONET element that terminates a SONET/SDH path is called Path Terminating Equipment (PTE). The element that terminates a SONET line or a SDH multiplex section (MS) is called a SONET Line Terminating Equipment (LTE), or a SDH Multiplex Section Terminating Equipment (MSTE), respectively. The element that terminates a SONET section or a SDH regenerator section (RS)] is called a SONET Section Terminating Equipment (STE), or a SDH Regenerator Section Terminating Equipment (RSTE). Note that "*a SONET/SDH PTE is also a LTE and STE*. Similarly, a *SONET/SDH LTE (MSTE) is also a STE (RSTE)*".

Example 2-11: Determine the functions of the SONET network elements in Figure 2-56.

Element at	*Served as (functionality)*
City A	Path terminating equipment, line terminating equip. & section terminating equip.
City B	Section terminating equipment
City C	Section terminating equipment
City D	Line terminating equipment & section terminating equip.
City E	Path terminating equipment, line terminating equip. & section terminating equip.

The functions of PTE, LTE and STE are summarized in Figure 2-57. For example, the SONET element (Figure 2-56) at City A serves as PTE, LTE and STE, therefore, it generates path overhead (POH), line overhead (LOH) and section overhead (SOH) bytes. Similarly, the element (Figure 2-56) at City E serves as PTE, LTE and STE. Therefore, it can interpret POH, LOH and SOH to perform network OAM&P functions. The results of these activities reflect SONET network performance status.

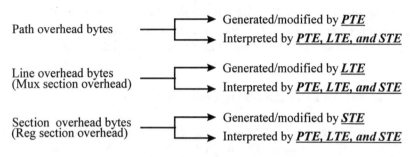

Figure 2-57   Functions of PTE, LTE (MSTE), and STE (RSTE).

2.6.3 SONET/SDH Network Topology

A SONET/SDH connection requires several network elements, that are classified as PTE, LTE and/or STE as defined in the previous section. However, from a network topology view, SONET/SDH network elements are classified as access layer elements, junction layer elements, or core (transport) layer elements (Figure 2-58).

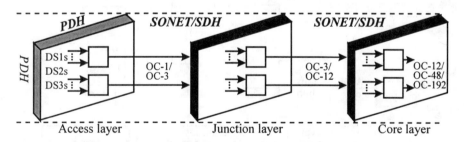

Figure 2-58   SONET/SDH Network Topology: Access, Junction and Core.

The access layer network elements (known as access products) convert PDH signals (DS1s, DS2s, etc.) into SONET/SDH signals. The SONET signals at the output port of access elements are typically OC-1 and OC-3. However, in many applications, such as SONET rings, the signals can be OC-12 or OC-48 signals. The output signals of the junction layer elements are typically OC-3 or OC-12, and for some applications, it can be OC-48. The core layer element has output signals of OC-12 and OC-48. In newer

SONET networks, the output signals of the core layer can be OC-192. The specific SONET network elements (e.g., add-drop multiplexers, digital cross-connect systems, etc.) used in these three layers are described in later sections of this chapter.

2.6.4 SONET/SDH Application Connections

Several commonly used SONET/SDH connections are described in this section: (1) overall end-to-end connections, (2) access connections (including business access applications), (3) intra-office connections, and (4) inter-office connections.

2.6.4.1 Access, Junction and Core-layer Connections

Figure 2-59 shows several applications (voice, data, video, and multimedia via ATM) served by SONET/SDH networks. On the PDH side, voice and video are digitized and multiplexed with data signals to form a digital bitstream (carried by DS1, DS1C, DS2 and/or DS3 signals) that is applied at the demarcation point of the SONET network. A SONET/SDH access multiplexer is used combine all the PDH signals and map them into a SONET/SDH signal format (e.g., OC-1 or OC-3 in general) for transmission. The multimedia signals are fed to ATM adapter which operates an ATM cell stream. This cell stream is input to an ATM mux/hub/router and is mapped into SONET/SDH signal frame (e.g., OC-3c: a concatenated OC-3 signal described in Chapter 6). These functions just are performed by SONET/SDH access layer.

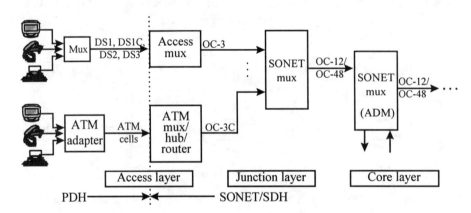

Figure 2-59   SONET/SDH Connections: Access, Junction and Core.

The SONET signals OC-3 and OC-3c are multiplexed to a higher rate SONET signal, such as OC-12 or OC-48, for farther transmission. This is known as the junction layer function. If the SONET signals are to be transmitted over a long distance, it is more economical to operate at the highest SONET rates (e.g., OC-48 or OC-192). This function is performed by the core (transmission) layer. In this example, only multiplexers, and add-drop multiplexers are used to illustrate the "access, junction, and core" connections. However, other SONET network elements (e.g., SONET switches and digital cross-connect systems) are used in practical applications.

2.6.4.2 SONET/SDH Access Network Connections

The SONET/SDH access applications can be grouped into: (1) residential and small business customers, and (2) large business and metro high-rise customers.

- Residential and/or small business connections: Figure 2-60 shows a typical SONET/SDH access connection for residential and small business applications. Voice and video signals are digitized and multiplexed with data signals, if applicable, to form PDH signals (e.g., DS1). Multimedia signals can be carried by ATM cells if appropriately provisioned. For small business applications, the customer's signals are first routed to a private PBX, and then carried by a digital signal (e.g., DS1 or PBX tie trunk). In some cases, widespread customers can first be fed to a Subscriber Loop Carrier (SLC) system, and formatted for transport by a common carrier signal (e.g., DS1). Some modern networks, a newer access SLC can transport SONET signals.

  All of the above customer signals (either PDH signals or ATM cells) are applied to a SONET access network element so that they can be mapped into SONET frame format (e.g., typically OC-3 or OC-12). These OC-3 and/or OC-12 signals are sent to a central office (wire center, or end office) where they are further multiplexed into higher rate signals for long distance transmission. Sometimes, they are fed to Digital Cross-connect System (DCS) for routing and concentration so that facility connections are more efficient.

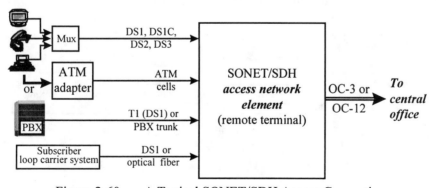

Figure 2-60   A Typical SONET/SDH Access Connection.

- Business and/or metro high-rise connection: This was actually the original SONET/SDH access application. The principles behind this type of application is exactly the same as described in Figure 2-60. In a business arrangement (especially in metropolitan areas), each floor in a high-rise office building may request voice, data, video, ATM multimedia, and PBX services (see Figure 2-61). These services can be in analog and/or digital formats (e.g., DS1 format or ATM cell stream). They are fed a SONET/SDH Business Remote Terminal (BRT) located in the basement of the building complex. The BRT is an access network element that maps/multiplexes incoming signals into OC-3 or OC-12 (currently; OC-48 may be used in the near future) for transport to service provider locations (e.g., a central office).

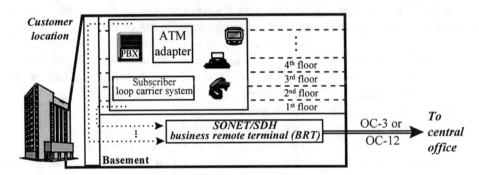

Figure 2-61   SONET/SDH Business or Metro High-rise Access Connection.

- SONET/SDH digital switch: A SONET/SDH switch can be applied at the access layer, junction layer, or core layer. The principles of the switch are similar for all three applications. Figure 2-62 shows the access switch application. In this case, the switch is always located at the local exchange carrier office (an end office wire center). The features of this switch are analog-to-digital conversion, ATM cell adaptation, packet (cell) switching, (traditional) digital signal switching, multiplexing and/or add-drop multiplexing, and digital signals cross-connection.

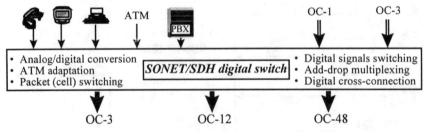

Figure 2-62   SONET/SDH Digital Switch for Access Network.

2.6.4.3 SONET/SDH Intra-office Connections

Four SONET/SDH network elements are typically used for intra-office connections: (1) digital switches, (2) Add-Drop Multiplexers (ADMs), (3) Digital Cross-connect Systems

(DCSs), and (4) digital (electronic) regenerators. The functions of a SONET/SDH digital switch shown in Figure 2-62 can also be used for intra-office connections. The typical signal interfaces within a central off ice are PDH signals (e.g., DS1 and DS3 in North America, and E1, E3 and E4 in other countries), ATM signals, and SONET/SDH signals (e.g., OC-1, OC-3, OC-12, ..., STM-1, STM-4, ...). The signal interfaces for digital switches, ADMs, DCSs, or regenerator depend upon the specific application (Chapter 6) and central office configuration (see Figure 2-63).

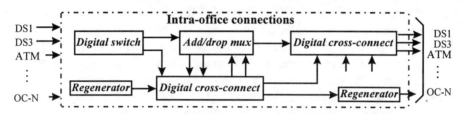

Figure 2-63   SONET/SDH Intra-Office Connections.

2.6.4.4 SONET/SDH Inter-office Connections

The SONET network elements shown in Figure 2-63 are used in almost every central office. Interconnecting two central offices may adopt one (or a combination of) the following three configurations: (1) point-to-point connections; (2) point-to-point with 1+1 or 1:n (1 by n) Automatic Protection Switching (APS) capability; and (3) ring automatic protection switching for self-healing capability.

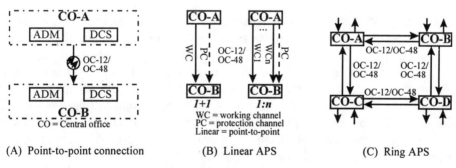

Figure 2-64   SONET/SDH Inter-office Connections.

The point-to-point interconnection without Automatic Protection Switching (APS) capability [Figure 2-64(A)] is found in some early SONET/SDH networks, in SONET/SDH networks without the needs for fast network restoration, or in some private SONET/SDH networks (e.g., between corporate headquarters and a branch office). This configuration has replaced many long-haul coaxial cable facilities because of several reasons: (1) using optical fibers as transmission media has considerably increased the available bandwidth [continues to increase with Wavelength Division Multiplexing (WDM) (Chapter 7)]; (2) using optical fibers improves the transmission quality compared to coaxial cable and microwave radio transmission; and (3) SONET protocol provides better network management capabilities because overhead bytes are used to implement/enhance Operations, Administration, Maintenance, and Provisioning (OAM&P) functionality.

When "fast" network restoration and/or self-healing capabilities are required, SONET/SDH networks can be configured as either linear (point-to-point) APS [Figure 2-64(B)] or ring APS [Figure 2-64(C)]. The APS function is described in Chapter 6. If linear APS is used in SONET/SDH network, and there are extra optical fibers available, the "1+1" APS provides better network restoration capability than the "1:n". This is because in the "1+1" architecture, each working channel is equipped with its own protection channel. Therefore, a multiple channel failure condition is no different from a single channel failure. In addition, the "head-ends" of each pair (working and protection channels) are "hard" bridged to provide a "faster" and "more reliable" protection switching. However, the "1:n" APS is popular when the number of available fiber cables is limited. Furthermore, if protection (spare) channels are used to carry low-priority traffic when everything is working, the "1:n" APS architecture is more cost effective.

The ring APS is the "*best*" configuration from a network restoration viewpoint. Ring APS networks (2-fiber or 4-fiber, unidirectional or bidirectional, line switched or path switched rings) are "*self-healing*". That is, service is automatically restored for the case or single or some certain multiple failure conditions.

2.7 WDM CONNECTIONS

Since the early 1980s, optical fibers have been widely deployed in telecommunications networks. In North America, the growth rate of optical fiber communications in the late 1990s has been almost exponential. Existing optical fiber systems a apply single wavelength per fiber to carry a digital bitstream with a fixed data rate. For example, in FT-G optical fiber systems one fiber carries a 1.7 Mbps digital bitstream at the wavelength of 1310 *nm*. Similarly, and in SONET OC-48 optical fiber links one fiber carries a 2.48832 Gbps digital bitstream at a wavelength of 1550 *nm*. An optical fiber link using this single wavelength architecture is shown in Figure 2-65(A). In general, optical fiber wavelength assignments [Figure 2-65(B)] are:

- For low-speeds and/or very short distance transmission, the wavelength adopted is around 820 nm [e.g., Local Area Network (LAN) applications].

- For medium-speeds and distances, 1310 nm is used (e.g., FT-G 417 Mbps, and some FT-G 1.7 Mbps systems).

- For high-speeds and long-haul transport 1550 nm is recommended (e.g., SONET OC-48 for long-haul applications).

It should be noted that the existing optical fiber technology is based on germanium (Ge)-doped silica materials. The optical fiber characteristics (attenuation versus wavelength) of Ge-doped fiber is shown in Figure 2-65(B). Additional information concerning this technology is provided in Chapter 7 of this book.

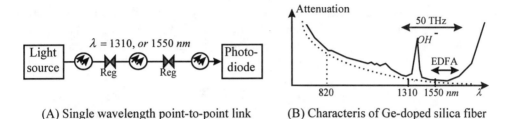

(A) Single wavelength point-to-point link   (B) Characteris of Ge-doped silica fiber

Figure 2-65   Wavelength Distribution of Ge-doped Silica Fibers.

To understand the basic principles of Wavelength Division Multiplexing (WDM) technology, it is important to note three characteristics exhibited by Ge-doped silica optical fibers [Figure 2-6B(B)]:

(1) An attenuation peak occurs around 1390 *nm*: This peak is the result of scattering loss caused by embedded hydroxide ions (OH⁻) in the fiber core, that are created during the manufacturing process. As a consequence, the operational wavelength has been broken into two parts: (1) around 1310 *nm*, and (2) around 1550 *nm* (besides the early 820 *nm* technology). This was the reason earlier 2-WDM systems were designed to work around 1310 *nm* and 1550 *nm*, but this peak has been nearly eliminated in "Allwave" fibers. Therefore, Allwave-fiber's "*available*" bandwidth is about 50 THz [see Figure 2-65(B)].

(2) Rapid attenuation occurs after 1700 *nm*: This attenuation is caused by the "intrinsic" absorption loss of Ge-doped silica material. Unless the basic optical fiber material is changed, it is believed that this degradation can not be improved using the present Ge-doped silica technology.

(3) Optical amplifiers have performance limits: Erbium-Doped Fiber Amplifier (EDFA) technology is the state of art. Optical amplifiers use "erbium" as the *active* material for the amplification function. These amplifiers can only provide acceptable optical signal gain around the 1550 *nm* window. Advanced research and development of other optical amplifiers is required to extend the WDM beyond the "1310 *nm*" and the "1550 *nm*" windows (i.e., fully utilize "Allwave" fibers).

Figure 2-66 shows an optical fiber link with/without applying WDM. The link in Figure 2-66(A) is a traditional optical fiber link without using WDM. The transmitter contains one light source (a LED or laser diode) that radiates "one" wavelength λ (around 820, 1310, or 1550 *nm*). Although every light source radiates more than one wavelengths, the central (nominal) wavelength is used to classify the operation of fiber link (see Chapter 7). The radiated light is transmitted over optical fibers, and as the optical signal travels along the transmission path (fibers), it is gradually attenuated. For long distance transmission, (electronic) regenerators must be deployed along the transmission path. When the light arrives at the receiver (i.e., the destination), a photodiode (photodetector) is used to convert the light energy into electrical energy in the form of electrical current (voltage). This electrical signal is applied to metallic wiring (e.g., twisted pair, coaxial cable, etc.), or converted into radio (microwave) signal for digital transmission, depending upon the network configuration.

Figure 2-66(B) represents a optical fiber link using WDM. The transmitter containers n light sources, each radiating a different wavelength (λ_1, λ_2, λ_3, ..., λ_n). The wavelengths are multiplexed together and transmitted over a single fiber. In modern optical fiber systems, several (5 to 8) Erbium-Doped Fiber Amplifiers (EDFAs) are used as optical amplifiers between adjacent regenerators. This is done to reduce the overall number of regenerators required in the network. Next generation optical networking technology is expected to replace electronic regenerators with photonic regenerators.

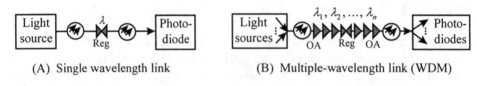

(A) Single wavelength link   (B) Multiple-wavelength link (WDM)

Figure 2-66   An Optical Fiber Link with/without WDM Technology.

2.8 FUTURE TRENDS

Chapter 2 has provided high level descriptions of several telecommunications services and connections (see Figure 2-67). Traditional, present, and future connection technologies are summarized as follows:

(1) Since the early 1960s, voice services have been implemented using digital carrier systems. Time Division Multiplexing (TDM) technology has been deployed worldwide using μ-law companding in North America, and A-law in other countries (Chapter 6). The digital signals (e.g., DS1, DS3, E1, E3, and E4) are typically transported over the Public Switched Telephone Network (PSTN). PSTN facilities (media) are twisted-pair metallic wiring, coaxial cables, microwave radio (analog and digital), satellite radio, optical fibers without Wavelength Division Multiplexing (WDM), and fibers with WDM. Eventually, the "backbone PSTN" will be an "all-optical network". In Figure 2-67, this traditional voice service technology is

indicated by the voice terminal "1", TDM technology "2", "PSTN" network "3, 4, and 5", wire/wireless "6", fibers without WDM "7", and fibers with WDM "8".

(2) Since the development of digital computers in the late 1950s, data communications have grown rapidly. To effectively transmit data signals over voiceband networks, data packetization using statistical multiplexing technology has been adopted. Data communications are typically provided by using Local Area Networks (LANs) in business and industrial applications. LANs are usually implemented with twisted pair wiring. Coaxial cables, digital radio, or other wireless systems are infrequently used because LANs traditionally localized networks. Recently optical fibers have been adopted for modern LAN applications. This line of data applications is indicated by data terminal "9", statistical multiplexing TDM "10", LAN "11 and 12", and facilities "6 or 7".

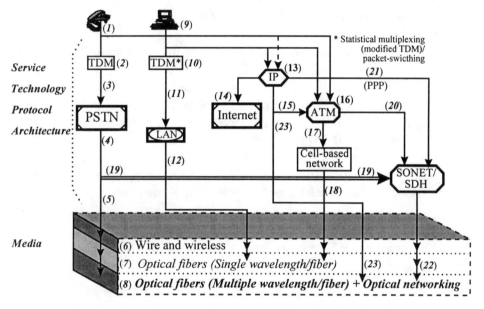

Figure 2-65   Various Call Connection Architectures (Options).

(3) Since the early 1990s, there has been an explosive growth of Internet services. TCP/IP (Chapter 11) has been the protocol suite adopted for Internet applications. Internet services have expanded from simple information downloading, to small business-oriented services, and finally into a highly competitive industry. Voice over IP (VoIP)is expected to become an important service that the Internet provides. Because of reliability and performance considerations, it has been proposed that Internet services be transported using ATM protocol [e.g., ATM adaptation layer (AAL) protocol 5, or 3/4]. It has also been proposed that SONET/SDH be adopted

as the physical layer protocol for Internet services. Data services over IP is indicated (in Figure 2-67) by the data terminal "9", IP protocol "13", and Internet "14". Voice over IP (VoIP) is indicated by the voice terminal "1", IP protocol "13", and Internet "14". IP over ATM is indicated by "15 and 16". From ATM, carrying IP or other signals, to transmission layer, and from IP to SONET/SDH physical layer facilities are described in the following paragraphs.

(4)   In 1988 ATM was recommended as the "3rd generation" packet switching technology for multimedia transport. Several different facilities have been used to carry ATM cell streams. In addition to PDH facilities, two commonly used facilities used for ATM transport are: cell-based networks, and SONET/SDH networks. These networks can be implemented using wire, wireless, and optical fibers. WDM is presently the most advanced technology available for ATM applications. The cell-based network ATM application is indicated by "17 and 18" in Figure 2-67. The SONET/SDH ATM transport is indicated by "20".

(5)   SONET/SDH networks have gradually grown into a "standard" physical layer protocol that is used worldwide. Since 1997, SONET/SDH networking has grown almost exponentially. It is expected by the year 2005, 75% of high-speed traffic over backbone networks will be carried by SONET/SDH-based optical fiber (both electro-optical and all optical networks). This is indicated by "19" for the applications of SONET/SDH carrying PDH signals; indicated by "20" for SONET/SDH networks to carry ATM cell streams; and indicated by "21", using Point-to-Point Protocol (PPP), for SONET/SDH to carry IP services. It should be understood that WDM can be used for SONET/SDH transport.

Note that the proposal of "*IP over DWDM*", indicated by "23" in Figure 2-67, may have the potential for bypassing bypass ATM and SONET/SDH altogether.

Review Questions II for Chapter 2:

(13) (True, False) An Internet is a network connecting many different networks which are owned, operated and managed by a single Internet Service Provider (ISP).

(14) TCP/IP is not a physical layer protocol, but it a layers _____ protocol suite. Similar to TCP/IP protocol are: _____ and _____ .

(15) The protocol for supporting applications that require reliable virtual circuits and flow control, and for detecting network congestion conditions, etc., is called the _____ (TCP).

(16) Since Internet connects many individual networks together, _____ and _____ are often required in an Internet connection.

(17) A _____ (i.e., _____) is a computer that contains pages of documents that are written in a standard format called _____ (HTML).

(18) Several Internet connections have been deployed for IP services: ____-IP-___, ___-IP-_____, _____-IP-_____, and _____-IP-_____ .

(19) (True, False) The PSTN-IP-PSTN service charge must be lower than the conventional switched service for voice applications for its competitive position.

(20) Internet users share a "common 32-bit" addressing scheme known as _____. Access to the Internet is usually via (1) _____ connections, or (2) a _____ link to an Internet Service Provider (ISP).

(21) For dial-up IP services, the customer must sign a _____ with the ISP to obtain a _____, and a _____ number used to access the Internet.

(22) For an office environment, the end-user stations are typically connected via _____ to the Internet or PSTN. If the connection is to be made to an ISP, the connection requires _____ .

(23) There are different packet switching technologies that have been deployed since the early 1980s such as _____, _____ and _____. X.25 is designed for _____ transport (up to 56 kbps) only; since it is a _____ technology. Even though the speed has increased up to 2 Mbps, the broadband _____ technology is still designed for _____ transport only. In contrast, the broadband ___ technology is suitable for multimedia services.

(24) The concepts of ATM Virtual Path (VP) and Virtual Channel (VC) allows physical channel bandwidth to be utilized _____ for customers with different _____ needs. ATM technology is known as a rate _____ packet technology.

(25) Between ATM network nodes, the maximum numbers of VPs and VCs are _____, and _____, respectively.

(26) A _____ is basically an "end-to-end circuit" for ATM services, and is a _____ communication link (capability) used for transporting ATM cells.

(27) From a provisioning viewpoint, ATM services can be grouped into two categories: _____ and _____ connections.

(28) ATM switched virtual connections and permanent virtual connections can be viewed as _____ public and private _____ network connections in the Public Switched Telephone Network (PSTN) environment.

(29) (True, False) The main advantage of ATM-based LAN over the traditional LAN services is the shared bandwidth the public network instead of expensive leased lines (e.g., T1 line in North America and E1 lines in other countries).

(30) It is possible to deploy LAM emulation by using ATM _____ Virtual Connections (VCs).

(31) Presently, there are five data rates defined for SONET signal transport: _____ (or _____), _____ (or _____), _____ (or _____), _____ (or _____), and _____ (or _____). The signals insides ( ) are SONET signals carried by optical fiber links.

(32) Similarly, there are presently four data rates defined for SDH signal transport: _____ (or _____), _____ (or _____), _____ (or _____), and _____ (or _____). The signals insides ( ) are SDH signals carried by optical fiber links.

(33) SONET Virtual Tributaries VT1.5, VT2, VT3 and VT6 are designed to carry sub-rate signals, that is, _____, _____, _____, and _____, respectively.

(34) SDH Tributary Units TU-11, TU-12, TU-2 and TU-3 are designed to carry sub-rate signals, that is, _____, _____, _____, and _____, respectively.

(35) Among all SONET various data rates, or among all SDH data rates, there is a very unique characteristic: _____. For example, STS-12 has a speed of ____ times the speed of STS-1, and STM-64 has a speed of ____ times the speed of STM-1.

(36) Even though SOENT/SDH protocol is equivalent to OSI physical layer (Layer 1), the end-to-end SONET/SDH connection is modeled as a three-layer protocol. The three layers are: the _____, the _____ (or _____ for SDH), and the _____ (or _____ for SDH) layers.

(37) For physical layer transport protocols, if SONET/SDH protocols are not applied, it is generally referred at as _____ digital hierarchy.

(38) (True, False) The proposal of "IP over SWDM" may have the potential for bypassing ATM and SONET/SDH altogether.

CHAPTER 3
Access: Traditional and Broadband -
HDSL, ADSL, VDSL & Wireless

Chapter Objectives

Upon the completion of this chapter, you should be able to:

- Describe general access networks, including wireline and wireless applications including: Base-Rate Interface ISDN (BRI-ISDN), Cable modem, Hybrid Fiber Coax (HFC), Integrated Digital Loop Carrier (IDLC), etc.

- Describe the traditional access networking: outside loop plant, outside plant planning, and typical loop facility tests.

- Describe several well-established broadband access systems: Integrated Digital Loop Carrier (IDLC) systems and applications, and Business Remote Terminal (BRT) for metropolitan area applications.

- Define channel capacity, spectral efficiency, Shannon-Hartley law, and their applications in traditional and broadband networks.

- Discuss Digital Subscriber Line (DSL): evolution from tradition to broadband access by using metallic local loops, xDSL classifications [High-speed, Asymmetrical and Very high-speed DSL (HDSL, ADSL and VDSL)], xDSL operational environment, frequency plane and frequency spectral bands, and different implementation methods [e.g., echo canceled hybrid, "carrier-less" AM/PM (CAP), and Discrete Multi-Tone (DMT) techniques].

- Describe wireless access technologies such as wireless local loop, satellite-switched broadband systems, and optical fiber/coax access schemes.

3.1 INTRODUCTION

The movement, management, and processing of large (and growing) volumes of information has become an important part of everyday life for modern society. To support these activities, a broadband "end-to-end network" with fast access capabilities is a necessity. Therefore, telecommunications system users are becoming increasingly interested in high speed access to various services:

- Interactive broadband services: Typical examples are high-speed Internet access, and video conferencing. These services require high data rates. Despite continuous progress in information compression techniques that have reduced the required bit rates for these services, new customer applications constantly require higher speeds and increasing capacities.

- Non-interactive services: Typical examples are video on demand, digital television, and databases uploaded.

As a result of these trends, telecommunications service providers must deploy both fast "**access**" networks and broadband "**backbone**" networks. The access network is referred to as the "last-mile network", and requires new/improved "last-mile" communications technologies. The demand for new technologies to support emerging broadband access communications is increasing rapidly. The present technologies available are:

- Wireline solutions:

  * Base-Rate Interface ISDN (BRI-ISDN)
  * Cable modem
  * Hybrid Fiber Coax (HFC)
  * Integrated Digital Loop Carrier (IDLC) systems
  * Business Remote Terminal (BRT)
  * Fiber To The Neighborhood (FTTN)
  * Asymmetric Digital Subscriber Line (ADSL)
  * High-speed Digital Subscriber Line (HDSL)
  * Very high-speed Digital Subscriber Line (VDSL)

- Wireless solutions:

  * Wireless Local Area Network (WLAN)
  * HomeRF
  * Satellite

A thorough understanding of: (1) the traditional twisted-pair wire loop, and (2) the Shannon-Hartley law used to determine the available channel capacity (rate) for computer and digital communications over a traditional copper twisted-pair wire is required as a foundation for further study of broadband and access technologies. In addition, Integrated Digital Loop Carrier (IDLC), Hybrid Fiber Coax (HFC), and Wireless Local Loop (WLL) solutions are discussed in this chapter.

3.2 TRADITIONAL TWISTED-PAIR ANALOG LOOPS

Traditional customer access to telecommunications networks is via "outside plant equipment". Although the outside plant was considered a mature technology, since 1980s it has gone through a series of upgrades and modifications to keep pace with the evolution of the telecommunications industry.

3.2.1 An Outside Plant

The conventional loop plant is designed to connect (interface) a customers' premises equipment to telecommunications networks. The loop plant has evolved from a "multiple plant design", to a "dedicated outside plant design", and finally to the present "serving area concept". In the "serving area concept", the outside plant within a serving area is called a "distribution network" (Figure 3-1). The distribution network is connected to the "feeder network" at a single point called the Serving Area Interface (SAI).

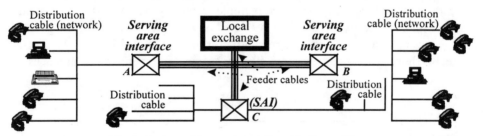

Figure 3-1   Conventional Outside Loop Plant.

The feeder network consists of a large numbers of twisted pair cables between the central office (local exchange) and the Serving Area Interfaces (SAIs). The SAIs are typically located within one-half mile of the customers. Several loop plant designs have been implemented, and are described as follows:

- Unigauge design: This configuration extends from the central office to the customer premises equipment, and uses only one size (gauge) of wire. It supports distance up to 52 *kft* (about 10 *miles*, or 16 *km*).

- Unified loop design: This configuration accommodates loop lengths up to 90 *kft* (about 17 *miles*, or 27 *km*).

- Long route design for replacing Carrier Serving Area (CSA): This configuration is designed for loops up to 105 *kft* (about 20 *miles*, or 32 *km*).

- Concentrated Range Extension with Gain (CREG) design: This configuration accommodates loop lengths up to 160 *kft* (about 30 *miles*, or 49 *km*) and contains active electronic components that insure proper signal levels.

Basically the outside plant is designed to meet transmission objects based on loop length and resistance. For customer lines longer than the distance limits, "loaded cables" and/or other active electronics are applied to extend the limits.

Definition 3-1: A loaded cable is a twisted pair wire with "loading coils" inserted at specific points along the transmission path.

For example, a 19H88 loaded cable consists of 19-gauge wire with an **88** mh coil inserted every 6,000 *ft*. The distances between two adjacent inserted coils is 3,000 *ft* for "B-type loaded cables", 4,500 *ft* for "D-type loaded cables", and 6,000 *ft* for "H-type loaded cables".

3.2.2 Outside Plant Planning

Planning of the outside plant for a local exchange begins with a geographic model that divides the outside plant into a primary system (feeder) and a secondary system (distribution). A serving switch can feed multiple secondary systems. The dimensions and density of the secondary system may vary based on the characteristics of the geographic area. For example, a secondary system can be further subdivided into smaller segments (Figure 3-2 shows a secondary system with four segments) if required for high density.

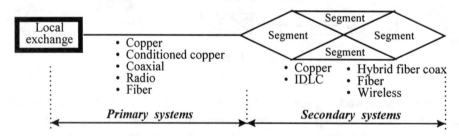

Figure 3-2   Primary and Secondary Systems of an Outside Plant.

The media for the primary system is typically (1) copper, (2) conditioned copper, (3) coaxial cable, (4) radio, or (5) optical fiber. The media for the secondary system is typically (1) copper, (2) Integrated Digital Loop Carrier (IDLC) system, (3) Hybrid Fiber Coax (HFC), (4) fiber (fiber to the neighbor), or (5) wireless.

Example 3-1: Describe a standard (typical) Serving Area Configuration (SAC).

Figure 3-3 illustrates the high level architecture and terminology used for a typical Serving Area Configuration (SAC). The servicing area interface is required to interface the feeder network and the serving areas. Distribution cables are used to connect all serving wires and street cables using distributing cabinets.

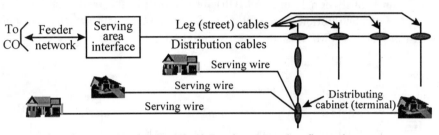

Figure 3-3   A Standard Serving Area Configuration.

The cost of materials, cables, terminals (a terminal is equipment that connects the distribution cable to the service wire entering a customer's premises), and interface (the interface is equipment that connects the feeder and distribution cables) components for a new service area can be very expensive. Therefore, the standard SAC model is designed to minimize outside loop plant costs.

The size of the distributing cabling is determined by the number of pairs provided per customer residential unit. The SAC model specifies a minimum of two pairs per unit. The extra pair may be used as a secondary line, or to implement ISDN. The extra pair may also be shared among several units. The SAC model is depicted as a single backbone cable that runs out from the serving area interface, and is connected to a number of street cables or "legs". Each leg connection point branches out in two directions. The spacing of leg cables is uniform along the backbone.

An important aspect of leg cables is that it is only available in "standard" sizes. For example, if a leg cable contains six terminals and each supplies four one-family houses with a minimum of 1.5 pairs per living unit, the number of pairs required would be 36 ($\equiv$ $6 \times 4 \times 1.5$). The "standard" cable size available contains 50 pairs, which equates to 2.08 pairs per residential unit.

Example 3-2: Describe a typical physical connection of a customer loop from the Customer Premises Equipment (CPE) to the end office terminating point.

Figure 3-4 illustrates the typical connection between the CPE and a local exchange (i.e., central office). Near the customer's premises, a distribution cabinet is required to distribute several "drop" cables to individual Customer Premises Equipment (CPE). Each drop cable is equipped with lightning protector via "grounding (GND)". The pair of feeder (subscriber cable) wiring are designated as "tip" and "ring" for indicating the wires connected to the positive side or the "ground" of the battery in the central office for detecting station's "on-hook" and "off-hook" conditions.

The distribution frame at the local exchange consists of two separate frames: (1) a Vertical Main Distribution Frame (VMDF) for interfacing the feeder cables, and (2) a Horizontal Main Distribution Frame (HMDF) for interfacing with switching machines or transmission systems. VMDF is equipped with lightning protector.

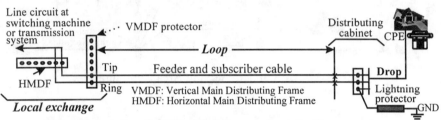

Figure 3-4   A typical Physical Connection of a Customer Loop.

Example 3-3: Describe the process used for wire gauge selection,   assuming a loop length of 34 *kft*.

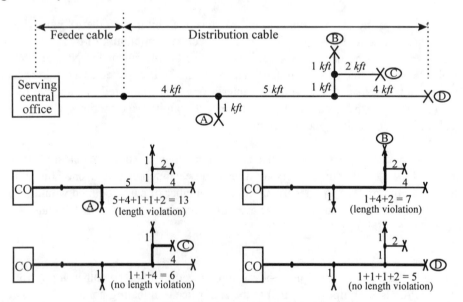

Figure 3-5   Distribution Cables and Bridge Taps.

A combination of 22-gauge and 24-gauge wiring is first examined and the bridge tap limit criteria (illustrated in Figure 3-5) is analyzed. It is usually most advantages to place finer gauge cable closet to the central office where a large number of connections is concentrated. The loop to be designed has a length of 34 *kft* [with a resistance of 1,500 ohms (Ω)], and a combination of 22 and 24 gauge. It is assumed that 5 loading coils must be applied along the loop since the length is 34 *kft* (i.e., assume H-type cable with a space of 6 kft between loading coils). Theoretically, the resistance for 24-gauge and 22-gauge copper wire is 51.9 Ω/*kft*, and 32.4 Ω/*kft*, respectively at 68°F. The resistance of each loaded coil is approximately 9 Ω. The design equations to determine lengths of each gauge wire, based on loop resistances, are given as follows:

$$\begin{cases} x + y = 34 \ \ (\text{kft}) \\ 51.9x + 32.4y + 5 \times 9 = 1{,}500 \end{cases} \tag{3-1}$$

where x is the length of 24-gauge wire, y is the length of 22-gauge wire, and 5 is the number of loading coils. Solving the simultaneous equation, Eq.(3-1), it can be determined that x = 18.1 *kft* (of the 24-gauge wire, which is used closed to the central office since it is the finer wire of the two given two gauges; 18.1 kft is assumed to be distributed as 4 + 5 + 4 + 1 + 1+ 1 +2) and y = 15.9 *kft* (for the 22-gauge wire; not shown in Figure 3-5 for simplicity).

   "***The acceptable cumulative bridge tap length***" is 6 *kft* or less. For the loop plant shown in Figure 3-5, the cumulative bridge tap lengths for four different cases have been calculated. With A or B as the working station, the bridge tap cumulative length requirement is violated (13 *kft*, and 7 *kft*, respectively). However, with C or D as the working station, the cumulative bridge tap length (6 *kft* or 5 *kft*) is within acceptable limits.

3.2.3 Typical Loop Facility Tests

In a customer loop, a subscriber can hear "noise" between the tip and ring. This noise is called the circuit noise (metallic noise) of the loop. Figure 3-6 illustrates three different ways of measuring loop circuit noise: (A) the overall loop circuit noise measurement, (B) the noise measurement toward the customer station, and (C) the noise measurement toward the serving central office. The requirements of these noise levels are beyond the scope of this book, and are not described.

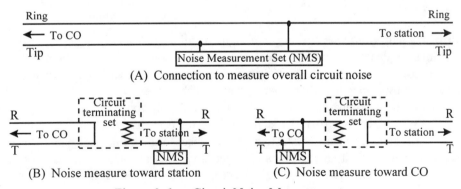

Figure 3-6   Circuit Noise Measurements.

3.3 INTEGRATED DIGITAL LOOP CARRIER (IDLC) SYSTEMS

Before the present (copper loop) broadband access systems (i.e., ASDL, HDSL, and VDSL) were available, Integrated Digital Loop Carrier (IDLC) systems were used for broadband access transport applications. Since the late 1960s, IDLC has been used to replace traditional twisted-pair local loops. The first well-known systems was called Subscriber Loop Carrier (SLC; e.g., SLC-96, SLC-5, SLC-120, SLC-240, and SLC-2000). This arrangement is also called a "pair-gain system" (described later in this section).

The development of Integrated Digital Loop Carrier (IDLC) was originally intended to provide basic telephone service for rural areas. However, applications for IDLC are also frequently found in metropolitan areas. Thus, IDLC has evolved into Business Remote Terminal (BRT). The reasons for applying an IDLC in a network are briefly described as follows:

- **Improves transmission quality**: An IDLC improves transmission quality over traditional long analog twisted-pair loops. As shown in Figure 3-7 (dashed curve), the system frequency response (signal attenuation versus frequency) for a long copper (twisted-pair) loop has a excessive attenuation in the high frequency range. This is an undesirable characteristic for speech signal transport, especially for female speakers with typically more signal power at high frequencies than male speakers. Referring to Figure 3-7 (solid curve), the IDLC characteristics provide uniform attenuation over the entire frequency spectrum.

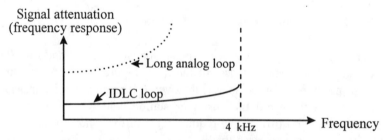

Figure 3-7   Transmission Quality Comparison.

- **Provides pair gain**: One advantage of digital facilities (especially digital optical fiber facilities) is that they require fewer physical wires compared to traditional twisted-pair loops. For example, a SLC-96 system (described in Chapter 6) can serve 96 voice customers by using four T1 carrier systems (each carrier system is a 4-wire digital connection). In comparison, a traditional analog loop approach would require 96 twisted-pair wires for 96 (i.e., a separate pair for each user). Therefore, a SLC-96 system offers a 'pair gain" of 12 to 1, with respect to the number of pairs needed to support an equivalent analog loops configuration. That is, each T1 signal carries 24 subscribers on two pairs of wires. Hence, this is a "pair gain system" with a characteristic of 12:1. If optical fiber pair (instead of multiple T1 carrier systems) is used, the "pair gain" characteristic will be even larger. For example, a 90 Mbps optical fiber provides a "pair gain" ratio of 1344:1 can be achieved. This is because a pair of fibers can carry 1,344 voice channels simultaneously.

- **Relieves duct congestion**: When an IDLC is deployed in a metropolitan business area, the advantage of "pair gain" can be used to reduce the number of wires in underground cable ducts, where congestion problems are common. For example, many Business Remote Terminals (BRTs) accept voice, data, and video signals within a metro high-rise building, and multiplex them to form a OC-3 or a OC-12 SONET signal (depending upon the business customers' needs; described in Chapter 6), that carries 2,016 or 8,064 voice (or "voice-equivalent") signals (in North American applications). In the case of OC-12 transport, instead of 8,064 separate pairs of wire, only two optical fibers are required (i.e., a "pair gain" ratio of approximately 4000:1 has been achieved).

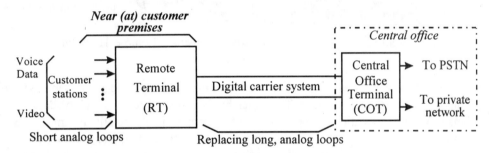

Figure 3-8   Major Components of An IDLC.

As shown in Figure 3-8, three components of a traditional IDLC system are:

1. ***Central Office Terminal* (COT)**: Digital signals transmitted by a Remote Terminal (RT) via digital facilities (e.g., T1 digital carrier, optical fiber system etc.) are typically delivered to different destinations. For example, some may go to switching machines, others may go to higher speed digital systems. Therefore, the Central Office Terminal (COT) is used to provide the necessary interfaces. The COT may be stand-alone equipment or part of other switching office equipment.

2. ***Remote Terminal* (RT)**: A RT is located on or near the customer premises, where voice signals are digitized and multiplexed with other signals (e.g., data, digitized video, etc.). Since the output of a RT is a digital facility, it effectively replaces long analog twisted-pair loops, and permits the use of very-short sections of analog loops for connections to customer stations (from the RT).

3. ***Digital facility***: Digital facilities are used to connect a RT and a COT. That is, "long analog" loops are replaced by a digital carrier system. In North America, T1 digital carrier systems or digital optical fiber systems are the digital facilities used to connect

RTs and COTs. In other parts of the world, the 2.048 Mbps digital carrier or optical fiber systems are used for this purpose.

3.4 BUSINESS REMOTE TERMINAL (BRT)

In metro-business applications, Business Remote Terminals (BRTs) have been widely used. BRTs are usually located in the basement of metro high-rise buildings (Figure 3-9).

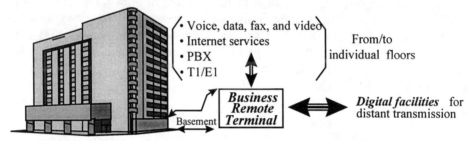

Figure 3-9   Increasing Needs for Business Remote Terminals.

In modern business, half of the telecommunications traffic is conventional voice signals and voice-mail services. The other half of the traffic is computer data and fax signals. Video applications are growing, but presently there is relatively limited demand for this service. Demand for Internet service is obviously strong, and indicates increasing growth. Traditional PBX and T1/E1 services are also still popular in the business arena. All these different services terminate on the Business Remote Terminal (BRT), and can be multiplexed together for long distance transport using high-speed digital facilities, thereby reducing cable duct congestion.

Various types of business remote terminals are available for different size business needs. In North American applications, the digital facilities are: T1 digital carrier systems, FT-G, and SONET OC-3, or OC-12. In other countries applications, the digital facilities are: E1 digital facilities, and SDH STM-1, or STM-4.

Typical modern IDLC systems, such as the type used in Next Generation Lightwave Networks (NGLNs), have functional block diagram shown in Figure 3-10. Two primary components in this IDLC are described as follows:

(1)   Remote Terminal (RT): Two functional components are contained within a remote terminal, and are described as follows:

    *   Access Resource Manager (ARM): This is a digital multiplexer used to combine digitized voice, data, and/or video signals, with DS1 signals to form a higher-speed signal (e.g., STS-3 or STS-12, depending on the business size). In addition, an ARM has the ability to add and/or drop traffic [i.e., like an Add-Drop Multiplexer (ADM)].

* Digital Channel Units (DCUs): A DCU performs the same functions as a digital channel bank (described in Chapter 6). That is, digitizing voice signals, performing time-division multiplexing of digitized voices, data, and other signals, and inserting control signals into the digital bitstream for network Operation, Administration, Maintenance and Provisioning (OAM&P) functions.

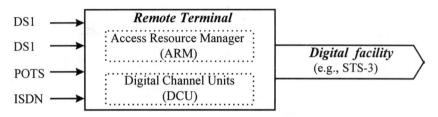

Figure 3-10 A Business Remote Terminal (Modern IDLC).

(2) Digital facility: The digital facility replaces the traditional analog loops for long distance transmission, and consists of:

* Multiple T1 carrier lines
* An STS-3 digital carrier system
* An OC-3/OC-12 optical fiber carrier system

For large business applications, T1 and STS-3 digital facilities have been replaced by high-speed OC-12 SONET or STM-4 SDH transport systems. It is possible that even OC-48 or STM-16 facilities may be deployed in the future. This example is for North American applications. Similar products are available to support global applications.

3.5 CHANNEL CAPACITY AND SPECTRAL EFFICIENCY

Interactive broadband access services require high data rate transmission. If a "noise-dominated" transmission channel (Figure 3-11) has unlimited bandwidth [e.g., Hybrid Fiber Coax (HFC) or Fiber To The Home (FTTH)], interactive services with high data rates can be implemented by simply increasing the signal rate. However, in most access environments, bandwidth is scarce, the channels are severely affected by crosstalk, and attenuation rapidly increases with frequency. In this noisy environment a compromise must be made between the number of signal levels and the signal rate. Spectral efficiency has been devised as a method for measuring the effectiveness of designs intended for high data rate broadband solutions.

Definition 3-2: The "*spectral efficiency*" (η_M) is defined as the ratio of the data rate (R in bps) to the bandwidth (W in Hz) required to transmit the signal with that rate [see

Eq.(3-20]. There are two factors that limit the spectral efficiency of a digital transmission system: (1) Inter-Symbol Interference (ISI), and (2) channel noise.

$$\eta_M = R / W \quad (bps / Hz) \tag{3-2}$$

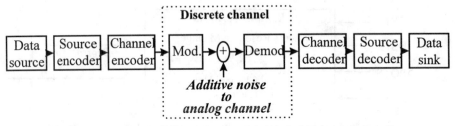

Figure 3-11   Major Building Blocks of A Digital Transmission System.

3.5.1 Nyquist Signaling Frequency or Nyquist Rate

During the early 1920s, H. Nyquist, an engineer at Bell Telephone Laboratories, published a series of technical papers ("Certain Factors Affecting Telegraph Speed", 1924; and "Certain Topics in Telegraph Transmission Theory", 1928) that formed a foundation for the theory of noiseless channel capacity. Nyquist derived the relationship between the signaling rate (R in bps) and the required bandwidth (W in Hz) for transmitting signals as expressed by Eq.(3-3):

$$R = 2W \tag{3-3}$$

The maximum spectral efficiency of a binary baseband link, according to Nyquist's theorem, is given as: $\eta_M = R/W = 2$ bps/Hz, for carrying speech signals. Example 3-4 is used to illustrate data transmission using Nyquist's theorem. However, Nyquist's theorem is not the only method available for determining the spectral efficiency of a given system.

Example 3-4: Assuming an existing copper loop plant has a bandwidth about 3,000 Hz, determine the data rate by using the Nyquist's theorem.

The spectral efficiency, according to the Nyquist's theorem, is 2 bps/Hz. The data rate can be obtained as follows:

$$\text{Rate} = 2 \text{ bps/Hz} \times (3,000 \text{ Hz}) = 6 \text{ kbps}$$

Note that this speed of 6 kbps is more than the data rate of 300 bps for a ITU-T V.21 modem. However, the data rate of 6 kbps is far less than a modern V.34 modem, which has a data rate of 28.8 or 33.6 kbps. In the V.34 modem, the spectral efficiency is 9.6 bps/Hz (for 28.8 kbps modem) or 11.2 bps/Hz (for 33.6 kbps modem). The difference between the Nyquist rate and the modern modem rate is:

"V.21 modems apply Non-Return-to-Zero (NRZ) binary transmission, while V.34 and newer modems use multilevel modulation or signaling."

Further details associated with the modem technologies is later in this chapter.

3.5.2 Shannon-Hartley Law

As previously discussed, Nyquist derived the spectral efficiency concept based on a simple Non-Return-to-Zero (NRZ) binary line code [Figure 3-12 (A)]. This line code has a poor bandwidth efficiency compared to many other line codes. The following is a partial list of line codes with better bandwidth efficiency (than simple NRZ binary line code). However, they all require more than two levels of transmission (described later in this chapter):

- Three-level or ternary codes: This technique is also known as an Alternate Mark Inversion [AMI, Figure 3-12(B)] code. Figure 3-12(B) illustrates a 50% duty-cycle Return-to-Zero (RZ) line code. Common AMI codes (described in Chapter 6) are BnZS (e.g., B3ZS, B6ZS, and B8ZS), HDBn (e.g., HDB3)] codes, and mBnT (e.g., 4B3T) codes. They are all three-level line codes, and are used to perform zero suppression (zero substitution) for timing synchronization purpose.

- Four-level or ternary codes: mBnQ (e.g., 2B1Q for Digital Subscriber Line (DSL) applications; described later in this chapter) codes are commonly used.

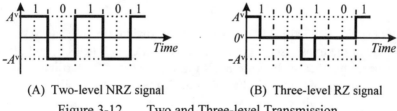

(A) Two-level NRZ signal   (B) Three-level RZ signal

Figure 3-12   Two and Three-level Transmission.

The power spectra for the three-level (ternary) and four-level (quaternary) line codes are narrower than the power spectrum for the NRZ binary code. Therefore, the spectral efficiency is increased as in modern modem design (i.e., a larger amount of data can be transported using the same system bandwidth).

In 1949 Claude E. Shannon (an engineer at Bell Telephone Laboratories) published a landmark paper titled "A Mathematical Theory of Communication" in the Bell System

Technical Journal (BSTJ). Shannon's paper expanded upon earlier work done by R. V. L. Hartley (titled "Theory of Information", 1928) at Bell Laboratories that confirmed the theory that "a given bandwidth x time" is required to transport a proportionate quantity of information. The impact of their contributions to modern computer and digital communications technologies was far beyond anyone's imagination. Shannon's concepts of information theory are:

- **Information** can be measured independently: The semantic aspects of information are not significant with respect to measurements.

- **Source coding theorem**: Every data source may be uniquely described with respect to its information content. That is, the required number of bits uniquely describes a data source, and can be coordinated with the corresponding information content as closely as desired.

- **Channel coding theorem**: Error-free data transmission is possible if the information rate is smaller than the channel capacity. That is, the error rate of data transmitted over a band-limited noisy channel (Figure 3-11) can be reduced to an arbitrarily small amount, provided that the information rate is lower than the channel capacity (examples are given in this section to illustrate the channel capacity theorem).

In multilevel (M level) baseband signaling (line codes), a sequence of m consecutive bits is represented by one of $M = 2^m$ distinct amplitude levels. In a noiseless environment the channel capacity increases linearly as M increases (i.e., if $M \rightarrow \infty$, then channel capacity $\rightarrow \infty$). However, in a noisy environment the maximum channel capacity is given by the Shannon-Hartley law as follows:

$$R \leq 2W \times \log_2 \sqrt{1 + \frac{S}{N}} = W \times \log_2 (1 + \frac{S}{N}) \qquad (3\text{-}4)$$

where R is the data rate in bps, W is the channel bandwidth (in Hz), and S/N is the signal-to-noise ratio of the transmission system. Note that in Eqs.(3-4), (3-5) and (3-7) the ratio S/N must be a numerical value (***not a dB value***). That is, S/N = (Signal power)/(Noise power) = P_{signal}/P_{noise} = a numerical value (e.g., 88, 120, etc.) $\neq$ 19 dB.

The maximum bandwidth efficiency (spectral efficiency), and the maximum signal amplitude are given, respectively, as follows:

$$\eta_{M-max} = \log_2 (1 + \frac{S}{N}) \qquad \text{(in bps/Hz)} \qquad (3\text{-}5)$$

$$\textit{Maximum signal amplitude} = \sqrt{P_{signal} + P_{noise}} \qquad (3\text{-}6)$$

The number of distinguishable levels for M-level coding is given by:

$$Number\ of\ distinguishable\ levels\ M = \sqrt{1 + \frac{S}{N}} \tag{3-7}$$

The minimum recognizable level difference is given by:

$$Minimum\ recognizable\ level\ difference = \sqrt{P_{noise}} \tag{3-8}$$

Example 3-5: Assuming a typical (copper) twisted-pair wire loop has a frequency spectrum (system frequency response) as shown in Figure 3-13, and a 3-dB bandwidth of about 3,000 Hz. Determine the channel capacity, spectral efficiency, and number of distinguishable levels (M) if multilevel transmission is adopted.

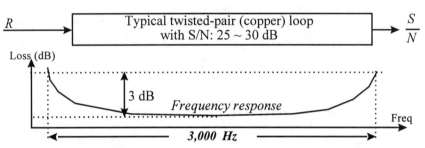

Figure 3-13 Typical (Copper) Twisted-pair Loop Characteristics.

First, assuming S/N = 25 dB, which must be converted into a numerical value using the "inverse dB" relationship as follows:

$$25\ dB \quad \Rightarrow \quad 10^{25/10} = 10^{2.5} = 316.2277$$

Substituting S/N = 316.2277, W = 3000 into Eq.(3-4), the data rate of this 3,000 Hz metallic cooper loop can be obtained as follows:

$$R \le W \times \log_2 (1 + \frac{S}{N}) = 3,000 \times \log_2 (1 + 316.227) = 24,928\ bps$$

Applying Eq.(3-5), the maximum bandwidth efficiency can be derived as follows:

$$\eta_m = \log_2 (1 + 316.227) = 24,928 / 3,000 = 8.309\ bps / Hz$$

This is known as the "***spectral efficiency***" ("***bandwidth efficiency***") of the system, and is sometimes called the "Shannon's predicted channel capacity limit".

By applying Eq.(3-7) the number of distinguishable level of a multi-level transmission system (M) can be calculated as follows:

$$M = \sqrt{1 + 316.2277} = 17\ levels$$

Next, consider the case for a loop that has a S/N = 30 dB. First, convert the S/N a ratio into a numerical value using the "inverse dB" relationship as follows:

$$30 \; dB \quad \Rightarrow \quad 10^{30/10} = 10^3 = 1,000$$

Next, substitute S/N = 1,000 into Eq.(3-4), to obtain the data rate as follows:

$$R \leq W \times \log_2 (1 + \frac{S}{N}) = 3,000 \times \log_2 (1 + 1,000) = 29,901 \; \text{bps}$$

Finally, derive the maximum bandwidth efficiency as follows:

$$\eta_m = \log_2 (1 + 1,000) = 29,901 \, / \, 3,000 = 9.967 \; bps \, / \, Hz$$

Thus, for a S/N of 30 dB, the spectral (bandwidth) efficiency of the system (i.e., Shannon's predicted channel limit) is calculated to be 9.967 bps/Hz. By applying Eq.(3-7) the number of distinguishable level of a multi-level transmission system as follows:

$$M = \sqrt{1 + 1000} = 31 \; \text{levels}$$

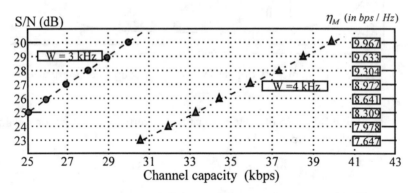

Figure 3-14   Channel Capacity for a Typical Twisted-pair Loop.

The same procedure can be used to analyze different S/N ratios, as indicated in Figure 3-14, where y-axis on the left-hand side is S/N (in dB); the y-axis on the right-hand side is spectral (modulation) efficiency (Shannon predicted capacity limit); and the x-axis is the calculated channel capacity. Note that two different system bandwidths (3 and 4 kHz) have been plotted in Figure 3-14.

Example 3-6: In Example 3-5, a typical loop is assumed to have signal-to-noise ratio of 25 ~ 30 dB and a length of 12,000 ~ 18,000 *ft* (365 ~ 548 *m*). If the loop length is shorter than the typical length, derive how the signal-to-noise ratio can be improved.

The S/N ratio has a higher value due to a shorter loop length (which has smaller noise power), hence the corresponding bandwidth efficiency is increased. Table 3-1 (W. Y.

Chen, "DSL: Simulation Techniques for Digital Subscriber Line Systems", Indianapolis, IN: Macmillan, 1998) illustrates the loop length versus capacity.

Table 3-1   Loop Capacity versus Loop Length (24-gauge twisted-pair).

Parameter	Bandwidth = 1 MHz				Bandwidth = 30 MHz		
Loop length (m)	50	200	1,800	5,500	500	900	1,400
Capacity (Mbps)	500	100	6	1.2	110	50	20
Efficiency (bps/Hz)	500	100	6	1.2	3.667	1.667	0.667

Example 3-7: Identify the ITU-T modem standards that support increased data rates.

Table 3-2 lists several ITU-T modem techniques, which show the effect of implementing modulation schemes, that approach the Shannon channel capacity limit.

Table 3-2   Spectral Efficiency of Data Transmission in a Local Loop.

ITU-T Rec. (year)	*Bandwidth (Hz)*	*Rate (kbps)*	*Modulation*
V.26 (1968)	1,200	2.4	4-PSK
V.27 (1972)	1,600	4.8	8-PSK
V.29 (1976)	2,400	9.6	16-QAM
V.32 (1984)	2,400	9.6	2D-TCM
V.34 (1984)	3,400	28.8/33.6	4D-TCM

PSK: Phase shift keying;   QAM: Quadratural amplitude modulation
2D-TCM: 2 dimensional trellis coded modulation

3.6 DIGITAL SUBSCRIBER LINE (DSL)

The bandwidth of backbone networks has been constantly increased since the late 1980s with the development of optical fiber technologies. A digital carrier system (in North American applications) using twisted-pair wire has a channel capacity of 24 voice channels, while the coaxial cable digital system has a capacity of 672 voice channels, and a single optical fiber digital system (e.g., FT-G) has a capacity of 24,192 voice channels. Most recently, several Dense Wavelength Division Multiplexing (DWDM) technologies have been deployed to further increase network bandwidth. DWDM technology is in the early stage of development, and the potential of higher network bandwidth (data rate) is conceptually unlimited.

Even if the bandwidth of backbone networks is increased, the end user services can not be improved unless the access technologies are advanced *accordingly*. Therefore, parallel efforts have been made to improve access technologies, especially the speed of access networks. As shown in Figure 3-15, two different approaches can be used to improve the access speed:

(1)   Install/deploy new access networks: A common technology, Hybrid Fiber Coaxial (HFC; discussed later in this chapter), is often used to increase access speeds.

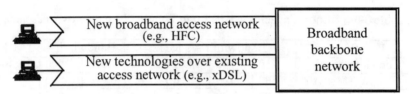

Figure 3-15   Two Approaches Used to Increase Access Speed.

(2)   Apply new technologies over the existing access networks: Use of the existing twisted-pair wire network is a popular approach for increasing access speeds. It is expected that competition in this area will be intense. The commonly used technology is Digital Subscriber Line (DSL). The generic name used for this technology is "*x*DSL", where "*x*" is used to indicate that the DSL technology can be Symmetric, Asymmetric, High-speed, or Very high-speed DSL. Several acronyms are generally accepted and are briefly defined as follows:

(a)   ADSL: Asymmetric DSL technology, with the speeds for the uplink (from user to network) and for the downlink (from network to user) are different. In contrast, symmetric DSL refers to the technology using the same speed for both directions. Without stating asymmetric, it is understood that it implies that a symmetric DSL is used.

(b)   HDSL: High-speed DSL technology, where the present speed (e.g., state of the art) is equivalent to the T1/E1 speed (1.544 Mbps, and 2.048 Mbps, respectively; discussed in Chapter 6) and the speed of uplink/down link are symmetrical.

(c)   VDSL: Very high-speed technology, where the speed is defined to be higher than T1/E1 speed. The speed of uplink/down link are also symmetrical.

3.6.1 Twisted-pair: Traditional and Broadband Access

Since 1881, when Alexander Bell first introduced twisted-pairs, this technology has been applied globally to reliably carry electrical phone signals to billions of subscribers. Recently, telecom operators around the world have moved towards including their existing (copper) twisted-pair infrastructure in next generation broadband access networks. Twisted-pairs have the potential to carry high-speed computer, television, and other digital signals in addition to Plain Old Telephone Service (POTS).

A frequently asked question is: "*How can DSL over a twisted-pair line carry data rates higher than 33.8 and 56 kbps signals supported by voiceband modems (originally considered the theoretical limit of a twisted-pair loop)*?"

It should be understood that "it is ***not*** the twisted-pair telephone line that prevents transport of broadband data signals from/to the customer". As indicated in Figure 3-16, the telephone switch in the central office allocates the bandwidth that is appropriate for voice calls, and thus limits the bandwidth of voiceband modems. That is, it is the associated ***filters*** used in the switching network that limit the access data rates.

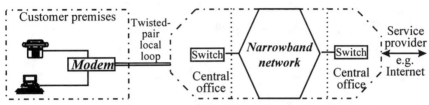

Figure 3-16   A Voiceband Modem Communication Model.

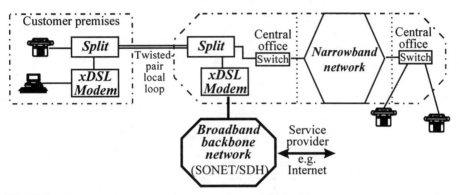

Figure 3-17   A DSL Modem Communication Model.

As previously stated, a twisted-pair line can carry data rates higher than the signal rates supported by voiceband modems. The maximum data rate depends heavily upon the loop length, but a large percentage of twisted-pair loops are capable of carrying very high data rate signals, provided the narrowband switch in the central office can be avoided. Figure 3-17 illustrates the architecture for this communication model. A splitter/combiner (shown as a "split" in Figure 3-17) is used at the customer premises and at the central office (before entering the switch and its associated filters). The data signals are extracted by the DSL modem at the central office and routed as broadband signals via an appropriate broadband network (e.g., SONET or SDH network) using packet technologies such as Asynchronous Transfer Mode (ATM) or Internet Protocol (IP).

Another commonly asked question is, "Why has the deployment of DSL technology been delayed?"

It is difficult to implement high-speed modems for use in the twisted-pair telephone loop plant: loops testing for qualification, narrowband filters bypassing, etc.

Several advanced techniques are required to support emerging broadband access solutions. These techniques are:

- High-speed signal processing
- High-bandwidth analog circuits (e.g., operating at hundreds of MHz)
- High-speed source encoders
- High-speed channel encoders
- High-efficiency modulator/demodulator equipment

3.6.2 Comparison Among Various Modem Technologies

Table 3-3 lists several ITU-T and xDSL (e.g., HDSL, SDSL, ADSL, and VDSL) modems with their corresponding data rates, bandwidth efficiencies, modulation techniques, and applications [IEEE Communications Magazine, May, 1999, Vol.37, No.5, p63].

Table 3-3   Various Modem Standards, Bandwidth Efficiencies, and Signal Rates.

Modem	Data rate (kbps)	η_M (bps/Hz)*	Modulation technique	Applications
V.21	0.3	0.10	FSK	Data transmission
V.22 bis	2.4	0.75	QPSK	Data transmission
V.29[1]	9.6	1.50	16-QAM	Data transmission
V.32[2]	9.6	3.00	TCM	Data transmission
V.34	28.8/33.6	11.00	TCM	Data transmission
V.90	56	18.00	TCM	Internet access
BR-ISDN	144	2.00	2B1Q	ISDN 2B+D
SDSL	768	2.00	2B1Q	HDSL over a single pair
HDSL	2,048	2.00	2B1Q	Server access, access to LANs/WANs, etc.
ADSL[3]	1.544-8.448[a] 16-640[b]	7.67	CAP/DMT	Access to Internet/multimedia databases; Video distribution
VDSL[4]	13-52[c] 1.5-2.3[d]	<4.00	CAP/DMT	ADSL + HDSL

(1) 4-wire leased lines   (2) Dialed-up lines   (3) Bandwidth up to 1 MHz   (4) Bandwidth up to 30 MHz
(a) Downstream (in Mbps)   (b) Upstream (in kbps)   (c) Downstream (in Mbps)   (d) Upstream (in Mbps)
* η_M : Bandwidth efficiency   BR-ISDN: Basic rate ISDN   TCM: Trellis Coded Modulation

The V.21 modem standard was originally established in the mid 1960s, and was still widely used until the mid-1980s. This modem applies Basic (Binary) Frequency Shift Keying (BFSK, or FSK) as its modulation/demodulation technique. FSK was adopted for its simplicity and immunity to noise (noise usually has more influence on signal amplitude than signal frequency). The FSK modem requires two frequencies (f_1 and f_2); one to carry a logical "1" and the other to carry a logical "0", for unidirectional transmission. Therefore, the V.21 modem was not efficient in bandwidth utilization.

The next modem standard was V.22, which applied Quadrature Phase Shift Keying [QPSK; which is also known as Quadrature Amplitude Modulation (QAM)] modulation as shown in Figure 3-18 (discussed in Chapter 6). Each pair of consecutive bits are carried by a sinusoidal wave. For example, "00" is represented by the carrier signal of $A \cdot cos$ (ωt + 45°), "01" is represented by $A \cdot cos$ (ωt + 135°), etc. Only one frequency is required, therefore, the bandwidth utilization is improved over the FSK (binary technique using two frequencies) modulation used in V.21 modems.

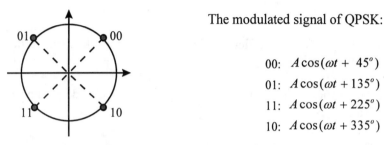

The modulated signal of QPSK:

$$00: \quad A\cos(\omega t + 45^o)$$
$$01: \quad A\cos(\omega t + 135^o)$$
$$11: \quad A\cos(\omega t + 225^o)$$
$$10: \quad A\cos(\omega t + 335^o)$$

Figure 3-18   Signal Constellation Diagram of QPSK.

An even more efficient modem (modulation/demodulation) technique is called Quadrature Amplitude Modulation (QAM), and can be 8-QAM, 16-QAM, 32-QAM, etc. Figure 3-19 shows 16-QAM used in V.29 modems, which uses four amplitudes ($\sqrt{2}$, 3, $3\sqrt{2}$, and 5) and eight phases (0°, 45°, 90°, 135°, 180°, 225°, 270°, and 335°) to represent sixteen ("0000", …, "1111") binary bit groups (i.e., "stars" in the 16-QAM constellation) each having a unique 4-bit code.

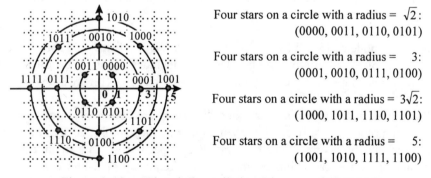

Four stars on a circle with a radius = $\sqrt{2}$:
(0000, 0011, 0110, 0101)

Four stars on a circle with a radius =   3:
(0001, 0010, 0111, 0100)

Four stars on a circle with a radius = $3\sqrt{2}$:
(1000, 1011, 1110, 1101)

Four stars on a circle with a radius =   5:
(1001, 1010, 1111, 1100)

Figure 3-19   Signal Constellation Diagram of 16-QAM.

<u>Definition 3-3</u>: The combination of QAM and convolutional coding is used to improve the noise performance of QAM. Modulation of this type is called Trellis Coded Modulation (TCM). For example, ITU-T V.32, V.34 and V.90 modems apply TCM technique to improve bandwidth utilization and noise performance.

As shown in Table 3-3, assuming the line quality is good, a symmetric full-duplex (FDX) data rate of 33.6 kbps is supported by a standard V.34 modem using TCM. V.90 with a rate of 56 kbps (but not symmetric) is a common modem used for accessing the Internet. The line quality (S/N) sets the spectral efficiency limit according to the Shannon-Hartley law, as illustrated in Example 3-8 and Figure 3-20. Typical loops with a length of 12,000~18,000 *ft* with a S/N of 30 dB have a theoretical limit of about 10 bps/Hz. However, if the loop is shorter and line quality is improved to a 40 dB S/N, then the limit can be extended to 13.288 bps/Hz.

Example 3-8: Plot the spectral efficiency limit, by applying Shannon-Hartley law, for various S/N performance ratios.

The spectral efficiencies plotted in Figure 3-20 are calculated by using Eq.(3-4) as described in Example 3-5.

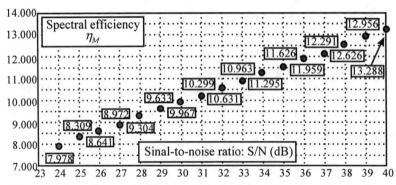

Figure 3-20   Spectral Efficiency Limit Using Shannon-Hartley Law.

The remaining modems (SDSL, HDSL, ADSL, and VDSL) listed in Table 3-3 do not utilize just the voiceband of a twisted-pair loop, but includes the entire bandwidth that is available (described later in this chapter) from a twisted-pair line (Figure 3-23).

Two different terms, "***baud rate***" and "***bit rate***" are commonly used to describe modem transmission speeds. Because of their similarity, they are frequently interchanged, but this is technically incorrect. The baud rate is the speed at which a signal can change during a one second interval. For example, if the frequency of a link connecting two devices changes 1,200 times every second, the interface is operating at a 1,200 ***baud rate***. In contrast, the bit rate (bits per second; bps) is the number of binary digits that a signal can carry during one second interval. For example, if a Bell 212A modem is operating at 600 baud (i.e., 600 different frequency values per second), and transmits two bits per each frequency value, the effective ***bit rate*** of this modem is 1,200 bps (= 600 × 2). Hence, in this case a 600 baud modem is operating at 1,200 bps. Obviously the bit rate (bps) is a more significant measurement of the overall performance of a device.

3.6.3 xDSL Loop Environment Considerations

The Public Switched Telephone Network (PSTN) currently has speeds as high as the OC-48 rate (2.48832 Gbps) without using the DWDM technology, and Internet links often have speeds of OC-12 (622.08 Mbps). Likewise, end user's PCs often in the Megabit range, and support interconnection via 100 Mbps Ethernet applications. However, bandwidth for the access links (Figure 3-21) is relatively limited.

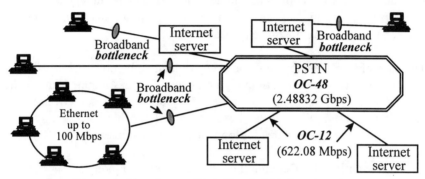

Figure 3-21   Access Link Bandwidth Bottlenecks.

Potential solutions (direction) for overcoming the bandwidth bottleneck of the access link (the first/last mile) are of great interest to Incumbent Local Exchange Carriers (ILECs) and Internet Service Providers (ISPs).

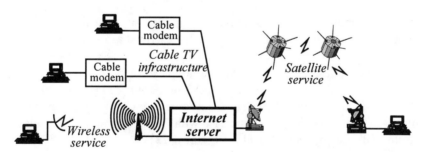

Figure 3-22   Various Options for Broadband Access Schemes.

Recently, customers have been offered alternatives to waiting for ILEC or ISP xDSL services. They can have megabit Internet access speed, especially in the downstream direction (from IP server to local PC client), using three schemes described as follows (Figure 3-22):

(1)   Local cable TV operators: Deploy cable modems to provide broadband access via the cable TV infrastructure.

(2)   Satellite services: These services can run at a higher speeds than cable TV.

(3)   Wireless services: Certain wireless services have been considered appropriate for broadband access technologies. For example, Multichannel Multipoint Distribution System (MMDS) and Local Multipoint Distribution System (LMDS) have been modified to provide broadband access.

As a result, ILECs are being compelled to use the existing infrastructure more efficiently for Internet access. If this is not accomplished, they may lose customers to newer broadband access providers. In addition, telecommunications deregulation has forced ILECs to shift their operations from "profits revolving around maximizing revenues" to "profits depending upon minimizing expenses".

The pace of xDSL deployment has been stimulated, by potential broadband access users, at a much faster rate than originally predicted. This has required pre-qualification of many existing copper loops to determine if they are suitable for xDSL services. An array of tools and test equipment has been developed for this purpose, and the ADSL Forum has proposed two procedures for loop qualification and testing:

1.   Pre-qualification: Determining whether a specific loop can support a specific xDSL service.

2.   Turn-up verification: Verifying that a loop is providing the proper xDSL speed and reliability in accordance with customer billing for these services.

3.6.3.1 Copper Loop Frequency Spectra

Figure 3-23 illustrates the various frequency spectra of typical copper loops. The applications for these frequency spectra are briefly described as follows:

- Since the late 1800s, copper twisted-pair loops have been applied for Plain Old Telephone Services (POTS) in the frequency range from 200 Hz to 3,200/3,300 Hz (a bandwidth of about 3,000 Hz). This configuration is illustrated in Example 3-5 (Figure 3-13). The switching machine used with POTS is optimized for this frequency spectrum [indicated by curve No. (1) in Figure 3-23].

- For the basic rate and primary rate ISDN interfaces, the copper loop spectrum between 200 Hz and 120 kHz is utilized [shown by curve No.(2) in Figure 3-23].

- For Asymmetrical DSL (ADSL) applications, two different frequency spectra have been used:

  * Upstream transmission [from the customer to the terminating central office] utilizes the frequency spectrum between 20 kHz and 120 kHz [shown by curve No. (3) in Figure 3-23].

  * Downstream transmission [from the central office to the customer] utilizes the frequency spectrum between 140 kHz and 552 kHz for transmission speeds up to

1.5 Mbps [shown by curve No. (4) in Figure 3-23]. The downstream also utilizes the spectrum between 140 kHz and 1.104 MHz for transmission speeds up to 8 Mbps [shown by curve No. (5) in Figure 3-23].

- For Very high-speed DSL (VDSL), the copper loop spectrum between 300 kHz and 30 MHz is used [shown by curve No.(6) in Figure 3-23].

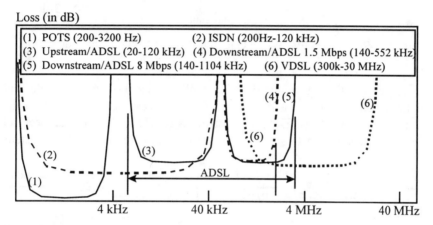

Figure 3-23   Various Frequency Spectra of Copper Loops.

As previously mentioned, the bandwidth limitation of 3 kHz for of voiceband lines is imposed by the customer access network interface filters used in the narrowband Public Switched Telephone Network (PSTN).

In addition, the copper loop bandwidth limitation can also be imposed by loading coils (see Definition 3-1), that are used to improve loop performance for analog speech. For xDSL applications, the filters and loading coils must be removed. With the filters and loading coils removed, the copper loop can pass frequencies up to several MHz (see Figure 3-23).

In the higher frequency spectral range, attenuation is substantially higher than int the voiceband spectrum (i.e., the attenuation increases with the square root of frequency).

Besides bandwidth limitations, copper loops have many other impairments. For example, thermal noise, reflection/echo, and crosstalk are intrinsic impairments. Likewise, components such as surge protectors, Radio Frequency Interference (RFI) filers, and bridge taps may exist in some loops. Impulsive noise (originating from lightning strikes), electric fences, power lines, fluorescent lighting, etc. are possible sources of environmental impairments (i.e., interference). The study, design and analysis of communication systems has adopted the use of the Additive White Gaussian Noise (AWGN) model shown in Figure 3-24. This model has been found to be adequate for approximating backbone network applications. From this model, system performance

[S/N for analog services, or bit error rate (BER; or bit error ratio) for digital services] can easily be evaluated. The assumption of Gaussian channel noise is based on the well-known "Central limit theorem".

Figure 3-24   Additive White Gaussian Noise (AWGN Channel) Model.

However, for DSL technologies the AWGN model is not suitable since the "Central limit theorem" can not be applied for the access networks. Instead, a model using Near-End CrossTalk (NEXT) and Far-End CrossTalk (FEXT) has been adopted. That is, crosstalk is the largest contributor of capacity limiting noise in DSL systems.

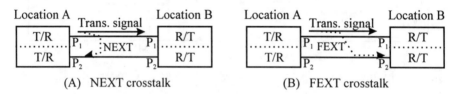

Figure 3-25   Near-end Crosstalk (NEXT) and Far-end Crosstalk (FEXT).

Figure 3-25 illustrates the two types of crosstalk found in "multipair" access network cables: (1) Near-End CrossTalk (NEXT), and (2) Far-End CrossTalk (FEXT). Two twisted pairs are shown; each pair (P_1 and P_2) has a transceiver (T/R) at location A and a transceiver (R/T) at location B.

A signal is transmitted from the transmitter of pair 1 (P_1) at location A, and creates interference on pair 2 (P_2) at the same location [Figure 3-25(A)]. It has been proven that NEXT affects any system that transmits in both directions simultaneously. Theoretically, NEXT could be eliminated by not transmitting in both directions in the same frequency band at the same time. That is, by separating two-way transmission either by using the same time and different bands, or the same band at different time intervals, would practically eliminate NEXT. This is illustrated in Figure 3-26:

(A)   Two directions of transmission are implemented in the same frequency band, but at two different time intervals, as shown in Figure 3-26(A).

(B)   Two directions of transmission are implemented during the same time interval, but in two different frequency bands, as shown in Figure 3-26(B).

(C)   Two directions of transmission are implemented in two different frequency bands and at two different time intervals, as shown in Figure 3-26(C).

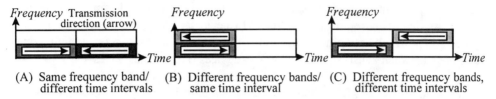

(A) Same frequency band/   (B) Different frequency bands/   (C) Different frequency bands,
      different time intervals   same time interval   different time intervals

Figure 3-26   Separating Two-direction Transmission.

The NEXT level is essentially independent of the cable length, but is highly dependent on signal frequency. The coupling factor (F_c) of NEXT interference, in terms of signal frequency, is given by the following equation:

$$F_c = -55 + 15 \cdot log \, (f/100 \text{ kHz}) \quad \text{dB} \tag{3-9}$$

Typically, NEXT increases as signal frequency increases, and can be modeled by Eq.(3-10) as follows:

$$\text{NEXT} \propto f^{1.5} \tag{3-10}$$

Definition 3-4: The procedure of systematically separating channels on the basis of directional characteristics is called "duplexing". This can be implemented either by Time Division Duplexing (TDD) or Frequency Division Duplexing (FDD). TDD assigns two different time intervals for each direction of transmission as illustrated in Figure 3-26(A). FDD assigns two different frequency bands for each direction of transmission as illustrated in Figure 3-26(B).

The analysis of Far-End CrossTalk (FEXT) is much more complicated. Besides frequency, FEXT is related to the loop length (L) and the loop insertion loss (see Section 3.6.3.3), as modeled in Eq.(3-11), as follows:

$$\text{FEXT} \propto L_{\text{insertion}} \times L \times f^2 \tag{3-11}$$

3.6.3.2 Frequency Plan for xDSL

A frequency allocation plan is an important aspect of xDSL transmission, from a system compatibility viewpoint. Presently, there is no global consensus, and few recommendations (see Section 3.6.3.4), for generating a worldwide xDSL technology. Different access networks have unique topologies and legacies of systems already deployed, therefore it is unlikely that a standardized frequency plan for different xDSL technologies (on a worldwide basis) will be established.

Example 3-9: ITU-T Recommendation G.922.2 (the G.Lite specification)

A major requirement for widespread xDSL deployment is to enable simultaneous voice and data service. This requires a "***splitter***" to be installed on the customer's premises (Figure 3-17), which is a time-consuming procedure. The splitter is used to prevent interference between xDSL signals and POTS devices (i.e., phones, faxes, etc.).

Another option is use an xDSL modem which has a built-in POTS splitter at the customer premises. However, this arrangement requires new wiring at customer premises. In either case, the implementation requires on-site support of telecom technicians, and implies a time-consuming/expensive procedure.

Therefore, an appropriate model for xDSL modem deployment is one that uses the present analog modem which is typically built-in inside Personal Computers (PCs). The user can simply plug in the traditional telephone jack without involving a telecom technician. ITU-T Rec. G.992.2 defines the "splitterless" ADSL specification commonly known as G.Lite, as having the frequency spectrum illustrated in Figure 3-27.

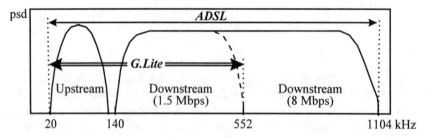

Figure 3-27   Frequency Spectrum (Power Spectrum Density) for G.Lite ADSL.

ITU-T G.Lite is one approach that can achieve wide consumer acceptance and rapid ADSL deployment, provided it meets the following requirements:

- Without requiring a splitter, G.Lite modem must support a long range [even exceeding 18 *kft* (Example 3-6)] to provide maximum coverage of users who are eligible for ADSL services. This "maximum coverage" is an important requirement, even if the trade-off of offering a lower data rate is unavoidable.

- G.Lite modems must be easily supported by ANSI ADSL standard (T1.413) equipment at the local exchange, with only a minor software upgrade.

- G.Lite modems must be low cost. It is proposed that the G.Lite modems be based on the T1.413 specification. That is, a T1.413-based ADSL Terminal Unit (ATU-C) can be easily upgraded to support G.Lite.

- G.Lite should be able to simplify the T1.413 specification to reduce system complexity, reduce cost, and promote multi-vendor interoperability.

The G.Lite ADSL is rate adaptive. Its data rate depends not only on the loop length, but also on "in-home wiring" conditions and the connected POTS devices (i.e., G.Lite does not have a splitter to prevent interference with POTS devices). It is possible to achieve 1.5 Mbps downstream and 512 Mbps upstream data rates for loops with good quality and a length of 18 *kft*. It is expected that G.Lite will operate beyond 18 *kft* at a lower data rate provided there are no loading coils in the loop.

It is important to understand that G.Lite modems cannot offer guaranteed bit rate services for the following reason. In order to avoid interference between the G.Lite modem and other POTS devices (i.e., phones or voice terminals), the G.Lite modem will cut back its power if it detects a POTS device going "off-hook". This requires the G.Lite modem to perform a "fast-retrain" procedure to operate at a lower data rate. The rate degradation, which depends on the interference with a POTS device, does not have a lower bound. That is, it is even possible to have a degraded data rate of zero. The retrain procedure may result in up to 1.5 seconds of no data transmission by the G.Lite modem. The original data rate is automatically resumed when the G.Lite modem detects the POTS device is "on-hook" again.

3.6.3.3 Loop Insertion Loss

VDSL can be used to transmit signals in the frequency range of 20 MHz via twisted-pair cooper loops. A huge variation in dynamic range may be the most significant impairment for VDSL transmission. Figure 3-28(A) illustrates a generic (not scaled) loop transfer function (insertion loss) with and without bridge taps. Since VDSL signals can operate in the higher frequency band, the effect of insertion loss must be taken into consideration.

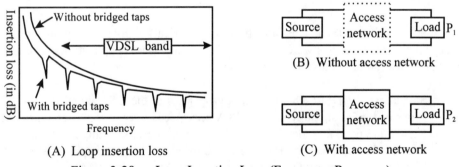

(A) Loop insertion loss

(B) Without access network

(C) With access network

Figure 3-28   Loop Insertion Loss (Frequency Response).

The insertion loss is an intrinsic loss of a metallic copper loop. It represents the loop impairment at various frequencies. To improve loop insertion loss characteristic, it is necessary to reduce loop impairments: bridge taps, loading coils, etc. Figure 3-28(A) indicates that a source is connected directly to the load without using a loop (access network); and assuming the received power at the load is P_1 (i.e., 8 μw). Figure 3-28(C) indicates that the source is connected to the load via a loop; and assuming the received power at the load is P_2 (i.e., 1 µw). These configurations are used to illustrate the procedure of measuring the loop insertion loss. By definition, this loop (access network) has an insertion loss (or gain; as some networks may have active regenerating devices) given the following equation:

$$Insertion\ loss = 10\ \log \frac{P_1}{P_2} = 10\ \log \frac{8\ \mu w}{1\ \mu w} = 9 \quad dB \tag{3-12}$$

3.6.3.4 The xDSL Family

There are several xDSL technologies that can be implemented/deployed for high-speed residential broadband access. For qualification purposes, these technologies can be classified into three groups as shown in Figure 3-29.

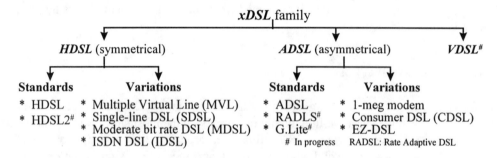

Figure 3-29   xDSL Family: HDSL, ADSL and VDSL

- High-speed DSL (HDSL): This branch of (symmetrical) high-speed digital subscriber line (HDSL) technologies were the first to broadband access applications. The HDSL technologies can be subdivided into two groups:

  (1)   Standardized HDSL: This technology operates at the DS1 signal rate of 1.544 Mbps or the E1 signal rate of 2.048 Mbps speeds over two twisted pairs out to 12,000 *ft* (3.6576 *km*) loop lengths. The line code used is usually 2B1Q (described later in this chapter), or (less often) Carrierless Amplitude/phase Modulation [CAP; a variation of Quadrature Amplitude Modulation (QAM) discussed in Chapter 6]. The local loops used for this technology must have a loop resistance of less than 900 Ω (ohms), and a total insertion loss less than 35 dB. If a HDSL "doubler" is used, the maximum loop distance can be extended to 24,000 *ft*.

  Another standard HDSL technology being developed is known as HDSL2 (Figure 3-29). This technology uses only one twisted pair for the same speed and distance supported by standard HDSL. HDSL2 applies a 16-level Pulse Amplitude Modulation (16-PAM) line code.

(2) HDSL technologies other than standard HDSL (i.e., HDSL variations): Four common HDSL variations have been deployed, and are described as follows:

* Multiple Virtual Line (MVL; a Paradyne solution): MVL can be scaled from 128 (2 × 64 kbps) to 768 (12 × 64 kbps), in 64 kbps increments, to serve up to 24,000 *ft* loop lengths. MVL allows POTS "lifeline" support, without requiring a premises splitter for isolating analog voice equipment from up to eight digital MVL devices.

* Single line DSL (SDSL): This is another variation from standard HDSL, that uses only one twisted pair for either 768 kbps (784 kbps including overhead) or T1/E1 speeds. An extremely high speed form may run at 2.3 Mbps. The local loops used for SDSL must have a loop resistance less than 900 Ω and a total insertion loss less than 35 dB (as in the standard HDSL).

* Moderate rate DSL (MDSL): This type of HDSL has a speed of 768 kbps on one twisted pair for up to 21,000 *ft* loop lengths.

* ISDN DSL (IDSL): This technology has the speed as basic rate ISDN: ("2B+D") for a speed of 160 kbps, and loop lengths up to 18,000 *ft*, but not to an ISDN switch. The local loops used for IDSL must have a loop resistance less than 1300 Ω and a total insertion loss less than 39 dB.

Noted that SDSL, MDSL and IDSL is not intended to support POTS on the same access line.

Among these high-speed DSL technologies, "standard HDSL" is the most popular choice. If a local copper loop qualifies for standard DSL service, it should also be qualified for the other HDSL technologies.

* Asymmetrical DSL (ADSL): ADSL is designed to optimize downlink operations, and can be subdivided into two groups (Figure 3-29):

(1) Standard ADSL: Standardized ADSL applies Discrete Multi-Tone (DMT) line encoding (an alternate line code is CAP), over a bandwidth of 1.1 MHz that is divided into 255 channels (*"bins"*). Some of these channels are not used, and some do not support analog POTS. The upper bound speeds are 8 to 9 Mbps downlink, and 800 kbps uplink for short loops. However, practical service data rates may be much lower. ADSL can be used for longer copper loops with much lower speeds. Currently, the minimum downlink speed that is acceptable to general customers is about 1 Mbps. The local loops used for ADSL must have a loop resistance of less than 1300 Ω, with variable loop insertion loss depending on the loop length.

Rate-adaptive DSL (RADSL; another standard ADSL): This standard is being defined, and uses CAP or DMT line coding. RADSL will adopt ADSL qualities and loop qualification. A DMT modem can choose up to 15 bits/symbol with QAM in each bin (i.e., channel). Some bins of RADSL can be "shut down" so that the downlink and uplink data rates can be "adaptive". The "rate adaptation" is supported by RADSL modems using CAP, but is inherent in the DMT bin structure. Presently, RADSL can also operate in a symmetrical mode, with a downlink and an uplink speed of about 800 kbps.

ITU-T G.Lite (see Example 3-9; another standard ADSL): This ADSL technology operates at a 1.5 Mbps downlink rate, and an uplink rate of up to 384 kbps on local loops. G.Lite uses a subset of the 255 DMT bins (channels), and is intended to inter-operate with "full" ADSL modems on either end.

(2) ADSL variations: They are vendor specific technologies, e.g., Nortel's 1-Meg modem, Rockwell's customer DSL (CDSL), and Cisco's EZ-DSL, etc.

ADSL, just like HDSL standard, will support RADSL and G.Lite once the local loop is qualified for standard ADSL technology.

- Very high-speed DSL (VDSL): This is intended to be an emerging ADSL technology for use in the areas of fiber loop carriers with short "copper tails". The copper loop can be up to 4,500 *ft* at variable speeds from 12 to 50 Mbps. Higher speed operations require a shorter loop length. The local loops for VDSL must have a resistance less than 325 Ω, with a variable loop loss requirement that depends upon loop length.

3.6.3.5 Copper Loop Characteristics/Impairments

Customers using access networks have recently become very demanding with respect to their requests for modern communication technologies. This is stimulated by the high-speed backbone (backhaul) networks that have been rapidly implemented using SONET/SDH standards. These backbone networks support data speeds as high as 2.5 Gbps (2.48832 Gbps for OC-48 or STM-16), or can be extended up to 10 Gbps (9.95328 Gbps for OC-192 or STM-64) without applying Wavelength Division Multiplexing (WDM) technology. In addition, if 80-WDM is used, the speed can be 80 times higher than 2.5 or 10 Gbps (e.g., in the range of 200 to 800 Gbps). For customers to utilize these high-speed backbone networks, the access network speeds must be increased accordingly.

Figure 3-30 illustrates a typical inter-connection for a backbone, Internet, and access network. The PSTN (backbone) has been implemented as a OC-48/STM-16 based SONET/SDH high-speed (2.48832 Mbps) network (typically in blocks of 64 kbps). The Internet has OC-12/STM-4 links operating at a speed of 622.08 Mbps. Local Area Networks (LANs) can run at speeds between 10 and 100 Mbps, and PC, can be equipped with cards/ports that operate in the megabit range. The access network is the set of links that desperately need (indicated by "?" symbols in Figure 3-30) much greater bandwidth and higher speeds than is currently available.

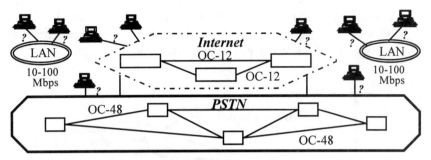

Figure 3-30   Backbone, Internet and Access Networks.

The access line to either the PSTN or Internet is typically provided by a Local Exchange Carrier (LEC) using twisted-pair copper loops. Various schemes [e.g., xDSL technologies deployed by Internet Service Providers (ISPs) and LECs] have increased the digital data transfer rate from 56/128 kbps to the megabit ranges for downlink (i.e., remote server to local client) direction. In some cases, the deployment of this service has been done on a grand scale at a relatively low cost.

The various alternatives (see Figure 3-29) that are available to implement higher-speed access networks has increased the pressure on LECs and ISPs to accelerate deployment of xDSL. The acceleration of xDSL deployment results in:

(1)   The need for copper loop (pre-) qualification: This procedure determines whether a particular loop is suitable for providing xDSL service.

(2)   The need for "turn-up" verification: This procedure verifies that the xDSL customer is receiving the full service which has been contracted and tariffed.

To perform loop qualification, the copper loop characteristics, limitations, and impairments must be completely understood. The impairments at xDSL frequency bands can be classified into two groups:

(1)   Physical impairments: These loop impairments arise from the techniques used for copper loop engineering. Originally loops were typically optimized for analog voice transport rather than xDSL applications. Primary physical impairments include:

   •   Loading coils: It has been previously mentioned that a typical loop exhibits a frequency response as shown in Figure 3-7. Due to the narrowband filters used in the local exchange office, the bandwidth of a typical loop rolls off rapidly after the frequency of about 3200/3400 Hz (Figure 3-7, or 3-31). When the loop length increases, the roll-off becomes steeper.

   Since the loops were originally designed for voice transport, loading coils have been used to flatten the frequency response across the voice band for longer loops. These loading coils extend the passband (0 to 4 kHz) voice range. That is, the spectrum has been flattened in the voice band, which results in a better voice (POTS) service quality. However, loading coils cutoff higher frequencies as shown in Figure 3-31, which can degrade xDSL performance.

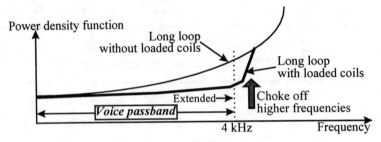

Figure 3-31   Frequency Spectra for Long Loops with/without Loaded Coils.

- Mixed gauges: A mixture of wire sizes (gauges) are used in many countries (typically outside the USA) which has the same effect as of loading coils, and creates problems for xDSL applications.

- Bridged taps: The effect of bridged taps on insertion loss has been illustrated in Figure 3-28. Bridged taps have been adopted in most of POTS loop plants to simplify adding or deleting users. They result in un-terminated impedance branches that reflect signals, and thus cause interference for xDSL applications.

- The use of proprietary line extenders (to the customer premises equipment) may pose additional interference problem for xDSL signals transport.

(2) Electrical impairments (interferers or disturbers): Electrical impairments are associated with signal interference that occurs in the copper loop. They include:

- Radio Frequency Interference (RFI): Potential sources of RFI are - (1) AM radio stations; (2) aerial wires; and (3) amateur radio broadcast.

- Crosstalk: In addition to NEXT and FEXT (illustrated in Figure 3-25), there are (1) Self crosstalk (SNEXT) caused by coupling between xDSL signals carried within a common cable binder; and (2) Foreign crosstalk (FNEXT) from other adjacent xDSL signals.

- Background noise: Thermal noise, semiconductor noise, and impulse noise can be created by any type of local exchange equipment.

3.6.3.6 Loop Qualification and Testing

It is desirable to have an array of local test equipment for xDSL loop qualification purposes. Since HDSL technology performs symmetrical operations at a modest data speed, its test equipment is more mature than the type required for ADSL. This is further complicated because ADSL technology intends to utilize the maximum available local loop capacity. In addition to the operational differences between HDSL and ADSL, the line code used (e.g., CAP and DMT) also requires different test procedures and test equipment. VDSL technology falls into a completely different category. It is much more difficult to test/qualify local loops for VDSL service. It is clear that FEXT interference will have an impact on VDSL technologies, and VDSL also operates in the frequency spectra started with AM and CB radio broadcast.

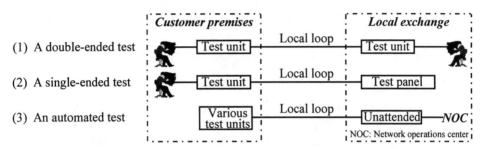

Figure 3-32   Various Local Loop Testing Approaches.

Three different approaches (Figure 3-32) have been adopted for loop testing: (1) a double-ended testing (which requires testing from the local exchange side and the customer premises side) is the most time consuming and least desirable approach even though it is widely used in the field; (2) single-ended testing, which is occasionally used in the field and is expected to become more popular; and (3) automated testing, which is the best approach, but is generally not available in the industry.

The Time Domain Reflectometer (TDR) is a tool that has been used for many years, but is suitable for testing/verifying xDSL loops. For example,

- The use of a "load coil counter/detector" and a TDR can determine the location of the first loading coil and/or mixed wire gauges in a loop.

- The use of a TDR and an "open meter" can measure loop lengths.

- A wideband frequency sweep can find bridge taps, and a TDR can determine their distances (i.e., locations in the loop).

Example 3-10: Describe the applications and operations of Local Loop Maintenance Operating System/Mechanized Loop Testing (LMOS/MLT).

LMOS is a family of integrated computer-based operations support systems that support the installation and maintenance of local telephone loops. Its primary function is to access, test, and manage the repair of loop troubles (i.e., failures). LMOS capabilities include: (1)

acceptance of trouble calls; (2) POTS and ISDN testing (e.g., analyze ac and dc current levels, detect hazardous electrical potentials, detect intercepts, test dialtone, measure capacitance and resistance, etc.), and (3) performing Mechanized Loop Testing (MLT).

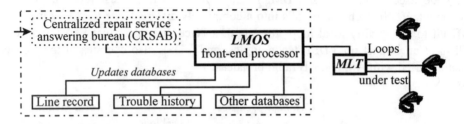

Figure 3-33   LMOS/MLT System Configuration/Applications.

MLT is a single comprehensive test system that is capable of analyzing both metallic and digital loop problems. The system is highly automated, relying on expert systems so that most tests are handled without intervention by technicians. The MLT test system allows a service provider to test loops routinely, frequently, accurately, and cost-effectively. Figure 3-33 illustrates the applications/configuration of LMOS/MLT. The Centralized Repair Service Answering Bureau (CRSAB) receives trouble reports from the customer, MLT performs loop tests (without human intervention), returns a real-time report back to CRSAB, and updates databases.

Sometimes, but not always, LMOS/MLT records (databases) can be used to predict xDSL success (i.e., qualify loops for xDSL applications). However, modifications must be made to LMOS/MLT so that the automated loop testing process is more reliable. In general, if a local loop supports T1 or E1 data rates, the loop should be able to support HDSL services as well. Compliance with Carrier Serving Area (CSA) loops helps the process of pre-qualifying loops for HDSL, with little or no modification of existing procedures. The process includes the detection and locations of loading coils, mixed wire gauges, bridge taps, loop resistance, loop insertion loss (typically 35 dB), Bit Error Rate (BER; typically 10^{-7}), and line power measurements.

Several test sets have been developed for xDSL prequalification and turn-up verification. It is believed that more advanced and cost-effective equipment will be available in the future. The following is a partial list of the new test equipment:

- Hewlett Packard (HP) manufactures a HDSL "Installer's Assistant (IA)" (formally marketed under the name of CERJAC) test system. It is a portable unit ("luggable"), that uses AC line power, and equipped with optional plug-ins for the specific type HDSL units being tested. The plug-ins can be used for field tests or to test the HDSL units themselves. The unit analyzes loading coils (number and locations), loop loss, and includes T1 or E1 test suites. This unit is designed for double-ended testing, but is also useful for some single-ended test applications.

- Sunrise Telecom introduced the SunSet xDSL test set for HDSL and ADSL testing. The unit is designed for double-ended testing, but also supports some single-ended applications.

- Tempo Research introduced the DSL2000 test unit for HDSL. This unit is designed for double-ended testing, with some single-ended testing applications.

- HP also introduced three test products for ADSL:

  * HP Service Advisor: It is a small handheld device that can be used to pre-qualify a loop for ADSL. The unit performs all the tests needed to provide a "go/no-go" decision for ADSL services.

  * HP Service Installer: It is a handheld "golden modem" with modules that supports several CAP and DMT chipsets, and can be used for turn-up verification. This unit is a "Windows-based" palmtop, with simple menus and downloadable test results. If the unit is used at the customer premises, it can verify the data speed, establish margins, and evaluate noise levels.

  * HP IP Service Installer software: This software package can be installed on the previous two units (described above) to measure IP layer connectivity and throughput.

- Turnstone Systems CX100 Copper CrossConnect™ test system can be used between the DSL Access Module (DSLAM) and Main Distribution Frame (MDF) of a local exchange. This unit is equipped with a "test probe" module to provide automated testing of xDSL loops, and manages the MDF to DSLAM interface. For example, if a subscriber upgrades from HDSL to HDSL2 or ADSL, no additional wires need to be installed in the local exchange office for this customer.

Definition 3-5: A "golden modem" is Windows-based handheld unit used for xDSL turn up verification. The unit is a battery, pocket-size device. Multiple test results can be stored and downloaded to a central database. It performs a test of the bit rate up to the customer premises. An Internet Protocol (IP) test suite or file transfer tests can be added to evaluate throughput and Bit Error Rate (BER).

3.6.4 High-speed Digital Subscriber Line (HDSL)

Currently, the HDSL systems transport bidirectional digital signals at the nominal rate of 1.544 Mbps (DS1 signal rate) or 2.048 Mbps (E1 signal rate). An option is also available for transporting 768 kbps (half of a DS1 payload). Standardization of HDSL will insure that compliant equipment. Standardization also encourages different equipment vendors to support interworking, which is a highly desirable characteristic. HDSL systems may eventually be able to operate on 99% of the copper twisted-pairs that conform to the Carrier Serving Area (CSA) guidelines. CSA guidelines were developed for local loops installed in telephone distribution plants that provide services to customer premises equipment via interoffice digital carrier systems terminating in a central office, or from Digital Loop Carrier (DLC) systems connected to an outside plant terminal.

When a transmission scheme uses two (or more) wire pairs, the characteristics of the wire pairs may differ in the following aspects:

- Loop make-up (configuration)
- Total loop length (working plus bridged taps)
- Wire gauges
- Bridged taps
- Loading coils
- Crosstalk (NEXT, FEXT, SNEXT, and FNEXT)
- Loop resistance
- Loop loss
- Delay
- Bit Error Rate (BER)

3.6.4.1 HDSL Rates and Payloads

The payload for HDSL systems in the North American digital hierarchy, is a bidirectional bitstream with a nominal rate of 1.544 Mbps (known as DS1 digital signal). The DS1 is described in Chapter 6 of this book, in various ANSI documentation (e.g., T1.102, T1.107, T1.403, T1.408), and in various ITU-T Recs. (e.g., G.704, G.732, and G.733).

The DS1 interface known as a DSX-1 signal (i.e., digital signal cross-connect point for a DS1 signal) is a bipolar or Alternate Mark Inversion (AMI) signal with a peak magnitude between 2.4 and 3.6 volts, alternating between "+" and "−" voltage to represent a logical "1", with a logical "0" represented by a zero volt level. For zero suppression (substitution) purposes, AMI may be replaced by the Bipolar 8 Zero Substitution (B8ZS) line code (both AMI and B8ZS codes are described in Chapter 6).

A DS1 signal may or may not be "framed". For framed DS1 signals, a SuperFrame (SF) of twelve 125-µs frames (i.e., a total of 1.5 ms), or an Extended SuperFrame (ESF) of twenty-four 125-µs frames (i.e., a total of 3 ms) structure may be used (see Chapter 6). Each frame contains an overhead bit with a rate of 8 kbps, and 24 channels (each channel contains an 8-bit byte at a rate of 64 kbps known as a DS0 channel). DS0 channels can be used to carry digitized voice, or any 64 kbps signal (e.g., digital data, FAX, etc.). Thus, the DS1 signal has a payload of 1.536 Mbps ($\equiv 24 \times 64$ kbps). The non-framed DS1 signal can be used to carry maintenance signals, fault conditions, and specialized services. A DS1 signal is designed to be carried by a T1 digital carrier system using metallic facilities (e.g., copper pair wires, coaxial cable, etc.). Digital regenerators are required along the transmission path for transmission up to 100 miles, but bridged taps are not allowed. A HDSL system also requires additional overhead bits above the DS1 1.544 Mbps capacity.

3.6.4.2 Loop Architectures for HDSL Systems

HDSL loops may adopt either of the following loop architectures:

(1) Dual simplex (also known as 2-pair simplex): This HDSL loop architecture is similar to the T1-line system architecture. Each pair carries a separate unidirectional signal at a nominal rate of 1.5 Mbps plus overhead bits.

(2) Dual duplex (also known as 2-pair full-duplex; Figure 3-4): This "dual-duplex" architecture is favored by the HDSL industry. Each pair uses an echo canceled hybrid transmission method, and carries the following data fields:

- 768 kbps payload ($\equiv$ ½ × DS1 payload)
- DS1 framing pattern at 8 kbps (in both pair)
- 8 kbps overhead

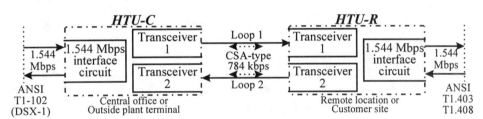

Figure 3-34   Dual Duplex HDSL Architecture.

A total of 784 kbps (768 + 8 + 8) transmitted in both directions simultaneously, is the binary equivalent line rate. Figure 3-34 shows a dual-duplex HDSL architecture. Two CSA-type loops (each carrying 784 kbps in both directions simultaneously), are connected to the HDSL Terminal Unit at the Central Office (HTU-C) or a network terminal site in the outside plant, and the HDSL Terminal Unit at the Remote (HTU-R) or customer end. The HDSL system extends from the 1.544 Mbps side of the HTU at one end (e.g., CO), to the 1.544 Mbps side of the HTU on the other end (e.g., customer premises). Note that the local loops used in this dual duplex mode are CSA-type loops. The two 1.544 Mbps interfaces are defined as follows:

- ANSI T1.102 interface: It defines the DS1 electrical interconnection requirements for frequency accuracy, frequency stability, line code, impedance, and pulse shape for DSX-1 cross-connect points.

- ANSI T1.403 and T1.408 interfaces: They define the requirements for DS1 signals at the network-customer interfaces.

3.6.4.3 Echo Canceled Hybrid Transmission Scheme

The transceivers in the "HTU-C" and "HTU-R" illustrated in Figure 3-34 are enlarged as shown in Figure 3-35. The transceivers are denoted as "T" and "R", and the hybrid with echo canceler connection is also shown.

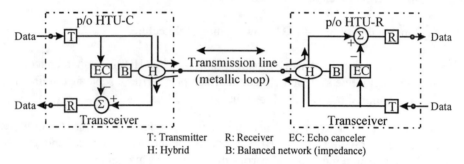

T: Transmitter   R: Receiver   EC: Echo canceler
H: Hybrid   B: Balanced network (impedance)

Figure 3-35   Echo-canceled Hybrid Transceiver.

A hybrid circuit is used to combine 2 separate two-wire facilities into a single two-wire facility at one end, and also to split the combined two-wire facility into 2 separate two-wire facilities at the other end. The hybrid is used with a balanced network [indicated by "B" in Figure 3-35] to achieve a better impedance match so that the echo is minimized. However, the balanced network (impedance) can never completely eliminate the echo. Thus, an Echo Canceler (EC) is used. The echo canceler produces a replica of the echo of the ***near-end*** transmission. The estimated echo (replica) is then subtracted from the actual received signal thereby canceling the echo. This arrangement allows full duplex operation over a two-wire subscriber loop.

The HDSL systems using echo cancelers with hybrids is intended for service on twisted-pair cables that meet the CSA guidelines for North American loop plants. This HDSL service will operate over loops up to 12,000 *ft* (4 *km*) in loop length, with a loop loss of 35 dB at 200 kHz into 130 Ω.

3.6.4.4 2B1Q Transmission Technology

Three technologies (line codes) can be used for HDSL transmission: (1) 2-Binary/1-Quaternary (2B1Q), (2) Discrete Multi-Tone (DMT), and (3) Carrierless AM/PM (CAP).

2B1Q is a 4-level code (see Figure 3-36) without redundancy. The symbol assignment of this 4-level line code is shown in Figure 3-36: the first bit is the sign bit and the second bit is the magnitude bit. It should be noted that +3, +1, −1, or −3 is a symbol only, not a numerical voltage value (see Figure 3-37). When the 1.544 Mbps digital bitstream enters the HTU, the bitstream is grouped into "pairs" of bits (called a "bit field"). Each bit pair is converted into one of the four standardized symbols (quats). The pulse mask for the four quats can be obtained by multiplying the normalized mask shown in Figure 3-37 by 2.64 volts, 0.88 volts, −0.88 volts, or −2.64 volts. That is, the voltages at various points for four quats of 2B1Q are shown in Table 3-4.

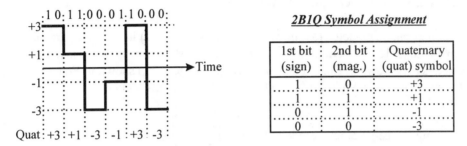

Figure 3-36   2B1Q Line Code/Quaternary Symbol Assignment.

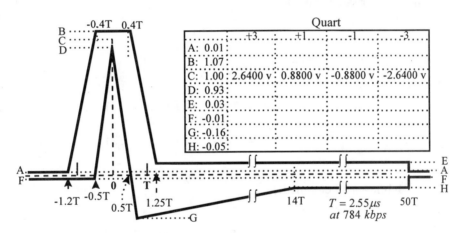

Figure 3-37   Normalized Pulse Mask for Transmitted Pulse.

Table 3-4   Voltages of 4 Quats Pulse Mask.

Normalized level		+3	+1	−1	−3
A:	0.01	0.0264 v	0.0088 v	−0.0088 v	−0.0264 v
B:	1.07	2.8248 v	0.9416 v	−0.9416 v	−2.8248 v
C:	1.00	2.6400 v	0.8800 v	−0.8800 v	−2.6400 v
D:	0.93	2.4552 v	0.8184 v	−0.8184 v	−2.4552 v
E:	0.03	0.0792 v	0.0264 v	−0.0264 v	−0.0792 v
F:	−0.01	−0.0264 v	−0.0088 v	0.0088 v	0.0264 v
G:	−0.16	−0.4224 v	−0.1408 v	0.1408 v	0.4224 v
H:	−0.05	−0.1320 v	−0.0440 v	0.0440 v	0.1320 v

The average power of a HDSL signal, consisting of a framed sequence of symbols with a frame word and equiprobable symbols in all other positions, should be between 13 and 14 dBm over the frequency band from 0 to 784 kHz.

3.6.4.5 Theoretical Loop Analysis Model

A local loop can be modeled (as shown in Figure 3-38) by: (1) impedance parameters (z_{ij}) shown in Eq.(3-12); (2) admittance parameters (y_{ij}) shown in Eq.(3-13); (3) scattering parameters (s_{ij}) shown in Eq.(3-14); (4) hybrid parameters (h_{ij}) shown in Eq.(3-17); or (5) ABCD parameters shown in Eq.(3-18) (described later in this section).

Impedance-parameter model:
$$\begin{bmatrix} V_1 \\ V_2 \end{bmatrix} = \begin{bmatrix} z_{11} & z_{12} \\ z_{21} & z_{22} \end{bmatrix} \begin{bmatrix} I_1 \\ I_2 \end{bmatrix} \tag{3-12}$$

Admittance-parameter model:
$$\begin{bmatrix} I_1 \\ I_2 \end{bmatrix} = \begin{bmatrix} y_{11} & y_{12} \\ y_{21} & y_{22} \end{bmatrix} \begin{bmatrix} V_1 \\ V_2 \end{bmatrix} \tag{3-13}$$

Scattering-parameter model:
$$\begin{bmatrix} b_1 \\ b_2 \end{bmatrix} = \begin{bmatrix} s_{11} & s_{12} \\ s_{21} & s_{22} \end{bmatrix} \begin{bmatrix} a_1 \\ a_2 \end{bmatrix} \tag{3-14}$$

where
$$a_i = 1/2(V_i + Z_i I_i) \quad (i = 1, 2) \tag{3-15}$$

and,
$$b_i = 1/2(V_i - Z_i I_i) \quad (i = 1, 2) \tag{3-16}$$

Hybrid-parameter model:
$$\begin{bmatrix} V_1 \\ I_2 \end{bmatrix} = \begin{bmatrix} h_{11} & h_{12} \\ h_{21} & h_{22} \end{bmatrix} \begin{bmatrix} I_1 \\ V_2 \end{bmatrix} \tag{3-17}$$

ABCD-parameter model:
$$\begin{bmatrix} V_1 \\ I_1 \end{bmatrix} = \begin{bmatrix} A & B \\ C & D \end{bmatrix} \begin{bmatrix} V_2 \\ I_2 \end{bmatrix} \tag{3-18}$$

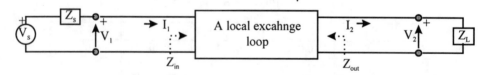

Figure 3-38   Two-Port Network Model for a Local Loop.

Figure 3-38 illustrates the two-port network model of a local loop: V_i and I_i = terminal voltages and currents ($i = 1, 2$); V_s is the source voltage; Z_s is the source impedance; and Z_L is the load impedance (typically, 110, 120 or 135Ω, resistive). The related performance measurements used for HDSL (ADSL, or VDSL) applications in a local loop configuration are given as follows:

$$Z_{in} = input\ impedance = \frac{AZ_L + B}{CZ_L + D} \tag{3-19}$$

$$Z_{out} = input\ impedance = \frac{DZ_s + B}{CZ_s + A} \tag{3-20}$$

$$V_{in} = input\ voltage = V_s \frac{Z_{in}}{Z_s + Z_{in}} \tag{3-21}$$

$$V_{out} = output\ voltage = V_L = V_s \frac{Z_L}{AZ_L + B + CZ_sZ_L + DZ_s} \tag{3-22}$$

$$P_{in} = input\ power = \frac{1}{2} |V_{in}|^2\ \text{Re}\ \{\frac{1}{Z_{in}}\} \tag{3-23}$$

$$P_{out} = output\ power = \frac{1}{2} |V_{out}|^2\ \text{Re}\ \{\frac{1}{Z_L}\} \tag{3-24}$$

where $|\ |^2$ is the magnitude, and Re{ } is the real part of { }. Two other important loop performance measurements are the loop insertion loss and the loop mean squared loss:

$$Loss_i = Loop\ insertion\ loss = \frac{Z_L + Z_s}{AZ_L + B} \tag{3-25}$$

$$Loss_{MSL} = MSL = Loop\ mean\ squared\ loss = \frac{1}{N} \sum_{i=1}^{N} \frac{P_{out}(f_i)}{P_{in}(f_i)} \tag{3-26}$$

Definition 3-6: The matrix {ABCD}, given in Eq.(3-18), is called the transmission matrix of a 2-port network, and the matrix elements are defined as follows:

       A = the open-circuit voltage transfer function
       B = the short-circuit transfer impedance
       C = the open-circuit transfer admittance
       D = the short-circuit current ratio

Example 3-11: Describe the insertion loss and the mean squared loss of the three typical local loops (with different loop lengths) plotted in Figure 3-39.

The insertion loss is expressed by Eq.(3-25), where Z_L is the load impedance and Z_s is the source impedance. Both Z_L and Z_s are functions of system operating frequency (f), and represents the impedance characteristic of the terminal devices. Likewise, the (access) network open-circuit voltage transfer function (A), and the network short-circuit transfer function (B) are also functions of network operating frequency. However, only A and B, not Z_L or Z_s, are functions of loop length.

From Eq.(3-26), it can be seen that the loop mean squared loss is clearly a function of network operating frequency. Since the output power, Pout, is related to the loop length, the loop mean squared loss is a function of loop length.

In summary, the insertion loss and mean-squared loss are related to the loop length. However, Figure 3-39 shows that the losses are within a few dBs for different loop lengths. Hence, the frequency variation has more effect on insertion loss and mean-squared loss than loop length.

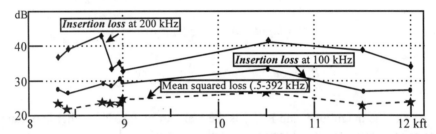

Figure 3-39   Insertion, Mean Squared Losses of Some Typical Loops.

3.6.4.6 Loop Performance Test Requirements and Test Procedures

The HTU-R requires a satisfactory performance of BER < 10^{-7} (with sufficient margin) when HTU-R receives a pseudo random sequence of pulses that are attenuated and distorted by loop impairments.

Near-End crosstalk (NEXT) is a potentially limiting impairment for two-way transmission of echo-canceler-hybrid based HDSL systems (Figure 3-35). Therefore, HDSL systems should achieve a performance level of BER < 10^{-7} in the presence of NEXT, at a level of at least 6 dB above the 1% worst case NEXT level that 49 other identical systems operating in the same 50 pair cable cumulatively present to the system.

Two HDSL transceivers are required when performing loop tests (i.e., one for each end of the loop under test). Figure 3-40 illustrates the HDSL transceiver NEXT performance test set up. The pseudo random binary pattern (sequence) generator provides a 784 kbps signal to the transceiver at point A. The sequence is transmitted via the loop under test, and is received at point C. An error counter is used to measure the error rate performance, but because the test is performed for only one direction, there is no error counter applied at point B. The pseudo generator applied at point D creates a realistic echo condition for the transceiver at that end of the loop.

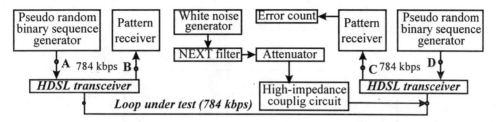

Figure 3-40   HDSL NEXT Performance Test Set-up.

For loop performance test purpose, the HDSL transceiver at the HTU-C end is locally clocked, and the transceiver at the HTU-R end derives its timing from the received bitstream (provided by the HTU-C transceiver). That is, there is no common clock between the HTU-C and HTU-R hardware. Therefore, the test can handle both transmission problem and timing problems imposed by the loop.

This test is performed in the full-duplex mode with both HDSL transceivers transmitting data simultaneously. The test set-up shown in Figure 3-40 is run twice: once from the network end, and again from the remote end.

3.6.4.7 Recommended Functional Characteristics

A HDSL Terminal Unit at the Remote end (HTU-R) should operate with a received signal baud rate in the range of 392 kbaud ± 32 ppm (parts per million). HDSL systems provide 1.544 Mbps transport by applying the dual-duplex method. That is, the binary equivalent line rate of a HDSL system is 784 kbps on each loop. This 784 kbps signal consists of:

- HDSL payload of 776 kbps of the following two parts:

  * 768 kbps payload ($\equiv$ ½ × DS1 payload)

  * DS1 "F" bit at a rate of 8 kbps: The "F" bit is duplicated and transported via both HDSL paths. A HDSL system should also transport unframed 1.544 Mbps signals [e.g., an Alarm Indication Signal (AIS) is a unframed all "1s" 1.544 Mbps signal].

- HDSL overhead of 8 kbps: The 8 kbps overhead bitstream carries: (1) the embedded operating channel, (2) Cyclic Redundancy Check (CRC), (3) synchronization word, and (4) an average of one stuffing quat per HDSL frame.

Example 3-12: Describe the signal flow of a 1.544 Mbps DS1 signal through a HDSL system operating in the dual duplex mode with a splitter for splitting/reconstructing the DS1 signal bitstream at the HTU-C and HTU-R (Figure 3-41 and Figure 3-34). The nominal DS1 data rate is 1.544 Mbps with a speed deviation of 32 ppm (parts per million) which is about 50 bps $[\equiv 32/(10^6) \times 1.544$ Mbps]. The DS1 signal (described in Chapter 6) consists of a "F", known as b_0, and 24 bytes (byte Nos. 1 to 24) of 8 bits each.

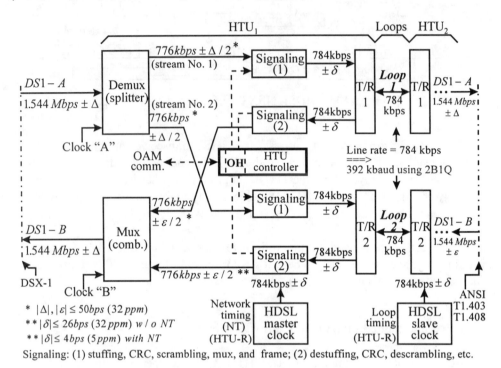

Figure 3-41   Dual-duplex HDSL Signal Paths for DS1 Signal Transport.

A demultiplexer (splitter) is used to split the DS1 signal supplied by the DSX-1 (DS1-signal cross-connect point) into two streams of 776 kbps $\pm \Delta/2$ each. One stream (stream No.1 in Figure 3-41) carries of bit b_0 and bytes No. 1 to 12, and the other stream (stream No.2 in Figure 3-41) carries bit b_0 and bytes No. 13 to 24. Note that the "F" bit, b_0, is duplicated in both streams. Each 776 bit data stream is fed to a "signaling" circuit that performs stuffing, CRC insertion, scrambling, multiplexing and frame pattern insertion. These functions are performed by commands from the HTU controller which adds 8 kbps HDSL overhead functional signaling data into each HDSL frame. The combined signal has a data rate of 784 kbps $\pm \delta$ [the speed deviation δ is 26 bps (32 ppm) if no network timing is applied, and is 4 bps (5 ppm) if network timing is applied]. This signal is fed to a transceiver for transport over a local loop (indicated as loop No. 1 in Figure 3-41) with a data line rate of 784 kbps (392 kbaud using 2B1Q line code). Network timing serves as the HDSL master clock at the HTU-C, while the clock source used at the HTU-R is "loop timing" (described in Chapter 14).

"Loop timing", other than "through timing" (described in Chapter 14), is recommended for HDSL systems. The High-bit rate Digital Subscriber Line operates in a master-slave mode, with the HTU-R slaved to the signal received from the network. That is, the signals transmitted to the network from the HTU-R are synchronized to a clock that is synchronized to the signals from the HTU-C.

A "steady-state" HDSL frame is illustrated in Figure 3-42. A HDSL frame, using 2B1Q line code, has a frame interval of $6 \pm 1/392$ ms (2,351 or 2,353 quats: Qs). Note that 2352 $= 392 \times 6$; $2352 - 1 = 2351$; and $2352 + 1 = 2353$.

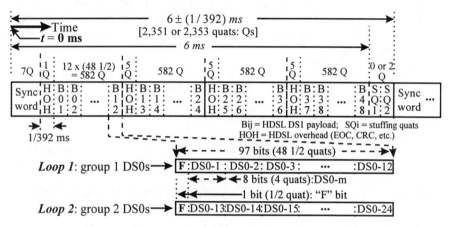

Figure 3-42   Frame Structure of 2B1Q HDSL Systems.

A HDSL frame is headed by a 7-Q (quat) synchronization word, that applies "Barker codes" (double Barker consisting of 14 bits). After this sync word, a 1-Q HDSL overhead word (embedded operations channel, CRC, etc.), is followed by 12 "DS1 byte blocks", each block leading with a single "F" bit. A splitter is used to split this DS1 payload into two groups (streams) of signals. Each group is led by an "F" bit (1/2 quat), followed by 12 DS0 signal blocks. Each DS0 block is an 8-bit block (4 Qs). Group No. 1 signal is carried, received, and transmitted via loop No.1. Likewise, Group No. 2 signals handled by loop No.2.

Note that "F" is duplicated in both groups. The data rate of these two groups of signals is 582 quats $[\equiv 12 \times (1/2 + 12 \times 4Q)]$. Following the first DS0 blocks, three more sets of bitstreams are transmitted. Each set is led by a 5-Q HDSL overhead, and another 24 DS0 blocks as before. By the end of a HDSL frame, 48 DS1 blocks (B01, B02, B03, ..., B12, B13, ..., B24, B25, B26, ..., B36, B37, ..., and B48). By this time if the 1.544 Mbps rate is fast or slow relative to the HDSL clock and the status of the data buffer, two or none stuffing sequences (SQi, i = 0 or 2) are inserted at the end of a HDSL frame.

From the frame structure, a HDSL frame is derived either $6 + 1/392$ ms (2,353 Qs), or $6 - 1/392$ ms (2,353 Qs), depending upon if 2Q or none stuffing sequences are used. The average HDSL frame is 6 ms. The receiver can determine the length of a given incoming frame by detecting the synchronization word in the following frame, and can adjust the demultiplexing procedure of the data stream.

The 7-quat sync word, leading a HDSL frame, serves as framing pattern for implementing HDSL frame synchronization. It is a Barker code sequence as shown in Table 3-5. The same binary values are assigned to these sequences as those used in Figures 3-36 and 3-37, and Table 3-4. The left-most quat in Table 3-5 is transmitted first (e.g., in loop 1, the "+3" quat is transmitted first). This mapping ensures that while peak signal levels are used to transmit the sync word quats, the corresponding bits constitute a "double Barker" pattern derived by repeating each bit of a 7-bit Barker code. This 14-bit pattern, that has excellent auto-correlation properties, may be examined for purposes of finding the HDSL frame synchronization. Note that the synchronization word shall not be scrambled by the HDSL system.

Table 3-5   Synchronization Quats.

Loop No.	Synchronization quats	Quat	1st bit	2nd bit
1	+3 +3 +3 −3 −3 +3 −3	+3	1	1
2	−3 +3 −3 −3 +3 +3 +3	−3	0	0

The same scrambling method used in BRI-ISDN (basic rate interface-ISDN) access DSL, described in ANSI T1.601, shall be used in HDSL systems. The generating (generator) polynomial used by the scrambler is $g(x) = 1 + x^{18} + x^{23}$ for the HTU-R to HTU-C direction, and $g(x) = 1 + x^5 + x^{23}$ for the HTU-C to HTU-R direction.

DS1 facilities (DS1-A and DS1-B in Figure 3-41) of a HDSL system may have two distinct and independent unidirectional signals in the two directions of transmission. They may or may not operate at exactly the same bit rate, depending on the choice of synchronization approach. Neither is there a prescribed phase relationship between received and transmitted signals.

The HDSL system must accommodate frequency differences between the two DS1 directions, different DS1 rates, and the HDSL system clock. This is implemented by using either "none or two" stuffing quats (SQ1 and SQ2; shown in Figure 3-42) at the end of each HDSL frame. The HDSL system should be able to accommodate DS1 signals when the two directions have frequency offsets relative to (1) each other, (2) the network, or (3) the HDSL timing average within a range of ±130 ppm. Note that the values of the stuffing quats are left as a choice made by individual equipment vendors, but they should *not* be scrambled, which is also the case for sync words.

3.6.4.8 HDSL Electrical Characteristics

The major electrical characteristics of a HDSL signal (which are the same as those specified for DS1 signals) are listed as follows:

- The nominal driving-point impedance: The impedance at the interface looking into the HTU-R shall be 135 ± 1% Ω.

- The longitudinal output voltage: Both the HTU-C and HTU-R shall present to the interface a longitudinal component whose voltage (in any 4 kHz bandwidth averaged in any 1 second period) is less than −50 dBV over the frequency range of 100 Hz to 400 kHz, and less than −80 dBV over the frequency range of 400 kHz to 1,000 kHz.

- The longitudinal balance: The "impedance to ground" is specified by the following equation:

$$L_B = 20 \times \log|V_1 + V_m| \quad dB \tag{3-27}$$

where V_1 is the applied longitudinal voltage (referenced to the building or "green wire ground" of the HTU-R), and V_m is the resultant "metallic voltage" appearing across a 135-Ω termination.

3.6.4.9 HDSL System OAM&P

Operations, Administration, Maintenance and Provisioning (OAM&P) processes are used by the HDSL service providers to achieve high quality service and a profitable operation of the HDSL network. The service providers must have appropriate "tools" to install new circuits, evaluate the performance of working circuits, repair defective/failed circuits, and create/maintain necessary operating records.

The overall operations approach used for HDSL networks must consider the existing network environment in which HDSL will be deployed. It should also consider the heritage of the technology from which it has evolved, and new opportunities that appear to be technically feasible. Since HDSL can serve as an alternative to T1 lines for transport of DS1-based services, HDSL must be compatible with the existing DS1 operations process and maintenance tools as much as possible. A full HDSL operations plan, that is cost-effective, should support the following tasks:

- Administration record keeping
- Pre-service provisioning
- Service provisioning
- In-service performance monitoring
- Out-of-service testing and repair
- Service restoration

HDSL system maintenance emphasizes a pro-active approach based on in-service Performance Monitoring (PM) using the CRC error detection/correction capability and Far End Block Error (FEBE) reporting function supported by the HDSL overhead bits. The maintenance objective is for the network (operator) to identify poor circuit performance before the customer does (i.e., prior to a customer complaint), and then remedy the situation. Processed CRC and FEBE information can generate autonomous alert/alarm indications to inform (warn) the network operator about marginal error performance.

Additional information based on DS1 error detection may also be available for use by HDSL system maintenance functions. Maintenance personnel can retrieve performance history data on demand to aid in verifying network problems. The six steps of the maintenance process are: (1) trouble detection, (2) trouble verification, (3) trouble sectionalization, (4) trouble isolation, (5) repair, and (6) repair verification. The following trouble-shooting sequence may be used in the HDSL system maintenance process:

- A customer trouble report, autonomous alarm, or alert message indicates a possible HDSL system problem.

- Maintenance personnel inspect/analyze HDSL system performance data and/or alarm indication signals to determine if the system problem is on the network side or the customer side of the interface.

- If the problem is on the customer side, it is referred to the customer maintenance agent.

- If the trouble is on the network side, it must be determined whether the problem is between the DSX-1 frame and the customer interface, or further back in the network.

- If the problem is between the DSX-1 frame and the customer interface, it may be within the HDSL system: (1) in the wiring between the frame and HDSL, (2) in the HTU-C loop(s), (3) at the HTU-R, or (4) in the wiring between the HTU-R and the customer interface.

- After the problem location is isolated as much as possible, a maintenance center will then dispatch the appropriate personnel to perform on site testing and repair.

- After the problem has been resolved and repair verification has been completed, the service will be restored to the customer.

3.6.4.10 Carrierless AM/PM (CAP) Transmission Technology

Another popular transmission technology (besides 2B1Q) used in HDSL systems is Carrierless AM/PM (CAP) modulation. HDSL systems using CAP transmission technology adopt a dual-duplex architecture (shown in Figure 3-34) with an Echo Canceler and Hybrid (ECH) function (Figure 3-35) in the transceiver sections. A functional block diagram of a CAP transceiver is shown in Figure 3-43.

The bitstream to be transmitted is scrambled and fed to the bit-to-symbol mapper and trellis encoder. This encoder generates two multi-level symbols: In-phase (I) and Quadrature components (Q). These symbols are represented as digital numbers and passed through a precoder. The CAP transceiver has two filters that have the same passband amplitude response, but with phase responses that differ by 90°. The filtered outputs are summed and fed to the hybrid before entering the local twisted-pair loop. The balanced network and the echo canceler perform the same functions, as previously described for Figure 3-35, when the loop takes the signal from the other end of the subscriber line and passes it to the receiver front end.

The received signal, after echo cancellation function is performed, is passed to the receiver which performs the necessary (1) equalization to reduce Inter-Symbol Interference (ISI) so that the acceptable Bit Error Rate (BER) can be achieved, (2) Viterbi decoding to decode the transmitted signal to it original format, and (3) descrambling the signal to the plain text.

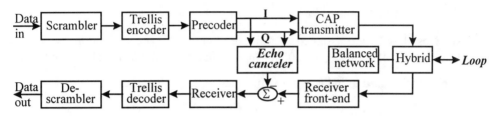

Figure 3-43   Functional Block Diagram of a CAP Transceiver.

The frame structure (Figure 3-44) of a HDSL signal using CAP is very similar to the frame structure of a HDSL using 2B1Q line code (Figure 3-42). The difference is that the CAP frame is in binary form instead of quaternary form.

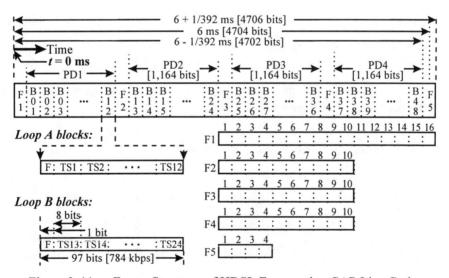

Figure 3-44   Frame Structure of HDSL Frame using CAP Line Code.

The CAP frame has an interval of 6 ms that is divided into four Payload Data blocks (PD1 ~ PD4) and five overhead framing blocks (F1 ~ F5). The 6 *ms* frame contains 4,704 bits, which corresponds to a nominal line bit rate of 784 kbps in each loop (loop A and loop B). Note that both loops A and B have the same frame structure but with different identifiers. Within each of the forty-eight payload block (B01 through B48) there are twelve sub-blocks, each containing 96 bits and an "F" bit (i.e., a total of 97 bits

= 1 + 12 × 8). As shown in Figure 3-42,each loop (A or B) contains 12 DS0s [i.e., (DS0-1 to DS0-12) for loop A, and (DS0-13 to DS0-24) for loop B]; and, they are carried by 24 different Time Slots (i.e., TS1, TS2, …, TS24).

The five HDSL overhead framing blocks (F1 to F5) are assigned as follows:

- F1 block - The first framing block contains 16 bits, that are assigned as follows:

    * Bit Nos. 1 ~ 14 are assigned as a frame alignment (synchronization) word (see Figure 3-45: "11111100001100" for loop A, and "00110000111111" for loop B).

    * Bit No. 15 is used to indicate a loss of input signal at the far end unit.

    * Bit No. 16 is used as the Far End Block Error (FEBE) indication.

Bit No.	1	2	3	4	5	6	7	8	9	10	11	12	13	14
Loop A	1	1	1	1	1	1	1	0	0	0	0	1	1	0	0
Loop B	0	0	1	1	0	0	0	0	1	1	1	1	1	1

Figure 3-45   Frame Alignment Word (FAW) for Loops A & B.

- F2 block - The second framing block contains 10 bits, that are assigned as follows:

    * Bit Nos. 1 ~ 4 are the first four bits of a 13-bit Embedded Operations Channel (EOC) [(EOC1through EOC4); see F3, and F4 for (EOC5 to EOC13)].

    * Bit Nos. 5 and 6 are the first two bits the six Cyclic Redundancy Check (CRC-6) bits (see F3 and F4 for the other CRC bits).

    * Bit Nos. 7 and 8 are assigned as the Power Status bits (PS1 and PS2) to report HTU-C power status (for the customer premises to central office direction only).

    * Bit No. 9 is used as Bipolar Violation Indication (BVI).

    * Bit No. 10 is the fifth bit of 13-bit EOC field (i.e., EOC5).

- F3 block - The third framing block contains 10 bits, that are assigned as follows:

    * Bit Nos. 1 ~ 4 are used as the 6th~9th bits of the 13-bit EOC field (EOC6 ~ EOC9).

    * Bit Nos. 5 and 6 are the 3rd and 4th bits of the CRC-6 field.

    * Bit Nos. 7 ~ 10 are regenerator associated bits. They are used to indicate "regenerator present", "regenerator remote block error indication", "regenerator central block error indication", and "regenerator alarm", respectively.

- F4 block - The fourth framing block contains 10 bits, that are assigned as follows:

    * Bit Nos. 1 ~ 4 are used as the last four EOC bits, that is, the 10th through the 13th bits of the 13-bit Embedded Operations Channel (EOC10 to EOC13).

* Bit Nos. 5 and 6 are used as the last two CRC bits of the CRC-6 field.

* Bit No. 7 is used as remote terminal alarm (for the customer premises to central office direction only).

* Bit No. 8 is used to indicate "ready to receive" (for both directions).

* Bit Nos. 9 and 10 are unspecified bits (unassigned).

* F5 block - The fifth framing block contains 4 bits, that are assigned as follows:

  * Bit Nos. 1 and 2 are "delete bits". A HDSL frame has a nominal length of 4,706 bits with a deviation of 1/392 ms (i.e., ±2 bits). Thus, a frame that has 4,708 bits (i.e., 4,706 + 2), uses bit Nos. 1 and 2 as "delete bits".

  * Bit Nos. 3 and 4 are "stuff bits". A HDSL frame has a nominal length of 4,706 bits with a deviation of 1/392 ms (i.e., ±2 bits). Thus, a frame that has only 4,704 bits (i.e., 4,706 − 2), uses bit Nos. 1 and 2 as "stuff bits".

3.6.4.11 Discrete MultiTone (DMT) Transmission Technology

In addition to the 2B1Q and CAP transmission techniques used for broadband access via copper twisted-pair local loops, another technology called "multicarrier" or "Discrete Multi-Tone (DMT)" transmission technology has been developed.

Even in the presence of crosstalk and Inter-Symbol Interference (ISI), the theoretical DMT range and performance is equal to or better than other modulation methods (line codes). DMT technology also offers strong impulse-noise rejection capability, and there are many ways of generating multicarrier line codes (signals).

Figure 3-46 illustrates the building blocks of a dual-duplex DMT HDSL transmitter. A 1.568 Mbps data stream and 32 kbps multicarrier HDSL control data stream (a total of 1.6 Mbps) are buffered in 500-µs blocks (each block has a length of 800 bits). The buffered blocks are fed to an encoder that converts the bits in these 500-µs blocks into (a maximum of) 512 complex amplitudes [i.e., ($X_{1,k}$, $X_{2,k}$, ..., $X_{256,k}$ and $\widetilde{X}_{1,k}$, $\widetilde{X}_{2,k}$, ..., $\widetilde{X}_{256,k}$) for dual-duplex operation, or ($X_{1,k}$, $X_{2,k}$, ..., $X_{256,k}$) for single-duplex operation]. Each of these complex signal is a two-dimensional signal with the specific magnitude and phase.

By using a 512-point Inverse Fast Fourier Transform (IFFT) function, the complex magnitudes and phases are combined into one (in the single-duplex case) or two (in the dual-duplex case) composite multicarrier signals. The resultant real (time-domain) sequences of IFFT output samples [$x_{1,k}$, ..., $x_{512,k}$ and $\widetilde{x}_{1,k}$, ..., $\widetilde{x}_{512,k}$] are converted to serial form using a Parallel-to-Serial (P/S) converter. In addition, a short (20-sample interval) cyclic prefix is appended to improve performance and assist system synchronization.

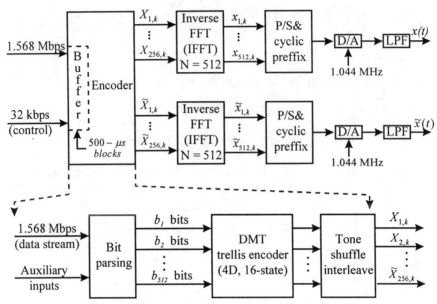

Figure 3-46   Dual-duplex Discrete Multitone (DMT) HDSL Transmitter.

Prior to passing through the hybrid line interfaces (see Figure 3-35 or 3-43), these sequences of samples are applied to Digital-to-Analog converters (D/As) operating at a clock rate of 1.044 MHz, and a Low Pass Filter (LPF). Note that the bottom branch (IFFT, P/S, cyclic prefix, D/A, and LPF) in Figure 3-46 is <u>not</u> required if the HDSL system is configured in the single-duplex mode.

The bottom half of Figure 3-46 illustrates the major functional blocks of the encoder. The input bit stream, composed of the data stream and auxiliary inputs, are parsed into (b_1, b_2, b_3, ..., b_{512} bits). These bits are applied to a four-dimension, 16-state trellis encoder (4D-16), that is designed for multicarrier applications. This code has an almost constant gain of 4.2 dB across all the sub-channels. The encoded amplitudes are shuffled in a manner that is determined on a per-line (per-pair of lines) basis that insures maximum protection against impulse noise.

For maximum range and performance of HDSL systems, the transmitter and receiver should both be optimized on a per-line basis. Multicarrier receivers perform this optimization by allocating a number of bits (depending upon the channel characteristic for the specific line in use at a particular frequency) to each sub-channel (sub-carrier). When multicarrier receivers perform this function exactly, and appropriate Trellis codes are used, the system's performance is optimized. Theoretically, no other modulation method is better. With the advent of digital signal processing and the use of the Fast Fourier Transform (FFT) algorithm, it is possible to approximate the optimum bandwidth in a cost effective manner. Likewise, small deviations from the "optimum bandwidth" reduce system performance only slightly in DMT systems.

Review Questions I for Chapter 3:

(1) Name five wireline solutions for broadband access (instead of the traditional analog loops): _____, _____, _____, _____, and _____.

(2) For modem outside loop plant, the "serving area concept" is widely adopted. The outside plant is called _____ network, which is connected to "feeder" network via _____ (SAI).

(3) Prior to xDSL, the commonly used broadband access system (especially for metro applications) is called IDLC. An IDLC system is used for several reasons; three primary ones are: _____, _____, and _____.

(4) (True, False) In North American, prior to SONET, the digital facilities commonly used in IDLC systems are T1 digital carrier systems.

(5) In modern SONET/SDH networks, Business Remote Terminal (BRT) interfaces with SONET's _____ or _____ signals; or interfaces with SDH's _____ or _____.

(6) Trellis Coded Modulation (TCM) is a technology that uses the combination of _____ and _____ to improve the noise performance of QAM.

(7) Copper local loop frequency spectra presently are classified into the following categories: (200~3200Hz) for POTS, (200Hz~120 kHz) for _____, (20 kHz~120 kHz) for _____ uplink, (140 kHz~552 kHz) for 1.5 Mbps ADSL _____, (140 kHz~1104 kHz) for 8 Mbps ADSL _____, and (300 kHz~30 MHz) for _____.

(8) From xDSL application viewpoint, besides bandwidth limitation local loops have the following primary impairments: _____, _____/_____, _____, _____, _____, _____ etc.

(9) Most xDSL technologies require "splitter" at the customer equipment and at the central office. However, ITU-T Rec. G992.2 defines _____ ADSL specification commonly known as _____.

(10) (True, False) HDSL systems may eventually be able to operate on 99% of the copper twisted-pairs that conform to the Carrier Serving Area (CSA) guidelines.

(11) The transceiver used in HTU-C and HTU-R (HDSL Terminal Unit at Central office, and at Remote site) uses _____ transmission scheme.

(12) Three technologies (line codes) can be used for HDSL transmission: _____, _____ and _____.

(13) The ADSL Forum has proposed two procedures for loop qualification and testing: _____ and _____.

3.6.5 Asymmetric Digital Subscriber Line (ADSL)

This section describes the electrical characteristics of Asymmetrical Digital Subscriber Line (ADSL) signals. The transport medium for the ADSL signals is a single twisted-pair copper wire that supports Plain Old Telephone Service (POTS), voiceband (voice-grade) data services, and high-speed duplex/simplex digital services. Mixed gauges of wiring or bridged taps are acceptable in most situations. However, the use of loading coils is not acceptable for ADSL applications.

"In the direction from the network to the customer premises, the digital bearer channels may consist of full-duplex low-speed signals and simplex high-speed signals. In the direction from the customer premises to the network, the digital bearer channels can only carry low-speed signals."

3.6.5.1 ADSL Network Architecture

The generic network architecture for a Digital Subscriber Line (DSL) shown in Figure 3-17 can be easily modified to implement an ADSL system (Figure 3-47). There is a "splitter" on each end of the loop, and the interfaces are U-C (Central office end) and U-R (Remote terminal end). The two major components of a splitter are a Low-Pass Filter (LPF) for handling the low-speed data stream, and a High-Pass Filter (HPF) for handling the high-speed data stream.

At the remote site, the customer can receive Plain Old Telephone Services (POTS) and voiceband (voicegrade) data services from the low-speed port of the splitter. In addition, a Network Terminal (NT) can be connected to the high-speed port of the splitter. Required interface are U-R (the interface between the ATU-R and the loop) and U-R2 (the interface between the Remote Terminal and the ATU-R) as shown in Figure 3-47. If the high-pass filter (Usually part of the splitter) is integrated into the ATU-R, then the U-R2 interface is the same as the U-R interface. In this example, the NT is connected to a customer premises network that provides services via Service Modules (SMs). The interface required is T/S [the interface between the NT and Customer Installation (CI) or home network] as shown in Figure 3-47.

At the central office end, the low-speed port of the splitter and LPF can be connected to a narrowband network (e.g., PSTN) to provide traditional voice or voiceband services. The high-speed port (and HPF) is connected to a broadband network that can be SONET or SDH. The current speed of these broadband networks is 622.08 Mbps (OC-12 or STM-4, see Chapter 6) for local exchange networks or 2.48832 Gbps (OC-48 or STM-16, also see Chapter 6) for toll/long haul networks. If WDM is used, a speed of 200 or 400 Gbps is feasible. The required interfaces are U-C (the interface between the ATU-C and the loop), U-C2, and VC [the logical interface between the ATU-C and a digital network element (e.g., switching system)]. If the high-pass filter (usually part of the splitter) is integrated into ATU-C, then the U-C2 interface is the same as the U-C interface. The VC interface may consist of interface(s) to one or more [Synchronous Transfer Mode (STM) or Asynchronous Transfer Mode (ATM)] switching elements. A digital carrier facility (e.g., SONET extension) may be interposed at the V-C interface.

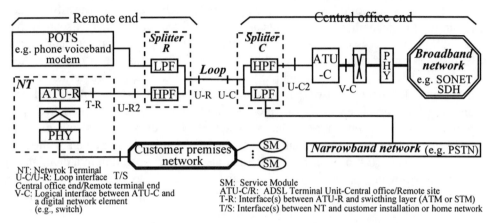

Figure 3-47   Network Architecture of An ADSL Network.

STM and ATM are two possible facility transport options. Thus, the ATU-C and ATU-R interface may be configured for either STM bit sync transport or ATM cell transport. Hybrid configurations (e.g., applications carrying STM over ATM) are not presently supported, but may be available in the future.

- If the U-C interface is STM bit sync based (i.e., no ATM cells on the U-C interface), the ATU-C is configured for STM transport.

- If the U-C interface is ATM cell based (i.e., only ATM cells on the U-C interface), the ATU-C is configured for ATM transport.

- If the U-R interface is STM bit sync based (i.e., no ATM cells on the U-R interface), the ATU-R is configured for STM transport.

- If the U-R interface is ATM cell based (i.e., only ATM cells on the U-R interface), the ATU-R is configured for ATM transport.

3.6.5.2 ATU-C/ATU-R Transmitter Reference Models

Figure 3-48 shows the "***downstream***" functional block diagram of an ADSL Transceiver Unit-Central office (ATU-C) transmitter for STM or ATM transport. Note the two dotted blocks marked "Cell TC" are only used for ATM transport configurations.

- Case 1 (downstream STM is supported): The basic STM transport mode is "bit serial" with no byte frames defined externally. Preservation of V-C interface byte boundaries in the ADSL data frame is optional. Outside ASx/LSx serial interfaces data bytes are transmitted with the Most Significant Bit (MSB) first, in accordance with ITU-T Recs. G.703 and G.707. All serial processing in the ADSL frame (e.g., CRC, scrambling, etc.) shall be performed with the Least Significant Bit (LSB) first (i.e., with the "outside world" MSB considered by the ADSL as being the LSB). As a result, the first incoming bit (i.e., "outside world" MSB from the external source) shall be the first bit processed as the ADLS (ADSL-LSB). ADSL equipment shall support both simplex and duplex bearer channels (AS0 and LS0). Other bearer channels, AS1, AS2, ..., LS1, LS2, ... are optional [ASx and LSx (x = 0, 1, 2, ...) are defined later in this section]. There are two paths connected to the mux/sync control and tone ordering function. The "fast" path provides low latency, and the "interleave" path provides a very low error rate but has a greater latency. An ADSL system supporting STM transport operates in a "single latency mode". That is, all user data are allocated to only one path (fast or interleaved). It shall also support operation in a dual latency mode downstream, in which user data can be allocated to both paths. However, support of operation in dual latency mode upstream is optional for ADSL equipment.

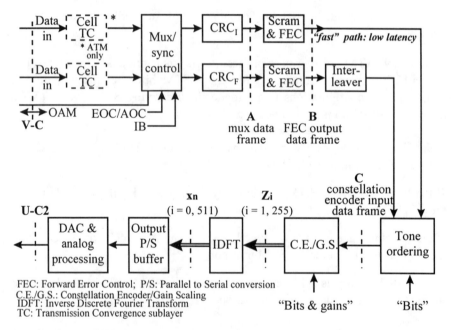

Figure 3-48   ATU-C Transmitter Reference Model for STM/ATM Transport.

- Case 2 (downstream ATM is supported): Byte boundaries at the V-C interface must be preserved in the ADSL data frame. Outside ASx/LSx serial interfaces data bytes are transmitted with MSB first, in accordance with ITU-T Recs. I.361 and I.432. All serial processing in the ADSL frame (e.g., CRC, scrambling, etc.) shall be performed with LSB first (i.e., with the "outside world" MSB considered by the ADSL as being the LSB). As a result, the first incoming bit (i.e., outside world MSB from an external source) shall be the first bit processed as the ADLS (ADSL-LSB), and the Cell Loss Priority (CLP) bit of the ATM cell header will be carried in the MSB of the ADSL frame byte (i.e., processed last). ADSL equipment shall support both the simplex and duplex bearer channel AS0. Other bearer channels, AS1, AS2, ... are optional. There are two paths connected to the mux/sync control and tone ordering function. The "fast" path provides low latency, and the "interleave" path provides very low error rate but has greater latency. An ADSL system supporting ATM transport shall operates in a single latency mode. That is, all user data are allocated to only one path (fast or interleaved). Support of operation in a dual latency mode, in which user data can be allocated to both paths, is optional. The ATM cell Transmission Convergence (TC) sub-layer provides an interface towards the ATM layer (V-C interface), and an interface towards the ADSL mux/sync control interface.

Figure 3-49 shows the "*upstream*" functional block diagram of an ADSL Transceiver Unit-Remote site (ATU-R) transmitter for STM or ATM transport (The two dotted blocks marked "Cell TC" are only used for ATM transport).

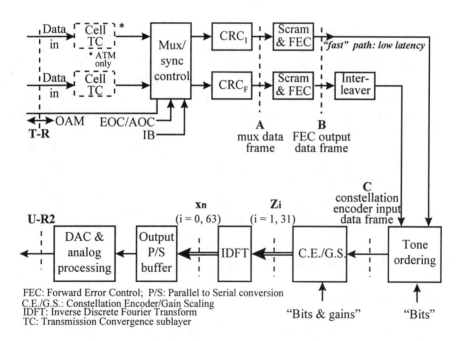

Figure 3-49   ATU-R Transmitter Reference Model for STM/ATM Transport.

- Case 3 (upstream STM is supported): The basic STM transport mode is "bit serial" with no byte frames defined externally. Preservation of V-C interface byte boundaries in the ADSL data frame is optional. Outside LSx serial interfaces data bytes are transmitted MSB first in accordance with ITU-T Recs. G.703 and G.707. All serial processing in the ADSL frame (e.g., CRC, scrambling, etc.) shall be LSB first (i.e., with the outside world MSB considered by the ADSL as being the LSB). As a result, the first incoming bit (i.e., outside world MSB from an external source) shall be the first bit processed as the ADLS (ADSL-LSB). ADSL equipment shall support the duplex bearer channel LS0 upstream. Other bearer channels LS1, LS2, ... are optional. There are two paths connected to the mux/sync control and tone ordering function. The "fast" path provides low latency, and the "interleave" path provides a very low error rate but has greater latency. An ADSL system supporting STM transport operates in a single latency mode, in which all user data are allocated to only one path (fast or interleaved). It shall also support operation in a dual latency mode downstream, in which user data can be allocated to both paths. However, support of operation in dual latency mode upstream is optional for ADSL equipment.

- Case 4 (upstream ATM is supported): Byte boundaries at the T-R interface must be preserved in the ADSL data frame. Outside LSx serial interface data bytes are transmitted MSB first in accordance with ITU-T Recs. I.361 and I.432. All serial processing in the ADSL frame (e.g., CRC, scrambling, etc.) shall be performed LSB first (i.e., with the outside world MSB considered by the ADSL as being the LSB). As a result, the first incoming bit (i.e., outside world MSB from an external source) is the first bit processed as the ADLS (ADSL-LSB), and the Cell Loss Priority (CLP) bit of the ATM cell header is carried in the MSB of the ADSL frame byte (i.e., processed last). ADSL equipment supports both simplex and duplex bearer channel AS0. Other bearer channels, AS1, AS2, ... are optional. There are two paths connected to the mux/sync control and tone ordering function. The "fast" path provides low latency, and the "interleave" path provides very low error rate but with a greater latency. An ADSL system supporting ATM transport in a single latency mode, in which all user data are allocated to only one path (fast or interleaved). Support of operation in a dual latency mode, in which user data can be allocated to both paths, is optional. The ATM cell Transmission Convergence (TC) sub-layer provides an interface towards the ATM layer (V-C interface) and an interface towards the ADSL mux/sync control interface.

3.6.5.3 ADSL Transport Capacity

An ADSL system can transport up to seven bearer channels simultaneously, including:

(1) Up to four independent "downstream" simplex bearer channels (AS0, AS1, AS2, and AS3) [unidirectional from the network to the Customer Installation (CI)].

(2) Up to three duplex channels (LS0, LS1 and LS2) (bidirectional between the network and the CI). As an alternative, these three channels can be configured as independent unidirectional simplex bearer channels. Note that the rates of the bearer channels in the two directions (the network toward CI and vice versa) do not need to be identified.

All bearer channel data rates shall be programmable in any combination of multiples of 32 kbps. The ADSL data multiplexing format is flexible enough to allow other transport data rates (e.g., channelizations based on the existing 1.544 Mbps data stream) but the support of these data rates (non-integer multiples of 32 kbps) are limited by the ADSL system's available capacity for synchronization. The ADSL system overhead and data rate synchronization provides sufficient capacity to support the framed DS1 data streams transparently. That is, the entire DS1 signal passes through the ADSL transmission path without interpretation or removal of the framing bits and/or other overhead bits. Part of the ADSL system overhead is shared among the bearer channels for synchronization. The remainder of each channel's data rate that exceeds a multiple of 32 kbps is transported using this shared overhead scheme. It should be noted that only framing structure 0 (other framing structures are described later in this chapter) can support non-integer multiples of 32 kbps.

The maximum net data rate transport capacity of an ADSL system depends upon the conditions/characteristics of the local loop, and certain configurable options that affect overhead. The ADSL bearer channel rate shall be configured during the "initialization and training procedure" to match the user data rate. One part of the ADSL initialization and training sequence estimates the loop characteristics. This is done to determine whether the number of bytes per data frame required to support the requested configuration's total data rate can be transmitted over the particular local metallic loop. The net data rate is computed as the total data rate minus the ADSL system overhead.

The transport capacity of an ADSL system is defined as only the high-speed bearer channels. However, when an ADSL system is installed on a line that also carries POTS signals, the overall capacity is that of POTS plus ADSL. If an ATU supports a particular bearer channel, it shall support it through both the "fast" and "interleaved" paths. The latency mode of an ADSL system may be different for "downstream" and "upstream" transmissions. A distinction must be made between the transport of STM and ATM signals, however bearer channels configured to transport STM signal can also be configured to transport ATM signal.

- Transport of STM signals: ADSL systems transporting STM data can be summarized as shown in Table 3-6. Support for data rates that are non-integer multiples of 32 kbps is optional. Besides transporting user data, an ADSL may optionally support a Network Timing Reference (NTR). The ATU-R may reconstruct the NTR, but this operation is independent of any timing that is internal to the ADSL system.

- Transport of ATM signals: An ADSL system can support ATM signal transport, using the single latency mode, at all multiples of 32 kbps up to 6.144 Mbps downstream and up to 640 kbps upstream. ATM data must be mapped into bearer channel AS0 in the downstream direction, and into bearer channel LS0 in the upstream direction using the single latency mode. However, the need for dual latency ATM operation depends on the service/application profile (this specification will be standardized in the near future). In summary, there are three different "latency classes":

(1) Single latency (not necessarily the same for each direction of transmission)

(2) Dual latency downstream and single latency upstream

(3) Dual latency for both downstream and upstream

Table 3-6   Required 32 kbps Multiple for STM Data Transport.

Bearer channels	Lowest required multiple	Largest required multiple	Highest required data rate (kbps)
AS0	1	$n_0 = 192$	6,144
AS1	1	$n_1 = 144$	4,608
AS2	1	$n_2 = 96$	3,072
AS3	1	$n_3 = 48$	1,536
LS0	1	$m_0 = 20$	640
LS1	1	$m_1 = 20$	640
LS2	1	$m_2 = 20$	640

- In addition to transporting ATM data, an ATU-C can support the transport of a NTR. The ATU-R may reconstruct the NTR, but this operation is independent of any timing that is internal to the ADSL system.

3.6.5.4 ATU-C Functional Characteristics

The ATU-C functional characteristics described in this section are: STM specific functions, ATM specific functions, network timing reference, framing, scrambling, Forward Error Correction (FEC), tone ordering, constellation encoding, gain scaling, modulation, and transmitter spectrum, etc.

(1) **STM specific functionalities**: The functional interfaces at the ATU-C for STM transport are shown in Figure 3-50. Input interfaces for the high-speed downstream simplex bearer channels are designated as AS0 through AS3 (see Table 3-6), and input/output interfaces for the duplex bearer channels are designated as LS0 through LS2. There is also a duplex interface for Operations, Administration, Maintenance (OAM) and control of the ADSL system, and a Network Timing Reference (NTR) provided by ADSL STM transport.

Note that LS0 is also known as the "C" or control channel. It carries the signaling associated with the ASx bearer channels, and may also carry some or all of the signaling associated with the other duplex bearer channels.

The one-way transfer delay for payload bits in all bearer channels (simplex or duplex) from the V reference point at central office end (i.e., V-C) to the "T reference point" at emote end (i.e., T-R) for bearer channels assigned to the "fast" buffer shall be no more than 2 ms. Similarly, bearer channels assigned to the "interleaved" buffer shall have a delay of no more than $[4 + (S - 1)/4 + S \times D/4]$ ms, where S is the mux data frames (S = 1, 2, 4, 8 or 16) and D is the interleave depth (D = 1, 2, 4, 8, 16, 32 or 64). Note that the same requirement also applies for the opposite direction.

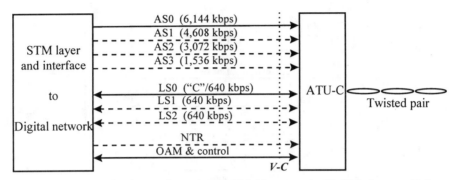

Figure 3-50   ATU-C Interfaces to the STM layer at the V-C Reference Point.

(2)   **ATM specific functionalities**: The functional interfaces at the ATU-C for ATM transport are shown in Figure 3-51. The ATM channel "ATM0" is always provided by ADSL systems, and the ATM channel "ATM1" is optional (i.e., may be provided for support of a dual latency mode).

Flow control functionality (i.e., Tx_cell_handshake and Rx_cell_handshake) shall be available on the V-C reference point to allow the ATU-C to control the cell flow to/from the ATM layer. Each ATM cell may be transferred from the ATM to the physical layer only after the ATU-C has completed the Tx_cell_handshake. Similarly, a cell may be transferred from the physical layer to the ATM layer only after the ATU-C has completed the Rx_cell_handshake. This functionality is important so that ATM cell overflow or underflow in the ATU-C and ATM layer can be avoided. There is also a duplex interface for OAM and control, and a network timing reference provided by ADSL ATM transport.

The one-way transfer delay, excluding cell specific functionalities, for payload bits in all bearer channels (simplex or duplex) from the "V reference point" at central office end (i.e., V-C) to the "T reference point" at remote end (i.e., T-R) for bearer channels assigned to the "fast" buffer shall be no more than 2 ms. Similarly, for bearer channels assigned to the "interleaved" buffer shall have a delay of no more than $[4 + (S - 1)/4 + S \times D/4]$ ms, where S is the mux data frames (S = 1, 2, 4, 8 or 16) and D is the interleave depth (D = 1, 2, 4, 8, 16, 32 or 64). Note that the same requirement also applies in the opposite direction. Several "ATM cell specific" functions (ITU-T Rec. I.432) are listed as follows:

*   Idle (unassigned) cell insertion

* Header error control generation
* Cell payload scrambling
* Bit timing and ordering
* Cell delineation
* Header error control verification

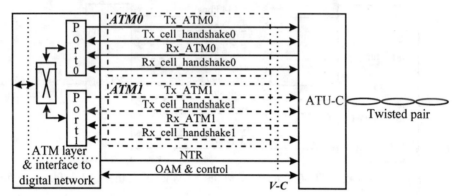

Figure 3-51   ATU-C Interfaces to the ATM layer at the V-C Reference Point.

(3)   **Network Timing Reference (NTR)**: Some services [e.g., Voice and Telephone over ATM (VTOA), or Desktop Video Conferencing (DVC)] require a reference clock be available for higher layers of the protocol stack (i.e., OSI layer 2, 3, 4, 5, 6 or 7). This is used to guarantee " end-to-end synchronization" of the transmit and receive sides. To support distribution of a timing reference over the ADSL network, the ADSL system transports an 8 kHz timing marker as Network Timing Reference (NTR). This 8 kHz timing marker may be used for voice/video playback at the decoder in VTOA and DVC applications. The ATU-C may have access to a source NTR, if it is provided by the network at the V-C reference point. The NTR signal may be derived from a Primary Reference Clock (PRC), or Primary Reference Source (PRS) in the ATM broadband network (e.g., SONET or SDH network).

Table 3-7   Four Framing Structures (ATU-C Transmitter).

Framing structure	Definition
0	***Full overhead*** framing with asynchronous bit-to-modem timing (enabled synchronization control mechanism)
1	***Full overhead*** framing with synchronous bit-to-modem timing (disabled synchronization control mechanism)
2	***Reduced overhead*** framing with separate fast and sync bytes in fast and interleaved latency buffers, respectively (64 kbps framing overhead)
3	***Reduced overhead*** framing with merged fast and sync byte, using either the fast and interleaved latency buffer (32 kbps framing overhead)

(4) **Framing**: The framing specification of the downstream signal (ATU-C transmitter) is given in this section, and the upstream framing (ATU-R transmitter) is described separately in Section 3.6.5.5. Two types of framing, and two versions of each type are defined, which results in four framing structures as shown in Table 3-7. If the ATU-C indicates it supports framing structure k, it shall also support all framing structures designated as "$(k-1)$" (k = 1 to 4).

Figure 3-48 illustrates functional block diagram of the ATU-C transmitter showing reference points (e.g., A, B and C) for data framing. Up to four downstream simplex bearer channels, and up to three duplex bearer are synchronized to the 4 kHz ADSL data frame rate, and multiplexed into two separate (fast and interleaved) data buffers as shown in Figure 3-52. A CRC, scrambling, and FEC coding are applied separately to the contents of each data buffer. In addition, the data from the interleaved buffer is passed through an interleaving function. Data from the two data buffers is then "tone ordered" (discussed later in this chapter), and combined into a data frame that is fed to the constellation encoder. Following the constellation encoder, a modulator is used to modulate the data frame that forms a DMT data symbol to produce an analog signal for transmission over the customer loop. Because of the addition of FEC redundancy bytes and data interleaving, the data frames have a different structural appearance at the three reference points in the transmitter:

* Reference point "A" (Figure 3-48: mux data frame): This is the multiplexed, synchronized data after CRC has been inserted. Mux data frames are generated at an average rate of 4 kHz.

* Reference point "B" (FEC output data frame): This is the output of the FEC encoder at the DMT symbol rate, where an FEC block may span more than one DMT symbol period.

* Reference point "C" (constellation encoder input data frame): This is the data frame input to the constellation encoder.

ADSL adopts the superframe structure shown in Figure 3-52. An ADSL superframe consists of 68 ADSL data frames (numbered from 0 to 67, which are encoded and modulated into 68 DMT data symbols), followed by a DMT synchronization symbol (sync frame, which carries no user or overhead bit-level data) that is inserted by the modulator to establish the superframe boundaries. From the bit-level user data perspective, the DMT symbol rate is 4,000 symbols/s (period = 250 μs), but to allow for insertion of the synchronization symbol, the transmitted DMT symbol rate is actually 69/68 × 4,000 symbols per second.

Each data frame [e.g., Frame No. 2, with a length of (68/69) × 250 μs as shown in Figure 3-52] within a superframe consists of data from the "fast" buffer and the "interleaved" buffer. The size of each buffer depends on the assignment of bearer channels which is established during initialization. The fast data buffer has a size of N_F bytes, used as the FEC output or constellation encoder input data frame. Within these N_F bytes, the K_F bytes (one byte is assigned as the "fast" byte) are used as the multiplexed data frame, and the R_F bytes used as FEC bytes. The "interleaved" data buffer contains N_I bytes used as constellation encoder input data frame.

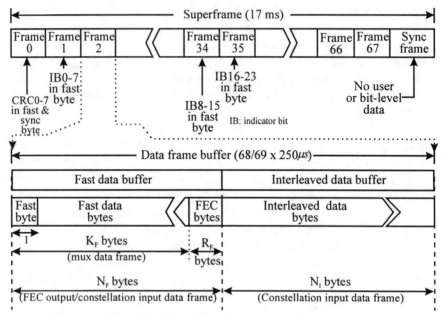

Figure 3-52 ADSL Superframe Structure (ATU-C Transmitter).

Eight bits per ADSL superframe are used for the CRC function on the "fast" data buffer (i.e., CRC0 ~ CRC7), and 24 indicator bits (i.e., IB0 ~ IB23) are assigned for OAM functions. As illustrated in Figures 3-52 and 3-53(A), the synchronization byte of the fast data buffer carries the CRC check bits in frame No. 0, and the fixed overhead bit assignment in frame Nos. 1, 34, and 35. The fast byte is assigned in "even-frame/odd-frame" pairs to either the EOC or to the Synchronization Control (SC) for the bearer channels assigned to the fast buffer.

As illustrated in Figure 3-53(B), for the "fast" data buffer (Figure 3-52) if the frames carry Synchronization Control (SC) information, bit 0 (i.e., the LSB) of an even-numbered frame (other than frames 0 and 34) and the odd numbered frame immediately following shall both be set to logical "0". The remaining 7 bits shall carry Synchronization Control bits (i.e., SC7 through SC1). If the frames carry an Embedded Operations Channel (EOC) message, bit 0 of an even-numbered frame (other than frames 0 and 34) and the odd-numbered frame immediately following shall both be set to logical "1". The first six bits (including the MSB) of the even-numbered frame, and the remaining 7 bits of the odd-numbered frame shall carry

EOC messages information. Note that bit 1 of the even-numbered frame is reserved [i.e., presently unused; designated as r_1 in Figure 3-53(A)].

(A) *"Fast"* sync byte

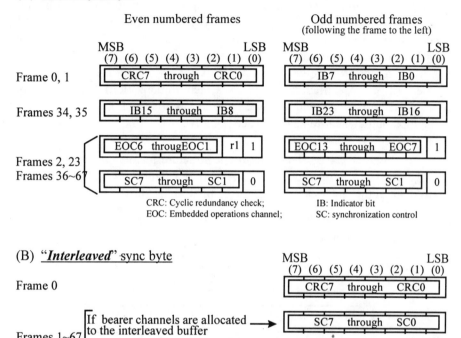

Figure 3-53   Sync Byte ("Fast" & "Interleaved") (ATU-C Transmitter).

Eight bits per ADSL superframe shall be used for the CRC on the "interleaved" data buffer (Figure 3-52) as CRC0 through CRC7 [Figure 3-53(B)] to perform the CRC check for the previous ADSL superframe in frame No. 1. In all other frames (1 through 67), the sync byte shall be used for synchronization control of the bearer channels assigned to the interleaved data buffer, or used to carry an ADSL Overhead Control (AOC) channel. In the "full" overhead mode (i.e., when any bearer channel appears in the interleaved buffer) the AOC data shall be carried in the LEX byte [Figure 3-54(B); described later in this section] and the sync byte shall designate when the LEX byte contains AOC data and when it contains a data byte from the bearer channels. When no bearer channels are allocated to the interleaved data buffer, the sync byte shall carry the AOC data directly.

Figure 3-54(A) shows the frame structure of the "fast" data buffer (Figure 3-52) for reference points A, B and C (Figure 3-48). At reference point A, the fast buffer shall always contain at least the "fast" byte, which is followed by $B_F(AS0)$ bytes of bearer channel AS0, B_F(AS1) bytes of bearer channel AS1, B_F(AS2) bytes of bearer channel AS2, and B_F(AS3) bytes of bearer channel AS3. Next come the bytes for any duplex (LSx; where x = 0~2) bearer channels allocated to the fast buffer. If any B_F(ASx) is non-zero, then both an AEX and an LEX byte follow the bytes of the last LSx bearer channel. Similarly, if any B_F(LSx) is non-zero, the LEX byte shall be included. AEX is the AS EXtension byte inserted in the transmitted ADSL frame structure to provide a synchronization function that is shared among ASx channels. LEX is the LS EXtension byte inserted in the transmitted ADSL frame structure to provide a synchronization function that is shared among LSx and ASx channels. The R_F (R_F is specified during ADSL system initialization) "Reed-Solomon FEC" redundancy bytes are added to the mux data frame (reference point A) to produce the FEC output data frame (reference point B).

Figure 3-54(B) shows the frame structure for the "interleaved" data buffer (Figure 3-52) for reference points A and B (Figure 3-48). Note that the frame structure of the "interleaved" buffer is similar to the "fast" data buffer.

(A) ***Fast** data buffer*

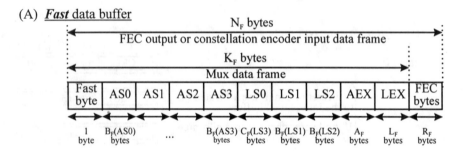

(B) ***Interleaved** data buffer*

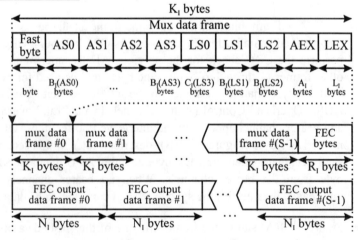

Figure 3-54   Fast and Interleaved Data Buffers (ATU-C Transmitter)

(5) **Scrambling**: The binary data stream outputs (in the order of LSB of each byte first) from the "fast" and "interleaved" buffers shall be scrambled using the following generator (generating) polynomial:

$$g(X) = 1 + X^{18} + X^{23}$$ (3-28)

The scrambler is applied to the serial data stream without reference to any framing or symbol synchronization.

(6) **Forward Error Correction (FEC)**: The ATU-C supports downstream transmission with any combination of the Reed-Solomon (RS) FEC coding (see Chapter 13) capabilities shown in Table 3-8. The R (R_F or R_I) redundancy check bytes c_0, c_1, c_2, ..., c_{R-2}, c_{R-1}, are appended to K (= K_F or S × K_F) message bytes m_0, m_1, m_2, ..., m_{K-2}, m_{K-1}, to form a Reed-Solomon codeword of size N = (K + R) bytes. The check bytes are computed from the message byte using Eq.(3-29):

$$r(X) = m(X) \, X^R \text{ modulo } g(X)$$ (3-29)

where   m(X) = message polynomial = $m_0 X^{K-1} + m_1 X^{K-2} + ... + m_1 X + m_{K-1}$

r(X) = check polynomial = $c_0 X^{R-1} + c_1 X^{R-2} + ... + c_1 X + c_{R-1}$

g(X) = the generator polynomial of the Reed-Solomon code

$$g(X) = \prod_{i=0}^{R-1} (X + \alpha^i)$$ (3-30)

That is, $$r(X) = \text{Re} mainder \text{ of } \frac{m(X)X^R}{g(X)}$$ (3-31)

The arithmetic [for Eqs.(3-29), (3-30), and (3-31)] is performed in the Galois field GF(256), where α is a primitive element that satisfies the primitive binary polynomial $1 + X^2 + X^3 + X^4 + X^8$. For example, a data bye of (d_7, d_6, d_5, d_4, d_3, d_2, d_1, d_0) is represented by the Galois field element of ($d_7\alpha^7 + d_6\alpha^6 + d_5\alpha^5 + d_4\alpha^4 + d_3\alpha^3 + d_2\alpha^2 + d_1\alpha^1 + d_0$) (see Chapter 13 for examples Reed-Solomon codes).

Table 3-8   Minimum FEC Coding Capability for ATU-C.

Parameter	Fast data buffer	Interleaved data buffer
Parity bytes per RS codeword	R_F = 0, 2, 4, 6, 8, ..., 16	R_I = 0, 2, 4, 6, 8, ..., 16
DMT symbols/mux data frames per RS codeword	S = 1	S = 1, 2, 4, 8, 16
Interleave depth	Not applicable	D = 1, 2, 4, 8, 16, 32, 64

R_F can be > 0 only if K_F > 0; R_I can be > 0 only if K_I > 0; R_I = an integer multiple of S.

(7) **Tone ordering**: A DMT time-domain signal has a high "peak-to-average ratio" [i.e., the ratio of the peak (of tone signal) voltage to the average voltage is required to be high], and large values may be clipped by the digital-to-analog converter. The error signal caused by clipping can be considered as an "additive negative impulse noise" during the time sample that clipping occurred. The clipped error power is almost equally distributed across all tones in a symbol that has been "clipped". Clipping is therefore most likely to cause errors on those tones that (in anticipation of a higher received S/N ratio) have been assigned the largest number of bits (and therefore have the densest constellation). These occasional errors can be reliably corrected by the FEC, provided the tones with the largest number of bits have been assigned to the interleaved data buffer. This procedure is known as tone ordering.

(8) **Constellation encoding**: The constellation can be implemented with or without Trellis coding. If Trellis coding is used, it is typically a 16-state 4-dimensional (4D-16) Trellis code.

(9) **Gain scaling**: Gain scaling of data symbols is applied to all data-carrying sub-carriers, as requested by the ATU-R, to equalize expected error rates. Gain scaling is not required for transmission of synchronization symbols.

(10) **Modulation**: Modulation using the Inverse Discrete Fourier Transform (IDFT) method is adopted in ADSL systems. The frequency spacing between sub-carriers is $\Delta f = 4.3125$ kHz $\pm$ 50 ppm.

(11) **Transmitter spectrum response**: Figure 3-55 shows the Power Spectral Density (PSD) mask for the transmitted signal.

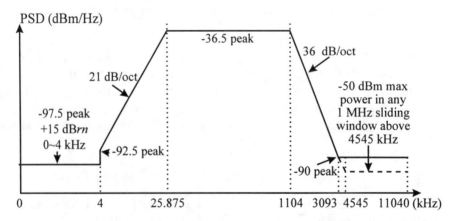

Figure 3-55   ATU-C Transmitter Power Spectral Density (PSD) Mask.

The "low frequency stop band" is defined as the voiceband. Likewise, the "high frequency stop band" is defined as frequencies greater than 1.104 MHz. In

Figure 3-55, all PSD measurements are at 100 ohms, but the POTS band aggregate power measurement is at 600 ohms. The measurements are made at the U-C interface (Figure 3-47). The breakpoint frequencies are exact, but the indicated slopes in Figure 3-55 are approximate.

3.6.5.5 ATU-R Functional Characteristics

The ATU-R functional characteristics include: STM specific functionalities, ATM specific functionalities, network timing reference, framing, scrambling, Forward Error Correction (FEC), tone ordering, constellation encoding, gain scaling, modulation, and transmission spectrum, etc.

(1) **STM specific functions**: Figure 3-56 shows the functional interfaces at the ATU-R for STM transport. Output interfaces for the high-speed downstream simplex bearer channels are designated as AS0 through AS3 (see Table 3-6), and input/output interfaces for the duplex bearer channels are designated as LS0 through LS2. The ATU-R also has a functional interface to transport OAM indicators from the Service Modules (SMs) to the ATU-R; this interface can be physically combined with the LS0 upstream interface. The data rate of the input/output interface is required to match the rate of the bearer channel connected to the interface.

The simplex bearer channels are transported in the downstream direction only; therefore their data interfaces at the ATU-R operate only as outputs. The duplex bearer channels are transported in both directions, hence the ATU-R l provides both input and output data interfaces. Note that the data rates are the same as the ATU-C.

An ATU-R configured for STM transport supports the full overhead framing structure 0 (Table 3-7). However, the support of full overhead framing structure 1, or reduced overhead framing structures 2 and 3 is optional. It can also support reconstruction of a Network Timing Reference (NTR).

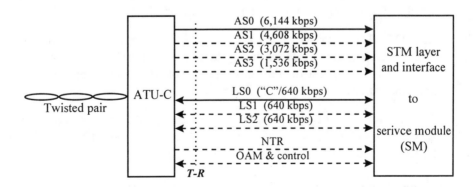

Figure 3-56   ATU-C Interfaces to the STM layer at T-R Reference Point.

(2) **ATM specific functions**: The cell specific functional interfaces at the ATU-C for ATM transport are shown in Figure 3-57. The ATM channel "ATM0" is always provided by ADSL systems, and the ATM channel "ATM1" is an optional used to support the dual latency mode. The ATU-R input/output T-R interfaces are identical to the ATU-C input/output interfaces.

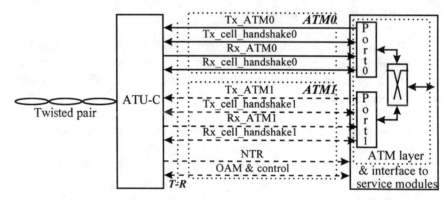

Figure 3-57   ATU-C Interfaces to the ATM layer at V-C Reference Point.

An ATU-R configured for ATM transport must support the full overhead framing structures 0 and 1 (Table 3-7). The support of "reduced overhead framing" structures 2 and 3 is optional. The ATU-R transmitter defines the T-R interface byte boundaries (either explicitly or implied by ATM cell boundaries) and the U-R interface, independent of the U-R interface framing structure. To ensure framing structure 0 interoperability between an ATM ATU-R and an ATM cell Transmission Convergence (TC) sublayer plus an STM ATU-C (i.e., ATM over STM), the ADSL system applies the following schemes:

* An STM ATU-C that transports ATM cells but does not preserve V-C boundaries at the U-C interface must indicate (during system initialization) that framing structure 0 is the highest framing structure supported.

* An STM ATU-C that transports ATM cells and preserves V-C boundaries at the U-C interface must indicate (during system initialization) that framing structure 0, 1, 2, or 3 is the highest framing structure supported (as applicable for the system implementation).

* An ATM ATU-R receiver operating in framing structure 0 can not assume that the ATU-C transmitter preserves V-C interface byte boundaries at the U-C interface, therefore it must perform the cell delineation "bit-by-bit".

(3) **Network Timing Reference (NTR)**: An ATU-R configured for ATM transport can support reconstruction of a NTR. If the ATU-C indicates that it has transmitted a

change of phase offset, the ATU-R delivers an 8 kHz signal to the T-R interface. Delivery of the NTR at the T-R interface is independent of the method of loop timing (see Chapter 14), as agreed upon by ATU-C and ATU-R during ADSL system initialization.

(4) **Framing**: The framing of the upstream signal (ATU-R transmitter) closely follows the downstream framing of the ATU-C transmitter with the following exceptions:

* There are no ASx bearer channels or AEX byte.

* A maximum of three bearer channels exist.

* The minimum Reed-Solomon FEC coding parameters and interleave depth are different (Tables 3-8 and 3-9).

* Four bits of the fast and sync bytes are unused.

* Four indicator bits for NTR transport are not used in the upstream direction.

(5) **Scrambling**: The data streams from both the "fast" and "interleaved" data buffers must be scrambled separately using the same algorithm as used for the downstream data buffers previously described in Eq. (3-28).

(6) **Forward Error Correction (FEC)**: The upstream data must apply the Reed-Solomon code previously described for the ATU-C downstream data.

The ATU-R must support upstream transmission with any combination of the Reed-Solomon FEC coding capabilities shown in Table 3-9.

Table 3-9   Minimum FEC Coding Capability for ATU-R.

Parameter	Fast data buffer	Interleaved data buffer
Parity bytes per RS codeword	$R_F = 0, 2, 4, 6, 8, ..., 16$	$R_I = 0, 2, 4, 6, 8, ..., 16$
DMT symbols/mux data frames per RS codeword	$S = 1$	$S = 1, 2, 4, 8, 16$
Interleave depth	Not applicable	$D = 1, 2, 4$

R_F can be > 0 only if $K_F > 0$; R_I can be > 0 only if $K_I > 0$; $R_I =$ an integer multiple of S.

(7) **Tone ordering**: The same algorithm used in for downstream direction is used here.

(8) **Constellation encoding**: The constellation can be implemented with or without trellis coding, as in the downstream direction.

(9) **Gain scaling**: For the transmission of data symbols, gain scaling shall be applied to all data-carrying sub-carriers, as requested by the ATU-C, to equalize the expected error rates. Gain scaling is not required for the transmission of synchronization symbols.

(10) **Modulation**: Modulation by the Inverse Discrete Fourier Transform (IDFT) is adopted in ADSL systems. The frequency spacing between sub-carriers is $\Delta f =$ 4.3125 kHz $\pm$ 50 ppm, which is the same as in the downstream direction.

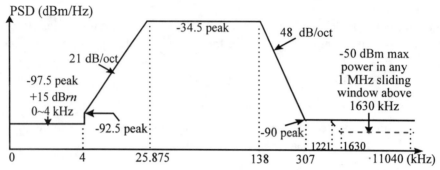

Figure 3-58   ATU-R Transmitter Power Spectral Density Mask.

(11) **Transmitter spectrum response**: Figure 3-58 shows the Power Spectral Density (PSD) mask for the transmit signal. The "low frequency stop band" is defined as the voiceband, and the "high frequency stop band" is defined as frequencies greater than 138 kHz. In Figure 3-58, all PSD measurements are at 100 ohms, but the POTS band aggregate power measurement is at 600 ohms. The measurements are made at the U-R interface (Figure 3-47). The breakpoint frequencies are exact, but the indicated slopes in Figure 3-58 are approximate.

3.6.5.6 ADSL System Operations and Maintenance

An Embedded Operations Channel (EOC) for communication between the ATU-C and ATU-R is used for ADSL system OAM (in-service and out-of-service), for the retrieval of a limited amount of ATU-R status information, and to obtain ADSL performance monitoring parameters. The EOC may also be used (in the future) to extend maintenance and performance monitoring function to the Service Modules (SMs) at the customer premises.

The EOC organization (e.g., protocols, message structure, address fields, data field, byte parity field, message/response field, information field, message sets, etc.) is beyond the scope of this book. They are specified in ANSI T1.413-1998, "Network and Customer Installation Interfaces - Asymmetric Digital Subscriber Line (ADSL) Metallic Interface."

3.6.5.7 ADSL Electrical and Physical Characteristics

An ADSL system has to specify several electrical characteristics such as Direct Current (D.C., DC or dc) requirements, voiceband, and ADSL band characteristics. Similarly, the system has to specify the following physical characteristics: wiring polarity integrity, connectivity, wiring requirements for an ATU-R with an integrated POTS splitter, and maximum distance for remotely located units.

- D.C. Characteristics: All the requirements specified for an ADSL system must be met in the presence of POTS loop currents ranging between 0 mA to 100 mA, and differential loop voltages as follows:

  * D.C. voltages of 0 volt (V) to −60 V

  * Ringing signals no greater than 103 V rms at any frequency between 20 to 30 Hz, with a D.C. component in the range from 0 V to −60 V

  * The input D.C. resistance of the ATU-C (at the U-C interface) and the ATU-R (at the U-R interface) must be ≥ 5 MΩ.

- Voiceband characteristics: The primary requirement is the input impedance. That is, the imaginary part of the ATU-x input impedance measured at the U-x (x = C or R) interface at 4 kHz must be:

  * In the range of $1.1 \sim 2.0$ kΩ ($\approx 20 \sim 34$ nF capacitor) with an integrated splitter.

  * in the range of 500 Ω to 1 kΩ ($\approx 40 \sim 68$ nF capacitor) with an external splitter

- ADSL band characteristics: The longitudinal balance (Figure 3-59 shows a measurement method) at the U-C and U-R interfaces shall be > 40 dB over the frequency range 30 kHz to 1104 kHz. If only the HPF part of the POTS splitter is integrated in the ATU, the measurement method shown in Figure 3-59 should exclude the part shown as dashed-lines. If both the LPF and HPF parts of the POTS splitter are integrated in the ATU, then the longitudinal balance measurement in the ADSL band shall be performed with the PSTN and POTS interfaces terminated with Z_{TR} and Z_{TC}, respectively. The longitudinal balanced is given by the following equation:

$$L_{Balanced} = 20 \times \log_{10} |\frac{e_1}{e_m}| \quad dB \tag{3-32}$$

where e_1 = the applied longitudinal voltage (reference to the building or green wire ground of the ATU), and e_m = the resultant metallic voltage appearing across a terminating resistor.

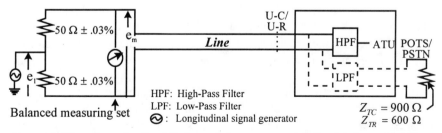

Figure 3-59   Longitudinal Balance Measurement Method (Above 30 kHz).

- Wiring polarity integrity: ADSL operation is independent of the polarity of the pair of wires connecting the ATU-C and the ATU-R transmission path.

- Network interface (RJ31X/RJ14C: For modems with internal/external POTS splitters the connection to the existing customer interface wiring shall be as specified in Table 3-10 using an 8-pin plug and jack (RJ31X/RJ14C) equipped with shorting bars. The cord connecting the POTS splitter and ATU-R unit is "hard wired" connection. The use of a separate POTS splitter (physically separate from the ATU-R) is not precluded by this standard, and other connection arrangements may be appropriate.

  The RJ31X/RJ14C is used an external POTS splitter that is typically mounted in the Network Interface Device (NID) or near the building entrance.

Table 3-10   Pin Assignments for 8-position Jack/Plug at Network Interface.

Pin No.	Assignment for jack	Assignment for plug
1	Tip or ring to POTS distribution	Tip or ring to POTS splitter (output)
2	No connection	No connection
3	No connection	No connection
4	Tip or ring from network interface	Tip or ring to POTS splitter (input)
5	Tip or ring from network interface	Tip or ring to POTS splitter (input)
6	No connection	No connection
7	No connection	No connection
8	Tip or ring to POTS distribution	Tip or ring to POTS splitter (output)

3.6.5.8 ADSL Loop Plant, Impairments and Testing

Laboratory testing is generally used to evaluate the compliance of an ADSL transceiver to performance requirements and standards. The testing methods are used to determine whether a system's ability to minimize digital bit errors caused by:

- Crosstalk coupling
- Background noise
- Impulse noise
- POTS signaling

These potential impairments are simulated in a laboratory set-up that includes loops, test sets, and interference injection equipment, that is connected to the equipment being evaluated. A typical lab arrangement for testing "downstream" and "upstream" is shown in Figure 3-60.

Two POTS splitters are shown in Figure 3-60. One for the remote (ATU-R) and the other for the central office (ATU-C) ends of the configuration. The splitters can be either internal or external to the ATU modem, and may contain a LPF and a HPF. Note that for the upstream direction, only an LPF is required, but the downstream direction requires both LPF and HPF. The purpose of the LPF is twofold: (1) For ADSL signals, protection from high-frequency transients and impedance effects that occur during POTS operation (i.e., ringing transients, ring trip transients, off-hook transients, and impedance changes); and (2) For POTS voiceband service, the low-pass filters provide protection from ADSL signals which may impact remote devices (e.g., handset, fax, voiceband modem, etc.) and central office operations.

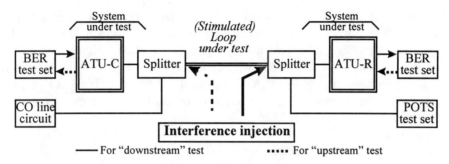

Figure 3-60   Overview of Laboratory Test Set-up ("Downstream" or "Upstream").

3.6.6 Very High-speed Digital Subscriber Line (VDSL)

The most recent xDSL technology is Very high-speed Digital Subscriber Line (VDSL), which is briefly described in this section. VDSL services provide very high data rates (tens of Gbps: see Example 3-13) over the metallic local loops to customers.

Example 3-13: As shown in Figure 3-23, the spectral allocation of VDSL is between 300 kHz and 30 MHz. Determine the highest theoretical speed of VDSL by applying the Shannon-Hartley theorem, or Eq.(3-4).

The bandwidth is given by 30 MHz − 300 kHz = 29.7 MHz. Assuming that the local loop has a S/N ratio of 30 dB (S/N = 1000), Eq.(3-4) representing the Shannon-Hartley law is expressed as follows:

$$R \leq 2W \times \log_2 \sqrt{1 + \frac{S}{N}} = W \times \log_2 (1 + \frac{S}{N}) \tag{3-4}$$

$$= 29.7 \times \log_2 (1 + 1000) = 296 \text{ Mbps}$$

Recalling R is the highest data rate in bps, and the spectral efficiency is 9.967 bps/Hz (Example 3-8) or as indicated in Figure 3-20. Higher data rates can be achieved (theoretically) if a higher spectral efficiency (Figure 3-20) is available.

VDSL is a standard that is actively being defined by ANSI Committee T1E1.4 in a coordinated effort with the ETSI Committee TM6. ITU-T SG 15 has recently assumed the responsibility from ANSI so that an international standard for VDSL can be created. VDSL modems eventually will be implemented for symmetrical or asymmetrical services. It is desirable to have a high data rate available to customers that desire broadband entertainment or data services, while prudently leveraging infrastructure costs of optical fiber system, avoiding wireless equipment placement, and bypassing unnecessary coaxial cable reengineering.

- Asymmetrical VDSL: Asymmetrical VDSL service is considered a residential DSL service that is introduced into the existing metallic local loops serving POTS and Basic Rate Interface-ISDN (BRI-ISDN) applications. VDSL will co-exist with POTS and ISDN signals on the same twisted pair of wiring, but will utilize separate frequency bands (Figure 3-61). A DSL signal splitter is required to perform the separation of functionality. The desired asymmetrical VDSL data rates over a distance of about 3,000 *ft* (1 *km*) are: 13 to 26 Mbps downstream, and 2 to 3 Mbps upstream. These data rates will allow delivery of Digital TV (DTV), High-Definition TV (HDTV), super-fast Web surfing, super-fast file transfer, and enable virtual office (at home) services. For shorter distances [1,000 *ft* (300 *m*) or less], the downstream data rate can be as high as 50 Mbps to allow simultaneous delivery of several DTV or HDTV channels.

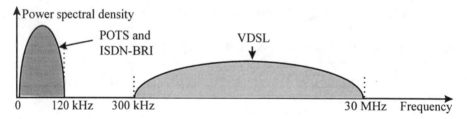

Figure 3-61   Spectral Allocations: POTS, ISDN-BRI and VDSL.

The connection required to provide asymmetrical VDSL services may not have to involve a central office. One possible scenario is HFC networking. That is, a connection from the central office to the street, where an Optical Network Unit (ONU) resides, utilizes optical fibers. However, the connection from ONU to the customer premises is implemented with existing metallic loops.

- Symmetrical VDSL: Symmetrical VDSL services are primarily intended for business applications. This service supports 13 Mbps connections over a twisted pair length of about 3,000 *ft* (1 *km*), or 26 Mbps for loop with lengths less than 300 *m*.

   Typical applications for symmetrical VDSL services are interconnecting ATM networks, and Ethernet/IP Local Area Networks (LANs) at data rates of 25.6 Mbps and 10 Mbps, respectively, over a twisted pair wiring. Buildings connected via LANs in a corporate campus environment (Figure 3-62) is an example of this type of application. In this environment, the main campus building is usually connected to the central office (service provider) via high-speed optical fibers. The signals carried by these fibers are OC-3 (STM-1 in SDH network) or OC-12 (STM-4 in SDH network). Copper twisted pair wires are typically used to connect all the other buildings in the corporate campus network, even though fibers and coaxial cables could be used for these connections.

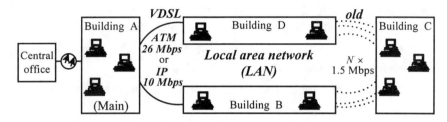

Figure 3-62   VDSL Applications in a Corporate Campus Environment.

Presently the services in corporate campus environments are typically provided by T1 (E1) digital carrier systems with data rates of 1.544 Mbps (2.048 Mbps). In these arrangements multiplexers and demultiplexers must be applied. Therefore, symmetrical VDSL offers the combined benefits of increased data rates and reduced overall cost.

3.6.6.1 Services Provided by a VDSL Link

The primary difference between VDSL and the services provided by other xDSL transport methods is the ability to deliver extremely high data rates. The high data rate enables several different services to be provided over a single VDSL link. One attractive feature is the use of ATM as a transport layer (ATM protocol consists of three layers: ATM adaptation layer, ATM layer and physical layer; to described in Chapter 5).

ATM protocol provides a universal transport interface for all types of services, including constant bit synchronous data such as T1, E1, or digital video. ATM transport also supports packetized data applications with variable bit rates (e.g., IP). In an ATM network, the ATM switch at the local exchange multiplexes different types of signals (e.g., voice, data, video, etc.) into a high-speed ATM cell stream. These ATM cells are typically transported over optical fibers (e.g., OC-3C, OC-12C, or OC-48C over SONET networks). The ATM service is carried by a specific Virtual Channel (VC). The services can also be bundled to form a specific Virtual Path (VP), that consists of several VCs. This scenario is typical for asymmetrical services in residential applications. Both single and dual latency modes may be applied.

Symmetrical services may be delivered by both ATM and STM transport protocols, usually in single latency mode. For telephone services the "fast" path (described later in this section) is typically used. ATM LANs and symmetrical IP services utilize the "slow" path.

The ATM implementation possesses the following advantages:

- ATM infrastructure is independent of the offered services.

- ATM interfaces are independent of the applications.

- ATM handles bursty, variable, and constant bit rates efficiently.

- ATM supports both "delay-sensitive services" (e.g., voice and interactive video), and "delay-insensitive services" (e.g., file transfer, etc.).

- ATM protocol provides a strong management capability over a single network infrastructure, and simplifies network operations.

- ATM ensures a path for evolution from existing services to new broadband service.

3.6.6.2 VDSL Transmission Environment

The metallic twisted phone line carrying VDSL signals may be connected directly from the local exchange office to the customer premises, or from the Optical Network Unit (ONU) of a Hybrid Fiber Coax (HFC) network to the customer premises (see Figure 3-83). The wiring is typically (24- or 36-gauge) unshielded and twisted pairs, either buried or aerial cable. The loop environment for DSL services s the same as the HDSL and ADSL applications that have been previously discussed. VDSL links provide powerful impulse noise protection by using effective error correction coding with "deep interleaving" (see Chapter 13).

To match the propagation delay requirements and benefits from the impulse noise protection for a wide array of services, the VDSL link transport capacity can be shared between "fast" and "slow" logical paths. The "slow" path operates properly in the presence of impulse noise up to 500 μs and produces a one-way propagation delay less than 20 ms. The "fast" path is more sensitive to impulse noise, but ensures a one-way propagation delay of less than 1.2 ms, thus it is suitable for all types of "delay sensitive services".

Definition 3-7: If only one path (either "fast" or "slow") is established, a VDSL link is said to operate in the "single" latency mode. If both the "fast" and "slow" paths are established, the VDSL link is said to operate in the "dual" latency mode.

3.6.6.3 VDSL DMT Transmission

The VDSL standards proposal draft document is based on multicarrier modulation. The proposed technology is Synchronized Discrete Multi-Tone (SDMT) which is a combination of two other technologies: (1) DMT modulation for line transmission features, and (2) Time Division Duplexing (TDD; commonly called "Ping-Pong"; Figure 3-63) for simple programmability of the downstream-to-upstream data rates.

Figure 3-63 illustrates the principles of TDD technology. For digital transmission, time is divided into many time slots (channels). For example, time slot No. 1 is assigned to customer No. 1, time slot No. 2 is assigned to customer No. 2, etc. Each time slot is subdivided into two halves. The first half (time slot) is used for communication from the service provider to the customer (i.e., the "downstream" direction, or "ping"), and the second half (time slot) is used for communication from the customer to the network (i.e., the "upstream" direction, or "pong"). TDD technology has been applied in several digital networks (e.g., digital cordless telephone systems).

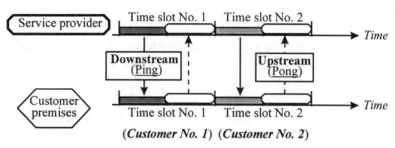

Figure 3-63   Time Division Duplexing (TDD) Technology.

The SDMT technique utilizes DMT in conjunction with TDD. All lines in the same cable binder are "loop-timed" (Chapter 14) and "locked" to the same network clock (i.e., all lines within the same cable binder "ping" and "pong" simultaneously) to avoid NEXT. SDMT-based systems support upstream and downstream transmissions within a single frequency band during different time periods [i.e., utilizing TDD; see Figure 3-26(A)]. This is implemented by the time-shared channel bandwidth of the "superframe" structure (Figure 3-64). A superframe consists of both "downstream" and "upstream" transmission periods separated by guard time intervals. The guard time is required to account for channel propagation delay and echo response time. Figure 3-64 illustrates a proposed 20-symbol superframe structure (ANSI contributions T1E1.4/98-265, August, 1998).

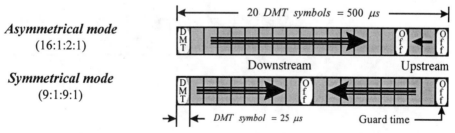

Figure 3-64   SDMT VDSL Transmission Superframe.

- Asymmetrical transmission in ratio 8:1 is implemented by using 16 symbols in the downstream direction, two symbols in the upstream direction, and two symbols between for the guard times.

- Symmetrical transmission in ratio 1:1 is implemented by using 9 symbols for both directions, and two symbols between for the guard times.

The 20-symbol superframe structure requires 500 µs (each DMT symbol interval is 25 µs), corresponding to 256 tones, each has a width of approximately 40 kHz. For longer length loops, the tone width can be narrowed.

Using DMT for VDSL modulation provides a number of advantages: (1) the ability to maximize the data rate, (2) adaptation to changing line conditions, (3) the ability to meet strict egress requirements, and (5) flexibility in the transmission band.

> *"The ability to maximize the transmitted data bit rate is provided by tailoring the distribution of information to the channel attenuation/noise characteristics."*

DMT technology partitions the channel into independent subchannels, and modems adapt to changing loop and noise conditions. Changes in a line's transfer function (e.g., noise) that do not reduce the line capacity substantially can be accommodated by *"bit swapping."*

The effects of some unpredictable noise sources (e.g., RF from AM or amateur radio transmissions) are mitigated because: (1) the tones are narrow, and only those near the interfering frequency are adversely effected, and (2) several RF interference cancellation techniques have been developed for DMT-based transmission.

Besides SDMT technology with TDD, there are proposals suggesting the use of CAP or QAM-based VDSL modulation techniques, in addition to Frequency Division Duplexing (FDD; Figure 3-65) for separating the upstream and downstream channels.

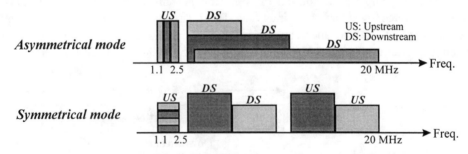

Figure 3-65   The Frequency Spectra Placement (FDD for VDSL).

3.6.7 xDSL versus Other Access Networks

Various access network technologies have been discussed in previous sections including: Subscriber Loop Carrier (SLC), Integrated Digital Loop Carrier (IDLC), Business Remote Terminal (BRT), and xDSL (i.e., HDSL, ADSL, and VDSL). Three more access technologies

[Wireless Local Loop (WLL), Hybrid Fiber Coax (HFC), and Cable modem] are briefly discussed in the following sections. Figure 3-66 summaries five of these access technologies: (1) Using metallic local loops as the media, the services can be POTS, BRI-ISDN, and low-speed ADSL if the loop length is between 3 and 4.5 *km*. For shorter loops (1.5 to 3 *km*), HDSL and high-speed ADSL technologies may be adopted. If loop is very short (0 to 1.5 *km*), VDSL technology can be applied to achieve very high-speed data transmission; (2) CATV industry has aggressively developed/implemented broadband access services using cable modems; (3) Hybrid-Fiber-Coax (HFC) technology has been implemented widely in North America. Optical fibers are used between the backbone network and Optical Network Unit (ONU). Connection between the ONU and customer premises can be implemented with metallic twisted pairs, coaxial cables, or optical fiber; (4) Wireless Local Loop (WLL) technology may apply Time-Division Multiple Access (TDMA) or Code-Division Multiple Access (CDMA) (Chapter 8), and has been adopted as a "fast deployment" method for accessing the backbone network, especially in countries where the metallic loop or cable plant has not been extensively deployed; and (5) Satellite access networks are still under study/standardization.

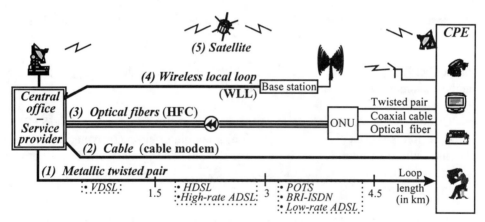

Figure 3-66   xDSL versus Other Access Technologies.

3.7 WIRELESS ACCESS TECHNOLOGIES

As shown in Figure 3-66, Wireless Local Loop (WLL) is a technology that can be implemented as a broadband access network. For fast deployment, WLL is the most attractive technique. WLL has several common names: (1) fixed wireless, because the "wireless" handset is located in a practically "fixed" location (i.e., the home or an office, rather than a car or cellular phone), (2) Wireless Subscriber System (WSS) because WSS performs the metallic local loop's functions, and (3) AirLoop™ since the system connecting the Customer Premises Equipment (CPE) to the backbone network is the airway (i.e., an <u>air</u> link; the substitute for a metallic <u>loop</u>). However, in the countries that traditional loop plants have been widely adopted, WLL may not as attractive in those underdeveloped (in telecommunications) countries. This is because of the bandwidth efficiency of WLL may not yet economically acceptable to compete with other technologies (e.g., cable modem, HFC, or even xDSL).

Figure 3-67 shows the wireless local loop system architecture that has been widely adopted for many wireless access networks, which utilizes mobile/cellular technology to supplement the metallic local loop plant. In some countries, WLL takes the place of metallic local loops, especially for fast deployment of telecommunications services. The major components of a WLL system are:

- Network Equipment (Element) [NE, also known as the Wireless Line Transceiver (WLT)], located at the local exchange office, consists of :

  * Switch interface [also known as the Central Access and Transcoding Unit (CATU)]: It provides an interface to the local exchange, and supports digital connections (E1, T1, or others) to the base station [Central TRansceiver Unit (CTRU)].

  * Base station CTRU: It performs speech encoding, encryption, scrambling, bit-interleaving, multiplexing, modulation, and radio (air) interface.

- Subscriber Equipment [SE, also called the Network Interface Unit (NIU)]: It is located at the customer premises, and is composed of:

  * Customer Transceiver [CT, also called the Subscriber TRansceiver Unit (STRU)]: It performs speech encoding, encryption, scrambling, bit-interleaving, modulation, and radio (air) interface (as in the base station). The primary difference between the base station CTRU and the customer STRU is the multiplexer is not required in the customer STRU.

  * Telephone Socket [TS, or Intelligent Telephone Socket (ITS)]: TS supports:

    ♦ Analog voice line interface
    ♦ BRI-ISDN interface (2B + D; B = 64 & D = 16 kbps)
    ♦ B + D interface

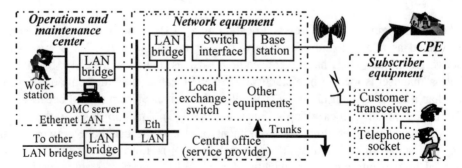

Figure 3-67   Wireless Local Loop System Architecture.

- Operations and Maintenance Center (OMC): An OMC is implemented by using an Ethernet LAN, that is interconnected via LAN bridges/routers. OMC servers and workstations are typically HP-9000 family equipment using HP-UX operating systems. The user interface is "X-Windows"-based Graphical User Interface (GUI). The management platform is HP OpenView Simple Network Management Protocol (SNMP, described in Chapter 12). An OMC interfaces the WLL system via a switched interface using an Ethernet LAN bridge/router. An OMC can provide OA&M functions up to 500 Network Equipment (NEs) entities, and probably more in the future. The major functions performed by an OMC are:

  * Configuration management: The OMC manages the configuration of WLL system hardware and software.

  * Provisioning: The OMC performs subscriber service provisioning.

  * Surveillance management: The OMC manages/monitors WLL system alarms.

  * Performance management: The OMC monitors and analyzes WLL system error performance management.

3.7.1 Wireless Local Loop System Interfaces

The required WLL interfaces are represented by: "A", "B", "C", "D", "E", "F" and "G" as shown in Figure 3-68.

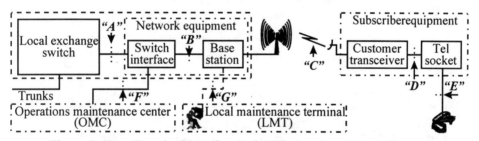

Figure 3-68   Required Interfaces of a Wireless Local Loop System.

- Point "A" is the interface between a local exchange switch and the switch interface in the Network Equipment (NE) of the WLL system.

- Point "B" is the interface between the base station and switch interface that is part of the NE of the WLL system.

- Point "C" is an air interface between the base station and the customer transceiver in the subscriber equipment in the WLL system.

- Point "D" is the interface between the customer transceiver and the Telephone Socket (TS) that is part of customer equipment of the WLL system.

- Point "E" is the interface between the TS and the customer equipment (e.g., voice terminal or PC) on the customer premises.

- Point "F" is the interface between the switch interface and the Operations Maintenance Center (OMC).

- Point "G" is the interface between a base station and a Local Maintenance Terminal (LMT).

The "G" interface is not permanent. The LMT is temporarily connected directly to the base station using the physical interface standard RS-232. It provides on-site maintenance and diagnostic capabilities (e.g., the ability to "read" and "change" base station configuration data).

3.7.2 Wireless Local Loop System Components

Figure 3-69 illustrates the functional building blocks in the downstream transmission path (i.e., from wireline network or base station to handset). The PSTN network accepts "N" calls from "N" users. Each user signal is fed to a Speech Encoder (SE), that is followed by an ENcryptor (EN), SCrambler (SC), and Channel Encoder (CE).

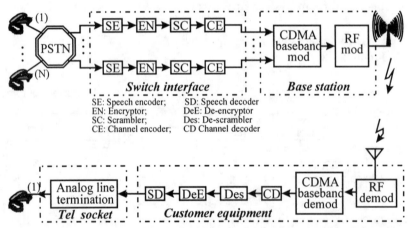

Figure 3-69   Functional Building Blocks In the Downstream Direction.

The WLL enabling technologies are described throughout this book, and are briefly summarized as follows:

- Speech Encoder (SE): It is used to encode speech signals to form a digital bitstream. There are numerable ways to implement speech encoding, but research in speech encoding is expected to continue forever. Table 3-11 lists some speech encoders, with their data rates and system performance characteristics[typically, Mean Opinion Score (MOS) is used to measure speech quality; MOS = 1 means "poor", MOS = 2 means "fair", MOS = 3 means "satisfactory", MOS = 4 means "good", and MOS = 5 means "excellent"; MOS is a "subjective", not an "objective", speech quality measurement; sometimes, Grade Of Service (GOS) replaces MOS for speech quality measurement].

Table 3-11   Comparison among various Speech Encoders.

Encoder		Data rate (kbps)	Performance (MOS)
PCM	(ITU-T G.711)	64	>4.00
ADPCM, ADPCM 2, ADPCM 3 encoding	(G.721)	32	>3.34
Low-delay CELP, LD-CELP 2, 3, 4 encoding	(G.728)	16	>3.38
GSM full-rate RPE-LTP		13	>3.47
IS-54 VSELP		8	>3.46
LD-CELP (4, 6ms frame)		8	>3.47
FS1016 CELP		4.8	>2.55

CELP: Code-Excited Linear Predictor;   RPE-LTP: Regular pulse Excited Long Term Predictor
VSELP: Vector Sum ELP.   MOS: Mean Opinion Score;   ELP: Excited Linear Predictor

- Encryptor (EN): It is used to implement system security. For wireless applications the encryptor usually requires a "key" with an algorithm to encrypt/de-encrypt the digital bitstream (from plain text to cipher text, and vice versa).

- Scrambler (SC): In general, a scrambler is used for security and to perform timing synchronization of the communications system. For digital systems (especially wireless systems), it is primarily applied for the purpose of timing synchronization (Chapter 14). That is, scrambling is used to guarantee ones density in the digital bit stream so that the receiver can implement timing synchronization easily.

- Channel Encoder (CE): When the digital bitstream is transmitted over the airway, every digit can be correctly restored with a very high probability, but, it can also be erroneously restored. Channel error control (detection/correction) schemes (described in details in Chapter 13) can be applied to reduce the error probability so that the Bit Error Rate (BER, bit error ratio) is low enough to yield acceptable quality. The most commonly used channel codes, in wireless systems, are Cyclic Redundancy Check (CRC) codes, Reed-Solomon codes, and convolutional codes.

- CDMA multiplexing: Direct Sequence (DS) spread-spectral multiplexing technique is used (described in Chapter 8). Other techniques [e.g., Fast Frequency Hopping (FFH) or Fast Time Hopping FTH); beyond the scope of this book] can also be used for wireless multiplexing.

- RF modulation: Modulation technique used in wireless systems are: Frequency Shift Keying (FSK), Phase Shift Keying (PSK), etc. (Chapter 8).

3.7.2.1 Channel Frequencies Used in Frequency Division Duplex (FDD)

For Wireless Local Loop (WLL) systems, the frequencies of operation are 1900, 2400, 3400 and 3900 MHz. Table 3-12 lists the channel separation, and the separation between downlink and uplink (forward and reverse) directions.

Table 3-12   Frequency Separation.

Operation frequency (MHz)	Separation (MHz) (forward and reverse)	Separation (MHz) (channel)
1,900	80	5
2,400	94	5
3,400	100	5
3,900	210	10

Example 3-14: Describe the channel frequencies for a WLL 1900 MHz applications.

For 1900 MHz operation band, the channel frequencies are shown in Table 3-13. The forward direction is the WLL system base station to the WLL subscriber equipment, and the reverse direction is from the WLL subscriber equipment to the base station. The central (carrier) frequencies for different channels are shown in Table 3-13. The frequency separation between adjacent channel is 5 MHz (Tables 3-12 and 3-13). The frequency separation between "forward" and "reverse" directions is 80 MHz.

Table 3-13   Channel Frequency Assignment for 1900 MHz Band.

Channel	1	2	3	4	5	6	...
Forward	1932.5	1937.5	1942.5	1947.5	1952.5	1957.5	...
Reverse	1852.5	1857.5	1862.5	1867.4	1872.5	1877.5	...
Δf (MHz)	80	80	80	80	80	80	80

Example 3-15: Describe the channel frequencies for a WLL 2400 MHz applications.

For 2400 MHz operation band, the channel frequencies are shown in Table 3-14. The central (carrier) frequencies for different channels are shown in Table 3-14. The frequency separation between adjacent channel is 5 MHz (Tables 3-12 and 3-13). The frequency separation between "forward" and "reverse" directions is 94 MHz.

Table 3-14   Channel Frequency Assignment for 2400 MHz Band.

Channel	1	2	3	4	5	6	...
Forward	2404	2409	2414	2419	2424	2429	...
Reverse	2310	2315	2320	2325	2330	2335	...
Δf (MHz)	94	94	94	94	94	94	94

Example 3-16: Describe the channel frequencies for a WLL 3400 MHz applications.

For 3400 MHz operation band, the channel frequencies are shown in Table 3-15.

Table 3-15   Channel Frequency Assignment for 3400 MHz Band.

Channel	1	2	3	4	5	6	...
Forward	3552.5	3557.5	3562.5	3567.5	3572.5	3577.5	...
Reverse	3452.5	3457.5	3462.5	3467.5	3472.5	3477.5	...
Δf (MHz)	100	100	100	100	100	100	100

Example 3-17: Describe the channel frequencies for a WLL 3900 MHz applications.

For 3900 MHz operation band, the channel frequencies are shown in Table 3-16.

Table 3-16   Channel Frequency Assignment for 3900 MHz Band.

Channel	1	2	3	4	5	6	...
Forward	3855	3865	3875	3885	3895	3905	...
Reverse	3645	3655	3665	3675	3685	3695	...
Δf (MHz)	210	210	210	210	210	210	210

3.7.3 Optical Wireless

Optical wireless technology has been developing rapidly, from early experimental prototype equipment to high-performance systems. It is now capable of delivering important benefits to modern telecommunications users. An overview of this technology is given in this section.

The benefits, limitations, and applications of the various available systems have different technological solutions that suit particular applications. These characteristics are described in this chapter.

3.7.3.1 Transition from Optical Fiber to Optical Wireless Communications

The first commercial optical fiber communications system was introduced in the early 1980s. The original data rate of 1.544 Mbps has been increased to tens (hundreds) of Gbps in modern optical networks. Optical fiber communication technologies are discussed in Chapter 7, but a simplified optical fiber link is illustrated in Figure 3-70. A typical optical fiber link consists of three key components:

(1) A light source [e.g., LED, Multiple Longitudinal Mode (MLM) laser, Single Longitudinal Mode (SLM) laser, Single Frequency (SF) laser, or tunable laser] which serves as the transmitter. Its function is to convert electrical signals (i.e., either voltage or current) into photonic signals (i.e., photons; light signals). LEDs are typically used for low data rate and/or short-haul systems, SF lasers are used of high data rate and/or long-haul systems. The special SF lasers, that are tunable for

adjusting to different wavelengths, are used in Wavelength Division Multiplex (WDM) or Dense WDM (DWDM) optical fiber systems.

(2)   Optical fibers [e.g., Step-Index (SI) Multi-Mode Fiber (MMF), GRaded INdex (GRIN) MMF, Single Mode Fiber (SMF), Dispersion-Shifted SMF, or special fiber] which serve as the transmission medium.

(3)   A photodiode (photo-detector) [e.g., Positive-Intrinsic-Negative (PIN) or Avalanche PhotoDiode (APD)] which serves as the receiver.

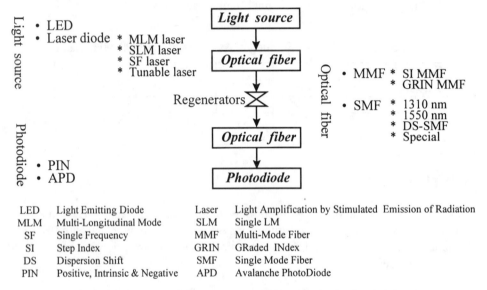

Figure 3-70   Major Components of an Optical Fiber Link.

The development of high-performance single mode fibers, and the reduced spectral width of narrowband laser diodes has drastically increased the data rate and low bit error rate of optical links. The attenuation characteristic of optical fibers is shown in Figure 3-71. Earlier optical fiber systems transmitted light at the wavelength around 820 *nm* (known as the 1st window; or 1G). The next generation of long-distance optical fiber systems utilized the wavelength around 1310 *nm* (known as the 2nd window; or 2G). Modern long-distance and high-speed optical systems utilize the wavelength around 1310 *nm* or 1550 *nm* (known as the 3rd window; or 3G) depending on the speed (super high-speed systems transmit at 1550 *nm*). The range of these wavelengths (820, 1310 and 1550 *nm*) is known as the "***infrared***" spectrum. Therefore, optical fiber communication technology is called often ***infrared technology***. Rapid advancements in infrared technology has "paved the way" for optical wireless technology, which basically removes "optical fibers" as transmission media and replaces them with "airways". Well-defined and high performance infrared technologies (including optical devices and optical wireless technologies) have been focused on improving "infrared" light transmission.

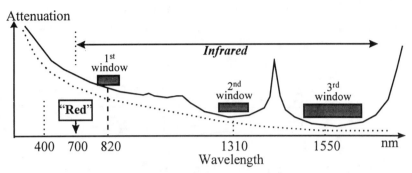

Figure 3-71   Attenuation versus Wavelength: Three Windows.

3.7.3.2 Issues and Limitations of Optical Wireless

Optical fibers serve as excellent "waveguides" in optical links, therefore removing and replacing them with "airways" loses the high-performance and low-loss advantage. Matching these advantages by developing optical wireless technology to achieve "acceptable" performance has been a challenge. That is, without fibers, conveying signals between terminal stations (transmitter and receiver) in a "controlled reliable manner" is a demanding task.

The light sources for high-speed optical fiber systems are typically "point-source" high power, and narrowband laser diodes. If these lasers are used for indoor optical wireless links, they pose a potential (eye) safety hazard. Therefore, a different light source (e.g., LED) must be used for indoor optical wireless applications. However, the highest speed using LEDs is about 10 Mbps, but may be improved to 50 Mbps in the future.

The high loss of the airway, and the poor coupling of the free-space and its receiver further reduces the received signal strength. Hence, optical wireless power budgeting is another challenge. For these reasons, transition from optical fiber to optical wireless communications is not straightforward, and new design solutions are needed.

3.7.3.3 Atmospheric Loss

It is important to understand atmospheric loss because power budgeting is one of the most important issues for optical wireless links. The atmospheric loss along the transmission path includes free-space loss, clear air absorption, scattering, refraction, and scintillation. Each of these topics is described separately as follows:

• Free-space loss: This is the portion of optical power arriving at the receiver that is not captured within the receiver's aperture. The free-space loss of a typical point-to-point system that operates with a slightly diverging beam (Figure 3-72) is about 20 dB. In comparison, the indoor applications with a wide-angle beam typically have a free-space loss of 40 dB or more. Free space loss exists in any optical wireless systems (i.e., either short-haul or long-haul indoor systems).

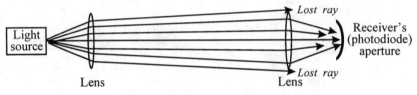

Figure 3-72   Illustration of Free-space Loss.

- Clear-air absorption: The clear-air absorption of an optical wireless system is similar to glass intrinsic absorption loss of an optical fiber system (Chapter 7). Clear-air absorption is wavelength dependent (Figure 3-73), and is critical for long optical wireless systems. Coincidentally, the smallest clear-air absorption occurs around 820 nm, 1310 nm, and 1550 nm (as in optical fiber applications). Therefore, the same opto-electronic devices used in optical fiber systems can be applied in optical wireless systems. This is crucial from an optical wireless commercial viewpoint.

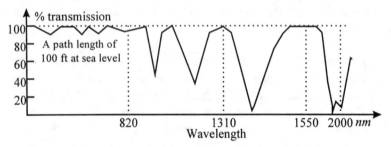

Figure 3-73   Clear-air Absorption at Various Wavelengths.

- Scattering and refraction: Scattering and refraction of optical wireless systems are due to water droplets in the air resulting from rain, fog, mist, or snow. This type of atmospheric loss depends on the season of the year, the geographical location of the optical wireless system, and the link length of the system. The suggested measurement of scattering and refraction is the percentage of "up time" (i.e., system is available) while the system bit error rate is better than an "acceptable" level. Various studies have shown that an availability of 99% can be achieved for an optical wireless link up to 1 km in length. With a suitable power budget, 99.5% availability may also be achieved. however, it is believed that a 99.9% availability may be very difficult to implement with the present technology.

- Scintillation: Solar energy heats small pockets of air to slightly different temperatures. This creates regions of varying refraction index along the light propagation path, and causes the optical signal to scatter at very shallow angles in the direction of propagation. Multiple optical signals that are phase-shifted (relative to each other) can arrive simultaneously at the receiver. This causes the amplitude of the received signal to fluctuate rapidly. The fluctuation can be as much as 30 dB when the atmosphere conditions are unfavorable, and the power spectral density can span 0.01 to 200 Hz. This scintillation effect can cause long bursts of data errors. In addition, it distorts the wavefront of the received signal, and causes the focused image at the photo-diode to "dance around" the surface (Figure 3-72) of the diode. Hence, a larger surface-area photodiode is required to ensure the signal is not lost.

Scintillation is usually not an issue for optical wireless systems with short links (i.e. < 500 *m*), however, it increases rapidly with link lengths greater than 500 *m*.

3.7.3.4 Infra-Red Data Association (IrDA)

The Infrared Data Association (IrDA) was established in 1993, and presently has more than 160 member companies. The task of IrDA is to develop an open standard protocol for wireless data communication using mature (commercially-available) infrared components [i.e., light sources (LED, or laser), regenerators, photodiodes (PIN, or APD)]. The agreed upon protocol provides a simple, low-power, low-cost, reliable means of wireless infrared communications for a wide range of computing, communications, and consumer devices.

3.7.3.5 Laser Power and Eye Safety

Optical wireless systems, like (microwave) radio wireless systems, may pose a hazard to human health if the system is not designed and/or operated correctly. One factor that may be a hazard in an optical wireless system is the use of semiconductor laser (laser diode). Therefore, it is important to classify optical sources in accordance with their emitted power. Figure 3-74 illustrates the principal classifications for a "point source" emitter such as a semiconductor laser. Note that the classification is a function of the wavelength of the emitted light. The vertical axis is the total power in a 5 cm lens, and the horizontal axis is the signal wavelength. The characteristics for three wavelength device categories is summarized as follows:

- **Short-wavelength** (around 880 *nm*) light source: A launch power of 0.2 mw or less is a Class 1 device. Class 3A devices have a launch power between 0.2 mw and 2.5 mw. Any device with a launch power larger than 2.5 mw is a Class 3B device.

- **Medium-wavelength** (around 1310 *nm*) light source: A launch power of 8.8 mw or less is a Class 1 device. Class 3A devices launch power between 8.8 mw and 45 mw. Any device with a launch power larger than 45 mw is a Class 3B device.

- **Long-wavelength** (around 1550 *nm*) light source: A launch power of 10 mw or less is a Class 1 device. Class 3A devices launch power between 10 mw and 50 mw. Any device with a launch power larger than 50 mw is a Class 3B device.

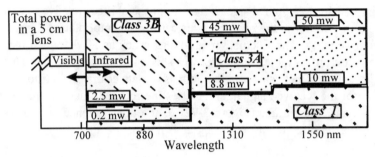

Figure 3-74   Laser (Point-source) Safety Classifications.

Example 3-18: Describe transmitter device for outdoor point-to-point systems.

Outdoor optical wireless systems are typically point-to-point systems that require high-power lasers operating in the Class 3B band to meet system power budget requirements. These transmitters should be located (as recommended by standards) where the light beam cannot be interrupted or inadvertently viewed by a human being. Roof-tops or tall masts are appropriate transmitter locations.

Example 3-19: Describe light source for indoor broad view systems

Industry standards recommend that Class 1 eye safety be used under any conditions implemented for indoor optical wireless systems. If short-wavelength lasers (for low-cost operation) are used, the launch power must be 0.5 mw or less (Figure 3-74). However, it is almost impossible to meet system power budget with this level of launch power. Therefore, to meet system power budget requirements with Class 1 eye safety, a Class 3B device having a "hologram" incorporated within the overall laser enclosure can be used. Another option is to use an array of LEDs that can launch sufficient power to meet the system power budget, yet still meet Class 1 eye safety recommendations. The reason an LED can meet Class 1 safety requirements is that it is not a "point source" device (Figure 3-75).

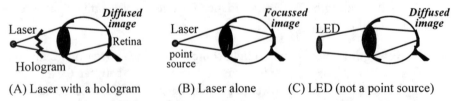

(A) Laser with a hologram   (B) Laser alone   (C) LED (not a point source)

Figure 3-75   Images on the Eye's Retina Produced by Lasers and LED.

Example 3-20: Illustrate a "point source" device using a hologram to meet Class 1 safety.

As the light from a "point source" passes through a hologram, the wavefront is broken up. Therefore, the image of the "laser spot" on the retina of the human eye is diffused [Figure

3-75(A)]. This is different from the focused image of a laser without a hologram [Figure 3-75(B)]. In contrast, an LED is not a "point source", and creates a diffused image on the retina of the eye [Figure 3-75(C)]. Therefore, a laser with a hologram is equivalent to an LED light source, with respect to eye safety.

3.7.3.6 Long Distance (100 *m* to 5 *km*) Optical Wireless Systems

Figure 3-76 shows the building blocks of a long-distance optical wireless system. Systems with this architecture can presently transport data at rates of 155 Mbps or higher for distances between 100 *m* and 5 *km*. The three components of this system are described as follows:

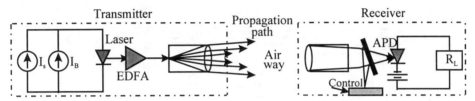

Figure 3-76   Building Blocks of a Long-distance Optical Wireless System.

(1)   **Transmitter**: The transmitter can be (1) a high-power Class 3B laser emitting a power around 100 mw (20 dBm), or (2) a lower-power device coupled to an Erbium-Doped Fiber Optical Amplifier (EDFOA; EDFA; OA). The latter requires the operating wavelength to be 1550 nm because EDFAs presently operate around 1550 nm. The emitting power is about 10 mw (10 dBm), which is a Class 1 eye safety operation. Using a laser as the transmitter requires a bias current (I_B) in addition to the signal current (I_s). Note the connection used at the transmitter is known as a "forward biased" configuration (Chapter 7).

(2)   **Propagation path**: To meet the system power budget requirement, it is important to minimize the overall loss of the propagation path. All atmospheric losses (clear-air absorption, scattering, refraction, and scintillation) can not be controlled. However, it is possible to reduce the free-space loss by using optical devices to (1) minimize light beam divergence from the transmitter, and (2) maximize the (numerical) aperture of the receiver. One solution is to use an astronomical telescope at each end of the propagation path (Figure 3-76).

(3)   **Receiver**: The receiver "sensitivity" parameter is most important (critical) for meeting system power budget requirements. Two receiver types that can be used are: PIN receivers, and APD receivers (Chapter 7). From a "sensitivity" viewpoint, an APD typically has a 10 dB better performance than a PIN receiver.

Example 3-21: Describe methods to improve the performance of optical wireless systems.

Because of the receiver's wide aperture (wide arriving beam angle at the receiver), it will take in stray light along with the optical signal beam. This stray light increases the noise level at the receiver, and thus reduces the S/N performance. This in turn, degrades the Bit Error Rate (BER). This is particularly significant for indoor systems that are subjected to ambient light interference. Figure 3-77 illustrates one way to improve the performance by applying an infrared filter. Another factor is that ambient light covers the power spectrum from DC to several hundreds kHz. Therefore, another solution is to apply a line code so that the encoded signal doesn't have power in the ambient light range. By combining these two schemes, the unwanted stray light can be entirely blocked, thereby improving system performance.

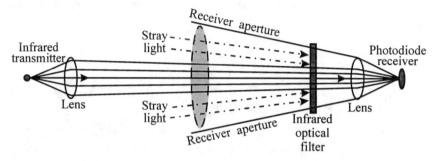

Figure 3-77   Stray Light Interference Control Scheme.

3.7.3.7 Short Distance (< 100 *m*) Optical Wireless Systems

All of the design and performance parameters for long-distance optical wireless systems are equally applicable to short-distance (< 100 *m*) systems.

One significant difference between short-distance and long-distance optical wireless systems is the effect of atmospheric loss. Atmospheric losses may have very little effect on (short-distance) system performance, especially for indoor systems. Short distance optical wireless systems may have different configurations, described as follows:

- Point-to-point systems: An outdoor point-to-point system can be used to link high capacity terminals located in two adjacent buildings. A high-power Class 3B emitter can be used since the systems are located away from people. It is quite easy to achieve the power budget requirement for this type of systems. In addition, designing, setting up, and aligning the system is much easier. System costs can also be reduced considerably. A system of 1 Gbps for a distance of 40 *m* is of particular interest to the industry, and may be installed using "protection-switched" links to temporarily "bridge" two buildings in the event a failure on the "working" link occurs.

  Point-to-point systems can also be applied indoors. For example, they can be used to extend a LAN port to a different part of an office. A system of 10 Mbps for a distance of 20 *m* has the highest potential for becoming popular. Two important differences between an indoor and an outdoor point-to-point system are : (1) Class 1 eye safety device must be used, and (2) weather proofing equipment is not required, thus system costs can be reduced.

- Telepoint systems: An indoor point-to-point system can be used for a telepoint application (Figure 3-78) if an emitter with diverging beams (rather than narrow beams) is used. One advantage of a high-speed (e.g., 10 Mbps) telepoint system is that it is accessible to any user within the cell diameter (0.5 m to 10 m) is possible. Likewise, roaming within the same cell is possible. The following design factor must be carefully considered: the uplink beam must "always" point toward the base station, therefore "roaming" (within a cell) is a greater technical challenge than uplink communications with stationary users. It should also be noted that "roaming" from one cell to another is impossible in this type of system.

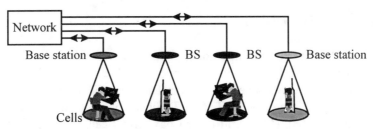

Figure 3-78   Optical Wireless Telepoint System Architecture.

- Diffused systems: Figure 3-79 illustrates the system architecture of a diffused optical wireless system. This is a multi-path system, rather than a point-to-point system. The key advantage is accessibility: either from the direct path or from any reflected path. However, several issues exist for diffused systems. Because of multi-path transmission, pulse spreading (delay spread, or pulse dispersion) may be severe enough to cause Inter-Symbol Interference (ISI). ISI degrades system BER performance. Interference from ambient light can also become a problem in this type of system.

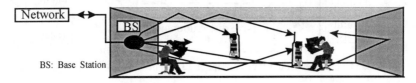

Figure 3-79   System Architecture of a Diffuse Optical Wireless System.

3.7.3.8 Very-short Distance (1 *m* or Less) Optical Wireless Systems

Very-short distance "point and shoot" optical wireless systems have been developed and are currently being manufactured for the applications that include: laptop computers, palmtops, printers, calculators, and mobile telephone. These wireless systems are known as IrDA systems in recognition of the IrDA standard that they follow. The IrDA standard specifies the following parameters for very short distance optical wireless systems:

- Data rates between 128 kbps and 4 Mbps (higher rates to several tens of Mbps may be possible in the future).

- Maximum operating range of 1 *m* (longer ranges up to 10 *m* may be possible in the future).

- A 30° cone pattern from the transmitter equates to a 50 *cm* beam diameter at the receiver positioned 1 *m* away.

- Very wide receiver "field of view" characteristics are required.

3.8 FIBERS IN THE LOOP (FITL)

Using optical fibers to replace traditional copper twisted-pair wiring to implement a broadband access network is not only technically achievable, but also cost effective. Several architectures are available for Fibers In The Loop (FITL): (1) Fiber To The Home (FTTH), (2) Fiber To The Office (FTTO), (3) Fiber To The Curb (FTTC), (4) Fiber To The Subdivision (FTTS), (5) Hybrid Fiber Coax (HFC) cables, or (6) fiber rings in the customer infrastructure. Figure 3-80 shows the major building blocks of fiber in the loop: a Host Digital Terminal (HDT), a Passive Distribution Network (PDN), and Optical Network Units (ONUs). Depending on the bandwidth requirements of the broadband access services, a PDN can be implemented by using OC-3 or OC-12 (STM-1 or STM-4) transport networks.

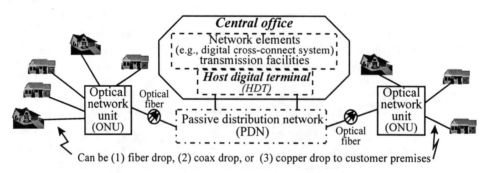

Figure 3-80   Fibers in the Loop: FTTH, FTTC, FTTS, HFC, etc.

A HDT supports the subtending Optical Network Units (ONUs) and facilities that interface the FITL system with the remainder of the transmission and operations networks. HDT may concentrate telecommunications traffic from all or part of its subtending ONUs for efficient feeder transport, and to present a highly utilized interface to local switching systems. The PDN may be implemented in one of the following ways: (1) point-to-point (i.e., one or two fiber to the ONU), or (2) multipoint (one or two fibers to the ONUs using passive splitters). The ONU provides the tariffed network interfaces, and future service interfaces for several residential and small business customers. Traditional telecommunications services may be provided over metallic twisted-pair drops and/or coaxial cables. The primary functions performed by the ONU are:

- Electrical-to-optical conversion, and vice versa.
- Analog-to-digital conversion for voice signals, and vice versa.
- Multiplexing (Time-Division Multiplexing; TDM).
- Maintenance for individual services and transmission facilities.

The last few feet between the ONU and the customer can be implemented with shielded or unshielded twisted pair metallic wiring, coaxial cable, fiber cable, or Hybrid Fiber Coax (HFC), based on the design requirements and decisions made in accordance with economic considerations.

Deregulation in the telecommunications industry allow cable TV operators to offer their customers many other services (besides cable TV) via the existing coaxial cables, that typically have a bandwidth as high as 700 MHz. Cable modems can also provide high-speed data services. The major components of a Cable TV (CATV) network (Figure 3-81) are listed as follows:

- Satellite (including earth stations)
- Microwave radio
- Head end (see Figures 3-81 and 3-82)
- Trunk systems (including coaxial cables and optical fiber cable; Figure 3-82)
- Automatic gain amplifiers (one-way or two-way)
- Passive devices (e.g., splitters, directional couplers, and subscriber taps)

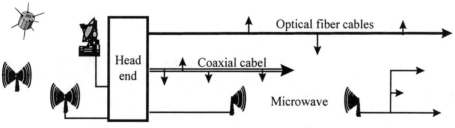

Figure 3-81   CATV Network Components.

It should be understood that coaxial cables are subjected to Radio Frequency Interference (RFI), which results in poor performance. Therefore, certain carrier frequencies with the available 700 MHz bandwidth may need to be avoided. In addition, coaxial cable attenuation increases rapidly for higher frequencies. Therefore, the best arrangement is to use optical fibers for main routes, and coaxial cables for the last few feet connected to the end user. The main advantages of this HFC arrangement are:

- Capability of integrating services
- Cost-effective (e.g., : fewer amplifiers required)
- Better system performance (less thermal noise due to heat reduction, and less RF interference: optical fibers are practically RFI free)
- Increased bandwidth for more users and different services

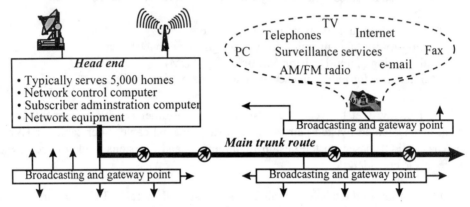

Figure 3-82   Broadband Access Services.

The switching points form the "hub" of a star switched network each serves a group of homes or business premises. In areas where it is too costly and difficult to bury network cables underground, the Multichannel Microwave Distribution Service (MMDS) may be a more cost-effective alternative for delivering TV, data, and voice signals (Figure 3-82). The computer not only manages "system housekeeping", conditional access, and customer billing, it also provides an interface for connection to the Public Switched Telephone Network (PSTN).

Example 3-22: Describe the HFC reference model shown in Figure 3-83.

As more telecommunications companies deploy "CATV-like" Hybrid Fiber Coax (HFC) architectures for providing both voice and video services, it is important to have a HFC reference model.

The HFC reference model consists of an "active" [working channel (WC)] and a an optional "protection" optical fiber path [Protection Channel (PC)] connected to the optical node [Optical Network Unit (ONU)]. From this optical node, the coaxial distribution cable is launched in four directions (branches 1, 2, 3 and 4) with each coax serving approximately 125 subscribers. The details of branch 1 are illustrated in Figure 3-83. At the head end (remote terminal or central office) HFC terminals are driven by an Uninterruptible Power Supply (UPS) with back-up battery. HFC terminals are connected to optical transmitters (WC and PC), and optical fibers, that are fed to the fiber-coax node. A launch amplifier is used to ensure the transmitted power is strong enough for transport, before the signals are distributed and dropped to customers. Other "express amplifiers" may be used along the path for signal amplification. The maximum number of active amplifiers, including the launch amplifier, is no more than 4 on any leg of coaxial cable.

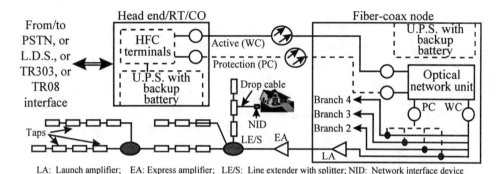

LA: Launch amplifier;   EA: Express amplifier;   LE/S: Line extender with splitter;   NID: Network interface device

Figure 3-83   HFC Reference Model.

Important considerations when implementing HFC networks are described as follows:

- Reliability: The reliability of the CATV network is typically engineered to meet the requirements of the broadcast video services (This may not be consistent with reliability requirements for telecommunications services).

- Powering: In a HFC network with centralized powering, it is sufficient to provide battery/generator backup at the fiber-coax node location serving approximately 500 subscribers. Centralized powering on the same coaxial cable that carries the telecommunications services is a feature of the HFC architecture that resolves most powering issues, and increases the reliability of the services.

- Architecture: The HFC architecture, as it exists today and will continue in the near future, uses the (5 ~ 30/40 MHz) band as the uplink (upstream: from subscriber to the head end) communications. This band is subject to ingress interference via the cable distribution plant from external sources. The downlink (downstream: from the head end to the subscriber) uses the (54 ~ 750 MHz) band for communications.

- Noise ingress: There are a number of steps that can be taken to either reduce the levels of ingress in the cable plant, or to mitigate the effects of ingress noise on the services offered. These actions are a combination of efforts by equipment manufacturers and service providers.

- Upgradability: Upgradability is a major strength of HFC networks. The fiber node can initially serve a large number of homes, or be scaled based on market needs.

- Services: The HFC network can support all of the switched and ATM/packet based services envisioned for the present and the near future. This includes: high-speed

Internet access, video on demand, "work at home", interactive services, games, transactional services, video telephony, tele-education, CATV services, etc.

3.9 SATELLITE BROADBAND ACCESS

The technology for satellite broadband access networking is still in the early research stage. It is believed that GEostationary Orbit (GEO) instead of Low-Earth Orbit (LEO) satellite systems may eventually be used for broadband access networks. This section briefly describes basic technical concepts and applications for satellite systems used in Wide Area Networks (WANs), that are also applicable for satellite broadband access networking. Generally, the following well-defined protocols are used to study/analyze new communications technologies. The particular characteristics that effect the performance of satellite communication systems are described as follows:

(1) **Transmission Control Protocol/Internet Protocol (TCP/IP)**: TCP/IP (described in Chapter 11) is the predominate protocol suite presently used for Internet applications. TCP is the data transport protocol from the "TCP/IP suite" that provides reliable data transmission and was designed to be robust in a variety of environments (including wireless and satellite networks). One feature of TCP is that it requires feedback to acknowledge successful data reception. Because of this feedback requirement, most TCP implementations do not allow efficient transmission over networks having a large bandwidth-delay characteristic. However, there are several parameters (e.g., segment size, timers, windows sizes, etc.) and congestion avoidance algorithms (e.g., slow start, selective retransmission, selective acknowledgement, etc.) that can be used to improve TCP performance. It should be understood that bandwidth-delay is an important issue for TCP satellite network applications. The theoretical throughput ($R_{throughput}$) for TCP, and the receiver buffer size (S_{buffer}) are given by the following equations:

$$R_{throughput} = \frac{S_{buffer}}{T_{round-trip}} \qquad (3\text{-}33)$$

$$S_{buffer} = W \times T_{roundtrip} \qquad (3\text{-}34)$$

where S_{buffer} is the storage size of the host computer's receiver buffer, $T_{round-trip}$ is the round-trip delay time, and W is the system bandwidth.

Example 3-23: Calculate the throughput and buffer size for a geostationary satellite with a round trip transmission (delay) time of 500 ms for various conditions.

(A) By applying Eq.(3-33), the throughput can be calculated for a receive buffer size of 8 kilobits as follows:

$$R_{throughput} = \frac{S_{buffer}}{T_{round-trip}} = \frac{8 \times 1024 \; bytes}{500 \; ms} = 16{,}384 \; bytes \, / \, s = 131.07 \; kbps$$

This procedure is repeated for other buffer sizes (16, 32, 84, and 128), and the corresponding throughput for each case is listed in Table 3-17.

Table 3-17   Theoretical Throughput for Geostationary Satellite.

Receiver buffer window size (kbytes)	Throughput (kbps)
8	131.072
16	262.144
32	524.288
64	1048.576
128	2097.152

(B) By applying Eq.(3-34), the required buffer size can be calculated for a bandwidth (W) = 64 kbps (DS0) as follows:

$$S_{buffer} = W \times T_{roundtrip} = \frac{64 \times 10^3 \times 500 \times 10^{-3}}{8 \times 1024} = 3.906 \text{ kbytes}$$

This procedure is repeated for other signals (DS1, DS3, STS-3, and STS-12), and the required buffer sizes are listed in Table 3-18.

Table 3-18   Required Buffer Sizes for Different Signals over Satellite.

Signal	Available bandwidth (Mbps)	Required buffer size (kbytes)
DS0	0.064	3.91
DS1	1.536	93.75
DS3	39.883	2,434.26
STS-3	135.102	8,245.97
STS-12	541.966	33,078.98

(2) **Asynchronous transfer mode (ATM) protocol**: The ATM protocol (described in Chapter 5) was originally conceived by the telecommunications industry to handle multimedia traffic over WANs. The features of ATM protocol are: connection-oriented (to reduce delay jitter, overhead bandwidth requirement, and buffer size), nearly-error-free fiber communication performance, guaranteed Quality of Service (QoS), ease of switching, and multimedia compliance.

(3) **Moving Picture Expert Group version 2 (MPEG-2)**: The MPEG-2 transport stream consists of 188-byte packets. The packets contain program specific information [e.g., Program Association Table (PAT), Program Map Tables (PMTs), Conditional Access Tables (CATs), Network Information Table (NIT), Program Clock Reference (PCR), and Program Element Stream (PES) packets. The transport stream is a multiplexed protocol that supports video, audio, mixed video and audio, and user specific data. The protocol is unreliable, and is insensitive to delay. If a reliable transport protocol is encapsulated into an MPEG-2 data stream, then the encapsulated protocol becomes delay sensitive.

3.9.1 Fixed Service Satellite Communication Network

A Satellite-Switched Code Division Multiple Access (SS/CDMA) network has been proposed for fixed services. Figure 3-84 shows a satellite network that utilizes an intersatellite link (ISL) to connect two adjacent satellites. These services use geostationary satellites, and require two-way communications between end satellite users having Ultra-Small Aperture Terminals (USATs) with a dish diameter of 26 inches. It must also have the capability to perform call routing onboard the satellite. Therefore, the satellite operates as a ***repeater*** as well as a ***switching center***. SS/CDMA service will only become economically feasible if the satellite system capacity and throughput are sufficiently high, and has a service quality comparable to wireline services. Therefore, the SS/CDMA system must provide a high spectral efficiency, which can be achieved by introducing multiple-beam satellite antennas that allow reuse of the available spectrum. Hence, the system must achieve an efficient utilization of the mass and power available on board the spacecraft. One factor that should be considered is that the power of the transceiver can be reduced by using new modulation methods.

Each satellite is linked to PSTNs via a satellite gateway office. In addition, each satellite has a direct link to CPE. The satellites can route calls (via the satellite gateway) between a PSTN and CPE, or between a CPE and another CPE.

This satellite network can offer fixed services for both circuit-switched (e.g., voice, video, and data) and packet-switched (e.g., data) signals. The signal rates for voice or data can be 16, 32, and 64 kbps; BRI-ISDN (2B+D) operates at 144 kbps (requires a BER of 10^{-6} or better), video at a rate of 384 kbps (requires a BER of 10^{-8} or better), DS1 signals operate at 1.544 Mbps, and E1 signals operate at 2.048 Mbps.

3.9.2 System Overview: Fixed Service Satellite System

As previously indicated, SS/CDMA is a satellite system intended to provide fixed services via geostationary satellites. It should provide both multiple access and switching functions for a multibeam satellite. The satellite must perform the switching function. That is, the user traffic channels will be switched from any "uplink" to any "downlink" beam. This can be implemented by using an "onboard" Code-Division Switch (CDS). Multiple access is achieved by using space, frequency, and code division techniques, which are briefly described as follows:

- Space-division multiple access: It applies spotbeam antennas so that the available frequency spectrum can be re-used.

- Frequency-division: It segments the available spectrum into several frequency bands, each having a predefined bandwidth of W Hz (presently, W is around 10 MHz).

- Code-Division Multiple Access: It provides access for each user within each frequency band (CDMA spreads user data over the bandwidth W) and within each beam.

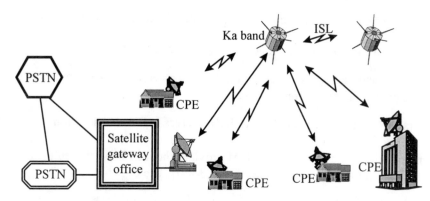

Figure 3-84   Fixed Service Satellite Network.

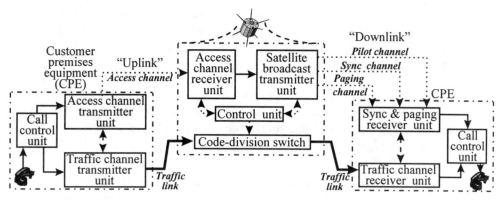

Figure 3-85   Functional Building Blocks of A SS/CDMA Satellite Link.

Figure 3-85 shows the functional block of a SS/CDMA fixed service satellite link. Only one direction of communication is illustrated. The four components of this link are:

(1) An "uplink" CPE: It is one of the two required CPE elements, and it consists of:

* Terminal Equipment (TE) functions as a voice terminal.

* Access Channel Transmitter Unit (ACTU) provides an access channel from the CPE to the satellite's Access Channel Receiver Unit (ACRU).

* Traffic Channel Transmitter Unit (TCTU) provides a traffic channel from the CPE to the satellite's Code-Division Switch (CDS)

* Call Control Unit (CCU) manages the interface between the end user, ACTU, TCTU. The CCU, ACTU and TCTU are known as the Subscriber Unit (SU).

(2) The satellite (onboard system) consists of the Code-Division Switch (CDS), an Access Channel Receiver Unit (ACRU), a Satellite Broadcast Transmitter Unit (SBTU), and a Control Unit (CU).

(3) The receiver CPE [in addition to the Call Control Unit (CCU)], consists of:

*   Terminal Equipment (TE) functions as a voice terminal.

*   A Sync and Paging Receiver Unit (S&PRU) receives and implements pilot, sync, and paging channels from the satellite.

*   Traffic Channel Receiver Unit (TCRU) receives traffic channels from the satellite.

(4) Common Air Interface (CAI): The CAI provides the interface between the Subscriber Units (SUs) [or the gateway office (Figure 3-84)]. The CAI consists of:

*   Traffic channels: They carry voice, data and signaling information between the subscriber units. The multiple access and modulation techniques are based on the Spectrally Efficient CDMA (SE-CDMA). The SE-CDMA will provide orthogonal separation of traffic channels within the beam. For some implementation schemes, satellite beams may also be separated by orthogonal codes. However, the onboard traffic channels are switched between the "uplink" and the "downlink" beams without data decoding.

*   Control channels: The control channel for the "uplink" is the access channel. Similarly, the control channel for the "downlink" are the pilot, sync, and paging channels. The sync and pilot channels provide timing and synchronization for the system. The paging channel provides signaling messages from/to the satellite.

The SS/CDMA system is one of several proposed satellite systems for fixed services. It is intended to be a high-capacity switching satellite system hat will operate at very low power while offering high-quality services. Multiple access and modulation schemes can achieve better frequency spectrum utilization, lower mass, and more efficient use of power onboard the spacecraft. In addition, it will achieve higher spectral efficiency with a low S/N ratio, and an excellent Bit Error Rate (BER). One other feature is that the onboard Code Division Switch (CDS) with its associated demand assignment control algorithm allows efficient access and routing of both circuit-switching and packet-switching traffic.

Review Questions II for Chapter 3:

(14) In an ASDL system, the digital bearer channels can only carry ____-speed signals in the direction from the customer premises to the network.

(15) At the remote end (the subscriber), a splitter is equipped with two filters: a _____ is used for providing POTS services, and the other is _____ for providing ADSL services. At the central office end, similar requirement is needed; a _____ is to route ADSL signals to a broadband network while a _____ is to route POTS signals to the traditional (narrowband) PSTN.

(16) ADSL downstream (downlink) supports: _____ and _____. ADSL adopts the superframe structure; a superframe consists of _____ ADSL data frames, followed by a synchronous frame, known as _____ synchronization symbol.

(17) (True, False) For Very high Digital Subscriber Line (VDSL) systems, the spectral (bandwidth) efficiency can be as high as about 10 bps/Hz.

(18) The spectral allocation for VDSL is in the range of ____ kHz _____ MHz.

(19) (True, False) VDSL can be adopted for both symmetrical and asymmetrical services.

(20) One of the attractive VDSL features is the use of _____ as a transport layer to deliver extremely high data rate.

(21) The proposed VDSL modulation is _____
(SDMT) which is a combination of _____ modulation and
_____ (TDD; also known as _____ technique).

(22) Wireless Local Loop (WLL) technology can be applied for _____ applications to replace the traditional analog local loop.

(23) (True, False) For multiple radio access, WLL can apply wideband-CDMA (WB-CDMA) or infrared technology.

(24) Besides multiplexing and modulation, WLL requires the following technologies (they are also used in cellular networks): _____, _____, _____ and _____.

(25) Presently optical wireless technology uses the _____ (short wavelength, or very high frequency) technology.

(26) (True, False) Unlike optical fiber systems, optical wireless systems have several issues and limitations, such as atmospheric loss and scattering, laser power and eye safety.

(27) The Hybrid Fiber Coax ((HFC) reference model consists of an _____, and a _____ optical fiber path; also known as the _____ and the _____.

Appendix 3-1

Modem Standards, Protocols, and Applications
(Listed in alphanumeric order)

Bell 103: Full duplex, 300 bps (300 baud), FSK, North American applications, supported, but no longer in general use.

Bell 212A: Full duplex, 1200 bps (600 baud), DPSK, North American applications supported, but has limited use.

ITU-T V.21: Full duplex, 300 bps (300 baud), FSK, based on Bell 103 (but not compatible), International applications supported, but no longer in general use.

ITU-T V.22: Half duplex, 1200 bps (600 baud), DPSK, based on Bell 212A (but not compatible), International applications supported, but has limited use.

ITU-T V.22bis: Full duplex, 2400 bps (600 baud), 16-QAM, North American and International applications. *Note that "bis" is a Latin adverb indicating "second or again", and in this context implies enhancement of the original ITU-T V.22 standard.*

ITU-T V.26: Full duplex, 2400 bps (1200 baud), 4-PSK, International applications, supported, but has limited use.

ITU-T V.27: Full duplex, 4800 bps (1600 baud), 8-PSK, International applications, supported, but has limited use.

ITU-T V.29: Half duplex, 9600 bps (2400 baud), 16-QAM, International applications, not considered an "approved standard", hence there are compatibility issues between different equipment vendors; typically used for FAX communications rather than modem, but limited use.

ITU-T V.32: Full duplex, 9600 bps (2400 baud), 2D-TCM, North American and International applications, considered a "de facto" standard.

ITU-T V.32bis: Full duplex, 14.4 kbps (2400 baud), 4D-TCM, North American and International applications, considered an adaptation of V.32.

ITU-T V.32fast: Full duplex, 28.8kbps (2400 baud), 4D-TCM, North American and International applications, supported, but generally replaced by V.34.

ITU-T V.34: Full duplex, 28.8 kbps (2400 baud), 4D-TCM, used for 28.8 kbps analog loop applications, with optional speeds of 31.2 and 33.6 kbps, North American and International applications, supported, considered a "de facto" standard.

ITU-T V.90: Full duplex, 56 kbps (2400 baud), 4D-TCM, relies on telecommunications service providers loop to support direct transmission of digital signals, typically implemented as 56 kbps for the downlink, and 33.6 kbps for the uplink direction; supported for both North American and International applications.

CHAPTER 4

Signaling Concepts and Applications

Chapter Objectives

Upon the completion of this chapter, you should be able to:

- Describe the relationships between signaling systems, switching systems, and transmission systems used for telecommunications.

- Describe signaling objects, signaling types and signaling applications (e.g., customer loop signaling, inter-office signaling, special service signaling, and network management signaling).

- Discuss call progress tones in North America (e.g., audible tone and call progress tones), and loop signaling.

- Discuss inter-exchange signaling: dial pulses and combinations of multi-frequency (MF) for address signaling between exchanges. Give examples of signaling techniques used for inter-exchange call signaling.

- Discuss various early signaling systems (e.g., ITU-T signaling system R1, signaling system R2, signaling systems No. 5 and No. 6) and explain their limitations.

- Describe common channel signaling systems including advantages, operational modes, network configurations, system interconnections, and ITU-T signaling system No. 7 (SS7) (i.e., its functional blocks, OSI layer relationship, message transfer parts, user parts, link-by-link signaling, and numbering plan).

4.1 INTRODUCTION

Information movement and management (IM&M) involves the application of transmission systems, switching systems, and signaling systems. Without the intelligent functions of a signaling system, IM&M over complicated telecommunications networks is not possible. Signaling controls for the transfer of information between switching system entities, which work together with associated transmission media and transmission terminals, to establish communication highways for voice, data, video, fax, etc. Signaling technology starts with the simple Direct Current (DC) transitions, and extends to the most sophisticated common channel signaling formats, signaling has kept place with the advancement of communications systems, and is no doubt the nerve center of every telecommunications network.

Figure 4-1   Relationship between Transmission, Switching and Signaling.

A simplified description of signaling is as follows: the process begins with the acceptance of information from an information source. The, additional functions such as signal encoding, signal generation (e.g., modulation), signal transmission, signal detection/demodulation, and signal decoding are applied to the original information. This ultimately results in the delivery of information to its intended destination.

According to ITU-T Recommendation Q.9, *signaling* is defined as follows: "*The exchange of information (other than speech, data, or video) specifically concerned with the establishment, release and other control of calls, and network management, in automatic telecommunications operation.*" The applications of signaling can be classified into two groups: (1) The subscriber signaling and (2) The trunk signaling (i.e., inter-switch or inter-office signaling). Signaling functions can be categorized into four basic functions:

1. **Supervision (supervisory) signaling**: It is used to initiate a call request on lines or trunks (known as line signaling on trunks), to hold or to release an established connection, to initiate or to terminate charging, or to recall an operator on an established connection.

2. **Address signaling**: It is used to convey the calling or the called subscriber's directory number, the area code, an access code, or a PBX tie-trunk access code. Address signaling also contains information indicating the destination of a call initiated by a customer, network facility, ..., etc.

3. **Call progress signaling**: It is used to convey call-progress or call-failure information to subscribers or operators. This type of signaling often utilizes audible tones or recorded announcements.

4. **Network management signaling:** It is used to control the bulk assignment of circuits, or to modify the operating characteristics of switching systems in response to overload conditions.

4.1.1 What is a Signaling System?

"A signaling system is a *language* that enables two switching equipment (element) to converse for the purpose of setting up calls. Like any other language, it possesses a vocabulary of varying size and varying precision, i.e., a list of signals which may also vary in size and a *syntax* in the form of a more or less complex set of rules governing the assembly of these signals." [A direct quote from Mr. Jouty, former Chairman of ITU-T Study Group XI: Telephone Signaling.]

4.1.2 Signaling Objectives

The objectives of signaling are to convey information between the subscribers and their associated switching systems (e.g., the originating and the terminating local exchange switches), and to convey information between the exchanges involved in the voice or data connection. The transfer of signaling information should occur at high speed. In the US, signaling transport utilizes the high-speed T1 digital carrier system instead of the voice-frequency channel [a T1 system has a capacity of 24 voice (VF) channels; discussed in Chapter 6.] Accurate signaling information transfer requires reliable signaling systems. The information transferred concerns the "setting up" and "disconnection" of voice and data connections as well as providing information for timing and billing for subscriber calls. Reliable and economical implementation of signal transfer technologies is essential.

4.1.3 Signaling Types

There are many types of signaling systems used in telecommunications networks. The various signaling types can be grouped into two basic categories as follows:

- **Channel associated signaling**: Customer-generated dial pulses used to notify the local exchange switch of the requested telephone number (or address) is the most common form of channel associated signaling. Another frequently used example of channel associated signaling are the multi-frequency tones (discussed later in this chapter) used between two switching systems. This type of signaling is also referred to as "in-band signaling" as defined in Chapter 1 (see Figure 1-17). In-band signaling information is carried on the same path set-up to convey speech or data information.

Figure 4-2(A) shows this type of signaling. Note that the channel is used to carry both the voice/data information and the signaling information simultaneously. In this scheme, the signaling information may occupy the same frequency band as the voice/data does. Similarly, signaling information may also be carried in the same channel that voice/data occupies, however, when it is carried above or below the range of voice/data frequency band. It is referred to as "out-of-band signaling" as shown in Figure 4-2(A) and (C).

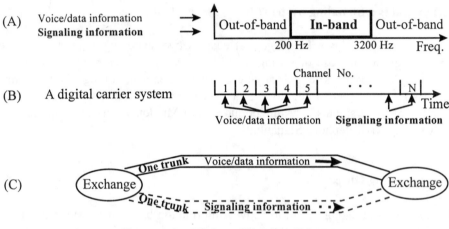

Figure 4-2   Various Signaling Types.

- **Common channel signaling**:   Typical examples of common channel signaling techniques are: Common Channel Signaling System (CCS7) used in the North American telecommunications networks, the Signaling System 7 (SS7) used in the other parts of the world, and the D channels of an Integrated Services Digital Network (ISDN) Basic Rate Interface (BRI) and the ISDN Primary Rate Interface (PRI). The D channel signaling method is discussed later in this chapter. In Common Channel Signaling Systems (CCS7 and SS7), a circuit that is not used for speech or data information transport can be used to carry the signaling information for several *speech circuits*. This type of signaling technology is also known as "out-of-band signaling" or "separate channel signaling". Separate channel signaling may adopt either of the two following implementations:

  * **Built-in separate channel signaling**: A typical digital carrier system has a capacity of many voice channels. One or more of these channels can be reserved to carry signaling information rather than the voice or data information as shown Figure 4-2(B). Some examples of built-in channel signaling are: the D channel of either ISDN BRI or PRI transport; the signaling channels (known as channels A, B, C and D; discussed in Chapter 6) of a DS1 signal, and channels 0 and 16 of an E-1 signal (discussed in Chapter 6).

* **Common Channel Signaling**: The signaling information is carried on a totally separate circuit that is not part of the carrier system used for the voice or the data information as shown in Figure 4-2(C). Some examples are: CCS7 signaling systems used in the North American networks, and SS7 signaling systems used in other networks (discussed later in this chapter).

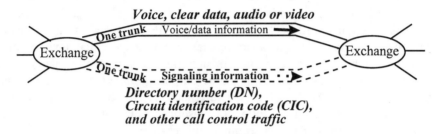

Figure 4-3   Common Channel with its Various Traffics.

In CCS7 (North America), the common channels are 56 kbps and full-duplex digital facilities (also known as T1 digital carrier systems; discussed in Chapter 6). In SS7 (ITU-T), the common channels are 64 kbps and full duplex digital facilities (also known as E-1 digital carrier systems; discussed in Chapter 6). For future digital networking, it is expected that other higher-rate digital trunks will be used for common channel signaling.

Common channel signaling results in a more elaborate control logic at the exchange offices, but eliminates the need for individual call control logic on a "per circuit basis". It also implies that common channel signaling technology has moved from pulse-based or frequency-based signaling to message-based signaling. Therefore, by applying additional encoding, the exchange offices can identify the services that the reserved circuit (trunk) will be used for voice, clear data, audio, or other types of information.

4.1.4 Modes of Common Channel Signaling

By conveying call control information over dedicated trunks, service providers can increase the revenue-producing utilization of the inter-office trunks which carry subscriber voice or data information. There are three common channel signaling modes that correspond to different topologies of the common channel signaling networks. They are listed below and will be discussed further in section 4.6.2 of this chapter.

* Associated mode
* Non-associated mode
* Quasi-associated mode

4.1.5 Signaling System Classification

There are many different types of signaling systems, but they can be classified into four main groups:

1. **Customer loop signaling**: This is the most elementary type of signaling, which is used between the customers (subscribers) and the serving exchange office. As described in Chapter 1, when a customer lifts up the handset or performs an equivalent action, an off-hook signal is sent to the exchange indicating a call request has been initiated. The exchange generates and sends a dial tone informing the customer to send the address information (i.e., dial the called party's telephone number). The address information can be sent using dial pulses, the dual-tone multi-frequency signal (DTMF; discussed in Chapter 1), or by operating push buttons. Other call processing signals follow (as described in Chapter 1). More detailed loop signaling is described in a later section of this chapter.

2. **Inter-exchange circuit signaling**: The signaling used for inter-exchange offices is more complicated than loop signaling. The implementation can be either one-way or two-way operation. For a two-way operation, the exchange of control signals in both directions must be symmetrical. There are many types of control signals. Some are mandatory (required) while others are optional. The minimum set of control signals for an inter-exchange office are:

   * Seizure of the circuit: it is used on the forward (calling) direction.

   * "Proceed-to-send" or "start-to-dial signal" on the reverse (backward) direction (i.e., a register has been assigned to receive the address information).

   * Address information on the forward direction: it can be dial pulses or multifrequency coded (MFC) pulses.

   * Answer signal (from the called party): It is used on the backward direction.

   * Clear back signal on the backward direction: It indicates that the called party has hung up the phone.

   * Clear forward signal on the forward direction: It indicates that the calling party has hung up the phone.

3. **Special service signaling**: These control signals are used to provide special services such as private (dedicated or "nailed up") lines or foreign exchange lines. These control signals are similar to loop signaling and inter-exchange signaling. However, they may be applied in a different time sequence and with different signal representations.

4. **Network management signaling**: They include traffic flow control signals for implementing network bandwidth management, signals for monitoring network performance, signals for reporting network performance, and network administration

functions. Two typical administration functions are: collection of billing information, and data routing.

4.2 Call Progress Tones in North America

There are many call progress tones adapted in North American network. Some of these tones are audible and the others are not.

4.2.1 Audible Tones

The fundamental audible tones are: the audible ringing tone, busy tone (from the station, or the customer equipment), "fast busy" tone (indicates network congestion), call waiting tone, dial tone, off-hook alert tone, recording warning tone, and ringing tone.

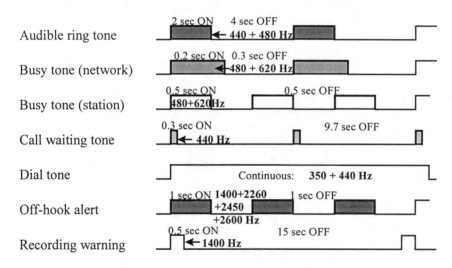

Figure 4-4   Audible Tones Used in North America.

Figure 4-4 shows the timing pattern of the commonly used audible tones in North American networks. The audible ring tone is a two-tone signal, which consists of 440 Hz and 480 Hz. The tone has a period of 6 seconds (it is turned on for 2 seconds and turned off for 4 seconds). The " busy tone" (network congestion or re-order tone) is also a two-tone signal with 480 Hz and 620 Hz tones. Its period is 0.5 seconds (the tone is on for 0.2 seconds and off for 0.3 seconds). The station busy tone is similar to the network busy tone except the period is 1 second (the tone is on for 0.5 and off for 0.5 seconds). A call waiting tone is a single tone signal with a frequency of 440 Hz, and it has a period of 10 seconds (the tone is on for 0.3 seconds and off for 9.7 seconds). A dial tone a continuous two-tone signal with 350 Hz and 440 Hz. An off-hook alert tone has a period of 2 seconds (the tone is on for 1 second and off for 1 second), and is a multifrequency tone with four frequencies (1400, 2260, 2450 and 2600 Hz). A recording warning tone is another single frequency tone (1400 Hz) with a period of 15.5 seconds (the tone is on for 0.5 seconds and off for 15 seconds).

4.2.2 More Call Progress Tones

Call progress tones can be classified into two groups as follows:

1. Customer tones: They include the low tone, the high tone, the dial tone, the audible ring tone, the line busy tone, the network busy tone, the re-order tone, and so on; and they are briefly described as follows:

   * Low tone: A two-tone signal (480 + 620 Hz). Its timing pattern is varied. The tone level is −24 dBm0 per frequency. (The unit of dBm0 has been discussed in Chapter 1: Appendix 1-2)

   * High tone: A single tone signal with a frequency of 480 Hz (the most common), 400 Hz, or 500 Hz. Its timing pattern is varied. The tone level is −17 dBm0.

   * Calling card service-prompt tone: A two-tone signal (941 + 1477 Hz) for 60 ms with a power level of −10 dBm0 per frequency (at −3 TLP). This tone is followed by another two-tone signal (440 + 350 Hz), which last for 940 ms. The power level of the second two-tone signal decays from −10 dBm0 per frequency (the same power level as the first two-tone signal) at −3 TLP with a time constant of 200 ms.

   * Audible ring tone: A two-tone signal previously described (Figure 4-4), with a tone level of −19 dBm0.

   * Data set answer back tone: A steady single tone signal with a frequency of 2025 Hz. It has a power level of −13 dBm0.

   * Deposit coin tone: A two-tone signal (see "Low tone") and is a continuous and steady tone.

   * Dial tone:: A steady (continuous) two-tone signal (see Figure 4-4) with a power level of −13 dBm0.

   * Line busy tone: A two-tone signal (see busy tone station and Low tone, as shown in Figure 4-4). This tone has a period of one second (the tone is on for 0.5 seconds and off for 0.5 seconds).

   * Local reorder: A two-tone signal (see "Low tone"). The period of this tone is 0.5 seconds (the tone is on for 0.3 seconds and off for 0.2 seconds).

* Recall dial tone: A two-tone signal (350 + 440 Hz). This tone's timing pattern consists of three bursts (0.1 second ON, 0.1 second 0.1 OFF, and STEADY ON). The tone's power level is −13 dBm0 per frequency.

* Receiver off-hook tone: A multi-frequency tone (1400 + 2260 + 2450 + 2600 Hz). This tone has a period of 0.2 seconds (The tone is on for 0.1 second and off for 0.1 second). The tone power level is (3.0 ~ −6.0) dBm0 per frequency.

* Reorder connected tone: A single frequency tone (400 Hz). The tone is on for 0.5 second burst every 5 seconds.

* Reorder warning tone: A single frequency tone (1400 Hz). The tone is on for 0.5 second burst every 15 seconds.

* Reverting tone: It's a two-tone signal (480 + 620 Hz). This tone has a period of one second (the tone is on for 0.5 seconds and off for 0.5 seconds). The tone power level is −24 dBm0 per frequency.

* Toll reorder tone: It's a two-tone signal (480 + 620 Hz). This tone has a period of 0.5 seconds (the tone is on for 0.2 seconds and off for 0.3 seconds).

2. Operator tones: These tones are used for network operations. They include the centralized intercept bureau order tone, the coin collect tone, the coin determination tone, the coin return tone, and so on; and, they are described as follows:

* Centralized intercept bureau order tone: A single frequency tone with a frequency of 1850 Hz (lasting 500 ms), and with a power level of −17 dBm0.

* Class of service: The tone used to distinguish the class of service can be (1) a two-tone signal (480 + 620 Hz), which lasts for 0.5 to 1 sec once (2) a two-tone signal as the "high tone", which lasts for 0.5 to 1 sec once, or (3) no tone signal.

* Coin collect tone: A steady two-tone signal (see low tone).

* Coin denomination tone (three-slot stations): The 5-cent slot is a single frequency tone (one tap, 1050-1100 Hz). The 10-cent slot is the same as the 5-cent tone except being two taps. The 25-cent slot is a single frequency tone (one tap: 800 Hz gong).

* Coin return tone: A two-tone signal (see high tone) that lasts for 0.5 to 1 sec once.

* Dial normal transmission signal: A steady two-tone signal (see low tone).

* Dial off normal tone: A steady two-tone signal (see low tone).

* Group busy tone: It is a steady two-tone signal (see low tone).

* Number checking tone: A steady two-tone signal (see low tone) or a steady single frequency tone of 135 Hz.

* Order (single-order) tone: A two-tone signal (see high tone) that lasts 0.5 sec.

* Order (two-order) tone: A two-tone signal (see high tone) that has two short spurts in quick succession.

* Order (triple-order) tone: A two-tone signal (see high tone) that has three short spurts in quick succession.

* Order (quadruple-order) tone: A two-tone signal (see high tone) that has four short spurts in quick succession.

* Permanent signal: A steady two-tone signal (see high tone).

* Proceed to send tone (International direct distance dial: IDDD): A steady single frequency (480 Hz) tone with a power level of −22 dBm0.

* Service observing tone: A steady (continuous) single tone signal (135 Hz).

* Vacant position tone: A steady two-tone signal (see low tone).

* Warning tone: A steady two-tone signal (see high tone).

4.3 LOOP SIGNALING

Signaling is the process of transferring (control) messages between any two point in a telecommunications network. It is also used to establish or to disconnect the communication paths between any two points. Other functions associated with call set-up and call connection release are also performed by signaling. The two major signaling messages sent over the network can be characterized as (1) customer signaling, and (2) inter-office signaling as shown in Figure 4-5.

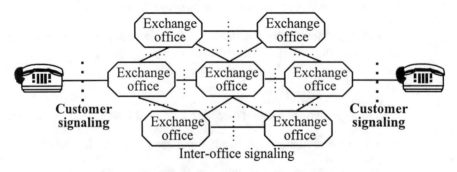

Figure 4-5   Customer Signaling and Inter-office Signaling.

Customer signaling refers to all control message between the customer and the exchange office (i.e., where the switching machines, transmission equipment, and terminals reside). An access line, (before the introduction of the PBX and the ISDN lines), is a facility between the customer station and the exchange office (or the switching network). The six main classes of control signal used for the access lines are described as follows.

1. **Call process signals**: Examples of "call process signals" are various audible tones (such as dial tone, ringing tone, audible ringing tone, audible busy tone, and congestion tones: previously discussed in Chapter 1) or announcements that inform the subscriber of the call progress status.

2. **Supervisory signals**: They are used by the customer to initiate a request for service, to hold a call connection, or to release a connection after the call has been completed. They are also used to initiate and to terminate charges for billing purposes.

3. **Control signals**: They are used for auxiliary functions with equipment connections to the Point of Termination (POT) or to the demarcation point. Examples are the toll connection restriction and the party identification functions.

4. **Address signals**: They provide information to the network to determine the destination of a call,. Typically the address signal is the called party telephone number, sometimes referred to as the directory number.

5. **Alerting signals**: They are supplied by the network to alert the customer equipment of an incoming call, and to the calling party for "call request confirmation".

6. **Testing signals**: They are used for measuring telephone circuit performance, such as noise bit error rate via a loopback arrangement, and also to "trouble shoot" circuit problem and/or equipment faults.

Loop signaling is also used as a method for the customer signaling since it consists of supervisory control messages. By lifting the telephone handset (or other equivalent action), the loop connecting the customer equipment to the switch at the exchange office is established. This off-hook signal alerts the switching machine to prepare for a call set-up request.

Similarly, at the end of a call, the handset is placed back in its cradle and the switchhook contact on the telephone set is opened. This action generates an on-hook signal that alerts the switching machine to release the connection because the call has ended (i.e., the subscriber has hung-up the phone).

Four aspects of loop supervision are briefly described as follows:

• The **on-hook signal** is continuously transmitted from an idle telephone (i.e., switchhook contact is open) to its local exchange switch so that the switch "knows" that this customer is available to receive an incoming call.

- The *off-hook signal* transmitted from the calling telephone is used to request service from the local exchange switch. Thus, this signal (i.e., switchhook contact closed) initiates the switching machine operations.

- The *off-hook supervision signal* is transmitted when the called party answers the call. This signal (i.e., called party switchhook contact closed) permits the local exchange to complete the call setup process. The off-hook supervision signal can also be used to indicate that the called party is busy and cannot accept the new incoming call.

- The *on-hook supervision signal* can be transmitted from either the calling telephone or the called telephone when either party has hung up the telephone. This signal (i.e., switchhook contact open) allows the local exchange switch to release the connection when the call has ended.

The signaling for the circuit (loop) connecting the customer and the local exchange switch can be either "loop-start signaling" or "ground-start" signaling. These concepts are described in the following sections.

4.3.1 Loop-Start Signaling

Loop-start signaling is the most common signaling scheme used in the North American Public Switched Telephone Network (PSTN). This signaling technology is applied in the network to provide two-way services between the customers and the network, and is used in the following four types of "two-way" service:

1. Public Telephone Service (PTS)
2. Message Telecommunications Service (MTS) for residence and business applications
3. Attendant-handled call service on manual Private Branch Exchange (PBX) or Automatic Call Distributor (ACD)
4. Manual or automatic data service

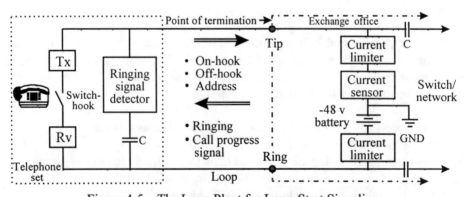

Figure 4-5   The Loop Plant for Loop-Start Signaling.

Figure 4-5 shows the loop plant connection between the customer's equipment (telephone set) and the local exchange switch. In the exchange office, a −48 v battery is used to power "loop plant". The positive end of this power supply is normally connected to ground (GND) as shown in Figure 4-5. Some digital switch systems (e.g., 5ESS) apply a "floating line supply" during a portion of the call. There is a current limiter on each end of the power supply to prevent excess currents. Current limiters are provided for the subscribers safety and to prevent equipment damage in failure conditions.

For the loop-start scheme, the "tip" conductor is connected to the positive end of this power while the "ring" conductor is connected to the negative end of the power supply. At the customer equipment, besides the transmitter (Tx) and the receiver (Rv) components, there is a switchhook (or equivalent) for initiating off-hook and on-hook signaling which is sent to the local exchange switch. There is also a dial pad (not shown in Figure 4-5) for generating the address signal. Another important element in the customer equipment is the ringing signal detector, which is used to detect the ringing signal generated by the network when an incoming call is intended for this customer.

The voltage across the subscriber equipment is nominally −48 volts DC. However, for ringing line conditions, the ac voltage can be as high as 105 v and as low as 0 v, with a nominal voltage of 88 v and 20 Hz superimposed on the −48 v. With "superimposed-voltage selective ringing at 60 Hz", the ringing signal is applied between tip and ring. Typically, the ringing and superimposed dc voltage is on the ring connection, and the tip is grounded.

The ringing signal is not a dc signal. It consists of a ringing interval of 2-seconds followed by a silent interval of 4-seconds (see Figure 1-26 of Chapter 1). With this implementation, a central office line can be seized for up to four seconds before this condition is recognized by the customer's equipment (i.e., the switch allocates the loop and sends the alerting signal before the subscriber hears the phone ringing).

4.3.2 Ground-Start Signaling

The "ground-start" signaling scheme was introduced in the early 1920s. Its purpose was to reduce the likelihood of seizure of the facility by both ends of the circuit during the 4-second silent interval of the ringing signal (Figure 1-26 in Chapter 1). Ground-start signaling is typically used on one-way or two-way PBX central office facilities with Direct Outward Dialing (DOD), and "attendant-handled" incoming-call service. In addition, ground-start signaling has been applied for Automatic Call Distributor (ACD).

In these applications ground-start signaling, instead of loop-start signaling, is specified for the following reasons.

- The ground-start lines provide a signal that can act as a "start-dial" signal.

- The ground-start lines provide a positive indication for a new call.
- The ground-start lines prevent unauthorized calls.
- The ground-start lines send a notification to the calling and the called parties to indicate "distant-end disconnect" (under normal operation).

In the idle state (i.e., on-hook) the network applies a negative DC voltage to the "ring" conductor while the "tip" conductor is held floating. The telephone set presents an open circuit "tip-to-ring" and "ring-to-ground". The telephone set also has a ground detector connected from "tip-to-battery" (off-hook) and a resistor from "ring-to-ground".

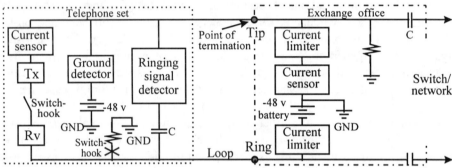

Figure 4-6   The Loop Plant for Ground-Start Signaling.

In the ground-start signaling configuration the telephone set grounds the "ring" side of the line by operating the switchhook contacts to initiate a call. The resultant current flowing through the ring conductor is detected by the switching equipment in the exchange office, and the "request for call" is recognized (Figure 4-6).

The telephone set remains in the loop mode for the duration of the addressing, call processing, and communications states. The line will revert to ground start mode only after the telephone set or the network goes to the "on hook" state.

The switching equipment connects a ringing circuit to the line when a call to the telephone set is initiated. This procedure applies "ground-to-tip", "negative battery-to-ring", and a 20 Hz "ringing-to-ring" signal. The ground makes the line busy at the telephone set. When the subscriber answers the call by closing the switchhook contact in the telephone set (i.e., goes "off hook"), the switching equipment will respond by "tripping" (disconnecting) the ringing signal and connecting the talking path.

4.4 INTER-EXCHANGE SIGNALING

The address signaling within the network (between any two exchanges) can be either dial pulses or combinations of Multi-Frequency (MF) tones. In modern networks MF tones are used to represent digits or special signals, which are referred to as inter-register signals. Table 4-1 lists examples of inter-register signal combinations.

Table 4-1   Multi-frequency Codes/Applications.

Code	Frequencies (Hz)	Signaling system	
		ITU-T system 5 and system R1	Expanded inband
0+1	700 & 900	1	
0+2	700 & 1,100	2	Coin collect
1+2	900 & 1,100	3	
0+4	700 & 1,300	4	
1+4	900 & 1,300	5	
2+4	1,100 & 1,300	6	
0+7	700 & 1,500	7	
1+7	900 & 1,500	8	Operator released
2+7	1,100 & 1,500	9	
4+7	1,300 & 1,500	0	Operator attached
2+10	1,100 & 1,700	KP, KP1*	Coin return
7+10	1,500 & 1,700	ST	Coin collect/operator released
1+10	900 & 1,700	KP2*	Ringback
0+10	700 & 1,700	Code 11*	
4+10	1,300 & 1,700	Code 12*	

Six frequencies: 700, 900, 1100, 1300, 1500 and 1700 Hz
* KP, KP1, KP2 = MF receiver gate opener (or start)
* ST = End of pulsing (or stop)
* Code 11 & Code 12 = Operator codes

There are six different frequencies are used in this multi-frequency scheme: 700, 900, 1,100, 1,300, 1,500 and 1,700 Hz (Note that 200 Hz separates any two adjacent frequencies). These MF tones are coded as "0, 1, 2, 4, 7, and 10", respectively.

The MF signals used for ITU-T signaling system No. 5 and system R1. The MF signals are transmitted in the forward direction only, and are used to transfer valid number (address) information over communication trunks. Each combination of two frequencies represents a pulse (dual-tone pulse as defined in Chapter 1) and each pulse represents an address digit. There are 10 digits (0, 1, 2, ..., and 9), and five other special control signals (see Table 4-1). Among these five signals, one is designated for indicating the beginning of pulsing (the key pulse, KP or KP1) for terminal traffic, one for transit traffic (KP2), one for indicating the end (stop) of the pulsing (ST), and Code 11 and Code 12 are used for operator applications. These pulses are sent over the transmitting channels. The receiver detects the pulses and processes them as digital information that is used to control equipment and processors that establish connections through the exchange switches. MF signaling is an application of in-band signaling technology.

Example 4-1: Signaling techniques for Inter-exchange call signaling.

In this example, a call is established by two or more exchanges. The description is given for a call that involves two exchanges: the one receiving the request for service is designated as the "originating exchange" and the one delivering the call to the called party is the "terminating exchange".

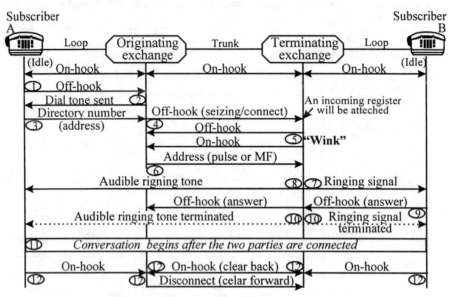

Figure 4-7   Inter-exchange Signaling (Call Process).

When there is no activity, both subscribers are said in the "on-hook state" as shown in Figure 4-7. It is assumed that subscriber A becomes active, and Figure 4-7 illustrates the inter-exchange signaling. The steps described below correspond to the steps [numbers enclosed by an ellipse] in Figure 4-7.

1.   Subscriber A (Figure 4-7) initiates a request for service by lifting the hand-set of the telephone. This action causes the loop to the originating exchange to be closed (described earlier in Figures 4-5 and 4-6). The signal sent to the originating exchange is known as the "off-hook" signal.

2.   The originating exchange switch's current sensor detects the off-hook signal. In response, the switch's dial tone generator sends the dial tone signal telling subscriber A to begin dialing the directory number, known as the address, which can be entered via dial pulses, Dual Tone Multifrequency (DTMF), or touch tone.

3.   Subscriber A dials the called party directory number which is sent to the originating exchange switch, that is located at a local exchange office.

4. Simultaneously, the originating exchange switch will seize an inter-exchange trunk (or circuit) by sending an "off-hook" (seizing/connect) signal to the terminating exchange. Upon the reception of the off-hook signal, the terminating exchange will request the attachment of an incoming register. This register is used to store the address information for the duration of call processing.

5. The terminating exchange, upon the reception of the off-hook signal from the originating exchange, will send an off-hook signal followed a on-hook signal to the originating exchange. The combination of an off-hook and on-hook signal is called a "wink" signal, which alerts the originating exchange switch that an incoming register has been attached at the terminating exchange to receive and store the address information.

6. Upon the reception of the wink signal, the originating exchange will start to send the address information to the terminating exchange.

7. After receiving the address of the called party, the terminating exchange will send a ringing signal to the called terminal (equipment) to alert subscriber B, provided the loop is not busy (i.e., subscriber B is on-hook).

8. After the terminating exchange has received the address information from the originating exchange, the terminating exchange switch may perform one of the following three actions:

    *   If subscriber B is not busy, the terminating exchange will send an audible ringing tone to subscriber A via the originating exchange.

    *   If subscriber B is busy, instead of an audible ringing tone, an audible busy tone will be sent to subscriber A via the originating exchange.

    *   If the call cannot be completed through either exchange, the affected exchange will send an audible congestion or "fast busy" tone to subscriber A.

9. If subscriber B is not busy, and willingness to accept the call is indicated by lifting the handset or performing an equivalent action.

10. The off-hook answering signal is sensed by the terminating exchange, which removes the ringing tone from subscriber B's loop and the audible ringing tone from subscriber A's loop.

11. The end-to-end connection is established and the communication between subscribers A and B begins.

12. If subscriber A hangs up the handset (i.e., returns to the on-hook state), the originating exchange will recognize the on-hook signal. In response, the originating

exchange will send a clear forward signal while the terminating exchange sends a clear backward signal to release the trunks between the exchanges. The network resources are now available to be allocated to handle a new call.

4.5 EARLY SIGNALING SYSTEMS

This section describes the evolution of inter-exchange signaling systems: from the early to the modern signaling systems. The development of inter-exchange signaling originally started with regional telecommunications network applications.

4.5.1 Examples of Early Signaling Systems

In this section, several examples of early signaling systems are described, including: ITU-T R1, ITU-T R2, ITU-T signaling system No. 5, and ITU-T signaling system No. 6.

Example 4-2: ITU-T Signaling System R1, which was standardized in 1968.

The R1 signaling system is primarily used in North American networks. It is formally called the "North American System". AT&T refers this signaling system as the 2600 Single Frequency (SF) signaling system. The R1 signaling system can be characterized by five parameters (summarized in Table 4-2):

- Line signaling: From a line signaling perspective, this R1 signaling system utilizes in-band signaling technology. The line signaling information is a continuous tone with a frequency of 2600 Hz. For end-to-end control signal transport, a "link-by-link" technique is implemented as shown in the above graph.

- Inter-register information exchange: Inter-register information is transferred by using multifrequency pulses generated by the combination of every two frequencies out of six pre-determined frequencies as given in Table 4-1. The information transport is also performed link-by-link, and is applied in the forward direction.

- Circuit operation: Implemented for both directions (i.e., two-way communications).

- Traffic: The traffic can either be terminal or transit [In Figure 4-8, if the intended traffic is from A to B, then (1) A $\Rightarrow$ B is terminal, and (2) A $\Rightarrow$ B$\Rightarrow$ C is transit].

- Transmission media: Any type of media can be used for telephone services except Time Assignment Speech Interpolation [TASI, also known as Digital Speech Interpolation (DSI)] lines. DSI lines share the media capacity between two subscribers by utilizing the speech pause periods of one subscriber to transport the speech signal of another subscriber.

Table 4-2   Major Characteristics of ITU-T Signaling System R1.

Line signaling	* An in-band signaling scheme * Continuous 2,600 Hz tone * Link-by-link transport
Inter-register	* Multifrequency pulses (2 out of 6 freqs: see Table4-1) * Link-by-link * Forward direction only
Circuit operation	* Two-way
Traffic	* Semi-automatic * Automatic working * Terminal or transit
Transmission media	* Telephone lines (except TASI/DSI lines)

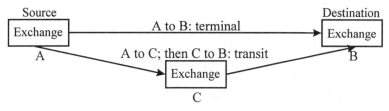

Figure 4-8   Terminal and Transit Transport.

Example 4-3: ITU-T Signaling System R2, which is standardized in 1968.

Table 4-3   Major Characteristics of ITU-T Signaling System R2.

Line signaling	* An out-of-band signaling scheme * Continuous 3,825 Hz tone * Link-by-link transport

Inter-register	* Multifrequency pulses (2 out of 6 freqs)
	   Forward: 1380, 1500, 1620, 1740, 1860 & 1980 Hz
	   Backward: 1140, 1020, 900, 780, 660 & 540 Hz
	* End-to-end: compelled sequences

Circuit operation	* One-way communications
Traffic	* Semi-automatic * Automatic working * Terminal or transit
Transmission media	* Telephone lines (except satellite circuits)

The R2 signaling system is used in both North American and international telephone networks throughout the world. It was formerly known as MFC Bem System. The main characteristics of this system is summarized on Table 4-3.

Example 4-4: ITU-T Signaling System No. 5, which is standardized in 1964.

Major characteristics of SS5 is given in Table 4-4.

Table 4-4   Major Characteristics of ITU-T Signaling System No. 5.

Line signaling	* An in-band signaling scheme * Continuous 2,400 and 2,600 Hz tones * Link-by-link transport
Inter-register	* Multifrequency pulses (2 out of 6 freqs: see Table4-1) * Link-by-link * Forward direction only
Circuit operation	* Two-way communications
Traffic	* Semi-automatic * Automatic working * Terminal or transit
Transmission media	* Telephone lines (including satellite and TASI lines)

Example 4-5: ITU-T Signaling System No. 6, which is standardized in 1968.

Major characteristics of SS6 is given in Table 4-5.

Table 4-5   Major Characteristics of ITU-T Signaling System No. 6.

Common channel signaling	* An out-of-band signaling scheme * Link-by-link transport
Mode of operation (to be discussed later)	* Associated * Quasi-associated * Non-associated
Transmission mode	* Serial data
Signal unit	* 28 bits in length (including 8 parity-check bits)
Transmission	* On analog circuit: 2.4 kbps * On digital circuit: 4 or 56 kbps * Per signal unit: each signal unit with 11 signal units, acknowledged and re-transmitted in case of error
Circuit operation	* Two-way communications
Traffic	* Semi-automatic * Automatic working * Terminal or transit
Transmission media	* Telephone lines (including satellite and TASI lines)

4.5.2 Limitations of Early Signaling Systems

Before the development of common channel signaling systems No. 6 and No. 7, the traffic flow over the telecommunications networks was much lighter than modern digital networks. The signaling systems used for "slow-traffic" networks were ITU-T Signaling Systems R1, R2 and No. 5. However, they are no longer adequate for modern digital telecommunications networks, which require higher-speed signaling capabilities. The limitations of the early signaling systems are summarized as follows:

- The early systems were designed for low speed signaling. The signaling speed is relatively low for the Plain Old Telephone Service (POTS). Modern communications networks require much higher speed signaling techniques. The complexity of networks and the physical distance of end-to-end connections for global services has resulted in demands for higher speed signaling.

- The early systems provided a limited number of signal types. The service features of modern networks are at least ten times greater than the features required for POTS, hence a greater number of signaling types are needed.

- In-band signaling applications, the signaling information shares the same path with the customer's voice signal. However, it is possible to have false network operations invoked by voice generated signals (e.g., "talk-off" conditions).

- The in-band signaling can also allow deliberate manipulation of the signaling system to circumvent service charges (e.g., long distance phone fraud). Therefore, special design techniques must be applied to minimize the potential effect of voice signals on signaling information.

- If the "tone on" signal is used to indicate the idle condition of a circuit, a carrier failure will cause multiple register seizures. This is because a carrier failure condition will interrupt the "tone on" signal, thereby affecting many circuits. That is, the interruption of the "tone on" signal will be interrupted as the "non-idle state", and will appear to be a large number of simultaneous seizures being applied to the switching exchange.

  As a result, the demand to attaching registers to the incoming seizures will be handled as if they were legitimate calls (i.e., cannot be distinguished from a carrier failure). Such failures can overload switching exchanges and exceed the traffic handling capacity of the switching equipment.

- Signaling equipment are required at both end of the connection, they contribute to a large portion of the system costs.

Because of the above limitations in early signaling systems, faster and more reliable signaling systems must be developed for modern telecommunications networks.

4.6 COMMNON CHANNEL SIGNALING SYSTEM

The common-channel signaling arrangement was designed to eliminate or reduce the limitations of early signaling systems. There are two types of signaling systems: per circuit signaling, and common channel signaling. The comparison of these two types of signaling systems is illustrated in Figures 4-9, which shows the network architectures of these signaling systems.

The main characteristics of the per circuit signaling architecture are listed as follows:

- The system per circuit requires a large number of relatively inexpensive line signaling equipment units, indicated by "signaling" in Figure 4-9(A).

- This system *does* require a large number of relatively expensive register signaling equipment units, indicated as "outgoing" and "incoming" registers.

- A large number of signaling states are available (via register signaling equipment units) only during the call set-up time.

- A few signaling states are available (via line signaling equipment units) at any time during the call.

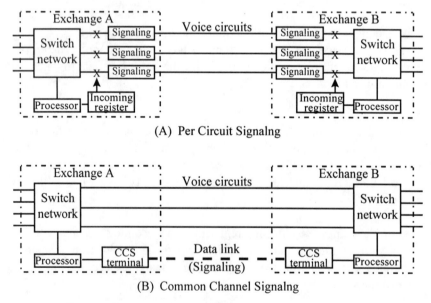

(A) Per Circuit Signalng

(B) Common Channel Signalng

Figure 4-9   Per Circuit and Common Channel Signaling Systems.

The main characteristics, rather different from those of a per circuit signaling system, of the common channel signaling architecture are listed as follows:

- The per circuit line signaling equipment is not required in the common channel signaling system.

- One relatively expensive common signaling channel element is required (indicated as "CCS terminal in Figure 4-9).

- The number of available signal types (e.g., call set-up related signal information and control signals) is practically unlimited.

- Signals can be sent any time during the call, without interference to the subscriber's information transfer because the signaling information travels over a separate data link.

4.6.1 Advantage of Common Channel Signaling

The common channel signaling architecture was implemented to meet the signaling needs of modern switching systems and customer services. From the network viewpoint, the common channel signaling system provides three major network functions:

(1) Routing control signal information by using Stored Program Control (SPC) switches (described in Chapter 5).

(2) Transporting signaling information over high-speed data links (e.g., T1 digital carrier systems, discussed in Chapter 6) between the processors in the SPC switching exchanges.

(3) Carrying all signaling and network control information on high-speed data links that are completely independent of the communications paths used to carry customers voice, data and/or video signals.

Based on the three major functions, the Common Channel Signaling (CCS) system has the following advantages over the per circuit signaling system:

- **Fast call set-up time:** CCS utilizes the high speed data link in association with modern SPC digital switching systems. The call set-up time is typically 1 to 2 seconds, or less.

- **High signal capacity**: The high speed data link and signal format adopted by CCS systems provides practically unlimited capacity for unique signals. The signaling information is not only used for call set-up and control, but also for non-circuit related functions.

- **Improved maintenance procedures**: The high speed data link, sophisticated signal formats, and unlimited signaling types allow CCS to also support maintenance procedures.

- **Expanded customer services**: The high-speed and wide range of signal types allow CCS systems to provide a variety of new sophisticated services.

- **Eliminates potential fraud**: Total separation of the signaling path from the customer communications path makes it is nearly impossible for customers to fraudulently manipulation communications networks.

- **Full communications bandwidth availability**: Since none of the voice circuit bandwidth is reserved for signaling purposes, a CCS system allows customers to utilize the full network bandwidth (i.e., separate paths for signaling and customer communications).

- **Reduced per-circuit signaling costs**: Because per circuit signaling equipment is eliminated (Figure 4-9), the per circuit costs of CCS systems is reduced accordingly.

In comparison, the in-band signaling architecture allows a circuit to be recognized as reserved by merely becoming active. The adjacent switching exchange offices associate this circuit with the directory number (DN) it carries, hence no explicit circuit identification is required.

In-band signaling requires dedicated connection control logic terminations for each individual circuit since the calls on various circuits can be in different phases (of the call). In addition, there are thousands of circuits between adjacent switching exchanges. In summary, the in-band signaling architecture reduces the revenue-producing utilization of circuits because:

1. The transmission time of dial pulses is in the order of tens of seconds, therefore in-band signaling typically takes 12 to 15 seconds for call set-up.

2. The end-to-end circuit must remain established to carry tones (e.g., ringing tone, audible ringing tone, busy tone and congestion tone) and announcements. With dual tone multiple frequency (DTMF), call set-up is much shorter (1 to 2 seconds or less).

4.6.2 Common Channel Signaling Operation Modes

Out-of-band signaling is carried over dedicated transmission facilities (channels) common to multiple inter-office circuits. There are three different modes of common channel signaling:

1. **Associated mode**: In this mode of operation, the dedicated signaling data link (indicated by the dashed line in Figure 4-10) between any two adjacent switch exchange offices carries the call control information that is **associated only** with the circuits between this pair of offices.

In the common channel signaling architecture, the call control information transferred between two exchanges over the signaling data link is formatted into **messages**, instead of pulse-based or frequency-based signals. In Figure 4-10, exchange A is assumed to be the originating office. Upon receiving the dial digits from the customer, exchange A attempts to establish the call by sending an Initial Address Message (IAM) over the data link to exchange B. This IAM provides exchange B with the following information:

* The Circuit Identification Code (CIC) of the reserved circuit

* The Directory Number (DN) of the called party

* The Connection Type (CT) to indicate if the service is voice, data, or video

* Other call associated information: for example, optional information that is becoming popular nowadays is the Calling Party Identification (CPN) displayed at the called party's equipment (i.e., "caller ID").

After receiving the IAM from exchange A, exchange B interprets the contents of the IAM and generates a new IAM for exchange C (assuming a third exchange is needed to establish the call). This newly generated IAM contains essentially the same messages as the received IAM (e.g., the CIC of the reserved circuit). Exchange B applies the Connection Type (CT) information received from exchange A to reserve the appropriate circuit connecting to exchange C. Note that specific connection types may require different circuit (e.g., for data transport, the echo cancelers along the transmission path must be disabled). Typically, exchange A derives this information from the bearer capabilities information element contained in the ISDN user's Q.931 setup message.

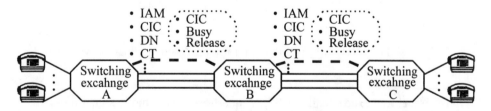

Figure 4-10   Functions (Example) of Associated Mode Operation

Another advantage of the common channel signaling architecture should be emphasized. Assuming that the called party is busy (an unsuccessful call), exchange C quickly notifies exchange B that the CIC circuit has been released. This makes the CIC circuit immediately available for another call. The release notification (message) is carried out-of-band via the signaling data link, in a "release message" that contains a "cause parameter" with the "busy information" encoded. Exchange B repeats this process in the backward direction (towards exchange A), thus allowing the "free-up" of circuit bandwidth throughout the network. In contrast, in-band signaling architecture, which uses the circuit for both customer traffic and signaling, must also send the "busy" tone throughout the connection. Thus, the circuit can not be "free-up" for another call immediately since it must carry the busy tone information.

2. **Non-associated mode**:   As shown in Figure 4-11, it can be seen that the out-of-band common channel (indicated by dashed lines) is not associated *exclusively* with circuits between a single pair of exchanges. For example, exchange C uses its common channel to transport call control information to exchange A when setting-up a call "from C to A". Similarly, exchange C uses its common channel to transport control information to exchange A for setting-up a call from exchange B (i.e., "B to C to A") or perhaps exchange D (i.e., "D to B to C to A").

The signaling (control) information is transported via a store-and-forward switch known as Signal Transfer Point (STP). Signal System 7 (SS7, discussed in detail later in this chapter) uses store-and-forward switches to reduce the number of common channels for connecting every node in a network to any other node in the same network (i.e., full inter-connectivity for all nodes in a network).

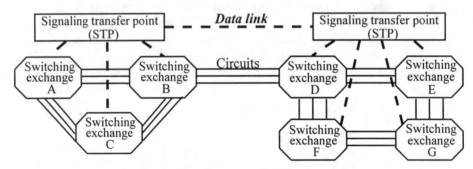

Figure 4-11   Network Architecture for Non-Associated CCS Systems.

3. **Quasi-associated mode (e.g., CCS7)**: A quasi-associated mode signaling system is a special form of non-associated mode. In this arrangement, predetermined and fixed common channel signaling routes are established between the endpoints (e.g., exchanges or databases).

4.6.3 Network Configuration for Common Channel Signaling

Figure 4-12 shows a two-level signaling network configuration using common channel signaling (CCS) links. This signaling network is based on quasi-associated signaling technology. There are two signaling transfer point (STP) serving a particular region via the "A" links (access links). Two STPs are used for reliability purpose. This architecture (using two STPs per a geographical region) is called a "mated pair STPs". For reliability reasons, the two STPs in each region are interconnected by "C" links (cross links). By doing so, the CCS network is highly redundant and can survive failures of multiple signaling links or an STP. Finally, all STPs are inter-connected by a "quad of B links" (bridge links).

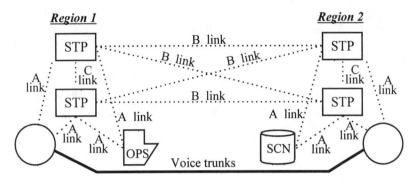

Figure 4-12   Two-level Signaling: Network Configuration for CCS.

Note that the quad STP configuration (shown in Figure 4-12) is an international standard for common channel signaling architecture. The quad configuration provides a multiplicity of independently functioning links (link sets) between any two signaling points. The two "mated STPs" share their region's workload under normal operation conditions. In the event of one STP fails, its mate STP is capable of handling the region's entire workload.

Definition 4-1: From a signaling applications perspective, *a signaling point* can be one of the following:

- The network's local and toll exchange switches

- Special nodes that are dedicated to switching signaling and call control related information. Generally, this type of signaling point is called a Signal Transfer Point (STP).

- Special systems that provide database services to various switching nodes. Examples are systems that handle "800" calls or credit card calls. These special system are called Network Control Points (NCPs).

Definition 4-2: The **users** of a signaling network (CCS7 or SS7) are the **applications** residing in the nodes of the common channel signaling network.

Responsibilities of a common-channel signaling (CCS) system are:

- The control of a call
- The control of certain switch resources
- The capability to query remote databases
- The capability to pass a specific volume of non-signaling data through the signaling network from an end user to another end user (e.g., from PBX to PBX).

Each signaling point (e.g., STP or exchange switch) that generates, interprets, or switches datagrams is assigned an ID (i.e., an address) known as a Point Code (PC). A datagram (see Figure 4-13) is a standardized groups of 272 octets used as a Protocol Data Unit (PDU) that contains a routing label and a message filed. The signaling point associated with the origination of a call is called the Originating PC (OPC), while the signaling point associated with the termination of the call is called the terminating or Destination PC (DPC). These PCs are part of the datagram (Figure 4-13). This particular example (datagram) is known as a sequenced datagram, the signaling route ID is included in the routing label.

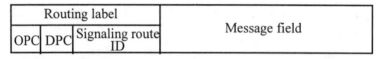

Figure 4-13   A Sequenced Datagram.

In addition to interconnecting exchange switches, a CCS network can also be connected to an Operator Position System (OPS) or Service Control Node [SCN, which is a database known as a Network Control Point (NCP)]. An SCN supports call-processing services, such as advanced messaging services, automatic call distribution, and customized announcements. Note that these nodes are connected to an STP via "A" access links.

The physical signaling channels ("A", "B", or "C" links) between any two signaling points have a speed of 56 kbps, and are configured in full duplex mode, and are called signaling data links. They transport all called related control messages. These 56 kbps ($\equiv 7 \times 8000$) channels are time slots of a digital carrier system. This speed is associated with digital carrier systems [T1 digital carrier systems with a speed of 1.544 Mbps used in North American networks, and E1 digital carrier systems with a speed of 2.048 Mbps used in other parts of the world],and are discussed in Chapter 6.

The configuration shown in Figure 4-12 is a standardized common channel signaling network, but variations from Figure 4-12 are possible. For example, in response to a particular traffic pattern, an exchange switch in one region can be connected to an STP in another region. In this configuration, the data link connecting an exchange switch to an STP in another region is no longer the "A" link, but, is called an "E" link (exchange link; not shown in Figure 4-12).

4.6.4 CCS Network Interconnection

Figure 4-14 shows the interconnection between a local exchange carrier's CCS network and an inter-exchange carrier (IXC or IEC) CCS network, and vice versa. The left-hand portion of Figure 4-14 shows the mated STPs [identified by the circles containing numerals "1" and "2"] of an originating LEC CCS network. The central part of Figure 4-14 is the full quad STPs [identified by the circles containing numerals 3, 4, 5 and 6] of an IEC CCS network. The right-hand port of Figure 4-14 is another mated STP pair [identified by the circles containing numerals 7 and 8] of another terminating LEC CCS network.

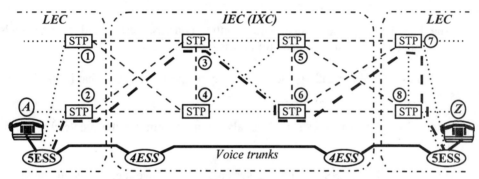

Figure 4-14   CCS Network Interconnection.

In this example, a call is made from customer "A" to customer "Z". It is assumed that two 5ESS exchanges serve as the originating and the terminating exchanges, respectively. Two 4ESS switches serve as the toll exchanges. Only two STPs of the LEC CCS network are shown. These mated STPs [STP Nos. "1" and "2"] serve the region of the originating 5ESS exchange. Two 4ESS exchanges are assumed to be required for this particular call. The two pairs of mated STPs are shown as Nos. "3", "4", "5" and "6 (which is a quad). Another pair of mated STPs (Nos. "7" and "8") support the terminating LEC 5ESS exchange. For this call, STPs "1", "2", "3" and "4" form a "quad". STPs "3", "4", "5" and "6" is the quad for the IEC CCS network. The third "quad" is formed by STPs "5", "6", "7" and "8". Note that the voice traffic for this call is carried over facilities that are separate from the signaling interconnections (shown as thick lines in the bottom section of Figure 4-14).

After the call is initiated and the called-party's number has been received by the originating 5ESS, the necessary call related control signaling is routed by using all three quads described earlier. In other words, the signaling message will be routed from the originating 5ESS to the terminating 5ESS as indicated by the *heavy dashed data links* (shown in Figure 4-14). That is, the signaling information is transferred from the originating 5ESS to STP "2", to STP "3", to STP "6", to STP "7", and then to the terminating 5ESS. Note that this is only one example of many possible routes since the STPs are mated for reliability purpose.

Review Questions I for Chapter 4

(1) (True, False) Signaling is the exchange of information (other than speech, data or video) concerned with the call-related and network management functions in a telecommunications network.

(2) From application viewpoint, signaling can be classified into two groups: _____ and _____ signaling.

(3) The four major types of signaling system are: _____, _____, _____, and _____ signaling systems.

(4) (True, False) Call process, supervision, control, address, alerting and testing signals are the major control signals used for access lines.

(5) The address signal typically is the called party telephone number, that is also known as the _____ (DN).

(6) (True, False) A supervisory signal can also be used to initiate/terminate charges for billing purposes.

(7) Prior to CCS7 and SS7 signaling systems, there have been several signaling systems specified by ITU-T as the early network signaling systems, for examples, _____, _____, Signaling system No. ___, and Signaling System No. ___.

(8) From signaling information and speech (video, or data) signal path flow viewpoint, signaling techniques can be _____ or common-channel signaling.

(9) Major advantages of common-channel signaling (over per circuit signaling) are: _____, _____, _____, _____, and so on.

(10) (True, False) There are two signaling transfer points (STPs) serving a particular geographical region. This architecture is called a "mated pair STPs".

(11) Since the mated pair STPs is used for a particular region for a common channel signaling system, for a particular call the such a signaling system often utilizes _____ STP configuration.

(12) (True, False) The users of a signaling network (e.g., CCS7 or SS7) are the applications residing in the nodes of the common channel signaling network.

(13) A sequenced datagram contains two parts: routing label and message field. In the routing label field, three components are needed: _____ (OPC), _____ (DPC) and _____.

4.7 ITU-T SIGNALING SYSTEM NO. 7 (SS7)

ITU-T Signaling System No. 7 was developed and standardized in 1980. The SS7 evolved from earlier signaling systems, and continues to be adopted as the worldwide signaling system for a digital trunk networks. SS7 is actually a set of protocols governing the transfer of call control information over a signaling network. SS7 protocol can also be applied to transfer information database oriented services. The SS7 network is a packet switching network (Chapter 5) within the Integrated Digital Network (IDN) or the Integrated Services Digital Network (ISDN).

Table 4-6   Major Characteristics of ITU-T Signaling System No. 7.

Common channel signaling	* An out-of-band signaling scheme * Link-by-link
Mode of operation	* Associated * Quasi-associated * (Non-associated)
Transmission mode	* Serial data
Signal unit	* Variable length
Transmission	* On analog circuit: 4.8 kbps * On digital circuit: (optimized) 64 kbps * On digital circuit: 56 kbps * Each signal unit is acknowledged and re-transmitted in case of error
Circuit operation	* Two-way communications
Traffic	* Semi-automatic * Automatic working * Terminal or transit * National or international networks * IDN and ISDN networks
Transmission media	* Telephone lines (including satellite and TASI lines)

According to ITU-T Recommendation Q.700, SS7's overall objective is to provide an international standardized general-purpose common channel signaling system.

- SS7 is optimized for operation in digital telecommunications networks [e.g., the μ-law and the A-law plesiochronous digital hierarchical (PDH) network, synchronous optical network (SONET) and synchronous digital hierarchy (SDH) network, etc.; PDH, SONET and SDH is discussed in Chapter 6].

- SS7 must be operated in conjunction with stored program control (SPC) exchanges to achieve optimum efficiency.

- SS7 can meets present and future requirements for transferring information used for inter-processor transactions within telecommunications networks (IDN or ISDN). The following topics are addressed by SS7 standards:

* Call control
* Remote control
* Management and maintenance of signaling

- SS7 provides a reliable means for transferring information in the correct sequence and without loss or duplication.

- SS7 is optimized to operate over 64 kbps digital channels. It is also suitable for operation over analog channels at lower data rates (i.e., 4.8 kbps).

- The SS7 signaling system supports the following applications:

  * Public Switched Telephone Network (PSTN)
  * Integrated Services Digital Network (ISDN)
  * Public land mobile network
  * Interaction with network databases
  * Service Control Points (SCPs) for service control
  * Operations, Administration and Maintenance (OAM) of networks

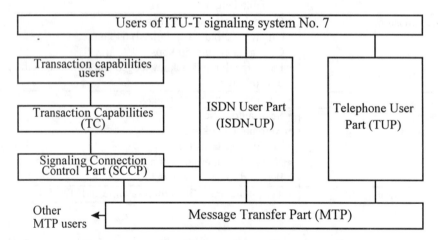

Figure 4-15   ITU-T Signaling System No. 7 Architecture.

4.7.1 ITU-T SS7 Architecture

As previously stated, Common Channel Signaling (CCS) is a telecommunications network composed of many switch processing nodes interconnected by transmission links (Figures 4-11, 4-13 and 4-14). Each node incorporates features of SS7, and forms a Signaling Point (SP) within the SS7 network. Messages between two SPs may be routed over a link directly connecting these points (known as the "associated mode" signaling; Figure 4-10). Messages may also be routed via one or more intermediate SPs that relay messages at the network layer (known as the "non-associated mode"; Figure 4-11).

Relaying signaling messages is carried out by Signal Transfer Point (STP). An STP is a point in the signaling network where a message received on a signaling link is transferred to another link, and is neither the source nor the destination of the user-part function.

The functional blocks of an ITU-T SS7 network specified in ITU-T Rec. Q.700 are listed as follows:

- Application Entity (AE)
- Application Service Elements (ASE)
- ISDN User Part (ISDN-UP)
- Message Transfer Part (MTP)
- Signaling Connection Control Part (SCCP)
- Telephone User Part (TUP)
- Transactions Capabilities (TC)

The fundamental principles of SS7 are the functional division into: (1) a common Message Transfer Part (MTP) that functions as a reliable transport system for transferring signaling messages between user functions, and (2) separation of user parts as shown in Figure 4-15. The term "user" refers to a functional entity that utilizes the transport capability provided by the MTP. As shown in Figure 4-15, the Signaling Connection Control Point (SCCP) also has users (ISDN-UP and TC users). The SS7 MTP user functions are listed as follows:

- Data User Part (DUP)
- ISDN User Part (ISDN-UP)
- Signaling Connection Control Part (SCCP)
- Telephone User Part (TUP)

4.7.2 Relationship between SS7 Functional Blocks and OSI Layers

SS7 can be regarded as a form of data communication (discussed further in Chapter 7) that is used for several types of signaling and information transfer between processors in telecommunications networks. Figure 4-16 shows the relationship between SS7 functional blocks and Open System Interconnection (OSI) layers. OSI layers 1 through 7 are the physical, data link, network, transport, session, presentation, and application layers (discussed further in Chapter 7).

The functions (usually performed by several data communications links in tandem) that transport information from one location to another are implemented by layers1, 2 and 3 (i.e., the physical, data link, and network layers). The functions defined by these layers provide the foundation on which the signaling communication network is built. The Signaling Connection Control Point (SCCP) with Message Transfer Part (MTP) provide services required (specified) by OSI layers 1, 2, 3 and 4.

The functions related to "end-to-end" communications are performed by layers 4, 5, 6 and 7. These layers (levels) must be defined so that they are independent from the internal structure of the signaling communications network. Transaction Capabilities (TCs) provide the services defined by OSI layers (levels) 4 through 7. Application Entities and Application Service Elements (ASEs) provide all required application layer protocols for layer 7. Layer 7 represents the semantics of a communications network, and layers 1 through 6 define the means by which the communication is carried out.

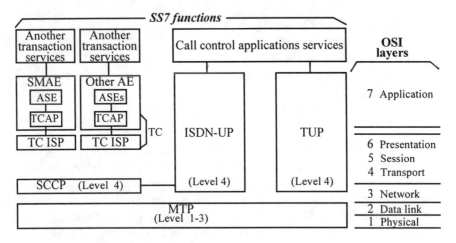

Figure 4-16   Relationship between SS7 Functional Blocks and OSI Layers.

4.7.3 Message Transfer Part (MTP)

The Message Transfer Part (MTP) of the SS7 signaling protocol is considered the "lower part" of protocol (the "upper part" is the user part). The MTP provides a routing service for its users. From an OSI perspective, MTP consists of three levels: the physical, data link, and network levels. The MTP applies level 3 header, level 3 address, and Signaling Points Codes (SPCs) or Point Codes (PCs) to forward "upper part" messages to their destinations in the signaling network. Note that in a digital network, every network entity [e.g., local exchanges, trunk exchanges, Signaling Transfer Point (STP), or Network Control Points (NCPs)] must be labeled with a unique SPC if the entity can generate or receive an SS7 message. The SPC can be an Origination Point Code (OPC) or a Destination Point Code (DPC), depending upon whether the entity is the origin or destination of a level 3 MTP message.

From a functional structure viewpoint, a signaling system consists of three elements: the signaling link, the message transfer part, and the user parts as shown in Figure 4-17. The four main components/functions of the Message Transfer Part (MTP) are signaling data link, signaling link functions, signaling message handling functions, and signaling network management. The signaling data link and signaling link functions are considered to be the signaling link. Similarly, the signaling message handling and signaling network management functions are considered to be signaling network functions. These functions are described in the following sections.

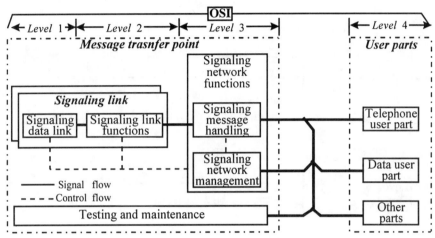

Figure 4-17   Functional Structure for SS7 System.

4.7.3.1 Signaling Data Link

The signaling data link corresponds to OSI layer 1 or level 1. It defines the physical, electrical, and functional characteristics of a signaling data link and the means to access it. It provides a bearer for signaling link. The signaling data link is typically a digital path with a speed of 64 kbps (for international applications) or 56 kbps (for North American applications). A signaling data link is a "bidirectional" transmission path used for transporting signaling messages. The data link comprises two data channels operating in opposite directions at the same signal rate. These channels may be derived from a 1.544 Mbps DS1 trunk, a 2.048 Mbps E1 trunk, or a 8.448 Mbps E2 trunk (defined in ITU-T Rec. G.704, and discussed in Chapter 6).   These channels can also be derived from a digital signal with the frame structure defined for data circuits in ITU-T Rec. X.50, X.51, X.50 bis, and X.51 bis. Note that a signaling data link consists of digital transmission channels, digital switches, and terminating equipment to provide an interface to SS7 (or CCS7) signaling terminals.

The signaling link can be accessed via a switching function, and the switch may also provide for automatic re-configuration of signaling links.

The standard bit rate on a digital bearer channel used as a signaling data link is 64 kbps, which are handled as switchable semi-permanent channels. The minimum signaling rate for call control and call processing applications is 4.8 kbps. For other functions, (e.g., signaling network management) lower bit rates can be applied. The signaling data link is dedicated exclusively for carrying signaling messages between adjacent signaling points. The signaling link cannot carry any other information.

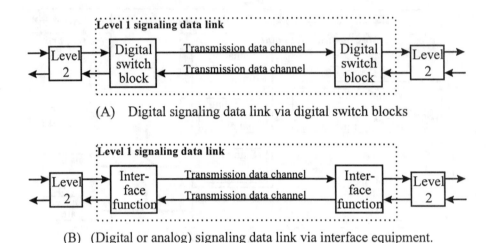

(A)   Digital signaling data link via digital switch blocks

(B)   (Digital or analog) signaling data link via interface equipment.

Figure 4-18   Signaling Data Link Transmission Channels.

- Signaling data links with channels derived from 1.544 Mbps (according to ITU-T Rec. Q.702, the standard for "signaling data links derived from the 1.544 Mbps path) is presently under further study.   However, if a signaling bit rate of 64 kbps is adopted, the values of bits within the signaling terminal or the interface equipment (Figure 4-18) must be selected to guarantee the ones density requirements of the ITU-T Rec. G.733 for Pulse Code Modulation (PCM) systems.

- For a signaling data link with channels derived from 2.048 Mbps, the interface requirements should comply with ITU-T Rec. 703 for electrical characteristics, and ITU-T Rec. 704 for functional characteristics (e.g., frame structure). The signaling bit rate must be 64 kbps, and the standard channel used as a signaling data link is time slot 16 (discussed in Chapter 6). If time slot 16 is not available, any other channel time slot available for 64 kbps transmission may be used, and bit value inversion is not required.

- For a signaling data link with channels derived from 8.448 Mbps, the interface requirements should comply with ITU-T Rec. 703 for electrical characteristics, and ITU-T Rec. 704 for functional characteristics (e.g., frame structure). The signaling bit rate must be 64 kbps, and the standard channel used as a signaling data link is time slot 67 to 70 in descending order of priority (discussed in Chapter 6). If time slot 67 is not available, any other channel time slot available for 64 kbps transmission may be used, and bit value inversion is not required.

It should be noted that the auxiliary equipment such as echo cancellers, digital pads, or A-law/μ-law A/D converters in signaling transmission path must be *disabled* to ensure **full duplex operation** and **bit count integrity** of the transmitted data stream.

4.7.3.2 Signaling Link Functions

The signaling link function level corresponds to the OSI layer (level) 2. It defines the functions and procedures for transferring of signaling messages over a single signaling link. This level 2 function (in conjunction with level 1) provides a signaling link for reliable signaling message transport between two network points.

A typical signaling message (delivered by the higher levels) is transferred over the signaling link in "signal units". A "signal unit" has a variable length. Each signal unit must consist of the following two parts: (1) the information content of the signaling message, and (2) the transfer control information used to insure proper operation of the signaling link.

The main signaling link functions are briefly described as follows:

- Flag (marker) insertion and signal unit delimitation and alignment: Since the signal unit (the signaling unit format is described later in this chapter) is variable in length, it is necessary to have flags for signal unit delimitation. The beginning and the end of a signaling unit are indicated by unique 8-bit patterns, known as the *flags*. It is important to have standards to insure the flag pattern cannot be imitated elsewhere in the signaling unit. A loss of signal unit alignment condition is declared when a bit pattern that is disallowed by the delimitation procedure (e.g., more than six consecutive ones) is received, or when the maximum length of signal unit is exceeded. Loss of signal unit alignment will cause a change in the mode of operation of the signal unit error rate monitor.

- Bit stuffing: To prevent flag imitation, dummy bits are stuffed (i.e., inserted) in signaling messages.

- Error detection: Cyclic Redundancy Check (CRC) forward error detection methods are applied, and parity-check bits are included in each signal unit. The parity-check function is performed by adding 16 parity-check bits at the end of each signal unit. This 16-bit parity-check field is generated at the transmitter by applying a specific CRC generator polynomial. The same algorithm is applied to generate parity check bits at the receiver. The CRC parity generated by the receiver is compared with the parity check field contained in the signal unit. If any differences are detected, an error indication is declared and the signal unit is discarded.

- Error correction: Error correction is implemented by applying the "preventive cyclic re-transmission method" or the "basic error control method".

  The "basic error control method" is applied to signaling links for non-intercontinental terrestrial transmission applications. It can also be applied to intercontinental signaling links where the one-way propagation delay is less than 15 ms. This method is a non-compelled, positive/negative acknowledgment, retransmission error control system. Each signal unit that has been transmitted must be retained at the transmitting signaling point until a **positive** acknowledgment for that particular signal unit is received. If a **negative** acknowledgment is received, transmission of new signal units must be interrupted. At the same time, signal units which have not yet received positive acknowledgments must be re-transmitted using the same sequence when they were first transmitted.

  The "preventive cyclic retransmission method" is applied to intercontinental signaling links where one-way transmission delay is equal to or greater than 15 ms. It can also be applied to signaling links carried via a satellite system, which can result in one-way transmission delays up to 250 ms. In the latter case, the preventive cyclic transmission method must be used for all signaling links of that set. This method is a non-compelled, positive acknowledgment, cyclic retransmission forward error control system. The signal unit that has been transmitted must be retained at the transmitting signaling point until a **positive** acknowledgment for that particular signal unit is received. During this holding time, no new signal units should be transmitted. That is, those signal units for which a positive acknowledgments has not yet been received must be re-transmitted cyclically. This forced re-transmission procedure is used to ensure that forward error correction does occur during adverse conditions [e.g., degraded bit error rate (BER) or high traffic loading]. If the number of retained and un-acknowledged signal units reaches a pre-determined threshold, the transmission of new signal units is re-transmitted cyclically until the number of un-acknowledged signal units is reduced.

- Signal unit sequence control: Explicit sequence numbers are applied in each signal unit, in addition to continuous acknowledgments.

- Signaling link error monitoring (failure detection): Signaling link error monitoring functions are applied for two different conditions: (1) during the power-up (i.e., the proving-in) state of the initial alignment procedure; called the alignment error rate monitor, and (2) when the signaling link is in service; called the signal unit error rate monitoring.

  The alignment error rate monitor is a linear count of signal unit errors. During loss of alignment, the signal unit error rate monitor is incremented in proportion to the duration (period) of the loss of alignment. In comparison, the signal unit error rate monitor procedure is based on a signal unit count that is incremented or decremented using the "leaky bucket" principle.

- Link state control function: This function provides directives to the other signaling link functions.

- Flow control: When a congestion condition is detected at the receiving end of a signaling link, the flow control function is initiated. The receiving terminal notifies the transmitting node of the congestion condition by sending an appropriate link status signal. At the same time, the receiving end withholds acknowledgment of all incoming signaling message units. This notification state must be continued as long as the congestion condition exists. If the condition lasts too long (i.e., exceeds a predetermined threshold), the transmitting end will be notified that there is a signaling link failure. As soon as the congestion condition abates, the receiving end will resume acknowledgment of all incoming signal units.

- Signaling link recovery

4.7.3.3 Basic Signaling Unit Format

The two main components of a signal unit are (1) a variable-length Signaling Information Field (SIF) carrying the information generated by a user part, and (2) a number of fixed-length fields for carrying information for message transfer control. For link status applications [i.e., Link Status Signal Unit (LSSU)], the SIF and Service Information Octet (SIO, described later in this chapter) are replaced by a status field generated by the signaling link terminal (instead of the user part). The three different signal unit formats. Figure 4-19 shows these three formats, shown in Figure 4-19, are differentiated by the Length Indicator (LI) field contained in each signal units.

The three signal units are:

- The message signal unit (MSU)
- The link status signal unit (LSSU)
- The fill-in signal unit (FISU)

The main components of a signal unit are described as follows:

- Flag: There are two types of flags: the opening flag and the closing flag. A signal unit is led by an opening flag (typically the closing flag of the preceding signal unit). In certain conditions (e.g., signaling link overload) a limited number of flags may be generated between two consecutive signal units. Therefore, a signaling link terminal must be able to receive consecutive signal units with one or multiple flags inserted between them. The closing flag indicates the end of a signal unit. The flag field has a bit pattern of "01111110". In SS7 applications, to ensure that the flag code is not imitated by any other part of the signal unit, the transmitting signaling link terminal inserts a "0" after every sequence of five consecutive "1s" before the flags are attached and the signal unit is transmitted. At the receiving end, after the flag has been delimited (i.e., detected and removed), each "0" following a sequence of five consecutive "1s" must be deleted.

A flag is considered to be an opening flag if it is not followed by another immediate flag. Whenever an opening flag is received, the beginning of a signal unit is declared. The next flag received is declared the closing flag, which indicates the end of a signal unit.

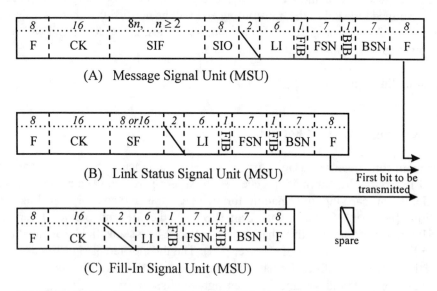

Note that the number of bit is indicated for each field of the signaling unit

BIB: Backward Indication Bit	FSN: Forward Sequence Number
BSN: Backward Sequence Number	LI: Length Indication
CK: Parity-check field	SF: Status Field
F: Flag	SIF: Signaling Information Field
FIB: Forward Indication Bit	SIO: Service Information Octet

Figure 4-19   Format Structure of Three Signal Unit Types.

After removing the 0's (inserted at the transmitting end), the receiver checks the length of the signal unit, which should be a multiple of 8 bits (at least 6 octets) including the opening flag.

*   If the length is not a multiple of 8 bits, the signal unit must be discarded, and the signal unit error rate monitor or alignment error rate monitor is incremented.

*   If more than M+7 (M ≤ 272) octets are received before a closing flag, the receiver will enter into the "octet counting" mode and the receiver will discard this signal unit. Once the receiver enters the "octet counting" mode, all the bits received between the last flag and before the next flag are discarded. The receiver exits the

"octet counting" mode when the next correctly checked signal unit has been received (that is, the parity-check field has the correct value for the signal unit).

* In either case described above, for a system using a basic error control method a negative acknowledgment will be sent in accordance with sequence numbering rules.

• Backward Sequence Number (BSN) and Backward Indicator Bit (BIB): The BSN is the sequence number of the signal unit which is being acknowledged. The BIB, BSN, Forward Sequence Number (FSN) and Forward Indicator Bit (FIB) are used in the basic error control method to implement the signal unit sequence control and acknowledgment functions.

• FSN and FIB: The FSN is the sequence number of the signal unit used in the forward direction. Both FSN and BSN are numbers in binary code with a range from 0 to 127 (i.e., a 7-bit field; "0000000", "0000001", "0000010", "0000011", ..., "1111111").

• Length Indicator (LI): This field has a length of 6 bits and is used to indicate he number of octets (8-bit bytes) following the LI field and preceding the parity-check field (CK). This LI field has a binary code in the range of 0 to 63, which is different for the three different types of signal units:

* For Fill-In Signal Unit (FISU):   LI = 0   (000000)
* For Link Status Signal Unit (LSSU):   LI = 1 or 2   (000001 or 000010)
* For Message Signal Unit (MSU)   LI > 2   (000011, 000100, ..., 111111)

It is mandatory to have the LI field set (by the transmitting signaling terminal) to the correct value corresponding to the specific signal unit type. However, there are conditions message units will be larger than 62 octets, and the LI field will be set to 63 (i.e., 111111).

• Service Information Octet (SIO): The Service Information Octet consists of two parts. The first part is the service indicator which is used to associate the signaling information message with a particular user part and is only present in Message Signal Unit (i.e., it is not provided for the Link Status Signal Unit or the Fill-In Signal Unit). The second part is the sub-service field, which is a four-bit field (bits A, B, C and D). Currently, bits A and B are for growth (not yet assigned). However, bits C and D provide the discrimination function: (1) between national and international signal messages, and (2) between two different national signaling networks.

• Service information field (SIF): Service information field has a field length of *8n* bits ($2 \leq n \leq 272$). This range allows a single message signal unit to accommodate information message lengths up to 268 octets (or bytes), which is accompanied by using a routing label (See Figure 4-20). The SIF format and SIF codes are uniquely defined for each user part.

- Parity-check field (CK): Each signal unit (MSU, LSSU or FISU) applies 16 parity-check bits for error control. Parity checking for SS7 applications uses a cyclic redundancy check (CRC) code with a generating polynomial given by

$$g(x) = 1 + x^5 + x^{12} + x^{16} \tag{4-1}$$

The details of CRC error control method are described in Chapter 13.

- Status Field (SF): The status field is only applicable for the Link Status Signal Unit (i.e., it is not provided for MSU or FISU). It has a length of 8 or 16 bits, depending upon the message length (i.e., the value of LI). When the field is an 8-bit code, 5 bits are currently spare, and 3 bits are used to indicate status. Six out of the 8 ($\equiv 2^3$) possible combinations have been assigned for six different "status indications" that represent unique alignment and out-of-service status indications.

4.7.3.4 Signaling Message Handling Functions

The signaling network functions (Figure 4-17) correspond to OSI layer 3 (level 3). It defines the transport functions and procedures that are common to, and independent of, the operation of individual signaling links. As shown in Figure 4-17, the functions defined in this level can be grouped into two categories:

- Signaling message handling functions: discussed later in this section.
- Signaling network management functions: discussed in the following section.

These functions are used to direct a signal message to its proper signaling link or user part during the actual transfer of messages. The signaling message handling functions guarantee that a signaling message originated by a particular user part (at an original signaling point) is properly delivered to the user part at the destination point (as indicated by the sending user part). The message delivery may be made through a signaling link connecting the originating signaling point directly to the destination signaling point. The message delivery may also be made through one or more intermediate Signaling Transfer Points (STPs).

The signaling message handling functions are based on the label contained in the signaling message, which identifies both the origination point and the destination point.

Definition 4-3: The routing label is the part of a signaling message that is applied by the message transfer part for signaling message handling (Figure 4-20).

The routing label is common to all telecommunications services and applications in a given signaling network. The standard routing label must be applied in international and national signaling networks.

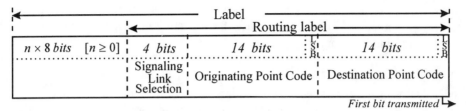

Figure 4-20   Routing Label Structure.

The routing label must be placed at the beginning of the Signaling Information Field (SIF, see Figure 4-19). Figure 4-20 shows the standard routing label, which has a length of 32 bits. The three parts of the routing label are briefly described as follows:

- The Destination Point Code (DPC): This code indicates the destination of the signaling message. The DPC is a simple 14-bit binary code. The least significant bit (LSB) occupies the first bit position and should be transmitted first.

- The Originating Point Code (OPC): This code indicates the signaling point where the signaling message is originated. The OPC is a simple 14-bit binary code. The least significant bit (LSB) occupies the first bit position and should be transmitted first.

- The Signaling Link Selection (SLS): The SLS field is used to perform load sharing wherever applicable. This SLS field contains 4 bits, exists in all types of signaling messages, and always occupies the same position.

  * For circuit-related messages of the telephone user part: The SLS field contains the least significant bits of the circuit identification code. These bits are **not** repeated elsewhere.

  * For circuit-related messages of the data user part: The SLS field contains the bearer identification code. These bits are **not** repeated elsewhere.

  * For all other user parts: The SLS field is independent, follows the signaling link selection on messages generated by any user part, and is used in the load sharing mechanism. As a result, if the user parts are not specified, the SLS field contains the same value for all messages belonging to the same transaction being sent in a given direction.

  * For message transfer part level 3 messages: The SLS field corresponds to the signaling link code for the signaling link between the destination point and the originating point (which the signaling message refers to).

The signaling message handling functions include the following:

- Message discrimination: The function of each Signaling Point (SP) is to determine (discriminate) whether the signaling message is applied at the signaling point. A message arriving at a signaling point may be destined for this particular SP, or may be routed to another SP. For the latter case, the signaling message is transferred to the message routing function.

- Message routing: This is the process for selecting the signaling link used for each individual signaling message. The routing procedure is based on analysis of the routing label of the signaling message and the pre-determined routing data at the signaling point. Message routing is destination code dependent. The load sharing element allows different portions of the signaling traffic sent to a particular destination to be distributed over two or more signaling links. The signaling links used (in succession) to carry a signaling message from its originating point to its destination point constitute a "message route".

- Message distribution: This function is used by a signaling point to distribute the signaling message to the appropriate user parts, if the message is destined for this particular signaling point.

4.7.3.5 Signaling Network Management Functions

The second category of the signaling network function is the signaling network management functions (Figure 4-17). These functions are used to control the current signal message routing and configuration of the signaling network facilities, based on the pre-determined data and status information for the signaling network. In the event of changes in the status of the signaling network, the management functions can also control re-configuration and other actions to preserve (to restore) the normal message transfer capabilities. The management functions and the testing/maintenance actions may include exchange of signaling messages with corresponding functions located at other signaling points.

The signaling network management functions consist of the following three functions:

(1) **Signaling traffic management (function):** This management function performs the following tasks:

    * Control of signaling message routing: Signaling message routing control is based on analysis of predetermined information for all "allowed candidate" routing possibilities and the status of the signaling network (i.e., the current availability of signaling links and routes). This information is supplied by the signaling link management and signaling route management functions.

    * Control of signaling message transfer: The transfer of signaling message (traffic) is controlled so there are no irregularities in the message flow. This function is performed in conjunction with signaling message routing.

    * Implement (traffic) flow control.

(2) **Signaling link management (function):** The SS7 signaling system can implement signaling links provisioning using different techniques. Some examples of implementation are:

* A signaling link may consist of a permanently assigned signaling terminal device and a data link.

* A signaling link may use an arrangement such that any switched connection to the remote end may be used in combination with a local signaling terminal device.

The main tasks of the signaling link management function are:

* Control of locally connected signaling link sets.

* Initiating actions to restore the normal availability of a signaling link that is experiencing changes in availability.

* Supplying information about the availability of local signaling links and link sets used by the signaling traffic management function.

* Interacting with signaling link functions at level 2, in accordance with indications of the of signaling link status.

* Initiating actions at level 2, such as initial alignment of an out-of-service link.

* Initiating reconfiguration of terminal devices and signaling data links (to the extent that such reconfiguration are automatic).

(3) **Signaling route management (function):** This management function is applicable only to a signaling system using quasi-associated mode. It is used to transfer information about changes in availability of signaling routes in the signaling network so that remote signaling points can take appropriate signaling traffic routing actions. Therefore, a Signaling Transfer Point (STP) can send messages that indicate the inaccessibility of a particular signaling point, thus enabling other signaling points to stop routing messages to an incomplete route.

4.7.4 Signaling System User Parts

The functional structure of the SS7 signaling system shown in Figure 4-17 can be simplified for describing the user parts as shown in Figure 4-21. There are several user parts in an SS7 signaling network: Signaling Connection Control Part (SCCP), Telephone User Part (TUP), ISDN User Part (ISDN-UP), Data User Part (DUP), and other user parts. Each user part is described separately in the following sections. The user parts correspond OSI layer 4 (level 4).They are functionally above signaling network functions, that are above signaling link functions and signaling data link functions.

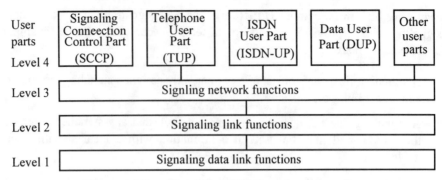

Figure 4-21   Simplified SS7 Functional Levels.

4.7.4.1 Signaling Connection Control Part (SCCP)

The Signaling Connection Control Part (SCCP) is a functional block above the Message Transfer Part (MTP), which is considered OSI level 4 of OSI (Figure 4-16). An example of a SS7 SCCP connection is shown in Figure 4-22. The combination of SCCP and MTP forms the Network Service Part (NSP) of a signaling network. Two objectives of the SCCP are:

(1) To provide a means for all logical signaling connections within a SS7 network.

(2) To provide a means for signaling data unit transfers, with or without the use of logical signaling connections.

The tasks performed by an SCCP are briefly described as follows. The SCCP provides additional functions to the MTP to transfer circuit-related or non-circuit-related information. It can also transfer other types of information between any exchange and specialized processing centers in a telecommunications network. All the information transfers can be applied to "connectionless" or "connection-oriented" network services via an SS7 signaling network.

The SCCP can be used to transfer circuit-related or non-circuit-related signaling information for the ISDN user part (with or without setup of an end-to-end logical signaling connection).

All services provided by the SCCP can be either "connectionless" or "connection-oriented" services. These services can be grouped into four classes (0-3) as follows:

• Class 0 (basic connectionless services): This is a basic datagram service between endpoints (MTP level 3 selects the outgoing link). The SCCP provides additional address information above the MTP level 3. Class 0 messages, which are datagrams, travel independent of each other across the signaling network.

- Class 1 (sequenced MTP connectionless services): In this service SCCP forces the arrival of several messages in their original order (sequence). This class of services applies a fixed route mechanism. The SCCP instructs the MTP level 3 to use the same Signaling Link Selection (SLS) for all messages associated with the same SCCP user.

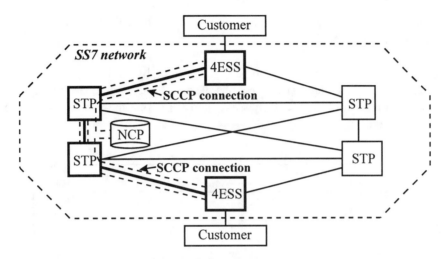

Figure 4-21 SCCP Connection for SS7.

- Class 2 (basic connection-oriented services): In this application the SCCP identifies specific messages as units of the same flow between endpoints. This type of service applies virtual connections which enable the receiving Signaling Point Code (SPC) to recognize that certain messages belong to the same flow. This technique requires incremental procedural capabilities: a common tag [known as the Local Reference Number (LRN)] and a 3-octet addition to the SCCP header. It also requires a connection establishment procedure prior to the user message flow, which instructs the destination SCCP to establish a state machine for the designated LRN flow.

  Note that when the SCCP function of an intermediary STP is invoked, it serves only to determine the ultimate Destination Point Code (DPC), and it doesn't act on the LRN. Although the intermediary SCCP function does not examine or act on the LRN, it does perform title translation, to provide a new DPC when necessary. Only the endpoints connected by the end-to-end LRN procedure examine and process the LRN.

- Class 3 (flow control connection-oriented services): The SCCP controls the LRN-related message flow by providing additional control messages. It does not impact other flows such as virtual connections or datagrams (i.e., connectionless flow). For data recovery, SCCP keeps track of messages on each virtual connection or LRN by numbering the messages. Based on the numbering, the receiver can detect gaps in the LRN sequence and recover missing messages.

Figure 4-22 shows the general SCCP message format. The SCCP messages are carried on the signaling data link (Previously described in this chapter) as a signal unit. The Signaling Information Field (SIF; Figure 4-19) of each Message Signal Unit (MSU) contains an SCCP message that consists of an integral number of octets. The five parts of an SCCP message are:

- Routing label: The routing label has been previously described in this chapter and is illustrated in Figure 4-20.

- Message type code: The message type is represented by an 8-bit binary code, and is mandatory for all messages. The functions of the message type code are: (1) uniquely define the function of an SCCP message, and (2) define the message format of the SCCP message.

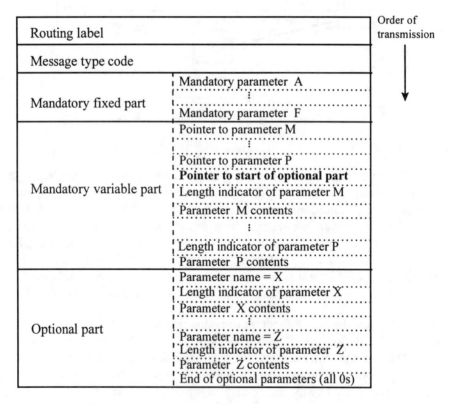

Figure 4-22   General SCCP Message Format.

- Mandatory fixed part: In this part of the SCCP message, all the mandatory parameters (i.e., mandatory parameter A, mandatory parameter B, mandatory parameter C, ..., mandatory parameter F) for a particular message type must be defined. All mandatory parameters have fixed number of bits. The position, the length, and the order of each mandatory parameter are uniquely defined by the message type.

- Mandatory variable part: All the parameters (i.e., parameter M, parameter N, ..., parameter P) with variable lengths are contained in this part. The parameter name and the transmission sequence are also implicitly contained in the message type code.

    Since the parameters have variable lengths, a pointer is required to indicate the beginning of each parameter (i.e., a separate pointer for parameter M, pointer for parameter N, ..., pointer for parameter P). A pointer to the start of the optional part is also provided if the SCCP message contains an optional part (Figure 4-22). Each pointer is an 8-bit codeword. The number of pointers is defined in the message type code. All pointers are sent consecutively at the beginning of the mandatory variable part as shown in Figure 4-22.

    The mandatory variable part also contains the actual parameter contents (i.e., parameter M contents, parameter N contents, ..., parameter P contents; Figure 4-22). Note that a length indicator precedes each parameter contents (e.g., length indicator of parameter M).

- Optional part: This part contains the parameters that are not mandatory. Therefore, an SCCP message may or may not contain this part. The optional parameters can be transmitted in any order since each parameter consists of the parameter name (8 bits), the length indicator, and the parameter contents. Note that if the optional part is contained in an SCCP message, an all-0s "end-of-optional parameters" octet must be sent after all the optional parameters have been transmitted. Obviously, this all-0s byte is not required if the optional part is not used in an SCCP message.

    The "message type code" that indicates whether the optional part is allowed or not. If the optional part is allowed, a pointer to the optional part will appear in the mandatory variable part. The pointer value will be assigned to "zero" if the message type code indicates the optional part is allowed, even though no parameters are sent in the optional part.

    It should be mentioned that all the parameters, including the routing label, in Figure 4-22 have a length of 8 bits. The sequence of transmission is from the top of the stack, as shown in Figure 4-22 (i.e., the routing label is sent first, followed by the message type code, mandatory fixed part, etc.).

4.7.4.2 Telephone User Part (TUP)

The telephone user part may carry the Initial Address Message (IAM), the subsequent address message, the end-of-pulsing signal, the continuity check of the telephone circuits, and the address-complete signals. They are briefly discussed as follows:

- Initial Address Message (IAM): The IAM is the first message sent during the call set-up process. The IAM must include all of the information required by the next international exchange (i.e., the gateway exchange) to route the call. The seizing function is implicit in the reception of the IAM.

First bit to be transmitted

8n	4	12	2	6	4	4	40
Address signals	Number of address signals	Message indicator	spare	Calling party category	Heading code (H1) 0001	Heading code (H0) 0001	Label

Figure 4-23   Initial Address Message Format.

The sending sequence of address information is: the country code (consists of 1 to 3 bits) first, and then followed by the national significant number [consists of trunk number and subscriber (telephone) number]. Note that this topic has been previously discussed in Chapter 1 (Figure 1-20).

Figure 4-23 shows the initial address message format. The first 40 bits are assigned as the "label" field. It is followed by two 4-bit heading codes (denoted as H0 and H1) each containing a codeword or "0001". A 6-bit field is used to indicate the calling party category. After 2 spare bits, 12 bits are assigned for the message indicator. The number of address signals field requires 4 bits. The end of the IAM is the address signals, which occupies an integral number (n) of octets. It should be noted that the transmission sequence of IAM is the label, followed by heading codes, and etc. as indicated in Figure 4-23.

- Subsequent Address Message (SAM): If additional address digits (i.e., a subsequent address message) need to be transmitted, they can be sent "digit by digit" or in groups of "multi-digits".

- End-of-Pulsing Signal: The end of pulsing (ST) signal will be sent: (1) if it is a semi-automatic call, (2) for any test calls, or (3) if the end of pulsing signal is received from a preceding circuit. For automatic cases, if the outgoing international gateway exchange knows (according to the digit analysis result) that the final digit has been seen, the exchange will send out the end of pulsing signal. Note that digit analysis may include the examination of the country code and counting the maximum number of digits in the national significant number field.

In all other cases, the ST signal will not be sent if the end of address information is determined by the reception of an address complete signal (described later in this section) from the incoming international gateway exchange.

- Continuity check of the telephone circuits: A SS7 system is a separate signaling network (i.e., separated from the traffic network used for carrying speech and other customer signals). It is important to check the continuity of the traffic network. It should be understood that a continuity check is **not** designed to replace the need for routine circuit testing of the traffic transmission path. It should also be noted that the continuity check of the transmission path is performed link by link on a per call basis. It can also be performed based on a predetermined statistical method prior to the beginning of communication.

  Some networks may have inherent performance and maintenance management capabilities, while others may not have these capabilities. Therefore, in transmission systems equipped with performance/maintenance features, the continuity check is not mandatory since the transmission systems can provide fault indication information to the appropriate switching systems. In general, most digital networks are equipped with these capabilities. However, if the fault indication information is lost, per-call continuity checking can be applied.

  If a continuity check for a digital circuit (which has an inherent fault indication) is received via an IAM, the request for continuity may be disregarded or a continuity check loop can be connected. The latter case will trigger an alert to the maintenance system, and may cause the call to fail since no continuity signal will be received from the distant end.

  There are many types of circuit configurations that must be checked for continuity. A continuity check loop must always be connected if: (1) the exchange has the capability to process IAMs with continuity check request, and the IAM has been received, or (2) the requests for continuity check are received. The following lists several types of circuit conditions:

  * The circuit type (analog or digital) is unknown to the SS7 system.
  * The exchange serves both analog and digital circuits.
  * The digital circuits have no inherent fault indication.

In four-wire traffic circuit applications that are served by an SS7 system, continuity checking is often performed. This continuity check can be applied to the circuits with or without Digital Speech Interpolation (DSI). However, for circuits equipped with echo control devices (e.g., echo cancelers or echo suppressors), these echo control devices must be disabled before performing the continuity check and re-enabled after the check is completed.

The check-tone transmitter and receiver are connected to the "go" and the "return" paths respectively, of the outgoing circuit at the first and each succeeding exchange (but not the last exchange of the circuits being checked) served by the SS7 system.

The check loop (loopback) is connected to the "go" and the "return" paths of the incoming circuit at each exchange (but not the first exchange).

The check tone generated by the transmitter is a single tone signal of 2000 ± 20 Hz with a signal level of -12 ± 1 dBm0 for international applications. The check loop should have a loss of 0 dB, with any difference between the relative levels of the two paths at the point of attachment being considered. A "recognition time" of the check tone at the receiver in the range of 30-60 *ms* is recommended. The continuity check is considered to be complete and successful when the receiver recognizes the check tone within the time limit.

- Address-Complete Signals (ACS): An address complete signal will be sent after the continuity check has been received and a cross-office check has been made. If the network has successfully provided the called party's line condition signals, the last exchange in the SS7 will originate and deliver an address complete signal. This is done when the end of address signaling has been determined and a possible General ReQuest (GRQ) message and General forward Setup information Message (GSM) cycle has been completed. The address complete signal will be sent after any of the following cases:

* The reception of an end of pulsing signal.

* The reception of the maximum number of the national significant number digits.

* The digit analysis of the national significant number is completed, and the result indicates a sufficient number of digits has been received so that the call can be routed to the called party.

* The reception of the end of selection signal from the succeeding network.

* In cases when the succeeding network user overlap of signaling and digit analysis is not possible, if an interval of 4-6 seconds has elapsed since the last digit was received and no fresh information has been received, an address complete signal can be sent. Before sending the address complete signal, the last digit received should not be transmitted over the national network to ensure a national answer signal *does not* arrive.

Note that when an incoming international exchange has sent a general request (GRQ) message, the system cannot send out an address complete message. Only after a GSM message has been received (in response to GRQ) can an address complete message be sent.

After the address complete message has been received, the first SS7 exchange will "through-connect" the traffic of the interconnected circuit. No other messages can be sent in the backward direction, except for the following signals:

* The answer or release guard signal, under normal operating conditions

* A call-failure signal

* The national network congestion signal

* The circuit group congestion signal

4.7.4.3 ISDN User Part (ISDN-UP or ISUP)

The ISDN User Part (ISUP) was developed as an interoffice protocol for circuit-related interworking with Q.931 protocol, and while supporting ISDN features. ISUP allows calls between ISDN subscribers, provides the signaling functions required for basic bearer services, and supports supplementary services for voice and non-voice applications in an ISDN.

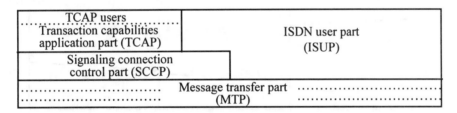

Figure 4-24   ISUP and TCAP, TCAP Users, SCCP & MTP.

ISUP also supports calls between non-ISDN subscribers. Therefore, ISDN callers can be served by an interoffice signaling system other than the ISDN user part. However, only the ISUP message repertoire and ISDN message structure provide the full set of ISDN features available for interoffice applications.

An ISUP message is generated and interpreted by the switching machines in a common-channel signaling network. The ISUP message is carried as user data in the MTP message, or in the SCCP message (when SCCP services beyond the MTP routing services are required as shown in Figure 4-24).

Example 4-1: (ISDN) Q.931 call control scenarios (i.e., establishment and clearing of a data call). *ISUP extends Q.931, which is a point-to-point network access protocol, over a store-and-forward message switching network that uses SS7 network services.*

A signaling data link must first be established on the "D" channel before the network can invoke the Q.931 procedure. After invoking the Q.931 procedure, the calling terminal (TE-A, which may be a PBX) will generate a "setup" message specifying various attributes of the requested service, known as the "bearer service". The following is a list of typical attributes specified in the "setup" message:

- Circuit mode service: The ISDN will provide a delay-free bit transport service, as is normally offered by long-distance circuit switched facilities.

- Access to a "B" channel: The basic rate or the primary rate interface selected to allow user data to enter the ISDN.

- Unrestricted digital information: Distinguishes digital data from speech or video signals.

- Transport rate: Specifies the speed that user data will be transported, end-to-end, across the network.

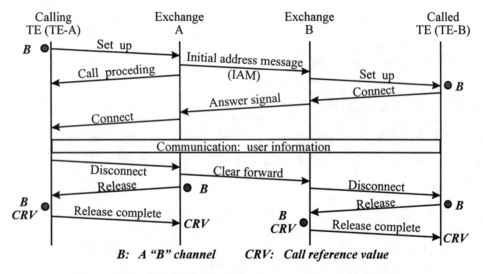

Figure 4-25   Establishment/Clearing of a Data Call (Q.931).

As illustrated in Figure 4-25, the "setup" message sent from TE-A reaches local exchange A via the "D" channel. Exchange A maps the relevant information of the "setup" message into an initial address message (IAM), and forwards the IAM on the common channel signaling network to establish the long distance facilities route for the data bit stream. Exchange B (a distant exchange) regenerates the "setup" message with the same attributes. This exchange selects a "B" channel (i.e., bearer channel) as the local attribute (i.e., the access attribute). Upon the reception of the call, TE-B generates and sends out a "connect" message. The (connect) message is regenerated at the calling interface. After the connection is established, the two TEs (TE-A and TE-B) can proceed with user data transmission, which flows across the selected B access channels and the circuit-switched long-distance facilities interconnecting the two B access channels.

When the data exchange between the two users is completed, the TE that initiates the "clearing" will issue a "disconnect" message on the "D" channel to its serving exchange (see Figure 4-25). This message contains the Call Reference Value CRV) which identifies the call to be cleared. The CRV causes the network to generate a "disconnect" message at the distant interface, by using the distant CRV. It also initiates a release procedure at both interfaces to release the B access channels and CRVs. This procedure makes both the physical and the logical facilities (i.e., "B" and "CRV" at each interface) available for a new call.

Example 4-2: Figure 4-26 illustrates various messages used for ISUP interworking between Q.931 Inter-Exchanges (IEs) . By comparing Figure 4-26 and Figure 4-25, it can be seen that the "set-up", "addressing", "alerting", "connecting", "facility request", "facility accepted", etc., are also required for ISUP interworking.

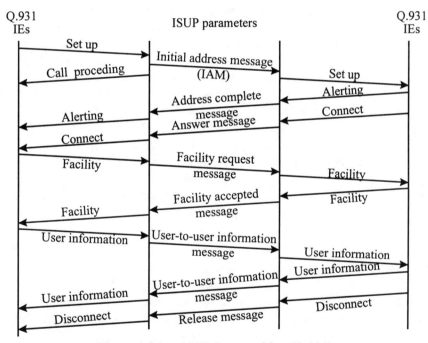

Figure 4-26   ISUP Interworking (Q.931).

Example 4-3: ISDN User Part (ISUP) versus Telephone User Part (TUP).

The Telephone User Part (TUP) is an interoffice protocol used to control of telephone calls emanating from analog subscriber lines. ISUP differs from TUP by having the capability for interoffice control of telephone calls emanating from both digital data subscriber lines (ISDN) and analog lines. In addition, ISUP can provide customer features not available under TUP. The following is a list of ISUP capabilities:

- ISUP supports data capabilities: ISUP allows the set-up of data calls. Although TUP has been enhanced to accommodate digital subscriber lines, it cannot interwork with Q.931 as fully as ISUP. In addition, SS7 provides another level 4 user part [i.e., data user part (DUP)] for interoffice control of circuit-switched data transmission.

- ISUP provides in-call modification: An active call can be switched between voice (speech) and data signals.

- ISUP provides interoffice end-to-end signaling to support the following:

  * The bearer services (for interoffice: set up, supervision and release)

  * Supplementary services (e.g., calling line identification, call forward, etc.)

  * The exchange of user-to-user signaling information

- ISUP provides supplementary services, such as Calling Line Identification (CLID), Call Forwarding (CF), Closed User Group (CUG), Direct Dialing In (DDI), User-to-User Information (UUI), Call Transfer (CT), Call Forward Busy (CFB), Call Forward No Reply (CFNR), Call Forward Unconditional (CFU), Line Hunting (LH), Call Deflection (CD: is under further study), Call Waiting (CW), Call Hold (HOLD), Completion of Calls to Busy Subscribers (CCBS: is under further study), CONFerence Calling (CONF), Three-ParTY service (3PTY), Closed User Group (CUG), Private Numbering Plan (PNP), Credit Card Calling (CRED), Advice Of Charge (AOC: is under further study), REVerse charging (REV: is under further study), and User-to-User Signaling (UUS). Note that TUP does not provide these services.

Example 4-4: ISUP message functional groups.

The main ISUP messages can be functionally grouped into two classes: the control message group and the circuit supervision group. It is important to recognize that circuit supervision messages are exchanged only between adjacent exchanges, while control messages traverse the transit exchanges along the path (although control messages may be changed in the transit exchanges).

- Call control message group (ITU-T Rec. Q.762): The control message group includes call set-up, call supervision, in-call modification, etc. (listed as follows).

  * Call setup: such as IAM, Subsequent Address Message (SAM), Address Complete Message (ACM), Call ProGress (CPG), Connect (CON), INformation Request (INR), INFormation (INF), and COnTinuity (COT).

*   Call supervision: such as ANswer Message (ANM), RELease (REL), FOrward Transfer (FOT), SUSpend (SUS) and RESume (RES)

*   In-call modification

*   End-to-end signaling: such as Pass-Along Message (PAM)

*   Supplementary services: such as FAcility Accepted (FAA), FAcility Request (FAR), Facility ReJect (FRJ), and User-to-UseR information (USR)

*   Circuit supervision message group: These messages deal with the management and supervision of single circuits, and circuit groups between two adjacent offices.

4.7.4.4 Transaction Capabilities Application Part (TCAP)

The two sublayers, collectively called the Transaction Capabilities Application Part (TCAP: Figure 4-27) of the SS7 stack, serve the non-circuit related users of the application layer.

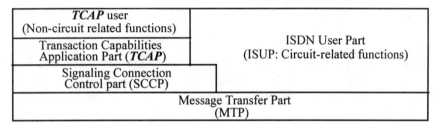

Figure 4-27   TCAP and TCAP Users.

Example 4-5: Circuit-related and non-circuit-related user functions.

*   A circuit-related user function: An example of a circuit-related function is the forwarding of the Initial Address Message (IAM) to toll exchange (switches) via the Common Channel Signaling (CCS) network. Since this SS7 message contains the routing information (i.e., the called party number) it enables the switches to set-up an end-to-end circuit between the local end offices.

*   A non-circuit-related user function: An example of a non-circuit-related function is the forwarding of a "toll free" 800 number query to a common channel signaling network database [known as the Network Control Point (NCP)] via a 4ESS switch to obtain routing information. Upon receiving this routing information, the 4ESS switch activates the circuit-related function described above (i.e., set-up an end-to-end circuit).

The user function of a TCAP consists of a request and response sequence as shown in Figure 4-28. In general, the users communicate by asking each other to perform actions (also, known as "operations") and by reporting the performance of the requested operations (i.e., a request followed by a response). This is identical to the fundamental paradigm of use communications: "ask questions, and get back answers".

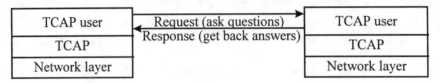

Figure 4-28   TCAP User Function: Request/Response.

In the SS7 environment, user requests can be grouped into four classes (i.e., there are four user classes of "operations").

(1) Class 1 "Report the result, either success or failure": An attempt is made to perform the requested operation, and a response is provided that is independent of success or failure. That is, a response is always supposed to be provided for class 1 requests. Failure to receive a response will cause the requesting user to take corrective action. An example of this class of operations is the translation of a toll free number into a valid subscriber number, and the return of the valid number if the translation can be performed. If the translation was not possible, the response should indicate why the operation could not be complete. The time allocated for this request/response is typically 2 seconds.

(2) Class 2 "Report only if failure": This type of request requires a response only if the outcome of the operation is unsuccessful. This reflects an optimistic approach, that considers unsuccessful outcomes to the exceptions. An example of a class 2 operations is "performing a routine test, and sending a reply only if an error is detected. The time allocated for this request/response is typically 1 minute. Note that the TCAP user is informed if no result has been received after a timer expires, which is interpreted as a successful outcome (even if the request was lost). This aspect should be considered when considering class 2 operations.

(3) Class 3 "Report only if success": This type of request requires a response only if the outcome of the operation (at the remote user) is successful. This reflects a pessimistic approach that considers failure to be the norm (i.e., even if an error is detected, it is considered to be the norm), and not requiring a reply. An example of a class 3 operations is "broadcasting a very specific question to a number of nodes, but only the node with the answer responds to the question".

(4) Class 4 "Report nothing": This type of operation does not require a response. An example of a class 4 operations is "sending an alarm, but not expecting a reply or acknowledgment of any kind". In this case, a response never arises from the

invocation of the operation. In this case the user relies upon TCAP and the network to deliver the invocation.

The above four classes of operations are referred to as "understandable" request/response messages. However, there are times that a request/response contains syntactic or semantic errors, and cannot be understood. Upon receipt of a "flawed" request or response for any class of operations, the remote user will respond with a message indicating " I don't understand", along with information. An example of this condition is "a protocol violation".

Example 4-6: TCAP contains two sublayers: (1) the Component Sub-Layer (CSL), and (2) the Transaction Sub-Layer (TSL) for serving TCAP users (Figure 4-29).

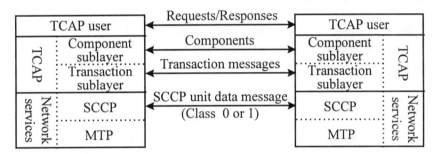

Figure 4-29   TCAP Sublayers for User Functions.

- The Component Sublayer (CSL): The CSL models the users' request and response behavior with a set of Protocol Data Units (PDUs), which are called "components". Each component carries one, and only one, request or response. CSL peers exchange components as illustrated in Figure 4-29. There are five TCAP components listed below and each component provides an envelope.

  * Component 1: This component is intended solely for containing a user request for any type of request in a particular users protocol.

  * Component 2: This component is intended solely for containing a response that indicates success for any type of response.

  * Component 3: This component is intended solely for containing a response that indicates failure for any type of response.

  * Component 4: This component is intended solely for containing a response indicating either success or failure for any of response.

  * Component 5: This component is intended solely for containing a "don't understand" response.

- The Transaction Sublayer (TSL): TSL is used to carry CSL components between CSL peers, using the connectionless services of the SCCP (note that connection-oriented services may be applied in the future). This connection is intended to be an efficient and low-overhead end-to-end connection (Figure 4-29).

Definitions 4-4: The Transaction Sublayer (TSL) protocol data unit is called a **_message_**. An exchange of related TSL messages is called a **_TSL dialogue_** or **_transaction_**.

TSL messages carry components exchanged by CSL peers. Each TSL message can carry several CSL components. Users set-up TSL connections to engage in dialogs or transactions.

4.7.5 Link-by-link Signaling versus End-to-end Signaling

A signaling implementation can be either "link-by-link" or "end-to-end" (The hypothetical signaling reference connections are specified in ITU-T Rec. Q.709). These two signaling scenarios are briefly described in the following sections.

Table 4-7   International Components: Max Number of STPs, SPs, etc. (Link-by-Link)

Connection countries	Connections	No. of STPs	No. of SPs	T_{una} (min/yr) [unavailable]
Large to large	*Mean*	*3*	*3*	*20*
	95%	*4*	*3*	*20*
Large to average-size	*Mean*	*4*	*4*	*30*
	95%	*5*	*4*	*30*
Avg-size to avg-size	*Mean*	*5*	*5*	*40*
	95%	*7*	*5*	*40*

4.7.5.1 Link-by-link Signaling

The maximum number of links for an international connection using "link-by-link" signaling is typically 12 links for average-sized countries or 14 links for large countries. Table 4-7 lists the parameters for a signaling network using a link-by-link signaling scheme for an international application. The first column indicates the countries involved in the international connection. The connection can be between one large country to another, between one large country and one average-size country, or between one average-size country to another. The parameters (Tables 4-7 to 4-9) are defined for two different connection cases: (1) the mean connection, and (2) the 95% connection as shown on the second column in the tables. Table 4-8, on the other hand, lists the same parameters for national applications in the international connections. Table 4-9 lists the overall signaling delay for link-by-link signaling.

- The maximum number of Signaling Points (SPs) - for both international and national applications

- The maximum number of Signaling Transfer Points (STPs) - for both international and national applications

- The maximum unavailability of the overall components of the signaling network in minutes per year - for both international and national applications

Table 4-8   National Components: Max Number of STPs, SPs, etc.

Country size	Connections	No. of STPs	No. of SPs	T_{una} (min/yr)
Large	*Mean*	*3*	*3*	*30*
	95%	*4*	*4*	*40*
Average-size	*Mean*	*2*	*2*	*20*
	95%	*3*	*3*	*30*

Table 4-9   Maximum Overall End-to-end Delay for Different Messages.

Connection countries	Connections	Delay (ms) for simple message (e.g., answer)	Delay (ms) for processing intensive message (e.g., IAM)
Large to large	*Mean*	*1170*	*1800*
	95%	*1450*	*2220*
Large to average-size	*Mean*	*1170*	*1800*
	95%	*1450*	*2200*
Average-size to average-size	*Mean*	*1170*	*1800*
	95%	*1470*	*2240*

Definitions 4-5 (from Q.709): When the maximum distance between an international switching center and a subscriber does not exceed 1000 km (i.e., about 935 miles) (or, exceptionally, 1500 km, i.e., about 1403 miles), and the country has less than $n \times 10^7$ subscribers (the value n is currently under further study), then the country is considered to be of "average-size". A country with larger distances between an international switching center (gateway exchange) and subscriber, or having more than $n \times 10^7$ subscribers, is considered to be of "large-size".

4.7.5.2 End-to-end Signaling

Another scheme used to implement international signaling is the "end-to-end" signaling method. The definition of country size remains the same as specified for the "link-by-link" signaling. The components of an end-to-end signaling connection are:

- Signaling End Point (SEP): The SEP includes the processing in the User Part/Application part (UP/AP), the Signaling Connection Control Part (SCCP), the Message Transfer Part (MTP), and the MTP-to-SCCP-to-UP/AP linkage.

- Signaling Point with SCC Relay functions (SPR): The SPR covers only the processing in the MTP-to-SCCP-to-MTP linkage.

- Signaling Transfer Point (STP): The STP covers only the processing in the MTP. Note that STPs are connected in series, using signaling data links, to implement an "end-to-end" signaling connection.

The maximum number signaling nodes (from the originating node to the destination node inclusively) is 18 in 50% of the national connections, and 23 in 95% of the international connections (for the connections between two average-size countries, the number is 24 instead of 23). The availability of a signaling connection for both the "link-by-link" signaling scheme and the "end-to-end" signaling scheme is approximately the same. That is, 0.99998 (10 minutes) downtime per year. Tables 4-10, 11 and 12 reflect the parameters for end-to-end signaling scheme, and are equivalent to Tables 4-7, 8 and 9 (used in link-by-link signaling scheme).

Table 4-10   International Components: Max Number of STPs and SPRs.

Connection countries	Connections	No. of STPs	No. of SPRs	T_{una} (min/yr) [unavailable]
Large to large	*Mean*	*4*	*2*	*20*
	95%	*4*	*3*	*20*
Large to average-size	*Mean*	*6*	*2*	*30*
	95%	*6*	*3*	*30*
Avg-size to avg-size	*Mean*	*8*	*2*	*40*
	95%	*8*	*4*	*40*

Table 4-11   National Components: Max Number of STPs, STPs, SEPs, etc.

Country size	Connections	No. of STPs	No. of SPRs	No. of SEPs	T_{una} (min/yr)
Large	*Mean*	*4*	*1*	*1*	*30*
	95%	*5*	*2*	*1*	*40*
Average-size	*Mean*	*2*	*1*	*1*	*20*
	95%	*4*	*1*	*1*	*30*

Table 4-12   Maximum Overall End-to-end Delay for Different Messages. (End-to-End)

Connection countries	Connections	Delay (ms) for simple message* (e.g., answer)	Delay (ms) for processing intensive** message (e.g., IAM)
Large to large	*Mean*	*900*	*1320*
	95%	*1270*	*1900*
Large to average-size	*Mean*	*900*	*1320*
	95%	*1180*	*1740*
Average-size to average-size	*Mean*	*900*	*1320*
	95%	*1200*	*1760*

* For simple message processing   ** For intensive message processing

It should be noted that the link-by-link signaling delay (Table 4-9) is applicable where messages are processed by each signaling point along the signaling data link connection. The purpose of using end-to-end signaling is to reduce the overall signaling delay (as shown in Table 4-12) where the normal message loading is assumed.

4.7.6 Numbering Plan for International Signaling Points

Within the signaling path of an international signaling network, an International Signaling Point Code (ISPC) is assigned to each signaling point. The ISPC is used to identify each individual signaling point, and has a maximum length of 14 bits (utilizing a binary code).

In certain network environments, a physical signaling network node may serve (i.e., perform functions) as more than one signaling point. Therefore, this type of signaling point may be assigned more than one ISPC.

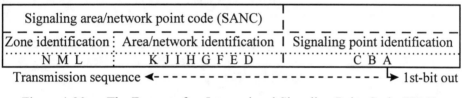

Figure 4-30   The Format of an International Signaling Point Code (ISPC).

An ISPC consists three fields as shown in Figure 4-30. These fields are described as follows:

- Zone identification: The three-bit (denoted as bits L, M and N in Figure 4-30) zone identification field identifies a geographical world zone. The zone identification codes 0 and 1 are currently reserved for future use (i.e., 000 and 001 are unassigned). Therefore, the 3-bit zone identification code consists of only six $[6 \equiv 2^3 - 2]$ binary codes that are available for zone identification assignments.

- The area/network identification: This field has a maximum length of eight bits (shown as bits D, E, ..., and K) and is used to identify a geographical area (not zone) or network within a specific zone. This filed and the zone identification field are referred to as the area/network signaling point code (SANC). Note that each country is assigned at least one SANC. There are a total of 256 $[\equiv 2^8]$ which available for SANC or ISPC assignment.

- The signaling point identification. This 3-bit code [shown as bits A, B and C] is used to identify the different signaling points within a particular geographical area or network. There are a total of 8 $[\equiv 2^3]$ which available for SANC or ISPC assignment.

The total number of binary codewords available for ISPC assignment is given by:

$$No.\ of\ codewords\ for\ ISPC = 6 \times 256 \times 8 = 12,288$$

The complete list of assignments for the zone identification codes and the area/network identification codes can be found in the ITU-T Rec. Q.708. Some examples are given as follows:

Example 4-7: The countries assigned zone identification codes 2, 3, 4, 5, 6 and 7 are listed as below.

- Zone 2 countries: Austria, Belgium, Bulgaria, Cyprus, Czechoslovak, Denmark, Finland, France, Germany, Greece, Hungarian Republic, Iceland, Ireland, Italy, Monaco, The Netherlands, Norway, Poland, Portugal, Romania, Spain, Sweden, Switzerland, Turkey, UK, USSR (old), Yugoslavia, ...

- Zone 3 countries: Barbados, Bermuda, Canada, Cayman island, Cuba, Dominica, Grenada, Guadeloupe, Haiti, Jamaica, Martinique, Mexico, USA, ...

- Zone 4 countries: Afghanistan, Burma, China, India, Iran, Iraq, Israel, Japan, Jordan, Korea, Kuwait, Lao, Lebanon, Mongolian People's Republic, Nepal, Oman, Pakistan, Saudi Arabia, Syrian Arab Republic, Viet Nam, Yemen, ...

- Zone 5 countries: Australia, Fiji, Indonesia, Kiribati, Malaysia, New Zealand, Philippines, Singapore, Thailand, Tonga, Tuvalu, Vanuatu, ...

- Zone 6 countries: Algeria, Central Africa, Congo, Egypt, Gambia, Libya, Mali, Morocco, Niger, South Africa, Sudan, Tanzania, Tunisia, Uganda, ...

- Zone 7 countries: Argentine, Belize, Brazil, Chile, Columbia, Costa Rica, El Salvador, Guatemala, Honduras, Panama, Peru, Uruguay, Venezuela, ...

Example 4-8: The assignment of Signaling Area/Network Number Codes (SANCs) is administered by the ITU-T. In contrast, the assignments of signaling point identifications (i.e., bits A, B and C in Figure 4-30) within each geographical area are made by each country. These assignments must be provided to ITU-T Secretariat. Both SANCs and signaling point identifications are published in the Operational Bulletin of the ITU.

4.8 DIGITAL SUBSCRIBER SIGNALING SYSTEM NO. 1

The ITU-T digital subscriber signaling system No. 1 was designed to support signaling between an ISDN user and the local serving exchange (which is equipped with ISDN). This signaling capability is essentially the D-channel as defined for ISDN. For digital subscriber signaling system No.1, only the first three OSI layers have been defined as follows:

* The physical layer (layer 1) is defined in ITU-T Rec. I.430 and Rec. I.431.

* The data link layer (or Link Access Procedure on the D-channel: LAPD, layer 2) is defined in ITU-T Rec. Q.920 and Rec. Q.921.

* Layer 3 is defined in ITU-T Rec. Q.930, Q.931 and X.25.

The D-channel is a 16-kbps channel used for basic rate ISDN (basic rate interface ISDN; BRI-ISDN) and is embedded in a multiplex structure of 192 kbps (for ITU-T networks), or 144 kbps (for North American networks). In comparison, the primary rate ISDN (primary rate interface; PRI-ISDN) has a 64 kbps D-channel which is also embedded in multiplex structure of 2.48 Mbps (E-1 signal for ITU-T networks), or 1.544 Mbps (DS1 signal for North American networks). These multiplexed signals will be described in Chapter 6.

This section is focused on the data link layer. The purpose of the data link layer (or LAPD) is to convey information between layer 3 entities across the ISDN user-to-network interface using the D-channel. In summary, LAPD supports: (1) multiple terminal installations at the user-to-network interface, and (2) Multiple layer 3 entities.

4.8.1 Main LAPD Functions

The data link layer function provides the capability of transferring information between multiple data link connection endpoints. The information transfer may be implemented via "point-to-point", "multicast", or "broadcast" data link connections. For the point-to-point case, a data link layer message frame (described later in this chapter) is directed to a single endpoint. In the multicast or broadcast cast, a frame is directed to more than one point. The LAPD functions are listed as follows:

• LAPD can perform flow control

• LAPD provides one or more data link connections on a D-channel. A Data Link Connection Identifier (DLCI) contained in each frame (discussed later in this chapter) is used to discriminate between multiple data link connections.

• LAPD performs frame delimiting, (frame) alignment, and transparency to allow a sequence of bits transmitted over a D-channel to be recognized as a "frame".

• LAPD implements sequence control so that the sequential order of frames across the data link connection is maintained.

• LAPD detects transmission, message format, and operational errors occurring on the data link connection.

- LAPD will attempt to recover detected (transmission, format, and operational) errors.

- LAPD will notify the management entity when unrecoverable errors are encountered.

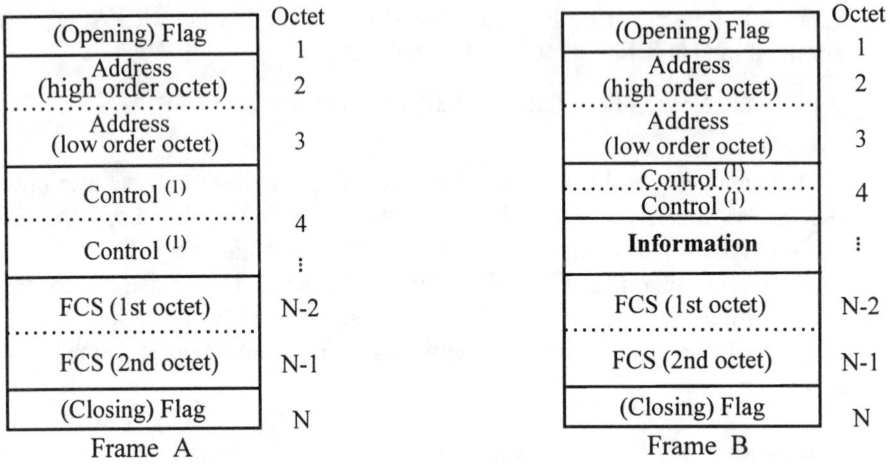

	Octet
(Opening) Flag	1
Address (high order octet)	2
Address (low order octet)	3
Control (1) .... Control (1)	4 ⋮
FCS (1st octet)	N-2
FCS (2nd octet)	N-1
(Closing) Flag	N

Frame A

	Octet
(Opening) Flag	1
Address (high order octet)	2
Address (low order octet)	3
Control (1) Control (1)	4
Information	⋮
FCS (1st octet)	N-2
FCS (2nd octet)	N-1
(Closing) Flag	N

Frame B

Note (1) Unacknowledged operation - One octet
Multiple frame operation - Two octets for frames with sequence numbers
One octet for frames without sequence numbers

Figure 4-31   LAPD Link Layer Frame Structure.

4.8.2 LAPD Frame Structure

All data link layer "peer-to-peer" information transfers are based on LAPD frames. The LAPD frame structure is shown in Figure 4-31. There are two types of LAPD frames, frames A and B. Frame A is used when there is no information field, and frame B is used when there is an information field. Both frames contains several fields, described as follows:

- Flags: All LAPD frames contain a flag heading, called the "opening flag", which has a standardized sequence of "01111110". The opening flag precedes the address field (see Figure 4-31). All frames are concluded with a closing flag, which has the same sequence as the opening flag. The closing flag follows the Frame Check Sequence (FCS). In some applications, the closing flag can serve as the opening flag of the next frame. However, in general a closing flag is usually followed by an opening flag. Therefore, all receivers must be able to detect (receive) one or more consecutive flags.

- Address field: There are two address octets; octet Nos. 2 and 3 of the LAPD frame, as shown in Figure 4-31. Octet No. 2 is known as the high order octet (Figures 4-31 and 4-32), and is led by one Extended Address bit (EA0). The EA0 bit is followed by one

Command/Response (C/R) bit and 6 bits of Service Access Point Identifier (SAPI). Octet No. 3, known as the low order octet, is also led by an extended address bit (EA1), which is followed by a 7-bt Terminal Endpoint Identifier (TEI).

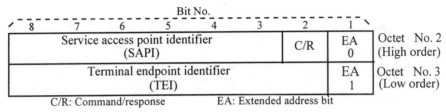

Figure 4-32   Address Field Format.

* The double-octet address field for LAPD operation has bit No. 1 of the first octet set to a logical "0" and bit No. 1 of the second octet set to a logical "1". The EA0 or EA1 bit is used to indicate the final octet of the address field. That is, if this bit is a "1", it is the final octet of the address field.

Table 4-13   Applications of Command/Response Bit.

Message	Direction	C/R value
Command	*Network to user*	*1*
	User to network	*0*
Response	*Network to user*	*0*
	User to network	*1*

* The Command/Response (C/R) bit indicates whether a frame is a "command" or a "response". If the **user** sends a command, this bit set to a logical "0", and when the **user** responses to a command, this bit is set to a logical "1". However, if the **network** sends a command, this bit is set to a logical "1", and when the **network** responses a command, this bit is set to a logical "0". (see Table 4-13).

Table 4-14   SAPI Code Assignments.

SAPI code	Related layer 3 or management entity
000000	Call control procedures
000001	Reserved for packet switching using Q.931 call control procedures
010000	Packet switching using X.25 layer 3 procedures
111110	Layer 2 management procedures
Others	Reserved for future standards

* The Service Access Point Identifier (SAPI) bits are used to identify the signaling point at which data link layer services are provided (by a data link layer entity) to

a layer 3 entity or a management entity. There are 6 bits (bit 3 is the least significant bit and bit 8 is the most significant bit of the SAPI field as shown in Figure 4-32) allocated for SAPI. Hence, there are 64 ($\equiv 2^6$) SAPI codes available, but currently only four SAPI codes have been assigned (see Table 4-14).

* The Telephone End-point Identifier (TEI) field in the frame octet 3 has a length of 7 bits (bit 2 is the least significant bit and bit 8 is the most significant bit of the TEI field as shown in Figure 4-32) allocated for TEI. Hence, there are 128 ($\equiv 2^7$) TEI codes available. There are two types of data link layer that require TEI, broadcast and point-to-point, described as follows:

  ♦ Broadcast data link layer connection: The TEI for this type of connection is associated with all user side data link entities having the same SAPI. The TEI value of 128 ("1111111") is assigned as the broadcast group TEI code.

Table 4-15   Point-to-point TEI Assignment.

TEI value	User type
0 to 63	Non-automatic TEI assignments (*user*)
64 to 126	Automatic TEI assignments (*network*)

  ♦ Point-to-point data link layer connection: The TEI for this type of connection can be associated with a single terminal equipment entity. A terminal equipment entity can possess one or more TEIs. The TEI values (codes) for point-to-point connections are shown in Table 4-15 (note that two different user types are shows). The non-automatic TEI values (codes) are selected by the user, thus the user is responsible for the TEI value (0 to 63) allocations. The network is responsible for the selection and the allocation of automatic TEI values (64 to 126).

• Control field: This field (Figure 4-31) is used to identify the type of frame as being either command or response. There are three types of control fields:

1. Numbered information transfer (known as I format), which is used to perform an information transfer between layer 3 entities.

2. Supervisory function (known as S format), which is used to perform data link supervisory control functions such as acknowledge, request re-transmission or request temporary suspension of I format transmission.

3. Unnumbered information transfers and control functions (known as U format), which is used to provide additional data link control functions and unnumbered information transfers for unacknowledged information transfer.

- Information field: The contents of the information field have a length that is an integral number of octets. The maximum length in this information field is a system parameter, but the default value is 260 octets.

- Transparency: The transmitter of the data link layer connection examines the frame content between the opening flag and the closing flag sequences and will insert a "0" bit after all sequences of five contiguous "1" bits (this includes the last five bits of the frame check sequence) so that a flag or an abort sequence does not exist within the frame. The receiver of the data link layer connection examines the frame contents between the opening flag and the closing flag sequences, and discards any sequence consisting of a "0" bit followed by five contiguous "1" bits.

- Frame Check Sequence (FCS): This field has a length of 16 bits. The generating polynomial used to produce this field is $g(x) = 1 + x^5 + x^{12} + x^{16}$. This generating polynomial will operate on all the bits in the frame, starting from the last bit of the opening flag to the first bit of the frame check sequence (exclusively), but excluding any bits inserted for transparency.

4.8.3 LAPD Operation Types

For layer 3 information transfer, ITU-T has defined two types of operations for the data link layer. They may coexist on a single D-channel, and are briefly described as follows:

1. Unacknowledged operation: Under this mode of operation, the layer 3 information is transmitted via Unnumbered Information (UI) frames. These UI frames are not acknowledged by the data link layer. Key operational characteristics are:

   * Even if the system has detected transmission or format errors, there is no error recovery mechanism available.

   * No flow control is applied.

   * Applicable to both the point-to-point and broadcast information transfer.

   * For broadcast applications, each endpoint has a specific Service Access Point Identifier (SAPI)

2. Acknowledged operation: In this mode of operation, the transmitted layer 3 data link frames must be acknowledged. Some characteristics of this mode of operation are:

   * The operation specifies the error recovery procedure, which is based on re-transmission of unacknowledged frames.

   * If the errors are un-correctable by the data link layer, a report is sent to the management entity.

* The operation also specifies flow control procedures.

* This operation is applicable for point-to-point information transfer.

* The layer 3 data link information is sent in numbered information frames, which may be outstanding at the same time.

* Multiple frame operation must be initiated by a multiple frame establishment procedure, which utilizes a Set Asynchronous Balanced Mode Extended (SABME) command.

4.8.4 Data Link Connection Identifier and Terminal Endpoint Identifier

The Data Link Connection Identifier (DLCI) must be used to establish the data link layer information transfer mode. In addition to DLCI, one has to understand the data link states and Terminal Endpoint Identifier (TEI) administration in order to understand data link layer information transfer.

The DLCI is applied to identify a data link connection and is carried in the address field of each frame. Note that the DLCI is associated with a connection endpoint identifier at two ends of a data link connection. The connection endpoint identifier is used to identify message units passed between the data link layer and layer 3.There are two elements in the DLCI: the Service Access Point Identifier (SAPI) and the Terminal Endpoint Identifier (TEI) (these functions have been previously discussed in this chapter).

Data link connection identifier is used internally by the data link layer entity and is not known by the layer 3 entity or the management entity. The layer 3 and the management entities apply Connection Endpoint Identifier (CEI), instead of DLCI, to address peer entities.

* A broadcast data link entity: Always remains in the information transfer state so that it is capable of in unacknowledged information transfer (known as a TEI-assigned state).

* A point-to-point data link entity: It may function in one of the following three states.

    * TEI-unassigned state: If the data link entity is in this state, layer 3 information transfer is not impossible since a TEI has not yet been assigned to the entity.

    * TEI-assigned state: In this state, layer 3 unacknowledged information transfer is possible since the TEI has been assigned by means of a TEI assignment procedure.

    * Multiple-frame-established state: In this state, both unacknowledged and acknowledged information transfers are possible.

4.9 SELECTED VOCABULARY OF SIGNALING

The most frequently used signaling terms will be defined in this section. Some of these terms are illustrated in Figure 4-35, 4-36, and 4-37.

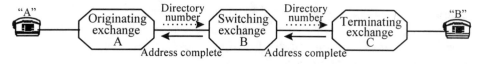

Figure 4-33   An End-to-end Connection.

- **Address complete**: A message that is sent in the backward direction, indicating all the address signals (Directory Numbers; DNs) required by the network for routing the call to the called party have been received (Figure 4-33).

- **Associated mode (of signaling)**: One of the three common channel signaling modes. In the associated mode, messages for signaling functions (which involve adjacent signaling points) are transported over a direct interconnecting link.

- **Call set-up**: One of the many states of call processing. The call set-up state establishes a communications path between: (1) the calling party and the called party, and, (2) between the calling party and the network entities.

- **Change-back**: Change-back is the inverse function of change-over. It is the procedure for transferring signaling information (traffic) from one or more alternate signaling links to a signaling link that has become available.

- **Change-over**: Change-over is the procedure for transferring signaling information from a signaling link (which is experiencing failure or is required to be cleared of traffic) to one or more alternate (different) signaling links.

- **Channel-associated signaling**: One of several signaling methods. In channel-associated signaling, the user traffic and signaling information is transported in the same channel (e.g., in-band signaling).

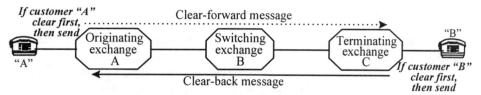

Figure 4-34   Clear-forward and Clear-back Messages.

- **Clear-back message**: The clear-back message is a signal (message) sent in the backward direction to indicate that the called party is no longer "off-hook".

- **Clear-forward message**: It is a signal sent in the forward direction to terminate a call (or a call attempt) so that the associated circuits can be released. This signal is normally originated when the calling party hangs-up the phone.

- **Compelled signaling**: It is one of several signaling methods. In compelled signaling, after one signal message (or element) has been sent, a second signal message cannot be sent in the same direction until the first signal message has been acknowledged (by a response in the opposite direction).

- **Continuity check**: To verify an acceptable path for transmission of voice, and/or data exists, a check must be made on each circuit in a connection.

- **Controlled re-routing**: Using a controlled methodology, the signaling information (traffic) is transferred from an alternate signaling route to the normal signaling route, when it becomes available.

- **Cross-office check**: To verify an acceptable transmission path exists, a check must be made on the circuits between two or more exchanges.

- **Dual seizure**: The condition when two exchanges attempt to seize the same circuit at approximately the same time (in a two-way operation mode) is called dual seizure.

- **En-block signaling**: En-block signaling is the method in which the address digits are encoded into blocks for onward transmission. Note that the blocks of code contain all the address information required to route the call to its destination.

- **End-of-pulse (ST) signal** (see Table 4-1): ST (STop) is an address signal sent in the forward direction to indicate there are no more address signals to follow.

- **End-of-selection signal**: The end-of-selection is a signal sent in the backward direction to indicate the successful completion of the call set-up process. It can also be used to indicate an unsuccessful termination of the call set-up process, and that may contain information on the called party's line condition.

- **Exchange**: The network consists of switches and many other network elements. The equivalent term for exchange is "office", "Central Office (CO)", "telephone switch", "telephone exchange", or "wire center". An exchange can be used as: (1) local exchange, (2) toll (trunk) exchange, or (3) gateway exchange.

- **Flow control**: Flow control is a function in a protocol used to control the flow of signaling messages between adjacent layers of a protocol, or between peer entities. For example, flow control allows a receiving entity to control signaling messages: (1) from a sending entity, (2) between or within different users and the Message Transfer Part (MTP).

- **Initial Address Message (IAM):** The IAM is a message sent in the forward direction at call set-up. The IAM includes the address and other information relating to routing and handling a call.

- **In-slot signaling**: It is a signaling information periodically sent (in a digital time slot) that is permanently allocated in the (voice or data) channel time slot.

- **Label**: It is one type of information contained in a signaling message that is used to identify the particular circuit, call, or management transaction to which the message is related.

- **Line signaling**: It is a signaling method for transmitting signaling traffic between equipment to terminate and continuously monitor part or all of the traffic circuit. For example: supervisory signaling between exchanges provides the busy, idle, or out-of-service status of circuits.

- **Link-by-link signaling**: In the link-by-link signaling scheme, signaling messages are transmitted "one link at a time" in a multilink connection (Figure 4-35). The network performs processing at each intermediate switching point for subsequent transmission. In the SS7 system, link-by-link signaling is the procedure used to exchange signaling messages between two Signaling Points (SPs) that are either directly connected to each other, or are connected via STPs.

- **Message Transfer Part (MTP)**: The MTP is a functional part of a common channel signaling system for transferring signal messages required by all the users. It also performs several subsidiary functions (e.g., error control & signaling security).

- **Multifrequency (Code) [MF(C)] signaling**: It is a in-band signaling method that uses multi-tone signals to represent the signaling traffic. A multi-tone signal contains two or more frequencies out of an "n" pre-determined frequency set (see Table 4-1).

- **Non-associated mode signaling**: Non-associated mode signaling one of many signaling methods. In this method, signaling messages involving two non-adjacent signaling points are conveyed (between those signaling points) over two or more signaling links in tandem that passes through one or more STPs.

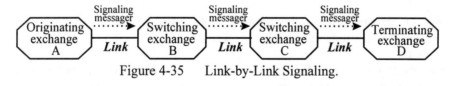

Figure 4-35   Link-by-Link Signaling.

- **Out-slot signaling**: In the out-of-slot signaling scheme, the signaling associated with a (voice or data) channel is sent over the same transmission path (channel) or circuit, but in a different time slot(s). An example of this method is the ITU-T 2.048 Mbps digital carrier system (described in Chapter 6).

- **Pass along method:** In the pass-along method, signaling traffic (information) is sent along the signaling path of a previously established physical connection.

- **Post-dial delay:** The post-dial delay is the time interval between "end of dialing " by the customer and the reception (by the same customer) of the call progress signaling generated by the exchange serving this customer. The call progressing signals can be the dial tone, a recorded announcement, or the abandon of the call.

- **Quasi-associated mode** (a special non-associated signaling operation): The signaling message route is determined (for each signaling message) by the signaling points between two or more signaling links (in tandem) passing through one or more STPs.

- **Release-guard:** After an exchange receives a "forward-clear" message when a circuit is placed in the idle condition, the exchange sends a "release-guard" message in the backward direction (in response to the forward-clear message).

- **Service indicator:** It is part of the information contained in a signaling message that is used to identify the user to whom the message belongs.

- **Signal unit:** The bit stream of a signaling message is grouped into "signal units" that function as a transferable entities used to convey information over a signaling link.

- **Signaling data link:** A signaling data link consists of two data channels that operate in opposite directions at the same data rate, and function as a single signaling system.

- **Signaling link:** A signaling link is the transmission media used for reliable transfer of signaling messages. A signaling link consists of a "signaling data link" and transfer control functions.

- **Signaling link group:** A signaling link group is several signaling links that directly connect two signaling points together and have the same physical characteristics (e.g., bit rate and propagation delay).

- **Signaling message route (or message route):** A message route is a signaling link (or consecutive signaling links connected in tandem) that is used to convey a signaling message from an originating point to its destination.

- **Signaling network:** A signaling network, that consists of many signaling points and signaling links, is used for signaling message transport.

- **Signaling Point (SP):** A SP is a signaling network node that is used to originate, receive, or transfer signal messages from one signaling link to another.

- **Signaling point code:** A signaling point code is used to uniquely identify each signaling point (network node) in a signaling network. This digital code can be used as either an originating point code or a destination point code, depending upon its position in the label.

- **Signaling point numbering plan**: It is a formal (standardized) description of the method (numbering plan) for converting "end-user-provided" address information into an address understood by the signaling network.

- **Signaling relation**: A signaling relation is formed by two signaling points, and involves the possibility of information interchange between corresponding user part functions.

- **Signaling route**: It is predetermined path, described by a succession of signaling points, that can be traversed by signaling messages (directed by a signaling point) toward a specific destination point.

- **Signaling time slot**: A signaling time slot starts in a particular phase of each frame (i.e., 125 μs interval) that is allocated for transmitting signaling messages.

- **Signaling traffic management functions**: They are used to control (and modify if necessary) routing information used for message routing function. These functions insure the transfer of signaling traffic is performed in a manner that prevents irregularities in message flow.

- **Speech digit signaling**: It is a type of channel-associated signaling where time slots (primarily used to transport of encoded speech) are periodically used to carry signaling messages.

- **Splitting**: Splitting is a switching function used for several purposes including: (1) disconnection (isolation) of a channel that precedes the point where signaling frequencies are injected, and (2) disconnection (isolation) of a channel that succeeds the point where the signal receiver is connected. Splitting can be used to prevent false operation of signaling equipment (caused by signal reflections and spillover) when signaling information is being received. Splitting can also prevent interference from a preceding circuit or nearby equipment (terminal, or device) when a signaling point is receiving a signaling message.

- **Subsequent address message**: A subsequent address message is sent in the forward direction after the initial address message, or after an address message that contains only one address signal.

- **User part**: User part is a function in the common channel signaling system. It is used to transfer signaling messages, via the message transfer part. There are different types of user parts for telephone, data, and ISDN services. Each user part is specific to a particular application in the signaling system.

Review Questions II for Chapter 4

(14) (True, False) SS7 signaling system evolved from earlier signaling system, and continues to be adopted as the worldwide signaling system for a digital trunk network.

(15) SS7 is optimized for operation in digital telecommunications networks: _____ (e.g., the μ-law and the A-law networks), _____ networks, _____ networks.

(16) SS7 can support the following applications: _____, ISDN, public and _____ network, _____ (OAM) of networks.

(17) In SS7 signaling system, the "upper part" of protocol is the _____, and the "lower part" of protocol is the _____.

(18) The signaling data link, typically a digital path with a speed of _____ or ____ kbps, defines the _____, _____, and _____ characteristics of a signaling link, and the means to access it.

(19) The main signaling link functions are: _____, signal unit _____ and _____, bit _____, _____ detection/correction, signal unit sequence _____, signal link _____ detection, flow _____, etc.

(20) Three basic signal units for SS7 signaling systems: the message signal unit (MSU), link status signal unit (LSSU), and fill-in signal unit (FISU). All of them starts with an 8-bit open _____ of "01111110", and ends with a closed _____. The CRC bit error control used for all three signal units is _____, with a generator polynomial of _____.

(21) The _____ [that contains three parts: the _____, the _____, and the _____ (SLS)] is the part of a signaling message that is applied by the message transfer part (MTP) for signaling message handling.

(22) The four major user parts of SS7 are: _____ (SCCP), _____ (TUP), _____ (ISDN-UP), and _____ user part.

(23) SCCP message contains routing label, _____ code, mandatory _____ and _____ parts, and optional part.

(24) Signaling implementation can be either _____ or _____.

(25) An _____ (ISPC) [having a maximum length of 14 bits (utilizing a binary code)], assigned to each signaling point within the signaling path of an international signaling network, is used to identify each individual signaling point of signaling network (e.g., SS7).

(26) (True, False) Currently, the world zone identification codes of "0" and "1" are reserved for future use.

CHAPTER 5

Switching Principles and Applications

Chapter Objectives

Upon the completion of this chapter, you should be able to:

- Describe the needs for switching, the random nature of traffic/service demands, and the evolution of switching technology.

- Describe circuit switching technology: switching functions, switching system architectures, switching control, switching interfaces, and switching applications.

- Describe exchange switching: local exchange switches (various terminations, billing and administration), trunk exchange and gateway switches (switch functions, features, evolution and gateway applications), and customer switches (basic functions, switching types, evolution, and future trends).

- Discuss packet switching technology: packet voice and data signal characteristics, differences between circuit and packet switching, and evolution of packet and ATM cell switching.

- Describe the 5ESS switching system: system architecture, ISDN and common channel signaling functions, wireless applications, and 5ESS-2000.

- Describe switching examples (other than 5ESS and 5ESS-2000): 4ESS digital switch (functional blocks, automatic message accounting, and software defined network), Datakit II virtual circuit switch, BNS-2000 broadband system, and GlobeView-2000 ATM public switch for SONET transport systems.

5.1 INTRODUCTION

The concept of switching is the selection of a transmission path for a signal, that is used to transport the signal over an extended distance (see Figure 5-1). The incoming lines (inputs) can be single-channel lines or a multiple channel line, and the outgoing (output) trunks are typically multiple channel line.

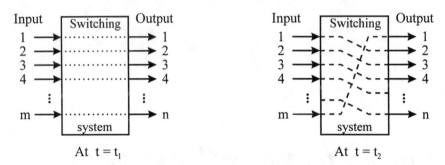

Figure 5-1   Concept of Signal Switching.

For example, during one interval (e.g., $t = t_1$), the signal on input-path No. 1 may be routed to output-path No. 1, the signal on input-path No. 2 is routed to output-path No. 2, …, and the signal on input-path No. m is routed to output-path No. n ($n \geq m$; in actual systems, n may be smaller than m). During a later interval (e.g., $t = t_2$), the traffic pattern may have changed. Now, the signal on input-path No. 1 may be routed to output-path No. 2, the signal on input-path No. 2 is routed to output-path No. 3, …, and the signal on input-path No. m is routed to output-path No. 1 (see Figure 5-1). The system, indicated by the rectangular box in Figure 5-1, is a switching system, that performs the "switching function" as described.

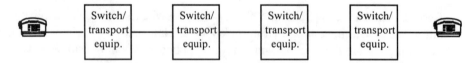

Figure 5-2   An End-to-end Connection.

5.1.1 The Need for Switching

On many occasions a typical call may involve several transmission facilities (connections), transmission (transport) equipment or terminals, and switching machines (see Figure 5-2). To implement an "end-to-end" connection, each switch must route its incoming signal to the intended routes (or trunks) for transport. The reason that switches to set-up the path for a signal to travel from its source to its destination is examined in the following sections.

Figure 5-3(A) shows a connection between two subscribers (A and B) without the need for selecting a different path (i.e., the two subscribers have a dedicated path).

Figure 5-3(B) is a case where each of the three subscribers (A, B and C) has the ability to talk to either of the other subscribers. If no switching machines are used, every subscriber must have two (= 3 − 1) phones to achieve the goal of full connectivity. If there are N subscribers, each must own (N − 1) phones. This arrangement is obviously not practical! Therefore, the network needs switches.

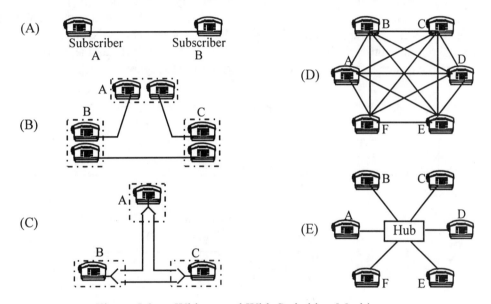

Figure 5-3   Without and With Switching Machines.

If distributed switches are applied to each individual subscriber [shown in Figure 5-3(C)], the required number of telephone is equal to the number of subscribers. However, the control of the switch becomes an issue. For example, if subscriber A needs to talk to subscriber B, both the switching machines at the subscriber A's and subscriber B's locations must be notified establish the connection. This control function is tedious, therefore, it can be concluded that the arrangements, without using a switching machine, shown in Figures 5-3(A), (B) and (C) are not practical.

A "somewhat" more practical arrangement is shown in Figure 5-3(D), which actually exist in certain small-size private networks. Figure 5-3(D) shows each of the six subscribers (A through F) has the ability to talk to the other five subscribers, without using a switching machine. If subscriber No. 1 needs to talk to subscriber No. 2, a simple switch (that has five different positions for the five other subscribers) at the subscriber will be activated (i.e., turned-on or pushed). The switch at subscriber No. 2 will be alerted (e.g., the phone rings). If the number of subscribers is not large, these simple switches are relatively easy to design/operate. The primary concern is the number of transmission paths connecting all the subscribers. It can be proven that the total number of transmission facilities (e.g., transport paths or connections) for N subscribers is given by the following equation:

$$N_c = C_2^N = \frac{N(N-1)}{2} \approx \frac{N^2}{2} \qquad \text{for very large } N \qquad (5\text{-}1)$$

To understand the required number of facility connections (N_C), Table 5-1 represents several values of N and the corresponding N_C. It can be seen that the number of connections increases rapidly for a large number of subscribers. This implies that the costs of transmission facilities (connections) will increase quickly as the network size increases [i.e., when using a network architecture similar to Figure 5-3(D)].

Table 5-1   Number of Connections vs. Number of Subscribers.
[If a switch is not applied; Figure 5-3(D)]

N	6	10	50	100	200	500	1,000	10,000
N$_c$	15	45	245	4,950	19,900	124,750	499,500	49,995,000

Table 5-2   Number of Connections vs. Number of Subscribers.
[If a switch is applied; Figure 5-3(E)]

N	6	10	50	100	200	500	1,000	10,000
N$_c$	6	10	50	100	200	500	1,000	10,000

If a switch is used as a hub device, as shown in Figure 5-3(E), the number of transmission facilities (connections) is given as follows:

$$N_C = N \qquad (5\text{-}2)$$

That is, only one (transmission) facility is required to connect each subscriber to the switch as indicated in Table 5-2. By comparing Table 5-1 and Table 5-2 (e.g., a network with 1,000 subscribers, the numbers of facilities with and without a switch are 1,000 and 499,500 respectively), it is clear that the economical tradeoff between the reduction of transmission facilities versus the addition of switching equipment can easily justify the decision to deploy switches in the network.

Figure 5-4 shows various locations where switches may be applied within a telecommunications network.

- Local exchange switches

For local exchange offices (i.e., central office), low-to-medium capacity switches are typically applied. In metropolitan areas, a large capacity switch may be used as the local exchange office.

A local call may involve only one central office. That is, the switch serves as both the originating and the terminating switch. This type of call service is known as an intra-office call (i.e., the call is switched by the same switch, rarely by two switches, within one central office). However, a local call may also involve two local exchanges; one serves as the originating switch, and the other serves as the terminating switch. A call of this type is known as an inter-office call. Both local exchanges are referred to as "end offices". Note that the transmission facilities connecting end users (subscribers) to the central office switch are called loops.

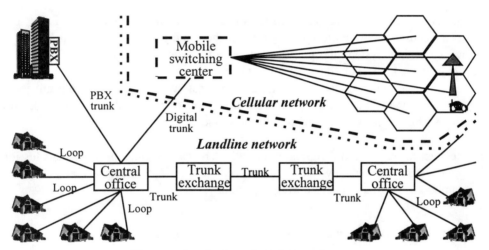

Figure 5-4   Various Switch Locations.

- Trunk exchange switches

For trunk exchange offices (also called toll offices; shown as trunk exchange in Figure 5-4), a medium-to-large capacity switch is often used. A typical long-distance call involves two or three trunk exchanges, and two end-offices. Note that the transmission facilities connecting any two adjacent switches are called trunks.

- Wireless switches

A switch is required to enable wireless networks to communicate with the Public Switched Telephone Network (PSTN; commonly referred to as "landline network"). The network dedicated for wireless applications is called a cellular network (discussed in Chapter 8) where the switch is located at the Mobile Switching Center (MSC or Mobile Telephone Switching Office: MTSO). This is typical a medium-size switch. One MSC may serve many cell sites, each having a cell antenna for communicating with several wireless subscribers. A serving area may consist of more than one mobile switching center. The typical transmission facilities connecting the cellular network to the landline network are digital trunks.

- For business locations, a Private Branch eXchange (PBX, i.e., a customer's switch) is often deployed within the business buildings. Their capacity is smaller than those used in the local exchange, trunk exchange, or MSC. The transmission facility connecting a PBX to the central office is called a PBX trunk.

5.1.2 The Random Nature of Traffic/Service Demand

The telecommunications service demand is usually random in nature. That is, typically not every subscriber wants to communicate at the same time. It is this randomness that allows switching system design to economically meet a given level of "Grade Of Service" (GOS). In summary, most switching system designs are based on the assumption that call origination is a random process, when the population universe of subscribers is viewed as a whole. The random nature of service demands includes both the request for service and the duration of the service.

- Request for services or origination of calls: The call origination rate fluctuates for a given hour, a given day, and a given year. Figure 5-5 shows a typical (relative) number of originating calls during a given day.

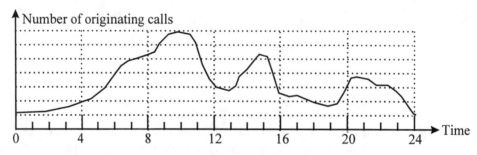

Figure 5-5   Number of Originating Calls During a Given Day.

Several factors may influence the call origination process. The following two factors are the key ones:

1. Customer behavior: This is by far the most important factor influencing the call originating process. For example, from Figure 5-5, it can be seen that the number of

calls during a workday peaks at mid-morning (i.e., around 10 AM). The second peak may occur in the mid-afternoon hours (i.e., around 3 PM). For off-hours the peak is the after-dinner (i.e., around 9 PM). The current Internet applications have triggered more network usage after hours. For future broadband ISDN applications (e.g., xDSL, HDSL, ADSL, or VDSL; described in Chapter 3), the usage peaks may change considerably. Another factor influencing usage peaks are special events.

2. The state of the system will also influence the call originating call process. For example, heavy system overloads can increase the number of customer's repeat call attempts. Special events can also definitely distort the randomness of traffic request patterns (e.g., holidays, natural disasters, etc.).

   A customer originates a call independently of other customers and independently of the state of the system. This results in the corresponding random nature of service needs.

- The duration of service (or the holding time): The during of service includes three main component, the customer dialing time ($t_{dialing}$), the ringing time ($t_{ringing}$) and the communication time ($t_{conversation}$) as expressed in Eq.(5-3).

$$\text{Duration of service} = t_{dialing} + t_{ringing} + t_{conversation} \qquad (5\text{-}3)$$

The communication (or conversation) time is the largest element of the holding time. The communication time for voice signal transport has a probability density that behaves as an exponential distribution (see Figure 5-6). The average holding time for the example shown in Figure 5-6 is 6.3 minutes.

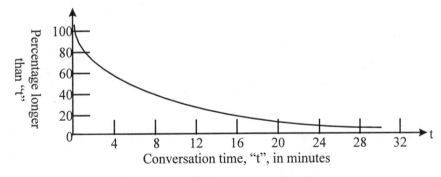

Figure 5-6   An Example of Conversation Time for Voice Transport.

The holding time may vary with customer types, charging method, and the geographic location. For Internet applications, the holding time is typically much longer than the holding time for the traditional voice communications shown in Figure 5-6. The expected average holding time for a typical Internet user is found to be at five to ten times longer than the voice applications (i.e., approximately an hour).

It is clear that the traffic characteristics of data communications are different from the characteristics of voice communications. This is because data signals occur in bursts. The holding time can be very short or very long (known as "bulk data" applications). Low priority traffic or traffic that can be preempted often has a longer holding time. Similarly, the latency requirement is different for data traffic (discussion of data communications and ISDN is given in Chapter 9).

5.1.3 Switching Technology Evolution

Switching technology has evolved from manual switches, to step-by-step switches, which were replaced by cross-bar switching technology. Finally, the Store-Program Control (SPC) switching technology was adopted for modern switching machines. The early SPC switches were analog designs that interfaced with analog signals. In the early 1970s, SPC switches started being replaced by digital switches.

The control methods used in switching machines also evolved from manual methods into direct progressive control methods. This was later replaced by the common control method, which eventually evolved into stored program control methods. For nearly 90 years, switching machines only interfaced analog signals. The first digital switch, which interfaced digital signals, was developed in the early 1970s. Table 5-3 and Figure 5-7 summarize the milestones of various switching technologies.

Table 5-3   Evolution of Switching Technologies.

Technology	Date introduced	Control method	Interface signal
Manual	1870s	Manual	Analog
Step-by-step	1890s	Direct progressive	Analog
Crossbar	1930s	Common control	Analog
SPC analog	1960s	SPC	Analog
SPC digital	1970s	SPC	Digital

SPC: Stored program control

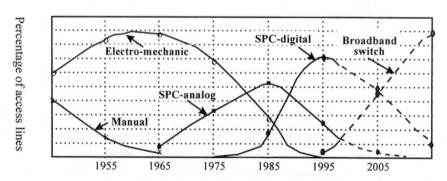

Figure 5-7   Switching Technology Evolution.

5.2 CIRCUIT SWITCHING TECHNOLOGY

The call process of an inter-office call is shown in Figure 5-8 and is used to describe the basic switching functions. A detailed description of call processing is provided in Chapters 2 and 4.

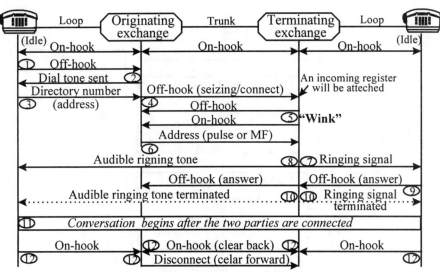

The steps shown in the graph have been discussed in Example 4.1, section 4.4.

Figure 5-8   Call Process of an Inter-Office Call.

Note that, in this inter-office call, the conversation begins at step No. 11, as indicated in Figure 5-8. The end-to-end connection involves two local loops and at least one inter-office trunk. The transmission facilities are dedicated to these two subscribers, and no other users are allowed to share these facilities during the duration of the conversation. This type of connection is known as a circuit switching technology (rather than the packet switching technology, which is described later in this chapter).

5.2.1 Switching Functions

In addition to interconnecting subscribers, the functions performed by a modern digital switch (using circuit-switching technology) can be grouped into four classes: (1) interconnection function (e.g., interconnecting two or more subscribers), (2) control-related functions (e.g., busy testing, interpretation, path establishment, path selection, and route selection), (3) signaling-related functions (e.g., alerting, attending, signaling reception, signaling transmission, and supervision), and (4) operations supporting functions (e.g., administration and maintenance). These functions are shown in Figure 5-9 and are described as follows.

- **Interconnection functions:**

  The interconnection function performed by switches provides transmission paths that are used to connect the calling party and the called party. This function is sometimes referred to as the "switching fabric" and is implemented by the functions within a switch known as the "switching network circuit".

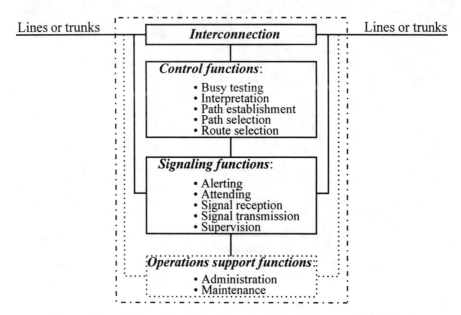

Figure 5-9   Basic Switch function of a Circuit Connection Switch.

- **Control-related functions**

  * Busy testing: Since the switch must seize an outgoing trunk or trunk group for routing incoming traffic, the switch must test the trunk group to determine whether or not it is available.

  * Signal interpretation: When a signaling message is received by the switch, the switch interprets the information contained in the signaling message. It then takes the proper action requested by the signaling message.

  * Path establishment: After the switch has selected a path, it must establish the physical interconnections for the path. This function typically requires a memory element that stores connection information for the duration of the call.

  * Path selection: The switching system selects an idle link (or a series of idle links) within the switch to establish the call connection.

* Route selection: All the switches involved with a call must select a trunk or a trunk group between any each pair of switches. This series of connections is called a "route".

- **Signaling-related functions**

  * Alerting: A switch must inform the distant office about the existence of an incoming call by transmitting the attending signal, if the call involves two or more switches. In addition, the called party must also be alerted about the presence of an incoming call. A ringing signal is usually sent for this purpose.

  * Attending: Each time after a subscriber originates a call (e.g., the off-hook condition), the originating switch must be able to recognize this request action. Similarly, an attending function is required from the trunk side as a "request for service" originated by other exchanges in the intended "end-to-end" connection.

  * Signal reception: After the switch has responded to the call originating request (i.e., attending the call request), it sends dial tone to the subscriber. The subscriber then sends out the called party's telephone number. The called party number is received by the originating switch, and then sent to downstream exchanges.

  * Signal transmission: This function applies to inter-office calls. For an inter-office call, the address signal must be transmitted to the distant office(s).

  * Supervision: The switch must detect when a connection is no longer needed. When this condition occurs, the switch requests release of the trunk so it is available for other calls.

- **Operations support functions**

  * Administration: There are several operations support functions used for administrative purposes described as follows.

    - Traffic engineering: This function provides traffic information used for planning growth needs. Traffic information can also be used to evaluate service quality.

    - Operations: There are numerous switching operation functions. The primary operations include: staffing of consoles, providing budgets, supporting business services to customers, providing assistance services, and interception of services.

    - Billing: Billing is the process of recording customer usage of services, and generating accurate billing statements for the services.

    - System administration: These operations include traffic measurement and analysis, load balancing, line and terminal assignments, and verifying quality of service.

Note that most operations support functions are based on collecting and processing of traffic data.

* Maintenance: Maintenance activities are used to verify switching systems are functioning properly. Two primary maintenance functional groups are:

   ◆ Performance Monitoring (PM): The PM function includes data collection for blocking probability computation, peg counts, call completion, incomplete calls, and re-directed calls.

   ◆ Alarm surveillance: The system must inform the appropriate network operations systems, about irregular and/or abnormal operations conditions. Based on these alarm signals, the system administrator can take proper actions to eliminate the error condition.

5.2.2 Switching System Architecture

Figure 5-10 shows the system architecture of a typical digital switching system. This system consists of four components that are briefly described as follows:

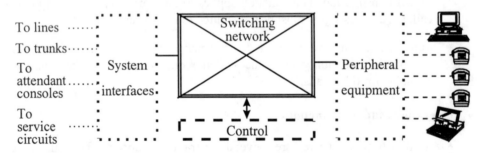

Figure 5-10   System Architecture of a Digital Switch.

* Switching network: This component provides communications links and the interconnection function for incoming and outgoing lines (or trunks). The switching network is also known as the "switching fabric" or "switching matrix".

* Controller: The controller is the "brain" of the switching system. The controller handles all call related functions/features. The control-related and the signaling-related functions described in previous sections of this chapter are functions of the controller.

* Peripheral equipment: Peripheral equipment includes voice terminals (telephones), data terminals, attendant consoles, and application processors. Note that some of this

equipment may not be required. For example, a trunk exchange may not need station lines and terminals. Peripheral equipment is particularly important for Private Branch eXchange (PBX).

- System interfaces: The system interfaces are circuits used to connect lines, trunks, attendant consoles, and service circuits to switching network.

5.2.3 Switching Networks - General Concepts

There are many unique characteristics of a telephone network. It is important to recognize and understand these characteristics when analyzing or designing a digital switching network. The main characteristics of a telephone network are:

- Most telephones are idle for a large percentage of the time.
- A subscriber does not talk with **_all_** other subscribers simultaneously.
- Call origination is a random process.

Based on these characteristics, digital switching networks are designed to:

- Provide full access for every customer to every other customer; but
- It is not necessary to insure every line and trunk can originate and terminate a call simultaneously (i.e., the switch has a finite probability of blocking).

The concepts of concentration, distribution, and expansion have evolved from switching studies. Various combinations of concentration, distribution and expansion stages make up the switching network for a digital switching system. Figure 5-11 illustrates the concepts of concentration [Figure 5-11(A)], distribution [Figure 5-11(B)] and expansion [Figure 5-11(C)].

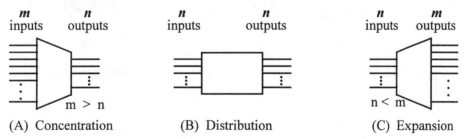

| m inputs | n outputs | n inputs | n outputs | n inputs | m outputs |

(A) Concentration   (B) Distribution   (C) Expansion

Figure 5-11   Concentration, Distribution and Expansion.

- Concentration: A concentration stage is typically applied between lines that originate calls and the other parts of a switching network. The number of inputs of a concentrator is always greater than the number of the outputs of the concentrator.

- Expansion: An expansion stage performs the inverse function of a concentrator. The function of expansion takes places between the switching network and the terminating lines. The number of inputs of an expander is always smaller than the number of the outputs [i.e., $n < m$ for n inputs and m outputs; Figure 5-11(C)].

- Distribution: A distribution stage (or several distribution stages) is applied between a concentrator and an expander. A distributor has an equal number of inputs and outputs [i.e., $n = m$ for n inputs and m outputs; Figure 5-11(B)].

Example 5-1: Figure 5-12 illustrates the concentration and the distribution stages for two different cases:

(1) Case 1: two switch networks have the same number of subscribers originating calls but with different call-hours [Figure 5-12(A)]. The two concentrators require the same number of inputs but different number of outputs ($n_1 < n_2$). Thus, the two distributors have different inputs/outputs ($n_1 \neq n_2$), and the expanders are the inverse of the concentrators.

(2) Case 2: two switch networks have different numbers of subscribers originating calls and the same call-hours [Figure 5-12 (B)]. Both the concentrators and expanders are different, but the two distributors are the same.

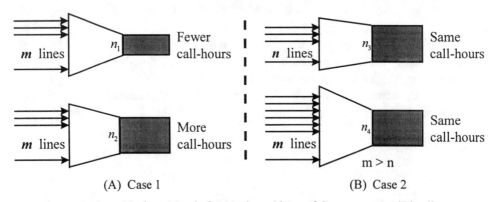

(A) Case 1   (B) Case 2

Figure 5-12   Various Needs for Various Sizes of Concentrators/Distributors.

Most modern digital switches are constructed from many individual switching elements. These individual switching elements are interconnected to form larger N x N switches. A typical switch fabric is a "16 x 16" or "32 x 32" in size (the meaning of switch fabric size is illustrated later in this chapter; for example, Figure 5-17 illustrates a 100 by 100 switch). A switch can be based on a single stage or multistage architecture. Similarly, large N x N switches can be formed by interconnecting several smaller switches. Multistage switching fabric is characterized by:

- A single path between an input and an output
- A multiple path between an input and an output

Definition 5-1 (Single-path and multiple-path switches): If there is one, and only one, path (connection) for connecting (by a switch) an incoming traffic line to its intended outgoing trunk, the switch is a single-path switch. If there are more than one connections that can be used for the input-output connection, it is a multiple-path switch (examples are given later in this chapter).

Definition 5-2 (Switch fabric): Switch fabric, also known as the switching network, is one of four switch system components as shown in Figure 5-10. It provides communications links and interconnection function for incoming and outgoing lines.

System performance, system capacity, "switch fabric" used, and costs must be considered when designing and interconnecting switching elements of a digital switch. Switching fabric can be classified into four categories: (1) shared memory time division, (2) shared medium time division, (3) single path space division, and (4) multiple path space division. Figure 5-13 illustrates the hierarchical relationship of these different types of switch fabric technologies, which are described in the following sections.

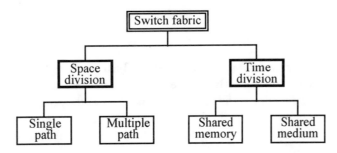

Figure 5-13   Various Switch Fabric Technologies.

5.2.4 Space Division Switching Network (Fabric) Technology

The main characteristics of space division switching networks are listed as follows.

- * A physical (spatial) link is established through the network.
- * The links are maintained for the entire duration of the information transfer.
- * External signals interfaces can be analog or digital.
- * Switching elements can be metallic (e.g., relays) or semiconductor devices.
- * Networks can consist of many stages

Example 5-2 Space division networks coordinate switches (also known as crosspoint switches), that can be used as concentrators, distributors or expanders.

There are many ways to implement "coordinate switch" arrangements, also called connecting networks or crosspoint switches. A connecting network is a system that establishes electrical paths between switch inputs and outputs. This type of connection network is capable of providing multiple "paths" for a large number of different connection patterns.

A coordinate switch arrangement can be viewed as a one-sided or a two-sided configuration. Figure 5-14(A) shows a one-sided triangular coordinate arrangement, Figure 5-14(B) shows a one-sided rectangular coordinate arrangement, and Figure 5-14 (C) shows a two-sided rectangular coordinate arrangement.

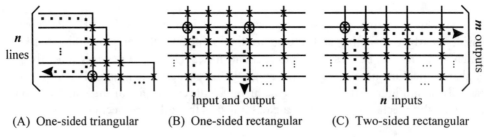

(A) One-sided triangular   (B) One-sided rectangular   (C) Two-sided rectangular

Figure 5-14   Various Coordinate Switch Arrangements.

- **One-sided triangular coordinate arrangement**: In this arrangement, the inputs and outputs are on the same side [In Figure 5-14(A), the total number of lines of the network is n]. The function of this connection network is to connect pairs of terminals attached to the input/output lines. The total number of crosspoints (N_c) [each crosspoint is shown by an "x" in Figure 5-14] is given by the following equation:

$$N_c = \frac{n(n-1)}{2} \qquad \text{where } n = \text{the number of lines} \qquad (5\text{-}4)$$

One crosspoint is required to connect each pair of lines as shown in Figure 5-14(A). A typical connection of this configuration is shown by the "bold dotted" line. The circled "x" indicated the required crosspoint for this connection.

- **One-sided rectangular coordinate arrangement**: In this arrangement [Figure 5-14(B)], the inputs and outputs are on the same verticals. Two crosspoints are required for each connection between a pair of inputs and outputs.

In Figure 5-14(B), a typical connection is shown by the "bold dotted" line, and the two crosspoints required for this connection are circled.

- *Two-sided rectangular coordinate arrangement*: In this arrangement [Figure 5-14(C)], the *n* inputs are on verticals while the *m* outputs are on horizontals. One crosspoint is required for each connection.

In Figure 5-14(C), a typical connection is shown by the "bold dotted" line, and the only one crosspoint required for this connection is circled.

Any of these three configurations (one-sided triangular, one-sided rectangular and two-sided rectangular coordinate arrangements) can be used as concentrators, distributors or expanders as follows:

* Serving as a concentrator: The inputs *(n)* and the outputs *(m)* are related as $n > m$.
* Serving as a distributor: The inputs *(n)* and the outputs *(m)* are related as $n = m$.
* Serving as an expander: The inputs*(n)* and the outputs *(m)* are related as $n < m$.

5.2.4.1 Advantages of Multiples and Stages of Smaller Switches

Grouping multiples and stages of smaller switches together is often advantageous when building a practical switch, rather than designing a single large switch. The reasons for using multiples and stages of smaller switches are listed as follows:

* More economical
* More flexible in design
* Less complicated than building a single large switch

Example 5-3: An implementation of a 3:1 concentrator using multiple stages to form a 4x4 (four inputs and four outputs) switch is shown in Figure 5-15.

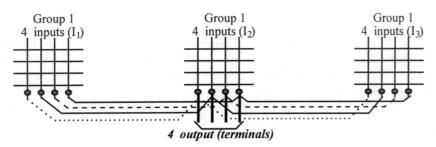

Figure 5-15   A 4-output 3:1 Concentrator Implementation
[Using Multiple Groups (Stages)].

The concept of using multiple switch stages is that a "group of output terminals" (4 outputs in this example) can be accessed by several (multiple) "groups of input terminals"

(3 groups in this example). As shown in Figure 5-15, there are three input groups (each having four input terminals) that can access four output terminals. This implementation can achieve a 3:1 concentration ratio since there are 12 inputs ($\equiv$ 3 input groups, each having 4 inputs) that can access four outputs (i.e., 12:4 = 3:1).

Example 5-4: (Staging switches): Connecting (staging) several small switches together to implement a larger switch is a common practice used to design modern digital switch networks. Figure 5-16 illustrate a two-stage switch implementation. Practical switching networks actually have several (not just two) stages.

As shown in Figure 5-16, instead of implementing a 4x4 switch as a single switch, a two-stage implementation can be used. The 1st stage uses switching elements of 2x4 while the 2nd stage uses switching elements of 4x2. These switching elements are easier (than a 4x4 switching element) and cost less than 4x4 elements.

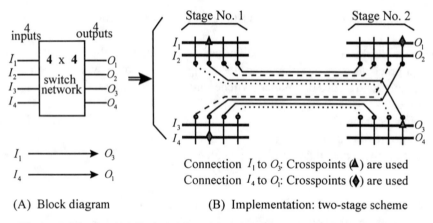

(A) Block diagram   (B) Implementation: two-stage scheme

Figure 5-16   A 4x4 Switch Network using Two-stage Implementation.

The connection from stage No.1 to stage No.2 is implemented in duplicated for reliability (solid connection for working (primary) channel while "dashed" or "dotted" connection for standby channel). Two examples of connection are shown in Figure 5-16: (1) from input port I_1 to output port O_3, and (2) from input port I_4 to output port O_1. It is clear that different crosspoints can be used to provide different (any input to any output) connections.

5.2.4.2 Network Blocking

Due to the nature of a call, connections within a network are never utilized 100% most of the time. For economics reason, it is not practical to design a digital switch that can simultaneously handle all incoming calls to the network. For example, if the total required bandwidth (data rate) for all subscribers is "*x*" Gbps, the network available bandwidth (data rate) is "*y*" Gbps, then switch design adopts $y < x$ allowing a certain percentage of call blocking. The term "blocking" must be defined.

Definition 5-3: When the simultaneous connection capacity of network is less than the total number of terminals, the (network) configuration is known as a "***blocking network***". In summary, if at least one of the links in each of the possible paths between inlets (inputs) and outlets (outputs) is unavailable (busy), it is a "***blocking network***".

Definition 5-4: For an *n* by *n* connecting network, there are n^2 crosspoints. If the network can perform all possible connections from its idle inputs to its idle outputs regardless of what other connections are currently in progress, this network is ***strictly non-blocking***.

Example 5-5 Designing strictly non-blocking networks: It is impractical to construct an "*n by n*" connecting network with n^2 crosspoints (if *n* is large) without applying a multi-stage implementation.

The example in Figure 5-17 (a two-stage network) is a "100 by 100" connecting network. Without multi-stage implementation, the total number of crosspoints is $100^2 = 10,000$. If a two-stage implementation is applied, the total number of crosspoints is 2,000 instead (for each stage, the total number of crosspoint is $10 \times 10^2 = 1,000$; thus, for two-stage configuration, the total number of crosspoints is $2 \times 1,000 = 2,000$). Therefore, a savings of 80% can be achieved.

This two-stage network provides full access from any inlet to any outlet. However, there is only one link from any of the first-stage "10 x 10" networks to the second-stage "10 x 10" networks. Hence, if a particular link between stages is unavailable (busy), other inlets in the "first-stage" attempting to communicate with the same "second-stage" outlet will be blocked.

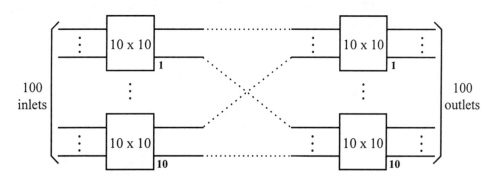

Figure 5-17   A 100 by 100 Switch Network.

Example 5-6 Clos switch network design: The Clos network is an early (developed in 1953) non-blocking multiple-stage network implementation. In a non-blocking Clos network, there is a unique relationship between the number of inlets (inputs) and the number of stages. If the number of inputs changes, the entire network must be rearranged to maintain the non-blocking condition. Therefore, this type of network is referred to as "rearrangeably blocking" network. It should be understood that these networks are not practical. A practical switch network normally has a finite probability of blocking. Figure 5-18 is used to illustrate the relationship between the number of inputs and the number of stages required for a three-stage non-blocking Clos network.

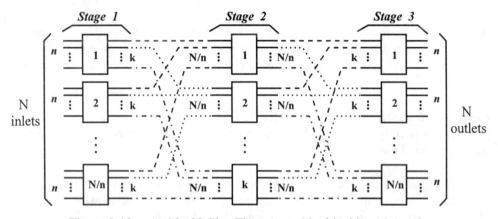

Figure 5-18   An N x N Clos Three-stage Nonblocking Network.

Assume the Clos network (shown in Figure 5-18) has the following parameters:

* N = The total number of inlets/outlets
* n = The number of inlets of each switch sub-network of the first-stage
* n = The number of outlets of each switch sub-network of the third-stage
* k = The number of the switch sub-networks of the second stage

For an N x N Clos network to be non-blocking, the relationship between the number of inputs and the number of stages must be held as follows:

$$k = 2n - 1 \qquad (5-5)$$

For a more general class of Clos network (other than the example shown in Figure 5-18), if the number of outlets of the third stage is m instead of n (i.e., the number of inlets n of the network is not identical to the number of outlets m), Eq. (5-5) is modified as follows:

$$k = n + m - 1 \qquad (5-6)$$

That is, $2n$ is replaced by $n + m$.

Table 5-4 shows the crosspoint comparison between a single stage non-blocking switch network and a three-stage Clos switch network [Note that the same notations are used in Table 5-4 as in Eq. (5-5)]. The numbers of crosspoints (N_c) for a single-stage and a three-stage Clos switch network are given by Eq. (5-7) and Eq. (5-8), respectively:

$$N_c = N^2 \; ; \qquad \text{for a single-stage switch} \qquad (5\text{-}7)$$

$$N_c = Nk\left(\frac{N}{n} + 2\right); \qquad \text{for a three-stage Clos switch} \qquad (5\text{-}8)$$

with
$$k = \frac{N}{n} - 1 \qquad (5\text{-}9)$$

Table 5-4 Crosspoints Comparison between a Single-stage switch and a Three-stage Clos Switch.

Network size			Number of crosspoints		% savings
N	n	k	Single-stage (1)	3-stage Clos (2)	(2) over (1)
16	2	3	256	288	−12.50%
36	6	11	1,296	1,188	8.30%
60	5	9	3,600	2,376	34.00%
105	7	13	11,025	5,655	48.70%
200	10	19	40,000	15,200	62.00%
8,192	64	127	67,108,864	4,161,536	93.80%

For example, if $N = 200$ and $n = 10$ for a three-stage Clos switch network, k can be obtained to be 19 [note that $k = 2n − 1$ as shown in Eq.(5-5)]. By substituting N, n and k into Eq.(5-8), N_c is calculated to be 15,200.

5.2.5 Time Division Switching Network (Fabric) Technology

Time-Division Switching (TDS) networks are based on two concepts:

1. Time-Division Multiplexing (TDM): TDM is a time-sharing scheme, that uses a common transmission medium for carrying input digital signals (see Figure 5-19). TDM technology is used many modern digital networks, is described further in Chapter 6.

2. Time-Division Switching (TDS): A TDM signal is transferred from the input points to the desired output points by using a TDS, as shown in Figure 5-19. Time-division switching is applied to establish communication paths between the input points and the output points. There are two methods used to implement time-division switching: (1) time division gates, and (2) Time Slot Interchangers (TSIs). They are explained in Examples 5-7 and 5-8.

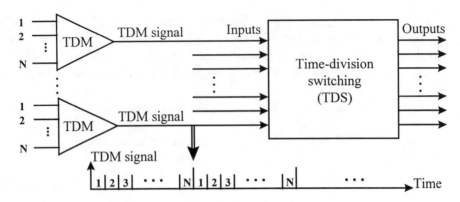

Figure 5-19   TDM and TDS of a Time-Division Switching Network.

Example 5-7 Time Division Switching (TDS) using time division gates: **This technique is used to interconnect multiplexed buses by "closing gates" for specific time slots.** The time-division gates method has been applied in systems where Pulse Amplitude Modulation (PAM is described in Chapter 6) is used. This type of switching system is also referred to as an "analog time-division switch".

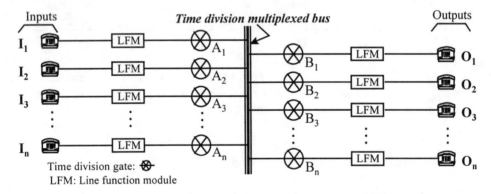

Figure 5-20   Time Division Switch Using Time-Division Gates.

In Figure 5-20, there are n input points (I_1 to I_n), and n output points (O_1 to O_n). This switch consists of a time-division multiplexed bus and many time-division gates, denoted by (A_1, A_2, A_3, ..., and A_n), and (B_1, B_2, B_3, ..., and B_n). If the input point I_1 is to be connected to the output point O_2, gates A_1 and B_2 will be **closed during the same time slot (interval)**, and a sample of the speech signal from the input point I_1 is transported from A_1 to B_2 via the time division multiplexed bus. Note that the speech signal is sampled at the Nyquist rate [8 kHz ≡ (2 × 4 kHz); 8,000 samples per second, which is equivalent to 125-μs interval between two adjacent samples] before entering into the switch (discussed further in Chapter 6).

Control of the switch connections is maintained by a time-division gate memory (sometimes referred to as the map of the switch controller), that stores every connection in the system.

Example 5-8 Time-division switching systems using a time slot interchanger: Time Slot Interchangers (TSIs) are commonly used in time division switching systems that handle Pulse Code Modulation (PCM; described in Chapter 6). This type of switching system is referred to as a "digital switch".

The time slot interchangers take in information (on the bus) and permute (i.e., re-arrange) the relative order of the time-slots. Figure 5-21 shows a switching system using one TSI with three input points and three output points. Assume that the signal from input point 1 is to be transferred to the output point 2, the signal from the input point 2 is to be transferred to the output point 1, and the input point 3 will be transferred to output point 3, the implementation is explained in the next two paragraphs.

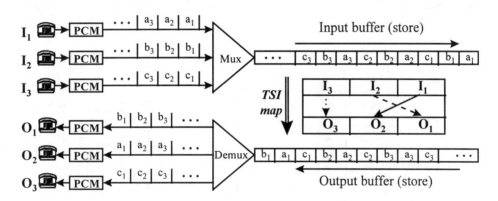

Figure 5-21   Time Division Switch Using a Time-Slot Interchanger.

The three input points (I_1, I_2, and I_3) are multiplexed and fed into a buffer (known as the input buffer) by applying bit- or byte-interleavingly multiplexing. The signal from the input point 1 carrying (a_1, a_2, a_3, ... bytes); the signal from the input point 2 carrying (b_1, b_2, b_3, ... bytes); and, the signal from the input point 3 carrying (c_1, c_2, c_3, ... bytes) are byte-interleavingly multiplexed and stored in an input buffer. Therefore, the storage sequence of the input buffer is (a_1, b_1, c_1, a_2, b_2, c_2, a_3, b_3, c_3, ...).

The storage sequence of the output buffer is (b_1, a_1, c_1, b_2, a_2, c_2, b_3, a_3, c_3, ...) in accordance with the TSI map. As a result of the demultiplexing function, the signal (b_1, b_2, b_3, ...) will be transferred to the output point O_1; the signal (a_1, a_2, a_3, ...) will be transferred to the output point O_2; and, the signal (c_1, c_2, c_3, ...) will be transferred to the output point O_3.

It should be noted that "*an n-channel (or n timeslots) Time Slot Interchanger (TSI) is equivalent to an 'n by n space division switch' because all of the n inputs have full non-blocking access to any of the n outputs.*"

5.2.6 Time Multiplexed Space Division Switching Network (Fabric) Technology

Besides space division and time division, a digital switching system can be implemented by applying "time multiplexed space division" technology. This type of switching network is often referred to as a Time Multiplexed Switch (TMS). It is a modified space division switch network that incorporates time sharing functions. The TMS provides communication paths for each time slot contained in the multiplexed signals. The "*m* by *m*" TMS shown in Figure 5-22 is used to illustrate the switching operations. There are *m* input multiplexers, m output multiplexers, *m* x *m* logic gates, and a controller that manages switching operations

Example 5-9 (TMS switch operations): Assume that input points (1, 2, 3, … and *m*) are to be connected to output points (2, 3, …, *m* and 1) respectively, for a specific time slot.

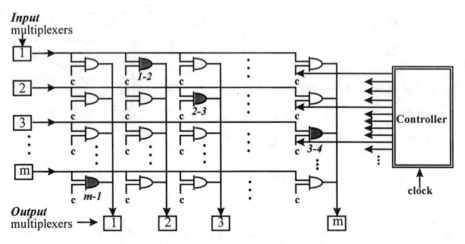

Figure 5-22   An *m* by *m* TMS Switch.

The clock signal provides timing information to the controller for identifying specific time slots. In this example, it is assumed that the input point 1 and the output point 2 are to be connected during a specific time slot. The controller delivers (at this specific time slot) a control signal to gate 1-2 [shaded] so that the gate is activated. The communication path between the input point 1 and the output point 2 is thereby established. A similar action will occur for the path between input point 2 and output point 3 by activating gate 2-3; between input point 3 and output point 4 by activating gate 3-4; and so on as directed by the controller.

5.3 DIGITAL SWITCHING NETWORK ARCHITECTURE

Although three different switching network technologies (space-division, time-division, and time multiplexed switching) can be used to design a digital switch, the space-division and the time-division methods are most commonly used. A digital switch may be implemented by using a space-division switch network(s) alone; a time-division switching network(s) alone; a space-division switching network followed by a time-division switching network; or a variety of similar combinations. Specifically, a digital switch network architecture can be any one of the following:

- Time (or time-division switching network) only (i.e., **T** only)
- Space (or space-division switching network) only (i.e., **S** only)
- Space-Time (i.e., **S-T**)
- Time-Space (i.e., **T-S**)
- Space-Time-Space (i.e., **S-T-S**)
- Time-Space-Time (i.e., **T-S-T**)
- Time-Space-Space-Space-Time (i.e., **T-S-S-S-T**)
- Other combinations

Example 5-10 A Time-Space-Time (T-S-T) switch: A "T-S-T" (or "S-T-S") switch is more complex, but more flexible switch than a T-S, S-T, T (alone) or S (alone) switch. A T-S-T digital switch is shown in Figure 5-23. Assume that the "channel A" signal from input point 1 is to be connected to the "B" channel of output point 2 as shown in the lower half of Figure 5-23.

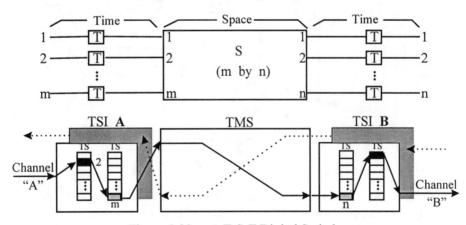

Figure 5-23   A T-S-T Digital Switch.

In this "T-S-T" switch, the Time Slot Interchangers (TSI A and TSI B) exchange information between the "external" channels and the "internal" (space array) channels. The space stage (TMS) provides connectivity between the TSI stages. In this example (lower half of Figure 5-23), the signal "A" (channel "A") from input point 1 is placed in time slot 2 of TSI A, the first "T" (switching) element of the T-S-T digital switch. The switch control processor establishes the path through TSI A and TSI B from external channel 2 to internal time slot m (i.e., TSI A transfers the channel A signal from time slot 2 to TMS time slot buffer "m"). During time slot "m", the TMS, the "S" element of the switch, connects the output buffer "m" of TSI A to the input buffer "n" of TSI B. TSI B transfers the input buffer "n" signal to channel B's time slot 1 (channel 1). Similarly, the other direction (i.e., from "B" to "A") of transmission progresses and simultaneously on the other half of the TSIs, using the same time slot of the TMS. That is, each TSI (TSI A or TSI B) has a mirror image (represented by the shaded squares in Figure 5-23) that handles the reverse direction of transmission shown by dotted lines.

5.3.1 Switching Control

The switching control element (also called the controller) is the "brain" of a switching system, and it manages all call-processing functions. Over time, the implementation of switching control has evolved (see Table 5-3; Section 5.1.3) from manual, to direct progressive, and finally to common control (note: common control is the only technology thoroughly described in this book).

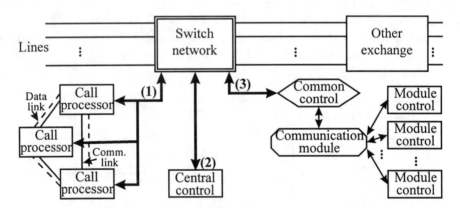

Figure 5-24   Switch Control Elements of a Switching System.

A switching system that uses a common control technology stores the called telephone number in a register. This (address) information is extracted as needed from the register to establish the desired path through the network. The registers and other control equipment are shared by all subscribers, thus the control method is called "common control". Three types of common control technologies have been developed over time, and are known as:

1. Electromechanical common control

2. Wired logic control (used for crossbar switching systems)

3. Stored Program Control (SPC; used for electronic switching systems): In SPC the logical steps involved in making connections reside in a stored program (software). An SPC controller can be designed by applying either of the following approaches:

    *   Distributed processing: Several methods can be used to implement distributed processing for switch control [see Figure 5-24, parts (1) and (3)]. In the approach shown as part (1) of Figure 5-24, each processor performs all the call-processing functions. The cluster of modules (three are shown) may communicate with each other via communication links that are managed by data links. This model can be implemented within a single system or a geographical area.

        In the second approach shown as part (3) of Figure 5-24, the distributed processing is a hierarchical structure. The common control includes global functions of the whole system (e.g., call processing, support functions, and system maintenance/administration). The module control elements contain the call-processing functions for the distributed unit. The communication module provides communications links between the common control and module control elements.

    *   Centralized processing: This arrangement is shown in part (2) of Figure 5-24. The central control element is an information processing unit that is capable of executing many different types of instructions. The control functions are governed by software programs, which are sets of instructions that are recorded in a memory element.

5.3.2 Switch Functions and Interfaces

For a digital switch to work properly with different types of lines and trunks, the switching system must provide various interfaces. To describe these various interfaces, it is necessary to understand switching system functions. These functions and interfaces are described in the following sections.

5.3.2.1 Switch Functions Related to Interfaces

The functions that a switching system performs for interface purposes are referred to as BORSCHT (**B**attery feed, **O**ver-voltage protection, **R**inging, **S**upervision, **C**odec, **H**ybrid, and **T**esting), and are described as follows:

*   Battery feed: The battery at the switching office is typically −48 volts (Figure 1-19 in Chapter 1). It provides power to the subscriber line for detecting the subscriber station's status [e.g., the off-hook condition; a current sensor (Figure 1-19) can sense if the subscriber is off-hook]. It also provides power for station signaling and the subscriber's signal transmission.

- Over-voltage protection: A switching system may be subjected to high voltage conditions such as lighting or power line crosses. The over-voltage protection is used to prevent damage to sensitive circuits in the switching system and for the subscribers personal safety.

- Ringing: The switching system provides the ringing signal for an incoming call (i.e., for the called party), and an audible ringing signal for call-in-progress confirmation (i.e., for the calling party).

- Supervision: The supervision functions include monitoring line status (e.g., on-hook, off-hook, and circuit release indications).

- Codec (coding and decoding): For a digital switch, the incoming speech signal is digitized by a Pulse Code Modulation (PCM) coder. At the receiving end, the digitized signal is converted to an analog speech signal by using a PCM decoder.

- Hybrid: A hybrid circuit (a three-port transformer) is used to separate the outgoing signal from the incoming signal, for both calling and called subscriber's circuit.

- Testing: Testing is applied to detect any fault condition on lines and/or trunks. Testing can be performed periodically or on demand.

These are common functions for all interface circuits. Specific functions depend on the type of interfaces will be discussed with specific interface in the next section.

5.3.2.2 Switch Interfaces

The switch interfaces illustrated in Figure 5-25 can be classified into three groups:

- **Station line interfaces**: Five typical station line interfaces are described here.

  * Analog voice terminal interface: The traditional telephone interface provides analog voice terminal interface (Figure 1-19 in Chapter 1), that is referred to as a tip/ring interface.

  * Digital voice terminal interface: For digital switches, incoming voice signals are processed by a PCM interface.

  * Data line interface: This interface supports signals generated by data terminals (i.e., not voice signals).

  * Hybrid voice terminal interfaces: A hybrid voice terminal uses an analog format for voice communication, but accepts digital control information. This interface is used in analog/digital PBXs, and key telephone sets.

* Basic rate ISDN interface: This signal consists of two B channels (information-*B*earing channels) that can be used for voice, data signal transport, or combined voice and data signals transport, and a D channel (Data link) used to carry control signaling information.

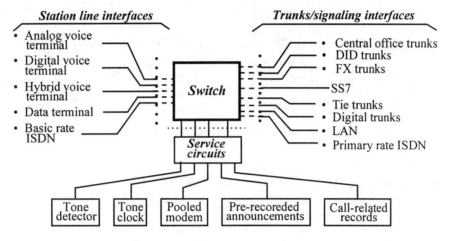

Figure 5-25   Switch Interfaces.

- **Trunk and signaling interfaces**: Eight typical trunk and signal interfaces are described as follows.

  * Central office trunk interfaces: The central office trunks include loop-start and ground-start trunks. A trunk is a transmission facility connecting two switch exchanges. Trunk can be categorized as: access trunk groups, toll-connecting trunks, inter-toll trunks, PBX trunks, etc.

  * Direct Inward Dialing (DID) trunk interfaces: Trunks that are connected to a PBX (Private Branch eXchange) or Centrex service are known as DID trunks.

  * Foreign eXchange (FX) trunk interfaces (Figure 5-26): Assume that service is connected to one or more foreign exchanges to cover a large geographic area, but a discount rate (rather than a toll rate) is applied.

  For example a call to 111-5454 (e.g., a national appliance repair center) from an area with an exchange of 333-xxxx is a toll call. As a result, customers in the "333" area are reluctant to request the service from 111-5454. However, if "111-5454" is a FX service subscriber, a call from any of its subscribed areas (e.g., 222-xxxx, 333-xxxx, ...) will be a local call. As shown in Figure 5-26, the customer in area C will dial 333-5454 (instead of 111-5454), that is clearly a locate call for the customer residing in the area with an exchange number of 333-xxxx. After the customer has dialed 333-5454, the originating switch serving 333-xxxx recognizes that this is a FX call, and routes via the FX trunk to 111-5454. As a result, the appliance repair center will receive more calls and business will increase because a larger geographic area is being served.

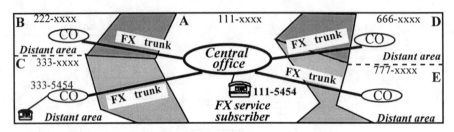

Figure 5-26   Foreign Exchange Services and FX trunks.

* Tie trunk interfaces: The trunk connecting a PBX or an equivalent device, to the central office is called a "tie trunk".

* Digital trunk interfaces: Digital trunks are used to connect digital terminals to a digital switch. For PCM terminals, trunks can be T1 or E1 (Chapter 6) facilities. For multiplexers [terminal mux/demux, or Add-Drop Multiplexers (ADM)] or Digital Cross-connection Systems (DCSs), trunks can be T3, E3, E4 or fiber facilities.

* Local Area Network (LAN) interfaces (described in Chapter 10): LAN can be connected via a switch to access to a PSTN or other LANs. The interface can be provided by using T1, T3, E1, E3 or other facilities.

* Primary rate ISDN interfaces: This is an ITU-T standard. It provides (23B + D) services for the μ-law (North American) digital networks, or (30B + D) services for the A-law (international) networks (see Chapter 9 for further details).

* SS7 signaling interfaces (SS7 has been described in Chapter 4): Since SS7 is an intelligent network performing call set-up related functions, SS7 must interface with local, trunk, or gateway switches. The interface is typically provided by T1 or E1 digital facilities.

• **Common resource service circuits**: Five typical common resource services are described as follows.

* Tone detector: Several tones are used to establish call connections (e.g., touch tone, call progress tones, and "modem answer" back tones).

* Tone-clock circuits: Touch tones, call progress tones, and modem "answer-back" tones, as well as the clock signals are used to achieve synchronization in digital networks.

* Pooled modem: Modem equipment is used in a digital switching system to interface outside analog facilities on a shared basis.

* Pre-recorded announcements: Many pre-recorded announcements are used in switching system. Two examples are: (1) "Your call cannot be completed as dialed, please check the number you dialed", and (2) "All circuits are busy, please try again later".

* Call related records: Many call-related records are collected based on traffic measurements. These records are used for network management, planning, and billing purposes.

5.3.3 Switching Software

Modern digital switching systems usually apply software based Stored Program Control (SPC; store and forward) technology to provide call processing, system administration, network management, maintenance, and special features. In addition to call processing (see Chapter 4; Example 4-1), other functions, such as system installation, testing, traffic engineering, billing, service changes, and management are described in the following sections. Before describing these functions, some definitions are provided first.

Definition 5-5: Switching software generally consists of a large number of (computer) programs, routines, and databases. Supervisory programs, used to schedule and supervise the execution of other programs, are generally referred to as Operating Systems (OSs). The collection (aggregate) of many small programs is called a "generic" program.

5.3.3.1 System Administration and Management Software

Besides call processing, system administration and management is an area that is important to the operations of a switching system. The objective is to ensure quality services are provided by the digital switching system. Because of its relative importance, software for system administration and management often comprises a large percentage of the total switching system software. The main system administration and management functions are listed as follows:

* System installation/testing
* Traffic engineering
* Service changes and management
* Billing
* Features selections
* Database management

5.3.3.2 Maintenance Software

Reliable service is the most important feature of modern digital switching systems. Therefore, these systems must have extensive software functionality for maintenance purposes. Maintenance software is generally under the control of an Operating System (OS).

The objectives of maintenance software is: *"detection, reporting, and trouble clearing as quickly as possible, with minimum disruption of services."*

Maintenance software functions can be grouped into four categories and described as follows:

1. Fault recognition and fault recovery software: Fault recognition and fault recovery functions are typically complementary. For example, if a system experiences a hardware failure or a software error, the fault recognition function identifies the source of the problem. Then, the fault recovery software proceeds to remedy the conditions (i.e., restore system stability and maintain customer service). In general, faults are detected through routine checking, continuous monitoring, or by specific tests performed during call processing.

2. System integrity software: System integrity software is designed to detect and respond fault conditions so that software errors have minimum impact on systems performance. In addition, system integrity software is responsible for detecting and handling system overload conditions.

3. System auditing software: Auditing is a technique used to verify consistency between the internal state of hardware and software system resources.

4. System recovery software: System recovery is the method used to restore normal service after a trouble condition has been identified. System recovery is categorized by levels of severity. The highest severity level (i.e., the most serious failure condition) may require a complete reload of system software from back-up storage (e.g., a system reboot).

There are other important considerations for maintenance software, such as:

* "Human-machine" software is required to support communications between technicians/operators and switch equipment so that administrative/maintenance tasks can be performed.

* The system must provide remote maintenance capabilities, which includes "dial-up connection" to a centralized maintenance center and automatic alarming.

* The system software must maintain a history or "error log" that reflects the alarm condition of the system.

5.3.3.3 Location Database

In addition to the generic program, the "location database" is an important software function in modern digital switching systems. The "location database" is used by the generic program to define unique attributes of a particular location. Two types of location database information is required by switching systems, and described as follows:

1. User data:

   - Identity of lines that terminate local subscribers
   - Identity of directory number trunks that are available to other switches
   - Specifications of available features
   - Methods used for call routing
   - Methods used for call billing

2. Equipment data:

   - Number of frames (i.e., equipment elements)
   - Number of circuits within each frame
   - Methods for software access identification of equipment
   - Configuration of the switch "center network"

5.3.3.4 Operating System (OS)

A modern digital switching system is a "real-time system" with lines and trunks as its input/output terminals. Each terminal is managed by stored-program control, and requires processing time, memory, software resources, and switching resources. Therefore, the operating system that administers the allocation of these resources must also satisfy overall system performance requirements. In addition, switching systems handle many concurrent calls, therefore the operating system must perform the following functions:

   - Implement concurrent processes
   - Provide process management
   - Provide scheduling and dispatching
   - Provide inter-process communications (i.e., communication between processes)
   - Provide memory management
   - Provide input/output control and support
   - Manage overall maintenance and administration functions

5.3.4 Switching Applications

A switching system can be designed for specific applications (see Figure 5-27). The application is based on the type of users and traffic patterns. Generally, switching systems can be classified into four categories as follows:

1. Local exchange switch: A local exchange switch is designed to serve the general public. Telephone lines for residential and business users are connected to a local exchange switch. Local exchange switches are typically connected to another local exchange switches or a trunk (exchange) switch.

2. Trunk (exchange) switch: Trunk switches are designed transport/switch network traffic. There are no "end users" (telephone lines, fax, or subscriber lines) directly connected to a trunk switch. Another name for a trunk switch is a "toll switch".

3. Gateway switch: Gateway switches are used to route international calls, and often perform functions similar to trunk exchange switches. Gateway switches provide interfaces to accommodate various signal formats used by different countries. A gateway switch also provides international operator assistance services, and is the point where international call billing is performed.

4. Customer switch: Customer switches are similar to local exchange switches, except they are designed for business applications. Specific users features, system management functions, and equipment costs are the primary concerns when designing customer switches. The most commonly used customer switches are called Private Branch eXchanges (PBXs).

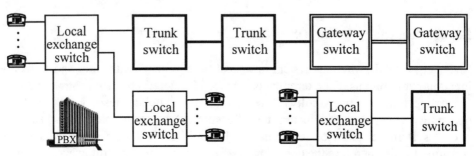

Figure 5-27   Switching System Applications.

5.4 EXCHANGE SWITCHING

From a functional perspective, all the switches in a telecommunications network can be considered to be "exchange switches" (see Figure 5-28). They can be used as local switches, or can serve as host/remote switches, tandem switches, trunk switches, and gateway switches.

Figure 5-28 shows a simplified telecommunications network. The network can provide: (1) intra-office calls, that require one and only one local exchange switch; (2) inter-office calls, that require at least two local exchange switches. In addition, the calls may require one or more trunk exchange switches; and, (3) international calls, that require two local exchange switches, one or more trunk exchange switches, and two gateway exchange switches. For business applications, PBXs are often used.

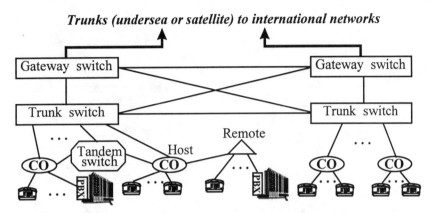

Figure 5-28   Various Exchange Switches.

5.4.1 Local Exchange Switches

The simplest application of a local exchange switch is to connect subscriber loops (within a geographical area) to a trunk switch (see Figures 5-27 and 5-28). This type of exchange switch is known as local exchanges, also called a Central Office (CO) or wire center. There are two special applications of a local exchange switch: (1) host/remote switch, and (2) tandem switch (shown in Figure 5-28).

The host/remote arrangement allows the local exchange switch (the host) to serve subscribers at distant locations via a "remote" unit. The remote unit can typically operate as a stand-alone switching element. This type of applications have been implemented in several digital switches such as 5-ESS, described later in this chapter.

A tandem switch provides a transit function by connecting several local exchange switches together. Its purpose is to provide greater traffic efficiency and routing diversity. The function of a tandem switch can be viewed as a switch serving as a hubbing device of many switches (including local exchange and trunk exchange switches) as shown in Figure 5-29. Note that Figure 5-29(A) is the application of a tandem switch while Figure 5-29(B) is the application of a local exchange switch serving many end users, as shown in Figure 5-3(D) and (E).

It has been discussed, in Figure 5-3, that if a hub switch is used to connect the end users in the same geographic area then considerable savings in equipment and facilities can be achieved. In addition, the network design is simpler. This is repeatedly shown in Figure

5-29(B). Therefore, from Figure 5-29(A), a "tandem switch" acts as a "hub switch" to many switches, and can achieve the same goal as a "hub switch" does. Note that trunk switches can be connected to a tandem switch although a tandem switch was originally developed to connect local exchange switches only.

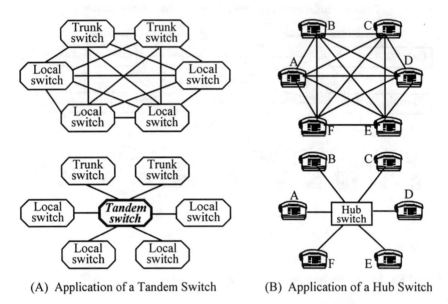

(A) Application of a Tandem Switch   (B) Application of a Hub Switch

Figure 5-29   Applications of a Tandem Switch.

5.4.1.1 Terminations of a Local Exchange Switch

The basic function of a local exchange switch is to provide "switched services", including call processing, routing, features, and operations. To understand switched services, familiarity with line terminations (Figure 5-30), trunk terminations (Figure 5-30), call types, call processing, and call scenarios is required.

- Line terminations: The switch interface connected to the subscriber's equipment is known as the line termination. In modern communications networks, there are four typical line terminations (note that there may be more types of line terminals in future applications): (1) analog voice frequency lines, (2) ISDN lines, (3) subscriber loops, and (4) integrated digital loops. They are described as follows:

    1. Analog Voice-Frequency (VF) lines: Analog VF lines are the oldest type of line terminations. However, they are still the most common type of termination in telecommunications networks. Although the local exchange is typically a digital switch, the subscriber's line must be connected to analog-to-digital conversion equipment (e.g., A-law or µ-law Pulse Code Modulation (PCM); discussed in

Chapter 6. In Chapter 3, it has been described that some of these VF lines can be used as Digital Subscriber Line (DSL; xDSL, HDSL, ADSL, or VDSL).

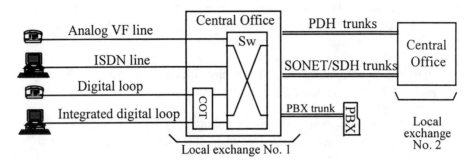

Figure 5-30   Line and Trunk Terminations of a Local Exchange.

2. ISDN line terminations: Currently, only the Basic Rate Interface ISDN (BRI-ISDN with 2B+D rate; described in Chapter 9) can be connected to the local exchange, but not Primary Rate Interface ISDN (PRI-ISDN and other ISDN rates; also described in Chapter 9).

3. Digital subscriber loops: Because of customer demands for improved analog VF line transmission, digital subscriber loops have been deployed to carry analog telephone traffic to the local exchange switch (i.e., the central office). A terminal combining several analog loops at a distant location from the local exchange switch is called a "Remote Terminal (RT)". In this arrangement, the connection between the remote terminal and local exchange switch is typically a digital carrier system (e.g., a T1 system; described in Chapter 6).

Note that this digital loop application was originally developed for suburban and rural areas in the late 1960s, and since then the applications have been extended to metropolitan areas. The equipment located in metropolitan business areas is often referred to as "Business Remote Terminals (BRTs)". The equipment terminating the digital loop applications is called "Central Office Terminal (COT)" (see Figure 5-30).

4. Integrated Digital Loops (IDLs): Over the past decade, digital loop applications have been extended to include ISDN capabilities to carry voice, data and video traffic to subscriber via digital facilities (e.g., T1 in North American networks). Similarly, IDL applications have been widely adopted for metropolitan business campus. In these applications, optical fibers have replaced the traditional twisted-pair wiring facilities such as T1 carrier systems, or coaxial cable facilities. The optical fibers used in: (1) SONET networks adopt OC-3 with a speed of 155.52 Mbps, or OC-12 with a speed of 622.08 Mbps, and (2) SDH networks adopt

STM-1 with a speed of 155.52 Mbps, or STM-4 with a speed of 622.08 Mbps. Higher rate optical fibers for IDL applications will soon be increased.

- Trunk terminations: The transmission facilities connecting two exchange switches are known as trunks. which may be implemented differently. In modern digital networks, the following three trunk terminations are commonly used:

  1. PDH (Plesiochronous Digital Hierarchy) trunks have been deployed since the early 1960s. Two regional PDH digital networking hierarchies exist:

     ◆ μ-law PDH network: The μ-law hierarchy is are used in the North American region, and some scattered parts in the world. There are several types of digital trunks in this kind of networking: the DS1 (carrying 24 voice channels or equivalent traffic), DS1C (carrying 48 voice channels or equivalent traffic), DS2 (carrying 96 voice channels or equivalent traffic), DS3 (carrying 672 voice channels or equivalent traffic), DS3C (carrying 1344 voice channels or equivalent traffic), and DS4 (carrying 4032 voice channels or equivalent traffic). The details of this topic are described in Chapter 6.

     ◆ A-law PDH network: The A-law hierarchy is used in areas other than North America. There are several types of digital trunks in this kind of networking: the E1 (carrying 30 voice channels or equivalent traffic), E2 (carrying 120 voice channels or equivalent traffic), E3 (carrying 480 voice channels or equivalent traffic), and E4 (carrying 1920 voice channels or equivalent traffic. The details of this topic are described in Chapter 6.

  2. Synchronous Optical NETwork/Synchronous Digital Hierarchy (SONET/SDH) trunks were developed in 1988 to implement modern high-speed networks, rather than continued deployment of PDH networks. Technically, SONET and SDH are consistent standards, even though they use different terminologies. There are also application differences, which are briefly described in the following sections (see Chapter 6 for additional details).

     ∗ SONET trunks are used to carry μ-law (North American) digital signals (e.g., DS1, DS1C, DS2, and DS3). Although SONET has five different data rates, there are ten categories of SONET trunks (more in future):

       (1) Synchronous Transport Signal (STS) level 1 Electrical (STS-1E): A 51.84 Mbps electrical signal (carrying one DS3 of 44.736 Mbps).

       (2) Optical Carrier level 1 (OC-1): A 51.84 Mbps optical signal (carrying one DS3 of 44.736 Mbps).

       (3) Synchronous Transport Signal (STS) level 3 Electrical (STS-3E): A 155.52 Mbps electrical signal (carrying three STS-1E signals).

(4)   Optical Carrier level 3 (OC-3): A 155.52 Mbps optical signal (carrying three OC-1, three STS-1E, or combinations of OC-1 and STS-1E signals).

(5)   Synchronous Transport Signal (STS) level 12 Electrical (STS-12E): A 622.08 Mbps electrical signal (carrying four STS-3E, 12 STS-1E, or combinations of STS-3E and STS-1E signals).

(6)   Optical Carrier level 12 (OC-12): A 622.08 Mbps optical signal (carrying four OC-3, four STS-3E, 12 OC-1, 12 STS-1E, etc. signals).

(7)   Synchronous Transport Signal (STS) level 48 Electrical (STS-48E): A 2.48832 Gbps electrical signal (carrying four STS-12E, 16 STS-3E, 48 STS-1E, etc. signals).

(8)   Optical Carrier level 48 (OC-48): A 2.48832 Gbps optical signal (carrying four OC-12, four STS-12E, 12 OC-3, 12 STS-12E, 48 OC-1, 48 STS-1E, etc. signals).

(9)   Synchronous Transport Signal (STS) level 192 Electrical (STS-192E): A 9.95328 Gbps electrical signal (carrying four STS-48E, 16 STS-12E, 48 STS-3E, 192 STS-1E, etc. signals).

(10)   Optical Carrier level 192 (OC-192): A 9.95328 Gbps optical signal (carrying four OC-48, four STS-48E, 16 OC-12, 16 STS-12E, 48 OC-3, 48 STS-3E, 192 OC-1, 192 STS-1E, etc. signals)

*   SDH trunks are used to carry A-law (non-North American) digital signals (e.g., E1, E2, E3, and E4). Although SDH has four different data rates, there are eight categories of SDH trunks (more in future):

(1)   Synchronous Transport Module (STM) level 1 Electrical (STS-1): A 155.52 Mbps electrical signal (carrying one E4 of 139.264 Mbps).

(2)   Optical Carrier level 1 (STM-1O): A 155.52 Mbps optical signal (carrying one E4 of 139.264 Mbps).

(3)   Synchronous Transport Module level 4 Electrical (STM-4): A 622.08 Mbps electrical signal (carrying four STM-1E signals).

(4)   Optical Carrier level 3 (STM-4O): A 622.08 Mbps optical signal (carrying four STM-1O, four STM-1, or combinations of STM-1 and STM-1O signals).

(5)   Synchronous Transport Module level 16 Electrical (STM-116): A 2.48832 Gbps electrical signal (carrying four STM-4, 16 STM-1, or combinations of STM-4 and STM-1 signals).

(6)   Optical Carrier level 16 (STM-16O): A 2.48832 Gbps optical signal (carrying four STM-4O, four STM-4, 16 STM-1O, etc. signals).

(7)   Synchronous Transport Module level 64 Electrical (STM-64): A 9.95328 Gbps electrical signal (carrying four STM-16, 16 STM-4, etc. signals).

(8)   Optical Carrier level 64 (STM-64O): A 9.95328 Gbps optical signal (carrying four STM-16O, four STM-16, 16 STM-4O, 16 STM-4, 64 STM-1, 64 STM-1O, etc. signals).

3.   Private Branch eXchange (PBX) trunks: Currently, PBX trunks are implemented by using PDH technology. PBX trunks are also considered to be "tie trunks" in some applications because they bring (tie) traffic directly to customer premises.

Example 5-11 Applications of local exchange switches for different call types: Figure 5-31 shows four common call types that involve one or more local exchange switches - (1) An intra-office call: a call with the same originating and terminating exchange; (2) An inter-office call: a call that involves two individual local exchanges; (3) A transit call: a "transit call" does not terminate at an exchange (e.g., passes through a toll/tandem exchange); and (4) An incoming call: a call that terminates in a local exchange, but was not originated by that exchange is known as an "incoming call".

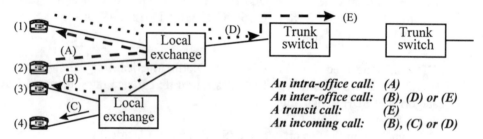

An intra-office call: (A)
An inter-office call: (B), (D) or (E)
A transit call: (E)
An incoming call: (B), (C) or (D)

Figure 5-31   Various Types of Call.

5.4.1.2 Billing and Administration

Billing is one of the most important functions performed by a local exchange switch. The billing and administration function of a local exchange are briefly discussed in this section.

*   Billing: The major steps of the billing process are listed below.

    *   Recording: The real-time collection of call data.
    *   Record collection: Manual or automatic record keeping performed by computer software.
    *   Sorting: Sorting of customers with respect to call types (local or toll calls).

* Rating: Application of tariff rates to individual call records.
* Disbursement: The results of call data collection is disbursed to various billing systems.
* Rendering: The process of preparing and issuing bills to subscribers.
* Collection: The process of collecting and recording subscriber bill payment.
* Enforcement: The process of taking action when subscribers fail to pay bills (e.g., involuntary service disconnection).

Modern Operations Systems (OSs) can be used to automatic recording functions. Two examples are briefly discussed as follows:

1. Periodic Pulse Metering (PPM): This function can be performed at the local exchange or at a higher level exchange. PPM is based on charging for a call in accordance with registering "meter pulses" on call meters in the local exchange. If the local exchange is the charging point, the pulses generated within the exchange are registered on the call meter of the "chargeable" subscriber. If the charging point is located in a higher level exchange, the local exchange analyzes the PPM information (received from the higher level exchange) and registers it on the subscriber call meter. In applications where this is impossible, the local exchange transmits the meter pulses to the preceding exchange where the PPM is registered on the subscriber call meters.

2. Automatic Message Accounting (AMA): An AMA system can be implemented in a local exchange to provide detailed call information for all telephone and non-telephone calls. The information can be used as either supplementary information to the PPM process, or separately for the purpose of call charging. If a local exchange is large enough, it is usually equipped with AMA capability, which is referred to as "Local AMA (LAMA)" as illustrated in Figure 5-32.

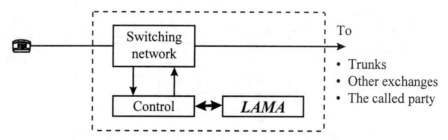

Figure 5.32   Local Automatic Message Accounting (LAMA).

The recording process for a local call can be grouped into three parts:

(1) After the calling customer completes dialing (i.e., enters the dialed digits), the class of service, calling-line equipment number, and call types

information is from the control function to the LAMA circuitry. The LAMA circuitry translates the calling-line information into a billing record. This billing record contains the digits dialed by the customer as the initial data entry.

(2)   When the call is completed (i.e., the called party answers) a second entry (known as the answer entry) is made in the billing record. This entry contains the "call identity index", a time stamp (i.e., time of day represented as minutes/tens of minutes, or seconds) that reflects the hour in which the call was initiated.

(3)   A third entry in the billing record (known as the disconnect entry) is made when the call is terminated. This entry contains the call identity index, and a time stamp (i.e., time of day represented as minutes/tens of minutes, or seconds) that reflects the hour in which the call was terminated.

In order to ensure correct call data processing numerous additional entries are recorded as part of the billing record.

Some local exchanges are not large enough to have their own LAMA. Centralized AMA (CAMA: Figure 5-33) is deployed at a large exchange, that serves a several small exchanges. This arrangement requires that the local exchanges to provide the "CAMA exchange" with information about the calling lines. In order to bill a customer properly, the directory number of the customer must be processed by the CAMA facility. The Operator Number Identification (ONI) or Automatic Number Identification (ANI) capability is required to obtain the calling customer's telephone number (see Figure 5-33). The billing process for two typical situations is described as follows:

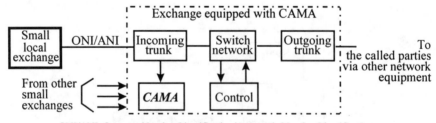

ONI/ANI: Operator Number Identification/Automatic Number Identification

Figure 5-33   Centralized Automatic Message Accounting.

(1)   A local exchange not equipped with ANI: In this case, the CAMA exchange switches the connection to an operator (ONI system) who requests the calling customer's telephone number before allowing the call to proceed. In local exchanges equipped with ANI, operator intervention is only required to handle calls from multi-party lines and for ANI system errors/failures.

(2) A local exchange equipped with ANI: In this case, the ANI equipment responds to signals generated by the CAMA equipment, and transmits the requested information (e.g., calling customer's number) to the CAMA system.

- Administration: There are many types of administration functions that a local exchange performs. Three primary functions are described as follows:

(1) Line administration: This function assigns and allocates lines in accordance with subscriber preferences.

(2) Recent change and verify: This function updates the subscriber database information with new line assignments, service types (e.g., single-party service), and features (e.g., call waiting).

(3) Load balancing: The concentration-distribution-expansion function within a local exchange switch are usually shared by groups of customers. In Figure 5-34, two groups (A and B) are shown. An above-average load in one group (e.g., group A) will result in more blocking than another group (e.g., group B) that has a below-average load.

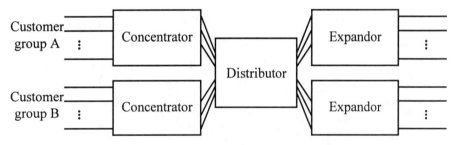

Figure 5-34   Load Balancing of a Local Exchange Switch.

From a traffic engineering perspective, a group with the same number of output lines and concentration ratio, but carrying a higher traffic load will have a larger blocking probability. To equalize the blocking probabilities among different groups, the concentration ratios can be changed. However, this is not practical because "non-uniform switches" are expensive. Therefore, another approach must be used. The most common method is called *"load balancing"*, which is used to improve blocking by equalizing the load for each customer group. The simplest and most economical way to implement and maintain load balancing is through control of routing line assignments. That is, *new customers are assigned to lightly loaded components (i.e., trunk groups), to avoid heavily loaded components*.

Note that if load balancing was not used, switches would have to operate at approximately 5% to 10% below their design capacities to maintain the same quality of service. In extreme cases, when loading is severely imbalanced, it may become necessary to disconnect active customers from overloaded components and transfer them to lightly-loaded components. Such corrective action is very costly, and should be avoided if possible.

5.4.2 Trunk Exchange and Gateway Switches

Typically, local exchange switches are used to connect telephone subscribers within small geographic areas, the serving charge is a fixed monthly bill (i.e., a flat rate). As serving areas become larger or the customer population increases, local exchanges require the aid of tandem switches (as shown in Figures 5-28 and 5-35) to effectively access the network's trunk, and gateway exchanges for international calls.

The cost of connecting telephone subscribers increases in direct proportion to the physical distance between serving areas. These "long-distance" calls are usually charged on an individual basis, not included in the fixed monthly flat rate. Trunk exchanges are designed to handle these types of calls (i.e., long-distance calls), and are often called "toll switches".

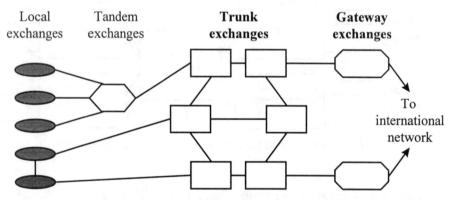

Figure 5-35   Trunk Exchanges Connection: Local, Tandem and Gateway.

5.4.2.1 Trunk Exchange Functions and Features

In addition to "trunk-to-trunk" switching, call routing, and signaling; a trunk exchange performs the following administration and maintenance functions:

- Network management: Enables traffic to be re-routed when there is congestion (e.g., during heavy demand).

- Traffic measurement: Collects data about the traffic load being carried by various components of the switching system.

- Billing: Records call-related information that is needed to determine the appropriate charges for service.

- Maintenance: Provides features that automatically detect, isolate, and locate failures (i.e., trouble conditions) to the circuit pack (plug-in) level.

Typical features of a trunk exchange switch are listed as follows:

  - Large capacity, with respect to trunk terminations, call attempts, and network load capacities
  - Network routing capability
  - Common channel signaling
  - Centralized Automatic Message Accounting (CAMA)
  - Clock synchronization
  - Gateway exchange access
  - Intelligent Network (IN) service capability
  - High reliability

5.4.2.2 Trunk Exchange Evolution

Trunk exchanges have evolved dramatically over the past seven decades. This section briefly describes the history of trunk exchange switch evolution.

- Step-by-step switch: Invented by A. B. Strowger in 1889, it was used extensively as a trunk switch in the 1920s by the Bell System. This design provided a separate stage of switching associated with each dialed digit. As the user dialed successive digits, the switch responded to the dial pulses and progressively selected a path through the network. This is referred to as "direct progressive control". Note that the dialed number could not be re-used because the dial pulse (digits) were not saved after being used to select the position of the switching devices.

- Crossbar switch (crossbar tandem of the 1930s and No.4A crossbar invented in 1943): Crossbar switches use common control. They are electromechanical devices with control equipment that is only used to establish the transport path. The dialed digits are stored in a register, and logic associated with register is used to select the connection. Once the connection is "made", the register is cleared and can be used to serve other calls. No. 4 crossbar switches were the backbone network elements of the original Direct Distance Dialing (DDD) network which allowed customers to place long-distance calls without the operator assistance.

- Stored Program Control (SPC) switch: The analog 1AESS SPC switch was first introduced in 1976. Later in the same year, the 4ESS switch was the first electronic switching system deployed that used digital SPC methods, which applied extensive software programs. With SPC, new services are implemented by changing rather than deploying hardware modifications. The 5ESS is another electronic SPC switch that was introduced in 1982.

5.4.2.3 Gateway Applications (Gateway Exchange)

An international call requires access to a gateway exchange. The international gateway switch provides connections between different countries that have a wide range of operating parameters. Specifically, a gateway switch must support conversions between different signaling formats.

Another issue for international calls is voice quality and echo control. Voice signal service degradation occurs under two conditions: delay and impedance mismatch; when these conditions are excessive (e.g., delays greater than 25 ms one way) an objectionable reverberation occurs in speech signals that is commonly called "echo". Both these conditions exist for typical international calls, hence, echo cancelers are used to eliminate or reduce the annoyance of echo.

Automatic and manual capabilities for detecting and correcting faults arising from hardware, software, and procedural errors are provided by the gateway switch as basic Operations, Administration and Maintenance (OA&M) functions.

Typical examples of switches that are used as the gateway exchanges are Lucent's 4ESS and 5ESS. The interface provided for national and international communications supports the following signaling systems: ITU-T SS5, SS6, SS7, and R1 or R2.

Revenues derived from international traffic are accounted for at the gateway exchange. Due to different administrations (e.g., countries, service providers within countries, regulations, etc.), the revenues derived from the international calls passing through a gateway exchange to different facilities have to be apportioned. The gateway exchange generates the call accounting information necessary to apportion revenue between these different administrations. Per-call records may be used directly, or an associated call accounting system may be provided. The gateway exchange also provides charging records for an Operator Service Position System (OSPS) when such services are involved. In addition, the gateway exchange supports a "proportionate bidding" feature, which makes it possible to allocate outgoing calls among competing carriers sharing traffic routes to the same destination. The call records of a typical international call consist of the following items:

- Incoming route
- Outgoing route
- Date and time (i.e., start of billing)
- Call duration
- Type of call
- Country of origin
- Country of destination
- Call class
- Service class
- Bridging indicator

Review Questions I for Chapter 5

(1) The incoming lines of a switch can be a single-channel lines or a multiple-channel lines, but the outgoing lines are typically _____-channel lines.

(2) If a switch is not used, in order to be able to connect any two subscribers, the required number of facilities N_c is very large for a moderate populated area, and is equal to _____, where n is the number of subscribers.

(3) The telecommunications service demand is random in nature. This random nature of service demands includes _____ and _____.

(4) (True, False) Switching technology evolves in the following sequence: from manual switches, step-by-step, crossbar, Stored Program Control (SPC) analog switches, and then SPC digital switches.

(5) An international call typically involves several toll exchanges, two _____ exchanges, and two _____ exchanges.

(6) Besides interconnection, switching functions can be grouped into three categories: _____, _____, and _____ functions.

(7) (True, False) The number of inputs of a concentrator used in a switch can be smaller than the number of its outputs.

(8) (True, False) A single-path space division switch means that there is only one possible connection between a particular input to a designated output line.

(9) The most important advantages of using multiple-stage than a single-stage for an "N x N" switch implementation are: _____, _____, and _____.

(10) (True, False) Theoretically it is impossible to design a non-blocking switch.

(11) A time division switch uses a Time-Slot Interchanger (TSI) to perform switching function, that is, routing an input to its intended output. An input and an output buffers are required to implement TSI _____, that contains the routing table.

(12) (True, False) A switching fabric can either be time-division (T) or space-division (S) switch. In practical switching implementations, it is rare to use only time-division or space-division alone, but combinations such as S-T, S-T-S, T-S-T, etc.

(13) Besides service circuits (e.g., tone detector, tone clock, etc.), the switch interfaces can be grouped into two categories: (1) _____ (e.g., analog voice terminal, digital voice terminal, data terminal, BRI-ISDN line interfaces, etc.), and (2) _____ (e.g., central office trunk, DID trunk, FX trunk, SS7, LAN, PRI-ISDN digital trunk interfaces, etc.).

(14) (True, False) Switching software is one of the most important components in a modern digital switch.

5.5 CUSTOMER SWITCHING SYSTEMS

Customer switching are equipment configurations that permit flexibility in providing telecommunications services that can be tailored to a customers' specific needs (e.g., a particular set of service capabilities and features). These services include both voice and data communications. Customer switching can be provided as follows:

- By systems that are owned or leased by the customer (usually business customers) and placed on the customer's premises. The equipment is referred to as Customer Premises Equipment (CPE), and includes Key Telephone Systems (KTSs) and Private Branch eXchange (PBX).

- By systems that are located in the local exchange (e.g., the Central Office; CO). A typical example of this approach is known as "Centrex" services.

5.5.1 Customer Switch Functions

Although the specific communications needs of business customers depend on the size and the nature of their businesses, a generic customer switch must perform the following functions:

- Intra-location calling: Most business activities depends upon employees communicating within the premises. For voice services, this consists of calls between people in different offices. For data services, this may involve communications between a FAX terminal, desktop or laptop PC, and a host computer (or another terminal) in the same premises.

- Incoming calls: This capability is important in business applications because incoming calls often represent new or additional business opportunities. Therefore, it is essential to have an adequate number of facilities and attendants to ensure incoming calls are promptly answered and efficiently routed to someone who can assist the calling party.

- Outgoing calls: Controlling communications costs is important to business owners. The customer's switch must limit call permission for selected stations, and also route calls over the most cost-efficient facilities. In addition, it is important to make the communications system "user friendly", which in turn will significantly improve employee efficiency. Station equipment (i.e., telephone sets) with button-operated features, and switching systems with automatic processing routines meet these needs.

- Communications systems management: Communications costs are a major expense and a necessity for any successful business. Modern customer switches must be reliable, and perform system management functions as well as protection switching to restore services in the event of system failure.

- Office automation and management: To improve business efficiency, voice/text message processing capabilities and directory services can be provided by the communications system (i.e., voice mail, call transfer, call conferencing, etc.).

5.5.2 Customer Switch Types

Six types (key telephone system, PBX, hybrid system, etc.) of customer switching systems are described separately in following sections..

5.5.2.1 Key Telephone System (KTS)

Generally, Key Telephone Systems (KTSs) are designed for business customers having 2 to 80 stations (i.e., telephone sets). KTSs are based on the concept that a business customer subscribes several lines from the local exchange, but the individual user has the flexibility to access and control particular lines from each terminal (i.e., telephone). For example, a user can press a button on the telephone set to close a switch point (Figure 5-36), and a particular line will be made available so the user can place a call (e.g., access an "outside" line).

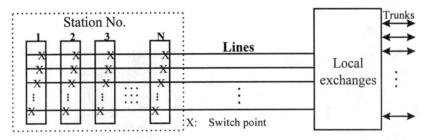

Figure 5-36   A Key Telephone System.

KTS is categorized as customer premises equipment. From a telecommunications hierarchy viewpoint, a KTS is one level below a local exchange. A KTS is a "user-controlled" switching system, with "keys" or "buttons" on the station sets for accessing central office (or PBX) telephone lines, or activating special features. A KTS arrangement allows users to operate the "buttons" (keys) to perform several functions. The primary KTS functions/characteristics are briefly described as follows:

- Multiline access: A KTS has telephone stations that can access several lines. Users are able to share these lines by pushing a button (key).

- Hold: A user can hold a call that is in progress on one line while another call is answered or originated on a different line. Note that by pushing a button, the station user receives a dial tone directly from the central office.

- Lamp signaling: This function is used to inform the user of the status of each line (e.g., such as line pickup, hold, busy, or ringing).

- Common audible ringing: All lines share the audible ringing function for alerting purposes.

- Intercom and paging: These optional functions are available for most KTSs, and are used to support business activities on the customer premises.

5.5.2.2 Private Branch Exchange (PBX)

A private branch exchange (PBX) is a switching system owned or leased by customer, and operated on the customer's premises. It is located remotely from the local exchange, and is dedicated to the exclusive use of one customer. The user's (usually the employees of a firm) telephone sets are connected to the PBX, which in turn is connected to the local exchange via PBX (tie) trunks as shown in Figure 5-37.

Like a KTS, a PBX is classified as customer premises equipment and is one level below a local exchange. A small PBX is designed to support up to 200 stations (i.e., telephone sets). A medium PBX can support several hundred stations, and a large PBX can serve several thousand stations.

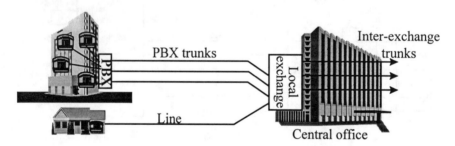

Figure 5-37   PBX and PBX (Tie) Trunks.

A PBX performs the following three functions:

(1) A PBX supports station-to-station calling on customer premises, using an abbreviated dialing plan (i.e., simplified extension numbers).

(2) A PBX concentrates customer station connections to central office trunks. That is, instead of connecting individual stations to the central office, several tens or hundreds of stations are concentrated into a single trunk that is connected to the local exchange.

(3) A PBX provides the capability for allowing an attendant to handle (i.e., intercept and direct) all incoming calls.

5.5.2.3 Hybrid (Customer Switching) System

A hybrid system looks very much like a KTS in that it uses multi-button telephone sets located on user desks. A station can handle several lines, put calls on hold, etc. Individual telephone sets also have many PBX features. The term "hybrid" or "hybrid key system" was first used in the 1970s for tariff filing purposes. That is, the tariff permitted telephone companies to charge more for "hybrid" than "standard" key telephone systems. Most modern hybrid systems have evolved from traditional key telephone systems because they are designed to access lines (rather than trunks) using push button operations. However, from a functionality viewpoint, a hybrid system offers features that are similar to a PBX (i.e., are more complex than standard KTS functions).

Table 5-5   Limitations/Solutions of KTS, PBXs, and Hybrid Systems.

KTS limitations	PBX solutions
• Line access limited by the number of line pick-up keys (buttons)	• Trunk access is provided via switching
• Intercom contention in applications having a large number of users	• Station-to-station calling is provided
• All stations may not be accessible via intercom	• Station-to-station calling is provided

PBX limitations	Hybrid system solutions
• Multitude of access codes	• Key press for accessing features
• Localized answering positions	• Key press for answering other station
• Need to dial "frequently called stations"	• Key press for intercom

Table 5-5 shows the limitations of a KTS compared to a PBX, and the solutions provided by using a PBX. In the bottom half of Table 5-5, the limitations of a PBX are listed with corresponding "hybrid" system solutions. That is, a hybrid system has multi-button telephone sets and offers characteristics of both PBXs and KTSs. Note that "Key press" (i.e., button entry) is used to access lines in a hybrid system.

5.5.2.4 Centrex Services

Centrex service provides functions that are similar to PBXs used by business customers. The telephone sets and attendant positions are located on customer premises, but all other switching associated equipment is located in the (serving) local exchange (see Figure 5-38). Centrex services provide station-to-station calling with abbreviated dialing, and access to the Public Switched Telephone Network (PSTN) without going through an attendant (e.g., switchboard operator). Similarly, incoming calls can be automatically routed directly to the appropriate station (i.e., extension). These features are called "Automatic Identified Outward Dialing" (AIOD) and "Direct Inward Dialing" (DID), respectively. The Centrex system also provides attendant services (from the attendant positions via connections to the local exchange. In the telecommunications network hierarchy, Centrex services are provided at the same level as a local exchange.

Twelve features of the Centrex service are listed below. Note that the first four are basic (Centrex service) features, while the eight others are supplementary (i.e., optional).

1. Abbreviated station-to-station dialing (basic)
2. Direct Incoming Dialing (DID; basic)
3. Direct Outward Dialing (DOD; basic)
4. Intercom (basic)
5. Call forwarding
6. Call transfer
7. Call waiting
8. Hunting
9. Abbreviated dialing
10. Conference service
11. Attendant
12. Administration, including:

    *   Account and authorization codes
    *   Call detail recording
    *   Station rearrangement

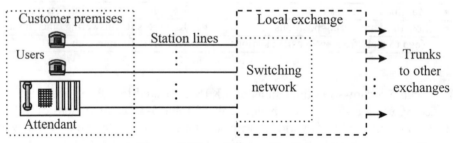

Figure 5-38   Centrex Services.

A comparison of features, equipment, customer control, maintenance, reliability and cost for PBX versus Centrex service is given in Table 5-6. PBXs and Centrex service provide approximately equivalent features, and functional differences are minor (depending upon the manufacturer). The major distinction between PBX and Centrex is the location (i.e., PBX equipment is located on customer's premises while most of Centrex equipment is located at local exchange) and ownership of the switch (thus, it is the customer's responsibility to perform PBX switch maintenance tasks, but local exchange performs maintenance task for Centrex equipment), which affects costs and the degree of customer control. Note that the PBX is Customer Premises Equipment (CPE; one level below local exchange switch in switching hierarchy), while Centrex is local exchange equipment.

Table 5-6   PBX and Centrex Service Comparison.

Parameters	PBX	Centrex service
Features	Equivalent	
Equipment	CPE based	Local exchange based
Customer control	Direct	Indirect
Maintenance	Customer's responsibility	Local exchange's responsibility
Reliability	Tailored to customer's specifications	Local exchange switch criteria
Cost	1. Requires capital investment for switching equipment if customer owns PBX 2. Customer pays trunk charges from PBX to local exchange	1. Does not require capital investment for switching equipment 2. Customer pays line charges from customer premises to local exchange

5.5.2.5 Automatic Call Distributor

The Automatic Call Distributor (ACD) function is an important feature for business customers. Many modern PBXs have ACD capabilities.

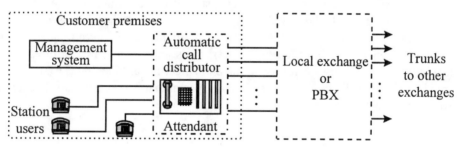

Figure 5-39   Automatic Call Distributor (ACD).

ACD is used to automatically switch large volumes of incoming calls to attendant (answering) positions. ACD applications include directory assistance and intercept operators (e.g., airline reservations, department store catalog sales, customer complaints, etc.). The characteristics of an ACD are listed below:

- Calls are evenly distributed among attendants to maximize the efficiency of the attendant group.

- When all agents (attendants) are busy, incoming calls are queued and answered in their order of arrival.

- Management information data is collected to administer the switching facilities and determine the number of attendants (agents) that are needed.

5.5.2.6 Telephone Answering System

A "telephone answering bureau" is a centralized location where the clients' telephones are answered by agents. This feature can be viewed as a two-part function. The telephone service provider provides equipment to do the necessary switching, while the bureau management provides sufficient operators (agents) to answer calls and record messages. Modern PBXs can provide integrated answering services and messaging features.

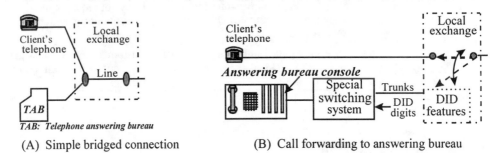

(A) Simple bridged connection   (B) Call forwarding to answering bureau

Figure 5-40   Telephone Answering Systems.

The concept of a telephone answering system is that a "simple bridge" is provided to connect the client's telephone and the answering bureau, as shown in Figure 5-40(A). In a Stored Program Controlled (SPC) system environment, another arrangement of the answering system is possible [Figure 5-39(B)]. This arrangement makes the answering system more flexible. It uses call forwarding and Direct Inward Dialing (DID) features to automatically route calls to a special switching system that is connected to the telephone answering bureau console.

5.5.3 Customer Switching Evolution

The first key telephone system was developed in 1938 by AT&T, known as 1AKTS (see Figure 5-41). Each telephone line connected to the system required a set of control relays, called Key Telephone Units (KTUs). The availability of low-cost, solid state, integrated circuit technology led to the development of Com Key KTS in the early 1970s. The Com Key KTS used Light Emitting Diodes (LEDs) to replace lamps, and speakers to replace bells. The analog Stored Program Control (SPC) technology was applied to the Horizon KTS in 1980. In the early 1980s, digital technologies were used to develop Merlin, Merlin Plus, and Merlin II, followed by Spirit and Digital TDM KTS in the late 1980s. Digital KTS systems have the following characteristics:

- Digital common equipment
- Enhanced features, including capabilities to support data
- Cost effective in providing both voice and data services

PBX development can be divided into three eras: (1) The first generation PBXs applied analog and space division technologies; (2) The second generation PBXs started to apply time division technology, and Stored Program Control (SPC) was introduced; and (3) The third generation PBXs applied digital, time division, and SPC technologies to support both voice and data services.

The first manual PBX was developed in the late 1880s. In the mid 1910s, step-by-step technology was used to develop 701 PBX. The 756 PBX was introduced in the early 1950s and used crossbar switch technology. The first PBX applying analog stored program control technology, No. 101 ESS, was deployed in the mid 1960s, followed by Dimension PBX in the mid 1970s. Digital PBX development started in the early 1980s, with the introduction of Systems 85, 75, and 25, followed by Definity which is widely used in the current applications.

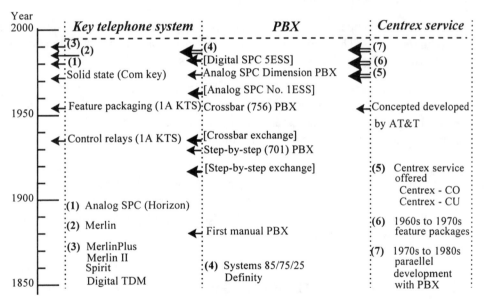

Figure 5-41   Customer Switch Evolution.

The Centrex service concept was developed by AT&T-Bell Laboratories in the early 1950s. It involved a set of services or features for a group of customers. The set of initial services/features included: (1) Direct Inward Dialing (DID); (2) Direct Outward Dialing (DOD); (3) station-to-station dialing; and, (4) Automatic Identified Outward Dialing (AIOD). Centrex services can be offered from a central office switch (local exchange; Centrex-CO), or the switching equipment can be located on the customer premises (Centrex-CU; Customer Unit). The development of electronic switching systems and stored program control technologies have resulted numerous Centrex services/feature offerings. From early 1970s to the present, parallel development in feature development exists in PBX and Centrex-CO.

5.5.4 Customer Switching Future Trend

The future trend in customer switching is briefly described as follows:

- In general, products and services are being offered with more features and better cost/benefit ratios.

- The proliferation of personal computers and data applications have forced PBXs, KTSs and Centrex services to provide data switching as a major feature offering.

- Digital switches typically handle data at rates up to 64 kbps, and have interfaces to Local Area Networks (LANs). As Integrated Service Digital Network (ISDN) interfaces are implemented, combined voice and data services will become more common, and customer switches will be the vehicle to carry office automation functions (e.g., message and directory services).

- Due to intense competition, cost is a major comparison factor in selecting customer switching systems. The technology trend is continuous improvement of the hardware and software being applied in customer switching products.

- Technology has advanced to a stage that KTSs and PBXs are applying the same technologies and architectures to implement similar features. It will become increasingly difficult to distinguish KTSs and PBXs. Only the user interface and methods used to control the network will remain unique for KTSs and PBXs.

- PBXs and Centrex systems definitely have common origins with respect to feature development. However, because of changes in the telecommunications industry, PBX and Centrex have become direct competitors. Since the telecommunications bill of 1996, the role of service providers has changed drastically, and long-distance carriers will eventually enter the PBX and Centrex business.

5.6 PACKET SWITCHING TECHNOLOGY

Two prominent forms of switching are used in modern telecommunications networks: (1) circuit switching, and (2) store-and-forward switching. Circuit switching is typically used to switch voice signals, while store-and-forward switching is not suitable for voice signals. To understand these switching technologies, the concept of voice and data communications is presented as background information.

5.6.1 Voice and Data Signal Characteristics

Voice signals are considered to be low bandwidth, continuous usage (fixed or constant rate), and have a greater acceptability of bit errors. In contrast, data signals are high speed, high bandwidth, but usage is intermittent (variable rate). For data signals, the acceptable Bit Error Rate (BER) must be close to zero, and is insured by applying error detection/correction technology. Voice signals are delay sensitive, but data signals may or may not be delay sensitive, depending upon the end user applications. A summary of voice and data signal comparison is provided in Table 5-7.

Table 5-7   Voice and Data Signal Characteristics Comparison.

Parameter	Voice signal	Data signal
Required bandwidth	Low	High
Use of bandwidth	Constant (uniform)	Variable (bursty)
Acceptable bit error rate	High	Low
Acceptable delay	Low	Low-high

5.6.2 Circuit Switching

Circuit switching technology is considered to be older one, because it was initially introduced to carry speech (voice-frequency) signals. Circuit switching allows the service provider to allocate network resources to provide a communication path between two users. This path is composed of resources dedicated to the users (i.e., the connection), and are not shared among different users. Three phases in a circuit switching connection are: (1) circuit establishment; (2) data transfer; and (3) circuit disconnect (release). The Public Switched Telephone Network (PSTN) applies circuit switching technology.

There are both advantages and disadvantages associated with circuit switching technology. Other than transmission equipment delay, there is no additional delay as information flows through the network. This technology is optimized for speech signals and interactive/real-time data signals. Other forms of constant rate data transmission is also possible over circuit switching networks, and no extra overhead is associated with transmission after the call has been set-up.

Circuit switching technology allocates dedicated resources, hence the user is billed immediately after a connection is set-up even if it is not use by the customer. If the customer's data application requires many short calls established in close succession, it is possible that the call set-up overhead will be significant compared to the duration of the data calls. In summary, circuit switching technology can be inefficient for bursty data applications.

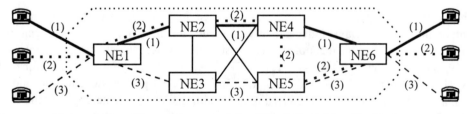

Figure 5-42   Connections used for Circuit Switching Technology.

Figure 5-42 illustrates the interconnection used to implement a circuit switching network. The network consists of six network elements (NEs). Each NE may consist of a switching machine and several transmission equipment [e.g., Analog-to-Digital (A/D) converters, multiplexers, Digital Cross-connect Systems (DCSs), etc., described in Chapter 6]. The transmission facility connecting two NEs is known as a trunk, which can carry many simultaneous voice signals. For the user pair No. 1, the circuit connection utilizes NE1, NE2, NE4 and NE6 as indicated by the bold lines in Figure 5-42. Note that the trunk between NE1 and NE2 has additional capacity, therefore, it is shared by users 1, 2 (as indicated by the dotted lines in Figure 5-42).

5.6.3 Store-and-Forward Switching

There are two types of store-and-forward switching networks: (1) message switching, and (2) packet switching.

__Definition 5-6__: Data communication typically involves the exchange of entities called messages. Telegrams, electronic mail files, and database files are common examples of messages.

1. Message switching: In a message switching network, a user assembles a message and passes it to the first node of the network. Each node stores the message, then passes it through the network (i.e., "forward" the message) until the message is delivered to the final destination. There are disadvantages to message switching. The delay associated with the delivery of messages makes this type of network inappropriate for interactive or real-time applications. Since message may vary in length, a buffering scheme may be needed to handle messages having different lengths. Slower secondary storage devices, such as tapes and disks, may also be needed to store messages.

2. Packet switching: This technique was first proposed in the early 1960s, and attempts to capitalize on the advantages of circuit and message switching while minimizing their disadvantages. The basic concept is that messages are segmented into fixed length packets, which are stored and forwarded at each network node.

5.6.4 Packet Switching Evolution

The principles of packet switching was invented in 1964 by Paul Nara of the Rand Corporation. Donald Davis independently originated a similar approach, and began to use the term "packet" for this technology. The first packet switching network was deployed in 1969, and was known as ARPANET. CCITT (now, ITU-T) adopted Rec. X.25 for packet switching protocol in 1976, and Rec. X.75 in 1978. This packet switching technology, with a speed up to 56 kbps, is referred to as "traditional packet switching". Technologies other than X.25 have been proposed and adopted (see Figure 5-43), but there are two basic packet switching categories: (1) traditional, and (2) fast packet switching methods. Fast packet switching can either be a frame relay (with speeds up to 2 Mbps) or cell relay. Cell relay is further divided into Distributed Queue Dual Bus (DQDB; with speeds up to 45 Mbps) and Asynchronous Transfer Mode (ATM), which can be applied to any speed the transport network supports.

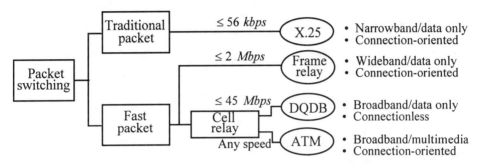

Figure 5-43   Packet Switching Technology Evolution.

The general approach of packet switching technology is to segment the customer data bitstream into "packets". The generic format of a packet is shown in Figure 5-44, and typically consists of the following fields:

- Flag (one each at the beginning and at the end of a packet): The flag fields are used to define (delimit) the beginning and the end of a packet.

- Header: Many types of information associated with a particular packet can be carried by the header field. Two common examples of header information are:

  1. Address: Address information is used to indicate the identity of two stations (i.e., origin and destination) in a point-to-point network, or the intended receiver (i.e., destination) in a multipoint network.

  2. Control: Control information is used to indicate the type of payload (i.e., information) that the packet is carrying. It may also carry sequencing and acknowledgment information.

- Payload (information): Depending upon the technology, the length of the payload field may or may not be fixed, and contains user information/data.

- Trailer: The trailer field is typically used to check packet parameters (e.g., length, sequence of the packets, and the error performance of the packet).

Flag	Header	Information (payload)	Trailer	Flag

Figure 5-43   A Typical Packet Format (Frame).

5.6.4.1 X.25 Packet Switching

ITU-T Rec. X.25 for packet switching is defined as the access protocol between the Data Terminal Equipment (DTE, a user machine) and the Data Circuit terminating Equipment (DCE, a network component). The X.25 Recommendation defines the lower three layers of the Open System Interface (OSI) model as follows:

1. Network layer: The protocol for this layer is called Packet-Level Procedure (PLP).

2. Link layer: The protocols for this layer are called the Link Access Protocol B and D (LAPB and LAPD).

3. Physical layer: The protocol for this layer is X.21 or X.21 bis.

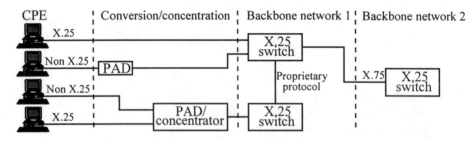

Figure 5-45   X.25 Network (Functional) Architecture.

In a typical X.25 packet network (see Figure 5-45), the major components are the Customer Premises Equipment (CPE, e.g., user terminals or host computers), Packet Assemblers/Dis-assemblers (PADs), concentrators, and X.25 switches. The X.25 standard specifies the network access protocol, but, the X.25 switches typically use vendor proprietary protocol to communicate with each other in the backbone network. A PAD is used to interface a non-X.25 CPE to an X.25 switch. A PAD with a concentrator is used to combine traffic from an X.25 CPE and a non-X.25 CPE to interface an X.25 switch. Figure 5-45 also shows that an X.25 backbone network can be connected to another X.25 backbone network using the X.75 network protocol.

5.6.4.2 Fast Packet Switching

Traditional packet switching (i.e., X.25 protocol) requires error control (error detection/correction) on each link. That is, at the transmitter end of each link (i.e., in the connection between two adjacent X.25 switches), error control algorithm is applied to every packet of the data stream. At the receiver end, the data stream is checked for errors. If errors have been detected in the data stream, re-transmission of the "un-acknowledged" packet (i.e., the packet with errors) is initiated by the transmitter.

Due to technology advances, modern transmission systems have achieved superior error performance. Likewise, at link-by-link error control methods are no longer required. This led to the introduction of "fast" packet switching technology. Two well-known fast packet switching technologies are frame relay and ATM (Figure 5-43).

5.6.5 Frame Relay

Frame relay (described in Chapter 9) is a relatively early fast packet switching technology. Conceptually, frame relay is similar to X.25 packet switching technology, with some minor differences. Specifically, frame relay's network nodes operate at the OSI layer 2, the nodes process frames faster (than X.25 nodes), the data speed for frame relay is 1.544 Mbps or 2.048 Mbps, and frame relay allows interconnection of local area networks (LANs).

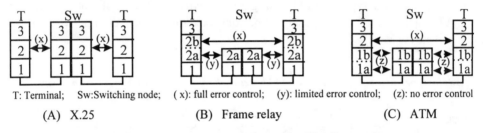

T: Terminal;   Sw:Switching node;   (x): full error control;   (y): limited error control;   (z): no error control

(A) X.25   (B) Frame relay   (C) ATM

Figure 5-46   X.25, Frame relay and ATM Comparison.

5.6.6 ATM Cell Switching

Figure 5-46 shows the major differences between X.25, frame relay, and ATM. Traditional X.25, the early packet switching methodology utilizes the transmission media (which has poor error performance). Therefore, an error control technique must be performed on every link of the end-to-end connection in order to achieve acceptable data quality. This error control is supported by a High-level Data Link Control (HDLC) protocol, which includes functions such as frame delimiting for identifying the packet frame boundaries, bit transparency, error checking [typically Cyclic Redundancy Check (CRC) technologies; described in Chapter 13], and error recovery (i.e., re-transmission) as shown in Figure 5-46.

With the advent of ISDN for narrowband services, the quality of the transmission facilities/systems, and the speed of switching systems increased. As a result, the errors within the network (for packet switched data transport) were reduce considerably. Therefore, this type of network only performs the core functions of HDLC protocol (frame delimiting, bit transparency and error checking) on a link-by-link basis. Other functions, such as error recovery, are performed on an end-to-end basis. (Note that this end-to-end operation in layer 2 is in contradiction with the OSI model, where the end-to-end operation is only performed at layer 4 or higher). Thus, layer 2 of the OSI model is divided into two sublayers: "2a" and "2b" to perform various functions: layer 2a supports the core functions and layer 2b supports the additional functions. This technology is called frame relay packet switching as shown in Figure 5-46(B).

For broadband ISDN or multimedia applications, the functional requirements of layers 1 and 2 are further extended. The core functions of layer 2a have been removed to the edge of the network and the error handling functions are no longer supported in the switching node inside the network as shown in Figure 5-46(C); this cell-relay technology is called ATM "fast" packet switching.

5.6.6.1 ATM Network and Network Interfaces

Figure 5-47 shows a typical ATM configuration, which consists of both private and public ATM networks. An ATM user can be connected to a public ATM network directly or to a private ATM network.

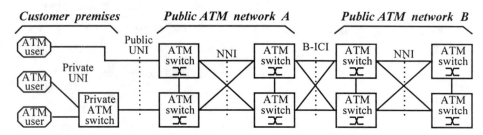

UNI: User to Network Interface; NNI: Network Node Interface; B-ICI: Broadband ISDN Inter-Carrier Interface

Figure 5-47   Typical ATM Network and Interfaces.

The interfaces defined in an ATM network are: (1) private User-to-Network Interface (UNI or NI; proposed by ANSI) for connecting ATM users to private ATM switches, (2) public UNI for connecting ATM users to ATM switches in public ATM networks, (3) Network Node Interface (NNI) for connecting ATM switches together in an ATM network, and (4) Broadband ISDN Inter-Carrier Interface (B-ICI) or Inter-Network Interface (INI; proposed by ANSI) for connecting two or more public ATM networks together. Note that private UNI and B-ICI standards are specified by the ATM Forum, while public UNI and NNI standards are specified by ITU-T. Differences between private and public UNI standards are distance limitation and maintenance capabilities. ATM connections can be provisioned as Permanent Virtual Connection (PVC) and Switched Virtual Connection (SVC), and correspond to private dedicated (also known as nailed-up, or leased lines) services and public switched services [e.g., Plain Old Telephone Service (POTS), and voice-grade (voiceband) data] in circuit switching networks.

5.6.6.2 ATM Channels

Many transmission systems have been proposed for carrying ATM cells, after the cells have been switched to their outgoing ports (facilities). In North American μ-law networks, clear-channel DS1, DS3, Fiber Distributed Data Interface (FDDI), and SONET facilities can be utilized to carry ATM cells. Similarly, in ITU-T A-law networks, E1, E3, cell-based, and SDH facilities have been proposed for transporting ATM cells. These different facilities are described in Chapter 6.

Independent of the speed of the physical facility, a transmission medium can be divided into many (a maximum of 2^{12}; see Section 5.6.6.3) Virtual Paths (VPs). Each VP can further be divided into many (a maximum of 2^{16}; see Section 5.6.6.3) Virtual Channels (VCs). Every VC within a VP, and every VP within a physical facility has a unique identifier (i.e., VCI or VPI) which is assigned at the originating node, and retrieved at the terminating node of an ATM link (joining two adjacent ATM switches). A detailed description of VPI and VCI is presented in later sections of this chapter. The definitions of VP and VC are given as follows:

- Virtual Channel (VC): A VC (see Figure 5-48) is a unidirectional communication link used for transporting ATM cells. It is a basic circuit (entity, or capability) for ATM services, and typically is found in a Switched Virtual Connection (SVC) service.

- Virtual Path (VP): A VP is a bundle of Virtual Channels (VCs). Note that all the VCs within a VP of a Permanent Virtual Connection (PVC) have the same endpoint (i.e., destination). In addition, the Quality of Service (QoS) of a VP must meet the QoS of the tightest QoS of all VCs.

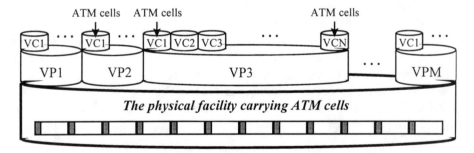

Figure 5-47   Relationship of ATM Physical Facility, VPs, and VCs.

An ATM user must subscribe to a VC capability, that is the smallest entity for ATM cell transport. However, for a user needs more than one VCs between two specific nodes, there are advantages for a customer to subscribe to a VP, rather than many separate VCs. The primary benefits are: discounted service charge, ease of management, and uniform transmission quality [i.e., Quality of Service (QoS)].

5.6.6.3 ATM Cell Structure/Organization

An ATM cell has a length of 53 bytes, which is divided into two parts. The first part is five bytes in length called the cell header. The remaining 48 bytes are assigned as the cell payload for information bit stream, that can be user information, network Operations, Administration and Maintenance (OAM) signals, etc. (see Figure 5-49).

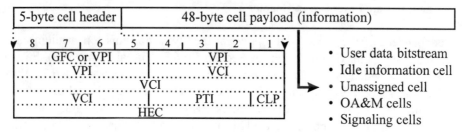

Figure 5-49   ATM Cell Structure/Organization.

The five-byte header consists of the following fields: GFI, VPI, VCI, PTI, CLP and HEC (See Figure 5-49). They are briefly described as follows:

- Generic Flow Control (GFC): This field has a length of 4 bits, and is used to establish connection between the ATM user and the ATM network node. The GFC signal is overwritten by the ATM switch, currently undefined, but has a default value of "0000". Note that this GFC field is not applicable within an ATM network.

- Virtual Path Identifier (VPI)/Virtual Channel Identifier (VCI): Both VPI and VCI are required for routing ATM cells from one ATM node (i.e., an ATM switch) to another, for Switched Virtual Connection (SVC) services. For Permanent Virtual Connection (PVC) services, only VPI bits are used for cell routing. VPI has: (1) a maximum length of 8 bits (4 bits each on the 1st and 2nd bytes of ATM header field) for connections between the user and the ATM switch node, or (2) a maximum length of 12 bits (8 bits on the 1st byte; including the 4-bit GFC field, and 4 bits on the 2nd byte of ATM header field) between two ATM switching nodes. VCI has a maximum length of 16 bits. Therefore, there are 256 ($=2^8$, between users and ATM networks), or 4096 ($=2^{12}$, between two ATM nodes) possible VPI values for carrying ATM VP users. Similarly, there are 65,536 ($=2^{16}$) possible VCI values for carrying ATM VC users. But, the first 32 VCIs (i.e., VC0, VC1, VC2, ..., VC31) within each VPI have been reserved for special applications, and cannot be assigned to carry VCs.

- Payload Type Identifier (PTI): This 3-bit field is used to indicate the type of information contained in the ATM cell. PTI can also be used to indicate when the ATM network is experiencing congestion conditions (available for ATM Adaptation Layer protocol No. 5 only).

- Cell Loss Priority (CLP): When the network is experiencing congestion and/or the customer's data rate is higher than the network capacity, ATM cells can be discarded. The CLP bit is used to indicate this condition: if CLP = 1 the cell is subject to discard; if CLP = 0 the cell is a high priority cell that should not be discarded. The ATM switches in ATM network manage the cell discard function.

- Header Error Control (HEC): This byte is used to perform cell delineation at the receiver (i.e., identify the cell boundary). It is also used to detect/correct bit errors in a received ATM cell header.

The 48-byte payload field is used to carry one of six different types of information as shown in Figure 5-49 and briefly described as follows:

- User data stream: The user data bit stream is segmented into bytes and placed in the 48-byte payload field for transport over the ATM network.

- Idle information cells: Since the line rate of an ATM link between any two ATM switching nodes must be maintained at a constant speed, "*decoupling*" must be applied when the incoming users' information rate is lower than the average rate. In ITU-T A-law (see Chapter 6) digital networks, idle information cells must be inserted along with the users' data streams to maintain ATM link speed.

- Unassigned cells: A "decoupling" algorithm is used in North American μ-law networks (see Chapter 6), which is the equivalence of A-law idle information cells.

- OA&M cells: Operations, Administration and Maintenance (OA&M) information cells need to be transported over ATM links. The main OA&M cells are "forward" error performance monitoring, "backward" error reporting, alarm surveillance, activation & deactivation, and loopback for testing. The frequency of transporting OA&M cells varies as specified in provisioning (e.g., one OA&M cell per 128 cells, one per 256 cells, one per 512 cells, etc.).

- Signaling cells: ATM cells can be used to transport signaling information for call-related processing functions.

5.6.6.4 ATM Virtual Path Switching and Virtual Channel Switching

Because an ATM connection can be either a Permanent Virtual Connection (PVC) or a Switched Virtual Connection (SVC), ATM switches must be able to perform both Virtual Path (VP) or Virtual Channel (VC) switching (see Figure 5-50).

In the case of VP switching, the ATM switch routes (i.e., switches) an entire VP link as a single entity. The Virtual Path Identifier (VPI) translation or routing table is shown in the first two columns of Table 5-8(A). Note that, in Figure 5-50(A), the Virtual

Channel Identifiers (VCIs) contained in each VPI remain unchanged. That is the reason why VCIs are not shown Table 5-8(A). In contrast, in the case of VC switching, both the VP and the VC links are switched as shown in Figure 5-50(B). Thus, both VPIs and VCIs must be translated as shown in the last four columns of Table 5-8(B).

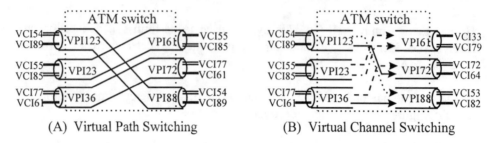

(A) Virtual Path Switching   (B) Virtual Channel Switching

Figure 5-50   ATM Virtual Path and Virtual Channel Switching.

Table 5-8   Translation Tables for VP and VC Switching Operations.

VPI$_{in}$	VPI$_{out}$	VPI$_{in}$	VCI$_{in}$	VPI$_{out}$	VCI$_{out}$
123	88	123	54	88	53
23	61	123	89	72	64
36	72	23	55	61	33
		23	85	61	79
		36	77	72	72
		36	61	88	82

(A) VP switching   (B) VC switching.

5.6.6.5 ATM Switching Technology

An ATM switch interfaces with several input trunks (input links) and output trunks (output links). The function of the ATM switch is to route the incoming ATM cells from any incoming link to any assigned outgoing link. Note that during the call setup process, an outgoing link is assigned to a particular customer (i.e., a specific incoming link) depending upon the value of the cell header. This switching function is performed by the header (VPI, VCI) translation since ATM is a connection-oriented technology. The switch performs internal routing between the incoming link and the outgoing link by using space-division switching. In ATM networks, there is no pre-assigned time slot for a given customer, hence it is likely that two or more cells from different incoming links will compete for the same time slot of an outgoing link. Therefore, a buffer must be provided to store cells so that queues can be used to ensure cells are not discarded.

An ATM switch has many features that are different from a conventional circuit switching switch, also known as Synchronous Transfer Mode (STM) switch. Table 5-9 lists differences between ATM and STM switches. In addition, an ATM switch provides a high throughput, typically in the Gbps range, and must also be able to switch a mixture of traffic types (voice, data, video and signaling) having different traffic priorities.

Table 5-9   Comparison between an ATM and an STM Switch.

Parameter	STM switching	ATM switching
Switching characteristics	• Circuit switching • Fixed data rate • Time slot based	• Cell (packet) switching • Variable data rate • Cell (header) based
Call setup	• Allocation of time slot	• Logical (virtual) channel
Queuing	• No buffer • Drop calls if capacity is exceeded	• Cells are queued if buffer (store) is available • Drop cells if buffer store overflows
Switching delay	• Delay is constant	• Delay jitter
Cell loss	• Cell loss is not an issue	• Cell loss/misinsertion can occur

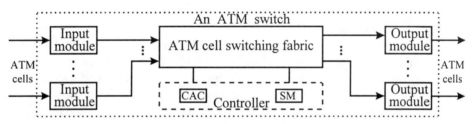

Figure 5-51   The Functional Blocks of a Typical ATM Switch.

1   ATM switch functional blocks:

Figure 5-51 illustrates the five major functional blocks of a typical ATM switch. They are the Input Module (IM), ATM cell switching fabric, Connection Admission Control (CAC), System Management (SM), and the Output Module (OM). Their functions are described as follows:

- Input module (IM): The IM is the entry point of ATM cell traffic into ATM switch. The "front end" of an IM is the circuitry required to perform the physical layer functions (e.g., SONET/SDH functions, header error verification, and cell delineation). Following the physical layer circuits are ATM layer's functional blocks. They are Usage Parameter Control (UPC), Network Parameter Control (NPC), and cell processing circuits. The ATM cells are demultiplexed, the cell header is extracted, and VPI/VCI is translated allowing user data to be routed through the switch fabric and passed to the output module for processing. Note that signaling cells and management cells are recognized by the cell processor in the IM, and are routed to Connection Admission Control (CAC) and System Management (SM), respectively, for processing.

- Switching fabric: The principle function of the switching fabric is to route cells from the input to the proper output. Cells that arrive simultaneously from different inputs and are destined to the same output are queued (stored) to prevent call loss. In addition, the switching fabric function selects cells to discard and monitors the congestion condition.

- Connection (or Call) Admission Control (CAC): CAC may be implemented as a centralized or distributed control. Its principle responsibilities are establishment, supervision, and release of switched virtual connections. CAC is implemented by the exchange of signaling information between the user and the network node, or between two network nodes. CAC interacts with another ATM switches via Network Node Interface (NNI) or a separate signaling network such as the SS7 system. CAC allocates resources to support QoS for a specific connection, and decisions are based on: (1) the connection-admission policy, (2) available resources, and (3) the existing traffic situation. Route selections are made to maximize the "long-term" network throughput by distributing traffic over the entire network, and to minimize the end-to-end delay.

- System Management (SM): SM is responsible for accurate and efficient internal operation of the ATM switching system. Its basic functions are: (1) carry out specific management responsibilities; (2) collect and administer management information; (3) communicate with users and network managers; and (4) supervise and to coordinate all management activities. The major management activities are: maintenance, performance monitoring, configuration, account, security, and traffic management.

- Output module (OM): The OM performs the inverse functions of IM. An OM receives cells from the switch fabric, CAC or SM and prepares the cells for physical transmission.

2. ATM buffering and queuing: When two or more ATM cells from different inputs arrive at the ATM switch simultaneously, and are in contention for the same output, these cells (except for one) must be queued in a buffer for later delivery to the same destination. The buffering options (see Figure 5-52) are described as follows:

- Input queuing/buffering: In this arrangement, each input port has its own dedicated buffer which allows the system to store the incoming cells until arbitration logic determines which queue is to be served next. The arbitration logic can apply a simple algorithm such as "round-robin", or a complex algorithm that takes into account the input buffer filling levels. There is one disadvantage of using input queuing technology, known as the "head-of-the line" blocking problem. At any given time, only one cell can be served.

  For example, assume that (1) the arbitration logic has decided to serve the cell from input No.1 going to output No.3; (2) at the same time the first cell from input No.2 is destined to go to output No.3; and, (3) the second cell from input No.2 is destined to go to output No.1, which is free for carrying a cell. It can be seen that the second cell (and subsequent cells) from input No. 2 will be blocked (i.e., "head-of-the-line" blocking).

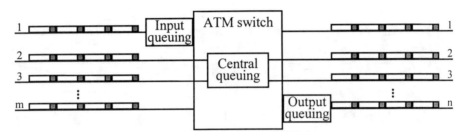

Figure 5-52   ATM Switch Buffering/Queuing Options.

- Output queuing/buffering: This arrangement allows cells from different inputs destined for the same output to be switched during one cell time. However, each output can only serve one cell at a time, hence output contention can occur. Therefore, each output needs a dedicated buffer for storing multiple cells. This scheme does not require arbitration logic since a simple queuing control algorithm such as "First-In-First-Out" (FIFO) can be applied. A disadvantage is that each output buffer must be large enough to accommodate up to "m" cells from "m" inputs. This is because in theory, all inputs could have a cell simultaneously destined for the same output (i.e., the worst case contention condition).

- Central queuing/buffering: Centralized queuing buffers are not dedicated to either inputs or outputs. Instead, they are shared by all inputs and outputs. This is an effective way to use buffer storage. However, a complex buffer management scheme is required. The management methodology consists of two linked lists: one "empty linked" list and one "serving linked" list for each output (described later in this chapter; Figure 5-71). The empty linked list provides the locations of all available buffer for storing any incoming cells. The serving linked list of each output provides the information of which cell to be served next to a destined output. Both linked lists consist of a "head" (i.e., the next buffer location to be taken out or to be used to store incoming cell), the sequence, and a "tail" (i.e., the last buffer location to be taken out or to be used to store incoming cell) of list.

5.7 5ESS SWITCHING SYSTEM

The principles of switching systems [including (1) circuit switching, such as local and toll switching, and customer switches, and (2) packet switching] have been described. Several specific switching systems, commonly used in North American digital networks as well as several parts of the world, are described in the following sections. The 5ESS switch was first introduced in 1982, and has evolved into a truly universal digital switch that meets all the telecommunications needs of the modern information age. Its versatility, flexibility, reliability, quality of service, and economic benefits are a direct result of using state-of-art designs and technologies for the system's hardware and software implementation.

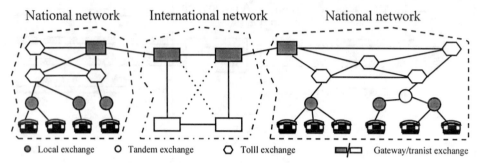

Figure 5-53 Global Telecommunications Network.

Figure 5-53 illustrates the architecture of a global telecommunications network (note that only voice terminals are shown in this example). The customers are served by local exchanges. If a call is within a local area, only one or two local exchanges are involved in the connection. In some cases a tandem switch is required to connect two local exchanges (see Figures 5-28, 5-29, and 5-35). For toll traffic, both local and toll exchanges are involved. For an international call, local exchanges, toll exchanges, gateway exchanges, and sometimes, transit exchanges are used to establish the end-to-end connection.

5ESS has a wide variety of applications. It can serve as a local, toll or international gateway exchange, or a combination of these functions. The following is a list of the major application of a 5ESS switching system:

- Local exchange
- Tandem exchange
- Toll or transit exchange
- Operator Services Position System (OSPS)
- Signal Transfer Point (STP) for a common channel signaling system
- International gateway exchange
- Wireless (cellular) switch (Mobile Switching Center/Mobile Telephone Switching Office; MSC/MTSO)
- Remote (switching) operations

5.7.1 5ESS Switch System Architecture

The architecture of a 5ESS switch emphasizes network flexibility through the use of distributed processing and a modular growth plan. The modular design allows the addition of switching capacity, system interfaces, network node applications, and increased call processing capacity.

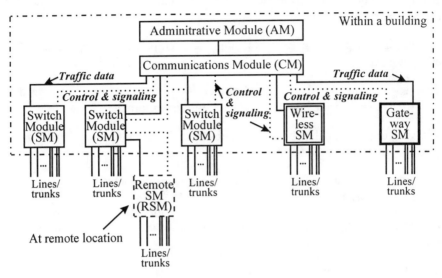

Figure 5-54   The Architecture of a 5ESS Switch.

The distributed 5ESS switch consists of Switching Modules (SMs), Remote SMs (RSMs), wireless SMs (for wireless networks), and global SMs (for global connections). The 5ESS hardware architecture shown in Figure 5-54 has the following characteristics:

- Provides non-blocking connections between unconcentrated terminations
- Integrates voice and data services
- Provides a direct interface to pulse code modulated digital facilities
- Hosts remote switching modules (away from 5ESS building)
- Hosts remote integrated services line units
- Integrates metallic access/testing facilities into SM/RMS terminating hardware
- Provides modular growth modularity to simplify the addition of new features and increased capacity

5.7.1.1 Switching Module

A Switching Module (SM) terminates subscriber loops (lines) and inter-exchange trunks interfacing the 5ESS switch. The SM is the basic "growth module" of the 5ESS switch. An SM performs the following functions:

- Scanning subscriber loops: Loop scanning has been described in Chapter 2.

- Routing calls: The SM is the switching network that provides a physical connection between the incoming facility and the outgoing facility.

- Making announcements: Example - "Your call cannot be completed as dialed, please check the number you dialed".

- Supervising call progress: Examples are - call setup, dial tone generation, alerting, call release, etc.

- Receiving analog signals from subscriber loops and inter-exchange trunks, and converting them into the internal digital time-division format

- Receiving digital signals from digital subscriber lines, and converting them into the internal digital time-division format

- Receiving digital carrier PCM facilities, and converting them into the internal digital time-division format

- Performing call-processing functions, such as recognition of requests and digit reception

- Supporting the Administrative Module (AM) in establishing intra-module and inter-module calls

- Providing service circuits, such as ringing circuits, tone generators, tone decoders, conference circuits, and test facilities

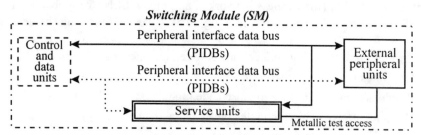

Figure 5-55   Functional Units of a Switching Module.

Figure 5-55 shows the three functional units of a Switching Module (SM) of a 5ESS switch. Their functions are described as follows:

1. External peripheral units: The external peripheral unit circuitry contains four sub-units, described as follows:

- Digital Line and Trunk Unit (DLTU): A DLTU functions as the system interface for PCM facilities. The facilities are either 1.544 Mbps (carrying 24 subscriber channels), or 2.048 Mbps (carrying 32-timeslot digital signal bit streams; but, carrying 30 subscriber channels). The DLTU's Digital Facility Interface (DFI) circuitry terminates the digital bitstream, converts the PCM format to the data time-slot format used within the switch, and performs idle time-slot stuffing. Currently, a DTLU can support up to sixteen 32-channel DFIs, or twenty 24-channle DFIs.

- Integrated Services Line Unit (ISLU): An ISLU supports analog line terminations. Each analog subscriber loop terminates on an individual "line card" (i.e., circuit pack), that provides Analog-to-Digital (A/D) conversion, battery feed, and over-voltage protection. An ISLU can also support digital line terminations, by using a Basic Rate Interface (BRI) card (i.e., circuit pack). Each BRI has two 64 kbps channels for voice, circuit-switched data, or packet-switched data; and one 16 kbps channel for signaling message and packet switched data.

- Analog Trunk Unit (ATU): An ATU terminates Voice-Frequency (VF) trunks connected to switchboards, operator positions, or local "test desks". Since an analog inter-exchange trunk carries a large volume of traffic, no concentration is provided.

- Line Unit (LU): A LU provides an interface for analog subscriber loops, and certain types of PBX trunks. LU performs analog space-division line concentration, origination scanning, battery reversal, battery feed, over-voltage protection, ringing, line supervision, and test access.

2. Service units: There are five service-related sub-units in a switching module, described as follows:

- Local Digital Service Unit (LDSU): The LDSU performs all tone generation and decoding functions required for call processing. Each switching module contains an LDSU.

- Global Digital Service Unit (GDSU): The GDSU provides conferencing capabilities and transmission test facilities, including generation and response to test tones and processing test results.

- Periodic Pulse Metering Unit (PPMU): The PPMU supports the analog line unit by providing PPM signals to subscriber loops.

- Modular Metallic Service Unit (MMSU): The MMSU provides metallic access circuits and automatic line insulation testing circuits for subscriber loop and inter-exchange trunks.

- Directly Connected Test Unit (DCTU): The DCTU provides low-frequency testing of subscriber loops, inter-exchange trunks, and special circuits.

3. Control and data units: The control and data unit circuitry contains two sub-units, described as follows:

- Packet Switch Unit (PSU): The PSU provides the 64-kbps interface that supports packetized signaling messages and packet data switching from digital subscriber lines.

- Module Control and Timeslot interchange Unit (MCTU): The MCTU performs the following functions:

  * Provides interfaces for the communication module, external interface units, and switching module processor

  * Provides call processing, call supervision, and maintenance functions

  * Provides time-division switching

  * Provides time-slot control bits

5.7.1.2 Remote Switching Module

As shown in Figure 5-56, a 5ESS switch can serve remote customers with the same features and services offered to local customers by using a Remote Switching Module (RSM). A RSM consists of standard SM hardware that augmented by circuits which terminate the digital facilities connecting the RSM to the host SM.

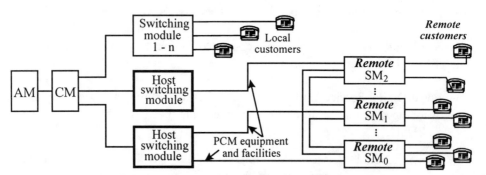

Figure 5-56   Remote Switching Module.

During normal operation (i.e., no failures), a RSM is connected to the Host Switching Module (HSM) via "control and data" links.

Definition 5-7: An RSM can process calls to lines directly connected to it via direct trunks, even if a total transmission failure occurs (i.e., failure of links to the host switching module). This processing is called "stand-alone operation".

During the transition from/to "stand-alone operation", intra-RSM calls will be maintained to minimize call interruptions, and normal dialing patterns are accepted.

When it is not possible to process a call request, the customer will be connected to a "reorder tone" or recorded announcement [e.g., "All circuits are busy, please try again later"].

A 5ESS switch can support operation of Multi-Module Remote Switching Module (MMRSM) configurations, which may consist of up to four RSMs. In addition to being connected to the host switching modules, MMRSMs are inter-connected to each other by dedicated PCM inter-RSM communications links (see Figure 5-56).

5.7.1.3 Communication Module

As shown in Figure 5-54, the Communication Module (CM) is the interface between the Administrative Module (AM) and Switching Modules (SMs). The main function of a CM is to provide a message interface between the AM and the SMs to control digital space switching. The CM contains the following four functional elements (see Figure 5-57):

1. Message Switch (MS): MS is the center for all communications occurring between the many processors within the SMs and AM of a 5ESS switch. It provides the switching function for various control, maintenance, and administrative (e.g., OA&M) messages are transferred between the AM and the SMs. The message switch consists of two sub-units:

   - Message Switch Control Unit (MSCU): MSCU performs the message transfer activities for the CM peripheral controllers, and provides an interface to the AM.

   - Message Switch Peripheral Unit (MSPU): It houses the Module Message Processors (MMPs), that performs message interpretation with the CM.

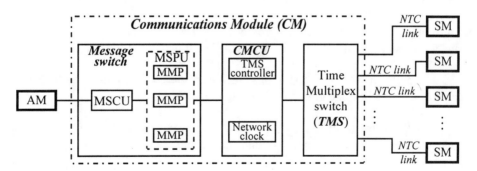

Figure 5-56 Communications Module (CM) of a 5ESS Switch.

2. Communications Module Control Unit (CMCU): It contains the network clock, and the metallic interface to the Time Multiplex Switch (TMS) and the Network, Control and Timing (NCT) links. The CMCU controls the operation of the message switch by

interfacing the MMPs and the TMS. The CMCU houses the TMS controller, the dual message interface, and terminates the message interface buses connecting the MMPs.

3. Time Multiplex Switch (TMS) unit: It provides the digital space-switching function in the 5ESS time-space-time switching architecture. In addition, the TMS provides the voice paths between the SMs, and the control paths between the message switch and the SMs. The switching fabric supports space switching of the time slots received over the NCT links. The single switching fabric configuration can serve up to 94 host SMs and local SMs (or 96 host SMs and RMs). This configuration provides a compact design for exchange switches not expected to grow beyond the 94 SM limit. For larger exchanges, a second switching fabric (i.e., dual switch fabric configuration) option is available to support up to 190 local SMs (or 192 host SMs and RMs), which is the maximum capacity of the 5ESS equipment. However, the single switch fabric configuration is the standard equipment arrangement. A "single to dual fabric conversion" procedure is available for switch exchanges that requires an in-service capacity expansion.

4. Network, Control and Timing (NCT) links: An NCT link is a full-duplex 32.768 Mbps serial link that carries digital control, timing, and network information. Each NCT link provides 256 time slots, each carrying 16 bits of information in a 125-μs frame interval. Optical technology is used to implement NCT links. Clock and frame timing is extracted from the user data stream to accomplish synchronous data transmission. Optical fiber technology was chosen because of the following advantages: electrical isolation, immunity to electromagnetic interference, high data rate capabilities, smaller physical size, light weight cables, improved security, low signal loss, and reduced cost.

5.7.1.4 Administrative Module

The Administrative Module (AM) provides the system-level interfaces required to operate, administer and maintain a 5ESS switch. The functions of the AM are:

- The AM performs common operations (such as resource allocation, maintenance, and control) that can be done globally and economically.

- The AM Processor (AP) is fully duplicated (i.e., two APs work in an active/standby configuration) to provide a highly reliable and stable system.

- The AM performs call-processing support functions such as:

    * System-wide maintenance access (craft interface)
    * Routine diagnostic/exercise control and scheduling
    * Software system initialization and recovery processes
    * Error-detection and fault-recovery functions (i.e., the AM contains error-checking circuitry used to detect and isolate faults)

- The AM performs administrative functions, and provides software access to external data links and disk storage.

- The call-processing functions of the AM consist of data routing and resource allocation. Data routing involves determining the SM on which a terminating line or trunk appears, and then selecting an available trunk in a trunk group to carry the traffic. The AM allocates and releases global resources such as TMS time slots to insure efficient operation of the switch.

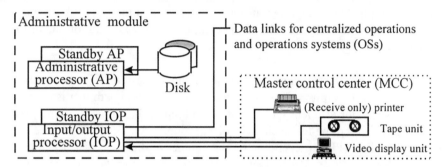

Figure 5-58   Administrative Module (AM) of a 5ESS Switch.

- A duplicated disk memory provides mass storage for programs and data. If needed (e.g., during initialization or error recovery), these programs and data can be transferred to the main memory in the AP or to the distributed memories in the SMs. In the event of a duplex system failure (a highly unlikely occurrence), the disk memory can provide rapid program and fixed-data restoration, as well as retention of billing data.

- The AM is the only processor in this distributed processing architecture, that keeps an "up-to-date picture" of the entire system.

- The Master Control Center (MCC) provides the primary interface between the 5ESS equipment and the personnel administering/maintaining the system. The MCC can also be equipped with video display units, test access equipment, printers, and/or tape units to assist personnel in performing administration/maintenance tasks.

5.7.2 ISDN and Common Channel Signaling Architectures

When the 5ESS switch is used for ISDN services, the SM has a different configuration (see Figure 5-59).   There are five functional elements in an SM used for ISDN applications: (1) Time Slot Interchanger (TSI), (2) Digital Line Trunk Unit (DLTU), (3) Integrated Services Line Unit (ISLU), (4) Packet Switch Interface Unit (PSIU), and (5) Switching Module Processor (SMP). Referring to Figure 5-59, it can be seen that:

- A data bus provides basic rate D-channel connectivity between the ISLU and the PSIU. This path can also be used to connect B-channels to PSIU for packet switching applications.

- The space switch in the Communications Module (CM) provides 64-kbps circuit-switched paths between SMs. The CM supports B-channel traffic between SMs, as well as packet switching between PSIUs in different SMs.

- The ISLU supports both non-ISDN analog and Digital Subscriber Line (DSL) terminations. The DSL supports four-wire interfaces, and the Network Terminal (NT) provides a two-wire interface. The ISLU can provide a concentration factor of "1-to-1" up to "16-to-1".

- The PSIU supports ISDN signaling messages and processed X.25 data packets. The packet bus is a fully arbitrated, unstructured communication path that is configured as a "star network" for efficient fault isolation and repair. For reliability, the packet bus is duplicated and provides each half of the duplicated SMP with access to any Protocol Handlers (PHs) (depending on the number of digital subscriber lines that the ISDN SM serves). The growth features offer small setup (startup) cost, but allows considerable flexibility for future capacity expansion.

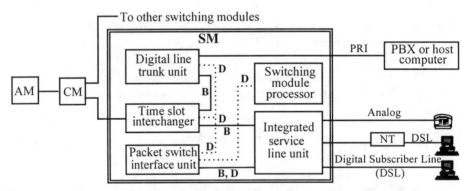

Figure 5-58   Switching Module Architecture for ISDN Services.

- The Switching Module Processor (SMP) operating system supports message communication between the PH and the SMP. The SMP operating system can also support the signaling interface between the D-channel messages and application level software in the SMP, when the PHs relay signaling messages to the SMP for delivery to ISDN application software.

- Operational software (for the ISDN application) is distributed among the PHs, SMs, and AM for ISDN circuit-switched calls. The SM and AM perform the same call processing functions as required for 5ESS circuit-switched calls.

For packet-switched calls, the PHs are responsible for establishing and disconnecting virtual calls. Software in the AM and SM performs routing,, traffic-reporting, and charging data reporting functions.

Example 5-12 Operator Services Position System (OSPS): An Application of ISDN within a 5ESS switch .

An Operator Services Position System (OSPS) is an optional function in the 5ESS switch. The OSPS requires 5ESS switch ISDN hardware and additional OSPS hardware [i.e., an Operator Position Switching Module (OPSM) equipped with an Operator Position Controller (OPC)]. OSPS is an application of ISDN within a 5ESS switch that uses the ISDN access protocol to support the integrated voice and data capability required for operator interaction. The OPC allows the operator's equipment to be remotely located [i.e., Remote Operator Service Center (ROSC); see Figure 5-60] with respect to the host 5ESS switch by using digital facility connections.

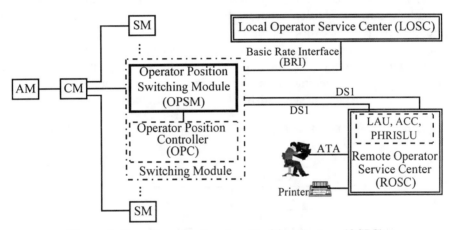

Figure 5-60   Operator Services Position System (OSPS).

The main hardware required to meet OSPS features are: (1) Asynchronous Terminal Adapter (ATA; see Figure 5-60), that allows communication with administrative processors, (2) Link Adapter Unit (LAU), that detects and reports remote alarm conditions and informs the maintenance personnel of active alarms by audio an visual indicators, (3) Alarm Conversion Circuit (ACC), (4) Protocol Handler and Remote Integrated Service Line Unit (PHRISLU), and (5) administrative printer.

5.7.3 Wireless Architecture

The 5ESS switch is designed to meet the needs of wireless services. The 5ESS switch may be used as the Mobile Switching Center (MSC), or Mobile Telephone Switching

Office (MSTO) in a Public Land Mobile Network (PLMN). The Wireless Switching Module (WSM) terminates facilities from the cell site Base Station System (BSS). The WSM handles wireless call processing and mobility management (Figure 5-61).

The Wireless Global Switching Module (WGSM) terminates signaling links and handles the routing of wireless messages to the appropriate WSMs. The WGSM also provides a BSS-Operation Maintenance Center (BSS-OMC) maintenance interface and access to information required for administration of mobile units. Chapter 8 of this book provides a detailed description of wireless operations.

5.7.3.1 MSC Administrative/Maintenance Functions

In addition to the functions described earlier, the digital cellular switch in the MSC also performs the following administrative and maintenance functions:

- *Collect billing records*: The call record includes calls originated and calls received by the mobile unit. In addition to usual billing information supplied for the public switching telephone network, billing records for cellular service should also include *radio channel seizure time*, *release time,* and *initial cell-site identification* for calls originated by the mobile unit. For calls received by the mobile unit, the billing record is created by the originating office to include the land telephone usage. The MSC usually records the radio usage statistics for the receiving mobile unit.

- *Collect traffic data*: Traffic data includes peg counts and call completion data.

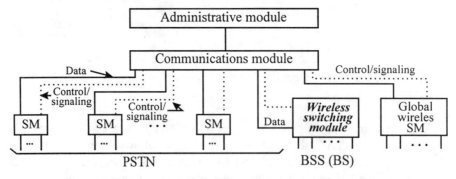

Figure 5-61   Wireless 5ESS Network Architecture.

- *Administer provisioning and service orders*: A service order is the MSC translation database that contains the records associated with each mobile subscriber (e.g., directory number, billing classification, service subscribed, etc.). Each time a subscriber is added to the system, deleted from the system, or changes service options, the MSC database must be updated by means of *"recent change messages"*.

Trunk assignment (i.e., provisioning) is based on the expected traffic loading. Trunks are also added or deleted in the MSC translation database by means of a *"recent change"* message. In most cellular systems, there is a direct correspondence between a cell-site trunk and a radio frequency assignment. When radio frequencies are changed at a base station, the corresponding trunk translations must also be modified by issuing *"recent change"* messages.

- *Maintenance*: The maintenance philosophy for an MSC is same as for the switching systems used in other applications. The maintenance functions include fault recognition, error recovery, diagnostics, fault sectionalization/isolation, routine testing, and periodically exercising the system to detect transient problems. In general, hardware and processors are duplicated to reduce the mean recovery time for failure. Typical "downtime" requirements are a maximum of 2 hours per 40 years.

5.7.3.2 Other Digital Cellular Switch

Obviously the 5ESS switch is not the only type of switch that is used as the Digital Cellular Switch (DCS) in mobile switching centers. Figure 5-62 shows the functional blocks of a *generic* digital switch in a DCS application.

The three functional blocks of a generic digital cellular switch are: (1) several Switching Modules (SMs) containing time-slot interchangers, microprocessors, and voice interface ports, (2) a Time Multiplexed Switch (TMS), and (3) various processors, and memory elements, such as Data Communications Interface Units (DCIUs) used for administration and maintenance activities.

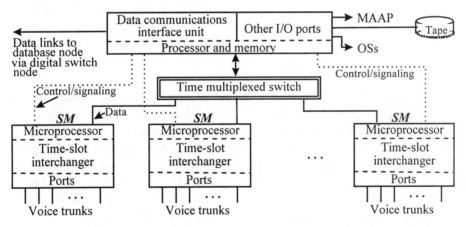

Figure 5-62   Functional Blocks of a Digital Cellular Switch (DCS).

As shown in Figure 5-63, all the voice trunks in a cellular network are connected to this DCS. Call processing software in Cellular Database Nodes (CDNs) send high-level orders to the DCSs instructing them to connect certain trunks together, thereby establishing connections for calls.

A typical DCS is equipped with two data links (usually one is active link, and the other is a standby link). Both data links are connected to duplicated Data Communications Interface Units (DCIUs). A DCS also has duplicated controllers (typically 16-bit processors). A tape system is usually provided to "boot" (i.e., initializing) the DCS with generic software and default translations. This tape system can also be used to deploy overwrites and system upgrades. In addition, the DCS is connected to several Operations Systems (OSs) that perform surveillance and alarming functions. A DCS is also connected to a Maintenance and Administration Panel (MAAP) used to perform system administration and maintenance functions.

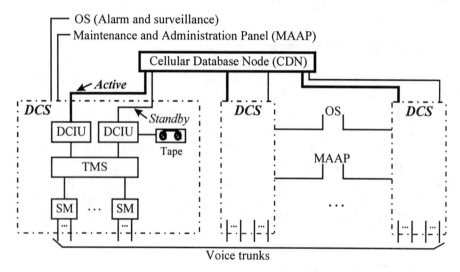

Figure 5-63   High-Level DCS Network Configuration.

A typical DCS is connected to approximately 5,000 trunk circuits. The actual switching fabric is grown in modules to accommodate the number of trunks that are needed for a particular serving area. The maximum number of Time Slot Interchange (TSI) modules is between 15 to 20. Each TSI module can physically terminate a maximum of 800 to 1,000 4-wire trunks. However, to implement a non-blocking switching network, the number of trunks connected to each TSI module should be limited to 400 to 500 terminations. Each TSI module is used to perform time division switching; and has separate duplicated controller and Time Slot Interchange (TSI) units. A TSI module can establish connections between trunks associated within its module, but cannot set-up connections between trunks that are in different TSI modules.

Another functional block in a generic DCS is a Time Multiplexed Switch (TMS), which is used to connect trunks from one TSI module to another. The TMS performs space division switching, and has a switching fabric that can be grown (i.e., expanded) to accommodate the number of TSI modules that are equipped.

Generic Cellular Digital Switch (DCS) equipment can be packaged in several cabinets. A common configuration has four cabinets (or frames) as follows:

1. *Auxiliary cabinet*: It is equipped with attenuators and signaling circuits to set the proper transmission levels on trunks, digital announcement circuits, cross-connect circuits, Local Area Data Sets (LADS) for data communications, Alternate Current (AC) power distribution unit, alarm distribution unit, and frequency generators.

2. *Common control cabinet*: It is equipped with duplicated control processors, a tape system for "booting" the DCS system with generic software and translations, an alarm panel, and power units.

3. *Module control cabinet*: It is equipped with duplicated TSI module controllers and duplicated time slot interchange units.

4. *Time Multiplexed Switch (TMS) cabinet*: It is equipped with duplicated controllers and switching fabric. To accommodate expansion of the number of TSI modules, the switching fabric can be "grown" by installing additional TMS carrier equipment in the TMS cabinet.

Voice trunks are connected to the TSI modules via voice trunk ports. The physical units for voice trunk ports are called "port carriers", which can be either analog or digital. Modern cellular switches have been implemented in compact physical configurations, which may consist of one or two frames.

5.7.4 5ESS-2000

5ESS-2000 is one of the most common switches used in North American digital networks. 5ESS-2000 is an improved version of the original 5ESS switching machine. It can be used for wireline and wireless network applications. This switch has three functional elements: administrative module, communications module, and switch module (typically more than one switch modules, and several remote switch modules are equipped in a 5ESS-2000 configuration). The characteristics of these three modules are described as follows.

1. Administrative Module (AM)

   * Centralized operations
   * Highly reliable
   * Duplex processor
   * Functions performed are: administration, control, and maintenance.

2. Communications Module (CM)

   * Centralized space-division switch
   * Switches data between the SMs
   * Switches control messages between the processors in the SMs
   * Switches control messages between the SMs and the AM
   * Provides system timing and synchronization

3. Switch Module (SM)

   * Distributed operations (multiple SMs and RSMs)
   * Provides line and trunk termination
   * Performs call processing
   * Provides time-division switching of data
   * Provides analog-to-digital conversion of user signals

The 5ESS-2000 has many features and benefits:

- High capacity non-blocking digital switch.
- Provides network interfaces required by both ANSI and ITU-T.
- Provides T1 and E-1 (CEPT-1) transmission interfaces.
- Performs robust trunk testing and maintenance exercises.
- Proven reliability in field applications.
- Architecture can meet emerging industry technology trends (e.g., ATM and SONET/SDH).
- Serves as a platform for multiple applications (i.e., local, toll, gateway and operator services).
- Mature network management platform.
- Flexible and scaleable switching system.

5.7.4.1 Switch Module (SM)

The Switch Module (SM) provides the required terminations for subscriber loops and inter-exchange trunks. It is the basic growth module of the 5ESS switch, and performs the following functions:

- The SM receives **analog** signals from the subscriber loops and inter-exchange trunks, and converts these signals into the internal digital time-division format used by the 5ESS-2000 system.

- The SM receives **digital** signals from PCM trunks, and converts these digital signals into the internal digital time-division format used by the 5ESS system.

- The SM performs call-processing functions, such as recognition of requests for service and digital reception.

- The SM supports the Administration Module (AM) for completing intra- and inter-module calls.

- The SM provides service circuits such as: ringing circuits, tone generators, tone decoders (detectors), conference circuits, and test facilities.

Since an SM is the basic growth module of a 5ESS, several different versions have been implemented to meet the needs of different applications. The most common SM configuration are described as follows:

- Single cabinet with Global Digital Service Units (GDSUs) and/or Digital Line Trunk Units (DLTUs): A GDSU can provide up to 3-port conference circuits for handoffs, with a transmission test facility. The number of GDSUs required for a 5ESS is determined by the traffic and handoff engineering estimates. The DLTU provides DS1 terminations in the switch module. A fully equipped DLTU can terminate up to 20 DS1 facilities (i.e. T1 carriers).

- Paired SMs: Within a cabinet, there are two functional SMs. Each functional SM is equipped with a fully loaded DLTU, capable of terminating 20 DS1 facilities (a total of 480 DS0 voice trunks). Therefore, a paired SM configuration can terminate up to 40 DS1 facilities (a total of 960 DS0 voice trunks).

- SM with Line Trunk Peripheral (LTP) cabinet: This configuration consists of two single cabinet SMs, each containing a GDSU and a DLTU. A LTP may contain a mixture of Line Units (LUs), Trunk Units (TUs), and Module Metallic Service Units (MMSUs). Assuming a 10:1 concentration, a LU can terminate up to 640 lines. A TU can terminate up to 64 analog trunks. The MMSU provides testing functions for analog lines and trunks.

- SM-2000: This SM has an architecture identical to the original SM. However, an SM-2000 can provide additional processing power and memory capacity. The switching fabric of the SM-2000 is supported by a faster Switching Module Processor (SMP).

5.7.4.2 Sizes (Types) of 5ESS-2000 Switches

A mobile or Personal Communications Service (PCS) switch provides the switching fabric that manages connectivity between: (1) the base stations, (2) the access manager, and (3) various types of networks [e.g., Local Exchange Carrier (LEC), Inter-Exchange Carrier (IEC), private, signaling, intelligent network, etc.].

The 5ESS-2000 switch is deployed worldwide and is a proven veteran in "landline networks". It offers superior reliability, every switching and control element is fully duplicated, and "hot-spared" to allow backup components to enter service instantly in the event of failure. The 5ESS-2000 switching platform utilizes **a common software** and

distributed hardware architecture, that simultaneously supports both wireless and wireline applications. The 5ESS-2000 platform is available in three sizes:

1. *Large switches*: The 5ESS-2000 Digital Cellular Switch (DCS) is the flag-ship platform. Switching components of the large, urban MSC, or PCSC (PCS Switching Center) consist of the following equipment:

   * Administrative Module (AM-2000)
   * Communications Module (CM-2000)
   * Up to 192 SMs, or 24 SM-2000
   * Serves up to 222 base stations

2. *Compact switch*: The 5ESS-2000 Compact Digital eXchange (CDX) is a reduced capacity version of the 5ESS-2000 DCS. Switching components of the compact suburban MSC, or PCSC consist of the following equipment:

   * Administrative Module (AM-2000)
   * Communications Module (CM-2000)
   * Up to 2 SM-2000
   * Serves up to 60 base stations

3. *Very compact switch*: 5ESS-2000 Very Compact Digital eXchange (VCDX) is a small capacity version of the 5ESS-2000 DCS. Switching components of the small, rural MSC, or PCSC consist of the following equipment:

   * Administrative workstation
   * Only one SM-2000
   * Serves up to 40 base stations

5.7.4.3 Functions of the 5ESS-2000 Switch

The 5ESS-2000 switch has a functionally distributed architecture that provides modular growth, duplicated hardware for reliability, and automatic failure recovery through reconfiguration. The MSC performs call processing, gathers data for network management, and provides interfaces to other MSCs, the PSTN, and external service centers (i.e., support services such as voice messaging, paging, and short message services). It has been previously indicated that there are three modules in a 5ESS-2000 switch: AM, CM and SMs. The individual functions of these modules have also been previously discussed. This section summarizes functions of a 5ESS-2000 switch as an overall system:

* Provides the switching function for base stations and various network elements.
* Exchanges of signaling information with other system entities.
* Provides advanced calling features via SS7.

- Provides call handling that copes with the mobile nature of subscribers (e.g., paging, physical movement of the mobile unit, etc.).

- Manages logical radio-link channels during calls.

- Manages of the MSC-base station signaling protocol.

- Handles location registration to ensure interworking between mobile stations and the Visitor Location Register (VLR).

- Controls inter-base station handovers (i.e., handoffs).

- Acts as a gateway MSC to interrogate the Home Location Register (HLR) for routing incoming calls to the appropriate mobile station.

- Supports both wireless and wireline applications simultaneously.

- Multiple applications are available on the same switch.

- Supports U.S. and international standards.

- Performs other typical switching functions associated with a local exchange in a fixed (i.e., wireline) network.

5.8 OTHER SWITCH EXAMPLES

Besides 5ESS switches, modern communications networks contain other types of switches that are used for different applications. In this section, the brief description of the 4ESS, 1PSS, Datakit II VCS, BNS-2000, and GlobeView-2000 switches are provided in the following sections.

5.8.1 4ESS Digital Switch

The 4ESS is another widely used digital switch in modern domestic and international communications networks. Its design and capacity is different from a 5ESS switch. However, the amount of traffic handled by 4ESS switches in North American digital networks is almost identical to the traffic load handled by 5ESS switches. The main design difference between 4ESS and 5ESS switches is that the 5ESS is modular. The following is a list of 4ESS switch characteristics:

- Applications: The 4ESS can used as a toll switch, tandem switch in local exchange carrier networks, and as an international network switch.

- Capacity: The 4ESS has a rated termination capacity of 107,520 trunks (maximum capacity, assuming 5% service circuits), 1,000,000 per busy hour call attempts, and a maximum load of 1.7 million "hundred call seconds" per hour.

- Signaling capabilities: The 4ESS can interface with any of the following signaling methods - Multi-Frequency (MF), Dial Pulse (DP), Dual-Tone Multi-Frequency (DTMF), CC6, CC7, SS5, SS6, and SS7.

- Services supported: The 4ESS supports advanced routing capabilities, Software Defined Network (SDN) services, advanced 800 services, and Automatic Message Accounting (AMA).

5.8.1.1 4ESS Functional Blocks

A 4ESS switch can interface analog trunks and digital trunks (as shown in Figure 5-64). Digital trunks (known as T1 digital carrier systems) arrive at the DS1 rate (twenty-four 64 kbps DS0 voice channels at a rate of 1.544 Mbps, known as a DS1 or digroup signal). Signals arriving to analog trunks are converted into a digital DS1 format using an A/D converter. A group of five DS1 streams are combined by the Digital Interface Frame (DIF) to form a DS120 format signal. The DS120 signal contains 120 data time slots that are multiplexed into a 128 time slot format (8 time slots are used for maintenance) per 125 μs frame. The DS120 signals are fed into the switching network, which consists Time-Slot Interchangers (TSIs) and a Time-Multiplexed Space switch (TMS).

The TSI and TMS elements are arranged in a "folded single-sided configuration" so that all trunks enter and leave the network on the same side.

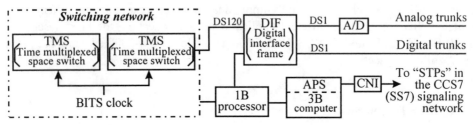

Figure 5-64   Functional Blocks of a 4ESS Switch.

The BITS (Building Integrated Timing Standard) clock provides the timing and synchronization signals for the 4ESS switch. A 1B processor is used as the primary controller for the 4ESS switch, and a 3B computer is used as Attached Processor System (APS) to handle specialized functions. The Common Network Interface (CNI) provides the connection between the 4ESS switch and Signal Transfer Points (STPs) in the CCS7 (SS7) signaling network.

5.8.1.2 (4ESS) Automatic Message Accounting (AMA)

The 4ESS Automatic Message Accounting (AMA) function is augmented by the APS for disk storage and database updating activities. A 4ESS can terminate up to 72 AMA operator positions, and a maximum of 8160 AMA trunks. The three primary AMA functions are:

1. Automatic Message Accounting Standard Entries (AMASE): This function enables billing data to be teleprocessed to a regional processing center (i.e., billing information is collected and sent to the regional processing center via the APS).

2. Automatic Message Accounting Recording (AMAR): This function has the capability of providing records for various types of calls entering the switch. Call records may be used for billing subscribers, determining access charges, and recovering transport costs from other carriers.

3. Automatic Message Accounting (AMA): This function produces records for incoming calls that are: utilizing equal access signaling, international calls, test calls originating outside the inter-exchange network, Wide Area Telephone Service (WATS) and 800 services, teleconferences, calls terminated due to "call denial", and Software Defined Network (SDN) calls.

5.8.1.3 Software Defined Network (SDN)

For decades large communications users have been searching for a cost effective way to implement "private overlay networks". A private overlay network utilizes leased circuits from a service provider (LEC or IEC) to meet the specific customer communications needs. Communications needs may vary weekly, daily, or even on an hourly basis. A common example of a private network is a Software Defined Network (Figure 5-65).

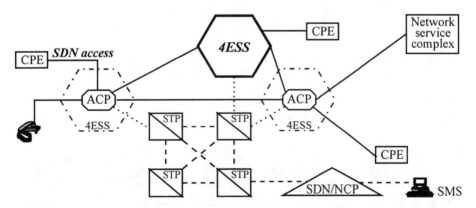

Figure 5-65   Software Defined Network (SDN) Architecture.

A SDN allows the customer to control the network configuration in ***real time*** via remote computers or workstation. This permits the user to configure the network in accordance with their specific application to optimize resource utilization. Figure 5-65 shows the SDN network architecture, which is a 4ESS-based network, managed by a quad STP signaling network. A Service Management System (SMS) allows the customer to access the signaling network via a Software Defined Network/Network Control Point

(SDN/NCP), or an Action Control Point (ACP) in the 4ESS network so the user can configure the network for optimum utilization.

5.8.2 Datakit II Virtual Circuit Switch (VCS)

The Datakit II VCS, shown in Figure 5-66, is a digital virtual circuit switch with a star topology architecture that interconnect local area networks, PBXs, data circuit switches, and X.25 packet switches. It applies virtual circuit switching technology to transport data in the network. A circuit is established and maintained for the duration of a call, but bandwidth is used only when information is actually transmitted. Functionally, the Datakit II VCS is a fast-packet switch that can be used as data networking vehicle.

Datakit II VCS can accommodate terminal speeds from 300 bps to 19.2 kbps. For inter-node applications, Datakit II VCS can accommodate speeds of 9.6 kbps, 56 kbps, 1.544 Mbps, and 8 Mbps. It is a network element that provides a vehicle for information exchange and resource sharing. Datakit II VCS is the central node (switch) of a data communication network. It thereby permits high speed data communications between computers and terminals, supporting both terminal-to-computer and computer-to-computer communications. The applications for Datakit II VCS are:

- Local Area Network (LAN): Used to connect terminals and computers in a specific area, such as office building, industrial park, factory complex, or university campus.

- Wide Area Network (WAN): Used to serve an extended geographic area (e.g., city to city, state to state, etc.) by connecting two or more LANs together.

- Computer-Based Operations Systems (CBOS): Used to provide connections between distributed computer systems for data networking applications.

- Central office LAN: Used to provide integrated data and voice services for telecommunications applications.

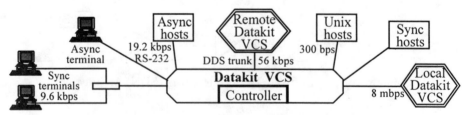

Figure 5-66   Star Topology of Datakit II Virtual Circuit Switch.

There are many benefits of applying Datakit II VCS. It can be deployed quickly, and is a secure and reliable, easy-to-grow network, that offers WAN compatibility. It is effective for data transport and networking small local businesses, widely dispersed businesses, or distributed processor systems, distributed satellite locations.

5.8.3 BNS-2000 Broadband System

BNS-2000 is a system that provides data and high-speed data networking services based on fast-packet switching (cell relay) technology. The core circuit of BNS-2000 is a cell relay switch that can simultaneously switch DS3, DS1, and subrate tributaries. Service providers using BNS-2000 can offer customer a full range of current and emerging data services, from asynchronous service to sophisticated LAN interconnected solutions. Typical BNS-2000 applications are:

- High-speed LAN-to-LAN connections
- Computer-Aided Design (CAD)
- Computer-Aided Manufacture (CAM)
- Point-to-point & point-to-multipoint video signals
- Video conferencing
- Computer generated graphics (full motion)
- Multimedia calls

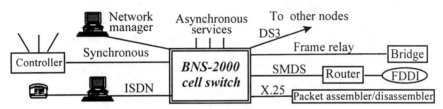

Figure 5-67   BNS-2000 Architecture.

Figure 5-67 shows the BNS-2000 architecture, that can simultaneously interface asynchronous services, synchronous services, frame relay services, Switched Megabit Data Service (SMDS), and LAN interconnect services (e.g., Ethernet, STARTLAN, token ring, FDDI, or TCP/IP). The BNS-2000 design is based on cell relay technology, which applies the concept of "bandwidth on demand". The benefits of using BNS-2000 are summarized as follows.

- From the service providers viewpoint, BNS-2000:

  * Offers a single state-of-the-art platform
  * Reduce the amount of equipment required
  * Reduces the number of personnel needed for operation/maintenance
  * Reduces the cost of training and ongoing network management

- From end user view:

  * Provides advanced SMDS functions, and flexible and customized services
  * Provides virtual private networking with security features, group addressing, customer network management and flexible billing arrangements
  * Supports high-speed public networking effectively and reliably

5.8.4 GlobeView-2000

Figure 5-51 shows the functional blocks of a generic ATM switch. The GlobeView-2000 is an ATM switch that has evolved from a Datakit system (speed of 8 Mbps), to the BNS-2000 system (speed of 200 Mbps), and finally into the GlobeView-2000 system (speed of 20 Gbps or higher). Note that the present GlobeView-2000 state-of-the-art speed is 20 Gbps, but future releases may have increased speeds.

Broadband service and network evolution can be simplified as shown in Figure 5-68. Customer services will ultimately be a form of advanced broadband services. The initial services capabilities are frame relay, switched multi-megabit data service (SMDS), and ATM. As time progresses, ATM bandwidth management functions will be added, followed by video, multimedia, interactive communication, and many other switched data services. Eventually advanced broadband services will be made available to a large population of subscribers.

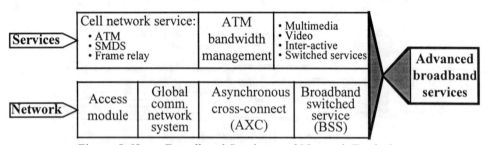

Figure 5-68   Broadband Services and Network Evolution.

GlobeView-2000 provides a high capacity ATM cross-connect for broadband provisioned services, Switched Virtual Connection (SVC) services, Permanent Virtual Connection (PVC) services, access services, broadband data, video, multimedia, and other switched data services. The GlobeView-2000 call processing and signaling functions have the following characteristics:

- Conforms to E.164 addressing standards
- Supports point-to-point calls
- Capable of mixing SVCs and PVCs
- Capable of mixing Constant Bit Rate (CBR) and Variable Bit Rate (VBR) traffic
- Supports between 5,000 and 10,000 simultaneous SVC calls

The functional block diagram for a GlobeView-2000 are shown in Figure 5-69. The primary function is performed by the Broadband Switch System (BSS), which consists of a service node, a control unit, several (growable) interface units, and a high-speed switch fabric. The GlobeView-2000's Switched Service Module (SSM) is connected to the BSS via ATM and control links. The SSM performs call-processing, signaling, and system administration functions. The BSS can be interfaced to various services via a BNS-2000 system, the access module, or direct connection. An important feature is GlobeView-

2000's interface to the Service Management Module (SMM) for network management functionality (e.g., operations, administration, maintenance and provisioning).

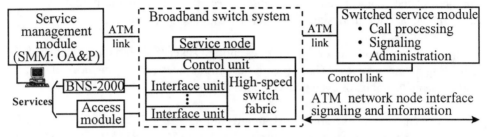

Figure 5-69   Main Functional Blocks of GlobeView-2000.

Figure 5-70 illustrates the growth/expansion feature of the GlobeView-2000 switch. During the design phase of a large "N x N" ATM packet switch, a partition is defined between the front-end (*memory-less*) Cell Distribution Network (CDN) and a column of Output Packet Switch Modules (OPSMs).

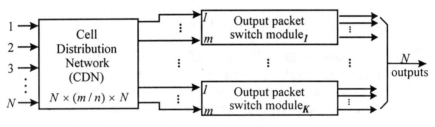

Figure 5-70   A Growable ATM Switch Architecture.

All incoming cells are routed through the CDN for instantaneous delivery, based on their destination output group address. The CDN performs its routing function for all the cells that arrive during each time slot. The outputs are grouped into "n lines" each, with a total being $K = N/n$ output groups. For each output group, the corresponding OPSM has m ($m \geq n$) inputs, therefore up to "m cells" can be accepted by an output group during each time slot. In an "N x N" switch, up to "N cells" can arrive simultaneously for a particular output group, so the maximum (conservative) design range of m is: $n \leq m \leq N$. Using the "knockout" design or another suitable algorithm, it is possible to derive a value for $m \ll N$ that yields an arbitrarily small cell loss probability that is acceptable under general traffic loading assumptions. The OPSM runs internally at 2.5 Gbps, therefore if more than m cells destined for the same output group arrive during a time slot, the excess cells are dropped (i.e., discarded). Reasons for doing so are manifold:

- The reduction of the expansion ratio ($N: mN/n$) required in the CDN translates into simpler hardware and reduced cost.

- Elimination of memory in the CDN avoids multiple stages of cell queuing delay, thereby achieving optimum output queuing in the OPSMs.

- Time-correlated (bursty) cell arrivals *only* affect the sizing of the output buffer in each OPSM, and not the design of the Cell Distribution Network.

Thus the key modular design of the entire packet switch is now reduced to implementing a "memory-less" CDN (i.e., a design that does not use memory elements, that can cause undesirable, or even unacceptable delay).

The present state-of-the-art OPSM architecture is "32 x 8" ($m = 32$, $n = 8$), and consists of a "32 x 8" concentrator with an "8 x 8" ATM Switch Module (SM). The *generalized knockout principle* (beyond the scope of this book) is used to size $m = 32$ corresponding to $n = 8$ output ports, which provides a relatively small cell loss probability. Thus, only 32 cells are accepted by the output packet switch module for 8 output ports during each time slot interval.

The "32 x 8" concentrator in the OPSM is a 32-input/8-output First In First Out (FIFO) buffer. All incoming cells are "funneled" into the 8-output buffer based on a FIFO. The "8 x 8" ATM SM stored all incoming cells in a shared RAM, that is later "read-out" according to the appropriate cell destinations.

The "m x n" dimensions of OPSM is asymmetric, namely $m > n$, because it may need to accept more than n cells in a cell period for an N x N switch (using the generalized knockout principle). A high-performance and efficient "small-size" OPSM is a key factor affecting delay-throughput performance which determines the overall switch performance.

The "8 x 8" ATM switch module of the ATM switch system architecture (Figure 5-70) has been prototyped by some switch equipment vendors as a "stand-alone unit". A full ATM switch architecture with a Cell Distribution Network (CDN or concentrator) has not yet been integrated in a national telecommunications network. It is a shared memory design. At 2.5 Gbps, an ATM cell period is approximately 160 nsec, and the write/read cycle time requirement for the RAM is 160 nsec/(m+n) where $m = n = 8$, which results in a requirement of 10 ns for writing or reading an entire cell. With a 2.5 Gbps "8 x 8" core switching fabric, the switching system can be configured in many ways depending on the line cards that are equipped (e.g., 155-Mbps line cards or 622-Mbps line cards) in accordance with standard SONET/SDH transmission rates.

Figure 5-70 shows a prototype arrangement for a "standalone 8 x 8" ATM switch. The entire "8 x 8" ATM SM (including the core fabric, line cards, and redundant circuitry) is designed to be packaged in one physical bay or frame. Most of the physical space is occupied by line cards. It should be noted that in addition to queuing and contention, additional issues must be considered: a congestion control priority scheme for real time services (e.g., voice, video, etc.).

An alternate "8 x 8" ATM switch architecture has been implemented using a shared-memory design. The hardware is divided into a data path and a control section. The data path is composed of a converter, a 28-bit-wide data bus and a large data RAM. The converter multiplexes the incoming and outgoing cells onto a wide data bus. The data bus is designed to write or read an entire cell in 10 ns, and the switching fabric operates at approximately 100 MHz.

An Address Control (AC) function maintains a FIFO queue for each output port of the switching fabric. Each FIFO queue contains the addresses of the cells stored in the data RAM that are destined for the respective output port. In addition, the AC function also maintains a buffer containing all the *available* (i.e., *empty*) locations in the data RAM. Both the data RAM and the address control functions are implemented in a custom device. The expansion module is "memoryless", and routes the cells to the concentration module. This module, in turn, requires a small amount of buffering, because it provides 32 inputs and eight outputs to the shared-memory switch fabric.

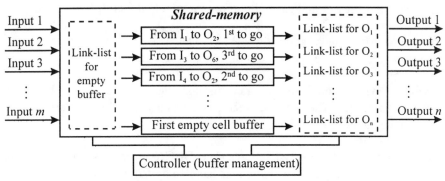

Figure 5-71   Buffer Store and Buffer Management of a Shared-Memory System.

Figure 5-71 illustrates the memory operation of a shared-memory system. For ATM applications, the buffer uses a "cell store" structure. The controller manages two lists: (1) the link-list for each output port, and (2) the link-list for empty buffer. The link-list for empty buffer determines the cell buffer location available for the incoming ATM (that can be from any input port), and then adds that particular cell location into the link-list of a specific output port. For examples, (1) the top cell location (Figure 5-71) stores the cell from input port No.1, destined for output port No.2, and is the first cell to go to output port No.2, (2) the second cell from the top stores the cell from input port No.3, destined for output port No.6, and is the third cell to go to output port No.6, and (3) the third cell from the top stores the cell from input port No.4, destined for output port No.2, and is the second cell to go to output port No.2. Once the cell has been shifted to the destined output port, the cell location is "free" for *any* incoming cell, and thereby is added to the link-list for empty buffer. The primary advantage of shared-memory system is the reduction of buffer size, and thus reduces the undesirable delay. However, the system requires a complicated controller.

5.9 SELECTED VOCABULARY OF SWITCHING

A key vocabulary of common switching terms and applications is briefly defined as follows:

- **Administrative Computer** (AC): An AC is typically located in the data center, and is used to provide security, track of job session history, and handle disk storage functions. It also supports the administration of the craft access network.

- **Adaptive Design Engineering** (ADE): ADE is the process of providing specialized design and development of switching, transmission, and customer-premises equipment, or modifications of standard products to meet the specific needs of a customer.

- **Administrative Module** (AM): An AM is an element in 5ESS (5ESS-20000) switch that is used for system-wide maintenance, administration, and resource allocation. It consists of a processor and general purpose input/output facilities.

- **Automatic Message Accounting** (AMA): The AMA function typically consists of processors used to store and retrieve of billing information.

- **Async**: Asynchronous (Async) transmission refers to the transport of unequally spaced "characters". The space between characters is controlled by start-stop elements at beginning and end of each character.

- **Busy Hour Call Attempts** (BHCA): The BHCA represents the total number of originating and terminating call attempts handled by a (switching) system in a typical one hour period (note that all call attempts, not just completed calls, are included in the BHCA calculation).

- **Central Automatic Message Accounting** (CAMA): The processing of utilizing centrally located equipment (associated with a tandem or toll switching office) for automatically recording billing data and customer-dialed exchange calls originating from several local central offices is known as CAMA.

- **Continuity Check Transceiver** (CCT): The CCT verifies outgoing trunk continuity for a common-channel signaling call and to report the results as well.

- **Centrex**: A set of customer services provided by a centralized switching systems. Examples are call forwarding, conferencing bridge, etc., provided by a central office.

- **Communications Module** (CM): The CM is a switch element that routes and distributes voice/data, control, and synchronizing signals.

- **Central Office** (CO): A CO traditionally referred to "the building housing the local exchange switch", that connects customers to the switch network. Today, the term CO refers a building that a service provider owns (rents), and contains any equipment used to implement a telecommunications network.

- **Central Office Terminal** (COT): The equipment located in the local exchange office that is used to terminate digital loop carrier systems (e.g., a subscriber loop carrier system; described in Chapter 6) is known as COT equipment.

- **Computer Subsystem** (CSS): The CCS used by a Switching Control Center (SCC), or by several SCCs, to interface the switching systems supported by these SCCs.

- **Direct Inward Dialing** (DID): The DID feature is available in PBX and Centrex systems, and allows PSTN callers to direct dial a particular extension served by a PBX without intervention by an operator.

- **Mobile Switching Center** [MSC, or Mobile Telephone Switching Center (MTSC)]: The MSC is the switching exchange used in a cellular network to connect a cellular user to other cellular or landline subscriber.

- **Network Call Denial** (NCD): The NCD function is a set of capabilities that a 4ESS uses to control subscriber access to the network served by the 4ESS, and utilizes screening of originating subscriber numbers for bill payment status.

- **Network Control and Timing** (NCT): A NCT is a control link between a Switching Module (SM) and the Communications Module (CM) in a 5ESS switching system.

- **Operator Services Position System** (OSPS): The OSPS is a feature in the 5ESS That provides processing for special toll calls and other calls/services that requires operator assistance.

- **Switching Module** (SM): An SM provides the termination for subscriber loops and interchange trunks connected to a 5ESS switching system.

- **Synchronous**: Synchronous transmission transports information (data) at a fixed rate with the transmitter and receiver equipment synchronized to a common clock.

- **Virtual Circuit (VC)**: A virtual circuit is represented by the association of the network addresses of two communications devices and a logical channel (i.e., programmable connection).

Review Questions II for Chapter 5

(14) A customer switch performs the following primary functions: handling _____, handling _____, handling _____, performing _____, and performing _____.

(15) The features provided by Centrex service are very much similar to PBX. But, there are several differences between PBX and Centrex. For example, the system maintenance: _____'s responsibility for PBX, but _____'s responsibility for Centrex; _____capital investment for switches if PBX, but not Centrex.

(16) Comparing voice signals to data, voice signals are ___ bandwidth, _____ usage, _____ and _____ rate, and have a _____ acceptability of bit errors.

(17) In circuit-switched service, the bandwidth utilization of voice transport is typically about _____%.

(18) Other than circuit-switching, store-and-forward switching has been developed for signal transport. Two schemes of store-and-forward switching are: _____, and _____ switching.

(19) Packet switching technologies have been divided into three categories: the first generation is ____, the 2nd generation is _____, and the 3rd generation is _____ (____).

(20) ATM cell (packet) size adopts a _____-length, and _____ cell structure.

(21) For ATM switching a physical channel (e.g., OC-3) can be divided into up to _____ VPs (within ATM network), and each VP can be divided up to _____ VCs.

(22) Two service types can be provisioned for ATM services: _____ connection, and _____ connection.

(23) A 5ESS switching system contains one _____ module, one _____ module, several _____ modules, and _____ modules.

(24) A switch module in a 5ESS system performs the primary functions: _____ subscriber loops, _____ calls, _____ announcement, _____ call progress, providing service circuits such as _____ circuits, _____ generators, _____ detectors, and _____ circuits.

(25) Resource allocation, system maintenance and control, craft interface, software system initialization, error-detection and fault-recovery are the primary functions of the _____ module of a 5ESS system.

(26) (True, False) 5ESS-2000 switch performs all the functions that 5ESS does, and is a SONET compatible system.

CHAPTER 6

Transmission Systems and Applications

Chapter Objectives

Upon the completion of this chapter, you should be able to:

- Describe voice digitization using waveform encoding technique: sampling voice signals, applying the Nyquist sampling theory, quantization (linear and nonlinear), and coding (binary and Gray).

- Describe the need for modulation and various modulation techniques: Amplitude Modulation (AM), Frequency Modulation (FM), Amplitude Shift Keying (ASK), Frequency Shift Keying (FSK), Phase Shift Keying (PSK), and Quadratural Amplitude Modulation (QAM).

- Describe the need for multiplexing and various multiplexing techniques: Frequency Division Multiplexing (FDM), Time Division Multiplexing (TDM), Wavelength Division Multiplexing (WDM), Code Division Multiplexing, TDM/FDM, and TDM/WDM.

- Discuss various digital signal hierarchies: the μ-law hierarchy (e.g., DS0, DS1, DS1C, DS2, DS3, and DS3C), the A-law hierarchy (e.g., E0, E1, E2, E3, and E4), the SONET hierarchy (STS-1, STS-3, STS-12, STS-12, STS-48, and STS-192), and the SDH hierarchy (STM-1, STM-4, STM-16, and STM-64).

- Describe different types of transmission equipment: Digital Channel banks (DCBs), Digital Cross-connect Systems (DCSs), Add-Drop Multiplexers (ADMs), Integrated Digital Loop Carrier (IDLC) systems, regenerators, and echo cancelers. Discuss interworking and applications of these equipment in digital networks.

6.1 INTRODUCTION

Telecommunications services, including voice, data, and video, have expanded rapidly from local (small-area) networks to regional, national and now global applications. Wireless (mobile/cellular) services (referred to as having the "first" and/or the "last" mile using airway transmission facilities) allow the once "impossible service-areas" to be reachable. Service downtime in hours is no longer acceptable, and self-healing network reliability has become a common design standard. Bandwidth capacity has increased from one voice channel to millions of voice (or voice-equivalent) channels.

Long-haul (back-haul) systems have evolved from twisted pair wire-based to broadband fiber-based systems. The transport protocols for an end-to-end connection have evolved from μ-law and A-law digital networking standards [known as Plesiochronous Digital Hierarchy (PDH)] into the improved SONET/SDH digital networking standard. The back-haul (backbone) systems used in a global digital communications environment now consist of the following four protocols (note that SONET and SDH may be considered to be the same technology in some digital networks):

(1) The μ-law Plesiochronous Digital Hierarchy (PDH)
(2) The A-law Plesiochronous Digital Hierarchy
(3) SONET (Synchronous Optical NETwork) digital hierarchy
(4) SDH (Synchronous Digital Hierarchy)

The signals: voice, audio, data, fax, and video, may be carried by one or more of the following transmission medium.

- Twisted-pair wires (two or four wires; described in Chapter 2)

- Coaxial cables (higher capacity than twisted pairs): "Coax" capacities range from several hundred to thousands of voice channels. They are popular for long distance applications in areas where optical fiber systems have not yet been implemented. Typical signals carried by coaxial cables are DS3 and E4 signals (discussed later in this chapter).

- Airway (Allow quick deployment in unusual environments; described in Chapter 8): Two (besides cellular) commonly used airway systems are briefly described as follows.

  * *Microwave radio transport system*: This technology is also known as digital radio, which is a "line of sight" high-capacity long-haul wireless system. The speed of these systems are in the range of a couple hundreds of Mbps (i.e., thousands of voice channels). They are used in areas where coaxial cables or optical fibers are too expensive to deploy due to geographic terrain conditions. The typical distance between a transmitter and its receiver is approximately 20 to 25 miles.

  * *Satellites*: Geostationary satellites are commonly-used to serve remote areas where conventional systems can not reach. In addition, satellites can be used to reach a wide area when broad coverage is required. They are also used as backup systems for conventional systems that have failed, or during maintenance conditions.

- Optical fibers (highest bandwidth and best transmission quality): Optical fibers are the most powerful systems with respect to high volume data, carrying signals over long distances without the need for regenerators, and delivering almost error-free performance. Chapter 7 provides additional details on this technology.

6.1.1 A Local or Toll Call

Call connections are typically classified as local calls, long-distance calls, or international calls. Figure 6-1 shows a simplified communication network composed of local exchange switching offices, toll exchange offices, and tandem offices.

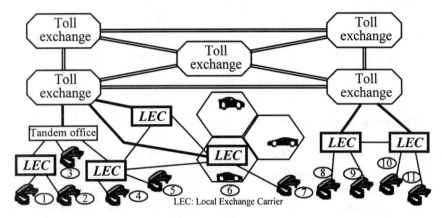

Figure 6-1   A Simplified Communications Network.

A local exchange office typically terminates end customers. That is, a switching office connected directly to the end customer is called a local exchange office. Toll exchange offices are required to support all toll calls, which include long-distance (within states or provinces), national (between states or provinces), or international (between countries) calls. In some applications, a switching office can serve as both a local and toll exchange. A tandem office can be considered either a local or a toll office, and is used to interconnect two or more local offices to reduce the number of connections between switches.

When a local switching office (which in the US, employs 1AESS, or 5ESS as described in Chapter 5) receiving a call request, it originates all call-related processes. This switching office is called the "originating office". The local switching office serving the called party of a particular call is called the "terminating office". If the connection between the *calling party* and the *called party* involves only one switching office (i.e., the switch serves as both the originating and the terminating office) it is an intra-office call. In contrast, when a call involves two or more switching offices it is an inter-office call. When the originating office and the terminating office are two different switching offices, they may be located in the same city or in different cities. Figure 6-1 illustrates the local calls, intra-office, inter-office, and toll (long distance) calls which are described as follows:

- Local calls: A local call can be a connection for any of the following cases: (1) between customers "1" and "2" (see Figure 6-1), which is also an intra-office call since both parties are served by the same office; (2) between customers "1" and "3", or between "2" and "3", which are both considered as inter-office calls since the calls are served by two different switching offices; (3) between customer "5" and wireless customer "6", or between wireless customers "6" and customer "7"; and (4) between customers "8" and "10".

- Toll (long distance) calls: A toll call can be a connection for any of the following cases: (1) between customer "1" and wireless customer "6" since the connection involves a toll exchange; (2) between customers "1" and "8" since the call involves at least two toll exchanges; and (3) between wireless customer "6" and customer "11" since the call involves at least two toll exchanges.

The cases where one or both the end users are mobile/cellular customers, the connection is actually slightly different from the diagram in Figure 6-1, and further details are provided in Chapter 8.

6.1.2 Brief Descriptions on Equipment for a Call Connection

Figure 6-2 illustrates a connection established for a toll (long distance) call. Assume end user "A" picks up the handset to request service via the local loop (typically a two-wire twisted pair line for carrying 4-kHz analog voice signals). In some areas (e.g., when an Integrated Digital Loop Carrier (IDLC) system is used) the analog loop is very short. A group of analog signals are converted into digital signals (typically 64 kbps) at the IDLC Remote Terminal (RT), and are then time division multiplexed together for transmission to the central office (shown as the local exchange in Figure 6-2). Another scenario is the arrangement when the originating (or terminating) end user is a mobile/cellular subscriber. In a mobile radio system, the airway media replaces the traditional wireline analog loop. This type of access system is known as a fixed wireless, airloop, Wireless Subscriber System (WSS), or Wireless Local Loop (WLL) Note that the industry uses many names for wireless loop applications.

Once the subscriber signal (either analog or digital) arrives at the local office the local exchange switches the signal to its intended outgoing trunk. If the incoming signal is analog, Pulse Code Modulation (PCM; discussed in a later section) must be applied to convert the analog signal into a digital format for multiplexing into a Time Division Multiplex (TDM) signal. The resulting TDM signal is further multiplexed so that the transmission facility can operate at its maximum speed to transport the signal to its final destination. The most common PCM system used in the North America digital network is a D4 digital channel bank. Later in this chapter it will be shown that a D4 channel bank (at the local exchange office) can accept several types of signals (e.g., analog and digital) and perform both analog-to-digital conversion and multiplexing functions. However, if the incoming signal to the local exchange office is in digital format, it will be multiplexed with other digital signals to form a higher-speed signal for further transmission. A popular multiplexer in the North America digital network is DDM-1000 (discussed later in this chapter).

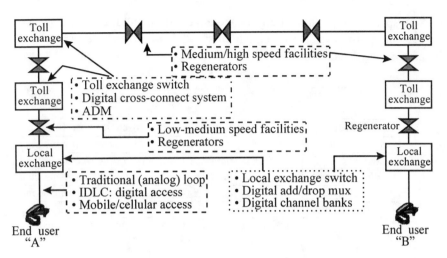

Figure 6-2 Components of an End-to-End Toll Connections.

In modern communications systems, the output signals of the local exchange office usually have a TDM digital signals. For short haul (≤ 200 miles) applications, these signals are carried by twisted-pair facilities [e.g., T1 or E1 (CEPT1) digital carrier systems]. For high-volume traffic, coaxial cable systems can be used for connecting the local exchange office to a toll exchange office. Examples of this configuration are T3 digital carrier systems in North America, and E3 or E4 carrier systems in other parts of the world. Regenerators must be used for any digital transmission spans that are longer than a few miles.

In the toll exchange office (besides high-capacity and high throughput digital switches) two transmission terminals are commonly used: Add/Drop Multiplexers (ADMs) and Digital Cross-connect Systems (DCSs). They are used primarily for routing individual digital signals to their intended destination. Depending upon the data speed, an ADM can be a low-speed, medium-speed, or high-speed element. Similarly, a DCS can be a wideband or a broadband DCS. The applications and functional differences between switches, ADMs, and DCSs is discussed later in this chapter.

The digital signals used to connect toll exchange offices are typically medium to high speed facilities, which can be transported by coaxial cables, digital radio, or optical fibers (multi mode fibers, single mode fibers, dispersion-shifted single mode fibers, dispersion flattened fibers, or low-loss fibers), depending upon the distance traveled and speed of digital signals being carried. In optical fiber systems (besides regenerators) optical amplifiers may also be used to increase transmission span distance (see Chapter 7). Thus, fiber systems are usually for deploying in high-speed, long distance applications. Table 6-1 is a comparison of the characteristics of various types of transmission media.

In summary, when digital signals are transmitted over a long distance, a high-speed transmission facility associated with a high-speed switching machine must be used. For example, a low-speed ADM can be used to handle (add, drop and transport) low speed digital signals for over short distances. However, a high-speed ADM is required to handle high-speed digital signals for long distance transmission.

Table 6-1   Comparison of Various Media.

Medium	Voice capacity	Applications/distance
Twisted pairs	*Tens*	*T1 carrier/short-medium*
Coaxial cables	*Hundreds*	*T3 carrier/medium-long*
Waveguides	*Hundreds*	*Radio*/short-medium*
Airway	*Thousands*	*Radio, satellite/short-long*
Optical fibers	*Tens of thousands*	*OC-N/short-long*

* Electrical signal conversion to radio (Presently, for waveguide applications)

In summary, when digital signals are transmitted over longer distance, a higher-speed transmission facility, associated with the high-speed switching machines and high-speed transmission equipment, must be used. For example, a low-speed ADM will be applied to handle low speed digital signals for shorter distance transmission while a high-speed ADM will handle high-speed digital signals for longer distance transmission. Therefore, to learn digital transmission systems, the first step is to learn digital signal (network) hierarchy.

6.2 DIGITAL HIERARCHY OVERVIEW

The digital hierarchy, which is the back-bone of digital networks used for global communications, is shown in Figure 6-3. This global digital hierarchy can be divided into two parts: (1) the SONET/SDH (Synchronous Optical NETwork/Synchronous Digital Hierarchy), and (2) the PDH (Plesiochronous Digital Hierarchy). SONET/SDH is a global communication standard that provides a framework for designing worldwide digital networks that have the following characteristics (i.e., advantages over PDH networks):

- Reliable and fully manageable
- Flexible and easy to grow
- Easy to adapt to new services

SONET hierarchy is primarily applied in modern next generation digital networks deployed in North America, and has five signal types (rates): STS-1 (Synchronous Transport Signal), STS-3, STS-12, STS-48 and STS-192. When these signal rates are carried by optical fibers, they are referred to as OC-1 (Optical Carrier), OC-3, OC-12, OC-48, and OC-192 respectively. In comparison, the SDH hierarchy has four signal types (rates): STM-1 (synchronous transport module), STM-4, STM-16 and STM-64. Note that STM-0 is a special signal for linking SONET and SDH. When these signals are carried by optical fibers, they are referred to as STM-1O (O for Optical), STM-4O, STM-16O, and STM-64O, respectively. SDH is the standard applied to modern digital communication networks outside the North American region.

STS-1 is the basic SONET signal, and STM-1 is the basic SDH signal. It will be shown later in this chapter that one STS-1 signal can carry one DS3 signal, seven DS2 signals, 14 DS1C signals, or 28 DS1 signals. Similarly, one STM-1signal can carry one CEPT-4 (E4) signal, three CEPT-3 (E3) signals, or 63 CEPT-1 (E1) signals. Note that it is **not** recommended that a STM-1 carry CEPT-2 (E2) signals, this is designated as " NA (Not Applicable)" in Figure 6-3.

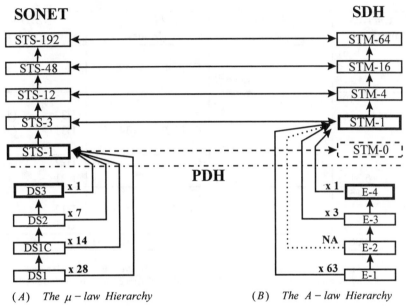

Figure 6-3   PDH, SONET, and SDH Digital Hierarchies.

PDH is a digital hierarchy that was implemented using standards different from SONET or SDH. PDH was first introduced in the telecommunications network during late 1960 to early 1961 in North America, and in the mid 1960s in Europe. There are three regional PDH standards as shown in Table 6-2: (1) North American regional, (2) ITU-T regional, and (3) Japanese regional. These three regional PDH hierarchies are implemented using two protocols: the *µ-law* and the *A-law* companding technology.

The µ-law PDH hierarchy is made up by four signal types: DS1 (digital signal-1st level), DS1C, DS2, and DS3. A single DS1, DS1C, or DS2 signal is typically carried by twisted-pair wires, DS3s are usually carried by coaxial cables, digital radio systems, or optical fibers. There are two other signals, DS3C and DS4, that are not as popular as the others. The PDH technology was introduced in 1960-1961, has been since widely deployed in Northern American networks, and is still in active operation.

In comparison, most areas outside North America have used the A-law standards. The A-law PDH hierarchy consists of four signal types: E1 [European digital signal-1st level, also known as PS-1 (PDH digital Signal-1st level), or CEPT-1 (Conference of European Posts and Telecommunications digital signal-1st level)], E2, (PS-2 or, CEPT-2), E3 (PS-3, or CEPT-3), and E4 (PS-4, or CEPT-4).

Table 6-2   Three Regional Digital Hierarchy and Associated ISDN Rates.

North America (Mbps)		*Japan* (Mbps)	*ITU-T* (Mbps)		*ISDN* (Mbps)	
DS4	(274.176)					
			CEPT4	(139.264)	H4	(139.264)
		(97.728)				
DS3C	(90)					
DS3	(44.736)				H32	(44.736)
			CEPT3	(34.368)	H31	(34.368)
		(32.064)				
			CEPT2	(8.448)	H22	(8.448)
DS2	(6.312)	(6.312)			H21	(6.312)
DS1C	(3.152)					
			CEPT1	(2.048)	H12	(2.048)
DS1	(1.544)	(1.544)			H11	(1.544)
DS0	(0.064)		CEPT0	(0.064)	B	(0.064)

Notes	DSn: nth -level digital signal	CEPTn: nth -level CEPT (digital) signal
	En: nth -level European (digital) signal	CEPT: Conference of European Posts & Telecom
	PSn: nth -level PDH (digital) signal	

Table 6-2 also lists the signal hierarchy for Integrated Service Digital Network (ISDN) standards: B-channel (information-*B*earing channel), H11, H12, H21, H22, H31, H32, and H4 signals. Table 6-2 also lists a 64 kbps (DS0) signal that typically represents a digitized voice signal. It should be noted that a DS0 signal can be either a ***digitized voice*** (speech) signal or a multiplexed data signal [from 2.4, 4.8, 9.6, 19.2 kbps, etc. sub-rate data signals].

6.3 VOICE DIGITIZATION

A customer initiates a request for service by picking up a hand-set or pressing a button, and then the call connection is set-up (described in Chapters 2 and 4). The voice signal (for voice service) is carried to the local exchange office as shown in Figure 6-2. The local exchange office has an analog-to-digital converter that is used to digitize the analog speech signal. The digitized voice signal is then combined with other signals by the TDM device. The local exchange office switching machine switches the TDM signals to the proper output trunk for transmission.

This section describes the details of speech signal digitization. A speech encoder that converts an analog speech signal into a digital speech signal does not take the many speech properties into consideration. It applies two speech properties: (1) the instantaneous speech amplitude at the sampling instances for coding an 8-bit codeword, and (2) the speech signal bandwidth for applying the Nyquist sampling theorem.

Figure 6-4 shows the building blocks of a Pulse Code Modulation (PCM) transmitter. The input signal to the system is an analog speech signal with a bandwidth of 4 kHz (described in Chapter 1). This speech signal is passed through a Low Pass Filter (LPF) to ensure that the signal bandwidth is limited to 4 kHz when applying the Nyquist theorem. The filtered signal is then "sampled and held" for further processing (i.e., quantization and coding). The output of the PCM system is a digitized speech signal with a speed of 64 kbps.

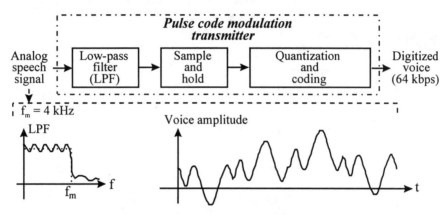

Figure 6-4   Major Building Blocks of a PCM Transmitter.

6.3.1 Sample and Hold Function

In a PCM transmitter, the analog speech signal is sampled at a rate known as the Nyquist sampling rate. The Nyquist sampling rate is given by the following equation:

$$f_s = 2 \times f_m \tag{6-1}$$

where f_s is the sampling rate in Hz, and f_m is the maximum frequency of the speech signal. A speech signal can extend its energy beyond 4 kHz, and there are cases when speech can extend up to 10,000 Hz. However, a typical speech signal has 90% of its energy distributed below 1000 Hz, and has 98% of its energy distributed below 3000 Hz. Thus it is not necessary to transport the portion of a speech signal that extends above 4000 Hz. Hence, the analog speech signal is applied to a low pass filter before entering the sampling circuit. There is another reason that a low-pass filter, with a bandwidth of 4 kHz so that the speech signal above 4 kHz is filtered out, is used. When applying the Nyquist sampling theorem, the analog signal must have a well-defined signal bandwidth (f_m). For speech signal applications, the low-pass filter has a bandwidth of 4 kHz, which is identical to the desired speech signal bandwidth (f_m).

Figure 6-4 also shows the characteristic (i.e., frequency response) of a Low Pass Filter (LPF). The dotted lines represent the characteristics of an ideal LPF, which would pass only the signal energy below f_m (the maximum frequency of the signal being sampled). A practical LPF (indicated by the curved line) is superimposed over the ideal LPF. Therefore, a speech signal will have its energy below f_m transmitted with very little filtering, but the speech energy above f_m will be heavily filtered. That is, only the speech energy of f_m and lower will be transmitted. **"What does a low pass filter do to an analog speech signal (in the time domain)?"** *Time domain speech (shown in Figure 6-4)that is processed by a LPF is "smoother" than an unfiltered signal.* A "sharp corner" in the time domain graph indicates high-frequency energy is possessed by the speech signal at that instance. Conversely, a smooth corner in the time domain graph indicates the speech signal has moderate to low frequency energy at that time instance. In summary, if a signal's corners have been smoothed, it will require less bandwidth to transmit the signal.

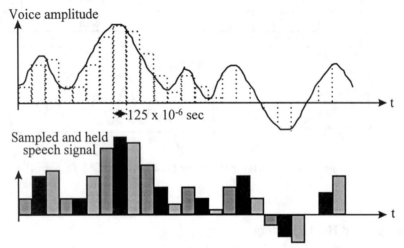

Figure 6-5   An Analog Speech Signal: Sampled and Held.

Figure 6-5 shows an analog speech signal in the time domain. The instantaneous voice amplitude is shown in the top graph of Figure 6-5. The speech signal is assumed to have a bandwidth of 4 kHz, and the Nyquist sampling rate is applied. Therefore, the analog speech signal is sampled at a rate of 8 kHz. That is, in order to restore the speech signal at the receiver with acceptable quality, 8000 samples must be taken during each one second interval by the transmitter. Therefore, sampling interval between two adjacent samples is calculated as:

$$\frac{1}{8000} = 125 \ \mu s \equiv \text{one frame} \tag{6-2}$$

Note that the **125-µs** sampling interval, between two adjacent samples, is defined as **"one frame"** by the telecommunications industry. Other time units are also based on the 125 µs frame, for example: (1) in North American digital networks a twelve (12)-frame time interval is defined as a **SuperFrame** (SF), and a twenty four (24)-frame interval is called an **Extended SuperFrame** (ESF), (2) in SONET/SDH networks a four (4)-frame interval of *500 ms* is defined as a VT superframe (multiframe) or TU multiframe [VT: Virtual Tributary; TU: Tributary Unit], and (3) in digital wireless communication systems, a 160-frame interval of *20 ms* is often used for speech encoding (i.e., in wireless communications, 125-μs, 5 ms, 10 ms, 20 ms, and 40 ms time intervals are used in the system design/analysis procedures).

After the speech signal is sampled, the (sampled) value must be held so that the sampled signal can be quantized and then coded. The sampled/held signal is shown as pulses in the bottom half of Figure 6-5. Each **pulse** represents the **amplitude** of the speech signal at the sampling point. It can be seen that the sampled/held signal waveform has much sharper corners than the original analog speech signal in the top half of Figure 6-5. From the definition of modulation (described later in this chapter), the original speech signal has been **modulated** to become the sampled/held signal. Therefore, the sampled/held signal (i.e., the bottom graph of Figure 6-5) is referred to as a *Pulse Amplitude Modulated* (**PAM**) signal. This P*A*M signal will be coded into an 8-bit *c*odeword to become a Pulse Code Modulated (P*C*M) signal. Note that the name P*C*M is derived from the P*A*M signal, in which the *A*mplitude of a sample is represented by an 8-bit *C*ode. At the output of the PCM terminal, the speech signal has a rate of 64 kbps ($\equiv$ 8 bits/sample × 8000 samples/sec), which is also referred to as DS0 or CEPT-0 signal (i.e., a voice channel, voice circuit, or voice trunk).

It should be recognized that not all telecommunications systems apply the Nyquist sampling rate. The Nyquist rate is twice the maximum signal frequency [see Eq.(6-1)], but there are applications where "*undersampling*" is used. For example, if a speech signal is sampled at 7 kHz, the resultant signal is undersampled and has a lower bandwidth (8 bits/sample × 7000 samples/sec $\equiv$ 56 kbps, compared with 64 kbps using the Nyquist sampling rate). In this example, the transmission quality will not be as good as the digital signal using the Nyquist sampling rate. This type of signal distortion is called signal **aliasing**. Some video signal transport systems (e.g., videoconferencing services) currently use sampling rates that are lower than the Nyquist rate because of the system bandwidth limitations. Likewise, there are other applications where "*oversampling*" is used to improve transmission quality. The consequence of oversampling (i.e. sampling a signal at a rate higher than 8 kHz) is that extra system bandwidth is required to transmit an "oversampled" digital signal.

6.3.2 Quantization of Sampled Signals

As shown in Figure 6-4, the sampled/held signal is quantized before it is coded into a digital codeword. It is important to know why quantization is needed in a PCM system, and, what quantization involves. An analog speech signal is eventually converted into digital codewords. For simplicity, assume that every sampled signal must be transmitted as an 8-bit codeword. Therefore, there will be a total of 256 ($\equiv 2^8$ for an 8-bit PCM system) *distinct* codewords that are available in an 8-bit PCM system. However, a speech signal can have any voltage value within its acceptable voltage range. For example, the speech signal sampled value can be 1.0

volt, 1.1 volts, 1.2 volts, 1.3 volts, 1.4 volts, 1.5 volts, 1.6 volts, etc. That is, a sampled speech signal has an ***infinite*** number of possible voltages which must be encoded into 256 different codewords. The following sections describe how this condition is handled through the quantization process.

Figure 6-6(A) represents an analog speech signal with its sampling points shown by the dotted vertical lines. This analog speech signal has a maximum voltage and a minimum voltage as shown on the y-axis in Figure 6-6(A). When the speech signal is sampled, the value can be any voltage level within the maximum and minimum voltage range. Therefore, individual samples must be ***grouped*** into 256 distinct categories. This grouping function is called ***quantization***. There are two different methods for grouping voltage samples:

(1) Linear or uniform equantization [see Figure 6-6(B)]
(2) Nonlinear or non-uniform quantization [see Figure 6-6(C)]

The advantages and disadvantages of each method and their functions are shown in Figure 6-6 and described in the following sections.

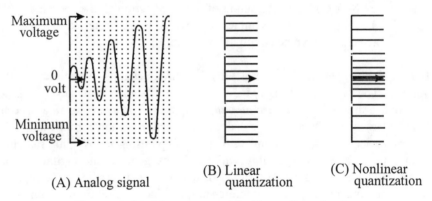

(A) Analog signal   (B) Linear quantization   (C) Nonlinear quantization

Figure 6-6   Linear versus Nonlinear Quantization.

6.3.2.1 Linear (or Uniform) Quantization

As shown in Figure 6-6(B), the voltage samples are grouped into 256 equally-spaced levels (known as quantization levels). This technique is called linear or uniform quantization. This method is further illustrated in Figure 6-7. The input signal to the linear quantizer is the sampled speech signal. The relationship between the input and the output of a quantizer is called the **quantization characteristic**, which is shown in the bottom graph of Figure 6-7.

The most negative sampled speech signal, having a voltage between x_0 and x_1 is quantized (i.e., grouped) into quantization level No.1, and encoded as an 8-bit codeword represented by "11111111". Similarly, a sampled speech signal with a voltage between x_1 and x_2 is quantized into level No.2, and is represented by "11111110". A signal with a voltage between x_{127} and x_{128} is in level No.128, and is encoded as "10000000". A signal with a voltage between x_{128} and x_{129} is in level No.129, and is encoded as "00000000". The most positive speech signal between x_{255} and x_{256} is grouped into quantization level No.256, and encoded as "01111111". This type of quantization (or grouping) is a linear or uniform quantization method since the input (i.e., the sampled speech signal), voltages are grouped into 256 levels $[(x_0, x_1), (x_1, x_2), (x_2, x_3), (x_3, x_4), (x_4, x_5), ..., (x_{255}, x_{256})]$, with the following relationship between any two adjacent levels (x_j, x_{j+1}).

$$x_1 - x_0 = x_2 - x_1 = x_3 - x_2 = x_4 - x_3 = x_5 - x_4 = ... = x_{j-1} - x_j = ... = x_{256} - x_{255} \qquad (4\text{-}3)$$

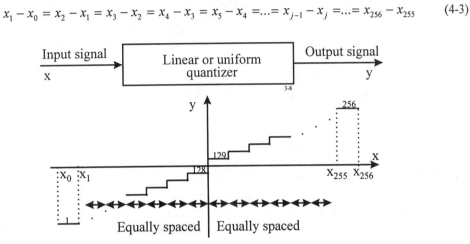

Figure 6-7   A Linear Quantizer Characteristics.

Can a linear quantizer be used in a PCM system to provide acceptable speech quality? This question is answered by examining Figure 6-8, which illustrates why the grouping function previously described is called "quantization". For simplicity, the following assumptions, (in addition to assuming an input signal range from –255 volts to 255 volts) are made when describing the functions of a linear quantizer:

* Quantization level No. 1 has a voltage level of –255 volts (and, a codeword of "11111111"), quantization level No.2 has a voltage level of –253 volts (and, a codeword of "11111110"), ..., quantization level No. 129 has a voltage level of 1 volt (and, a codeword of "00000000"), quantization level No. 130 has a voltage level of 3 volts (and, a codeword of "00000001"), ..., quantization level No. 255 has a voltage level of 253 volts (and, a codeword of "01111110"), quantization level No. 256 has a voltage level of 255 volts (and, a codeword of "00000001") (Figure 6-8).

* Any speech signal (sample; sample Nos. 1 and 2) that has a voltage between 0 volt and 2 volts (two threshold voltages as shown in Figure 6-8) is "grouped" to level No. 129 and transmitted as codeword "00000000".

* Any received signal with codeword "00000000" is restored to a voltage level of 1 volt, corresponding to quantization level No. 129.

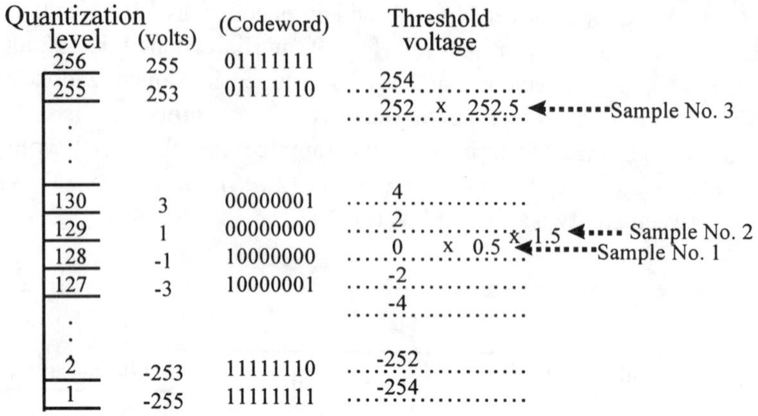

Figure 6-8   A Linear Quantization: Levels, Codewords, and Thresholds.

Two of the most frequently-asked questions are:

- **Why is the name "quantization" used?** Figure 6-8 shows a speech sample (sample No. 1) with a value (voltage level) of 0.5 volts, which is transmitted as codeword "00000000". When this signal is received , the restored voltage will be 1.0 volt, instead of 0.5 volts. Therefore, a quantization error (noise) of 0.5 volts has been introduced in the restored signal. Similarly, sample No. 2, with a voltage of 1.5 volts, is transmitted by the same codeword "00000000". Therefore, the restored signal for sample No.2 is 1.0 volt, and another error of 0.5 volts has been introduced. Hence, both samples (i.e., 0.5 volts and 1.5 volts) have been "*quantized*" to 1.0 volt.

- **Is a linear quantizer suitable for all signals?** To answer this question, two samples will be examined: sample Nos. 2 and 3 shown in Figure 6-8. For sample No. 2, the signal level is 1.5 volts and the quantization noise level is 0.5 volts. Fore sample No.3, the signal level is 252.5 volts and the quantization noise level is 0.5 volts. The signal-to-noise ratios for both samples are calculated as follows:

$$\text{For sample No. 2:} \qquad \frac{S}{N} \; = \; 20 \times \log \frac{1.5}{0.5} \; = \; 9.5 \quad dB$$

$$\text{For sample No. 3:} \qquad \frac{S}{N} \; = \; 20 \times \log \frac{252.5}{0.5} \; = \; 54 \quad dB$$

From this example, it can be seen that a linear quantizer provides a better signal-to-noise ratio for large analog signals (i.e., the signal-to-noise performance for small signals is relatively low). It should be noted that a practical PCM system does not process quantization voltages as high as 255 volts, and a S/N ratio of 54 dB is not typical. In actual PCM linear quantizers the voltage levels are different from those in this example, but the relative results for the signal-to-noise ratio are similar. Hence, the same conclusion regarding S/N ratio applies.

There are two commonly-used methods (rather than a linear quantizer) that can be applied to improve the signal-to-noise performance for small signal samples:

(1) **Increasing the number of quantization levels**: Instead of 8-bit codewords, 13-bit codewords can be used in a linear quantizer. This will provide 8,192 (instead of 256) quantization levels. In 13-bit linear quantization PCM systems, the value difference between two adjacent levels is much smaller than that of an 8-bit system. Therefore, the quantization error (noise) will be lower for small signals, and the signal-to-noise ratio is improved to become acceptable.

(2) **Applying a nonlinear quantizer**: A nonlinear quantizer utilizes a *"compandor"* to improve the signal-to-noise ratio for small signals. This technique is discussed in the following section. This approach is often more desirable than increasing the number of quantization levels, which requires more system bandwidth (i.e., 13 bits/sample × 8000 samples/sec = 104 kbps, compared to 8 bits/sample × 8000 samples/sec = 64 kbps).

6.3.2.2 Nonlinear Quantization

Figure 6-9 shows the quantization characteristic of a nonlinear (non-uniform) quantizer. The input signal (x) is the sampled speech signal, and the output signal (y) is fed to a PCM coder. The PCM coder converts the quantized speech signal into an 8-bit codeword.

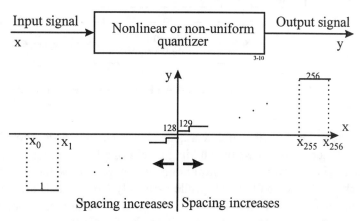

Figure 6-9   A Nonlinear Quantizer.

The major difference between a linear and a nonlinear quantizer is that the grouping of the input speech signal samples are no longer equally spaced. The voltage difference between x_1 and x_0 (i.e., quantization level No. 1) is larger than the voltage difference between x_2 and x_1 (i.e., quantization level No. 2). The voltage difference between x_{128} and x_{127}, between x_{129} and x_{128} (i.e., quantization level No. 128 or 129) is the smallest. The voltage difference between x_{256} and x_{255} is identical to the voltage difference between x_1 and x_0. The following notation is used to express the spacing of the 256 quantization levels, and the nonlinear quantization characteristic is expressed by Eq. (6-4).

$$x_1 - x_0 = x_{256} - x_{255} \;\succ\; x_2 - x_1 = x_{255} - x_{254} \;\succ\; \cdots \;\succ\;$$
$$x_{127} - x_{126} = x_{130} - x_{129} \;\succ\; x_{128} - x_{127} = x_{129} - x_{128} \tag{6-4}$$

where $x_j - x_{j-1}$ ($j = 1$ to 256) represents the spacing of quanitzation level "j". From Figure 6-9 and Eq.(6-4), it can be seen that when the speech signal is small, the corresponding quantization value around the 128th and 129th level are applied. The level spacing is small in this region, therefore a small quantization error is expected, and the speech quality is acceptable since the signal-to-noise ratio is acceptable. Recall that the signal-to-noise is relative low (e.g., 9.5 dB) when the speech signal voltage is low in a linear system. Now, as shown in Eq.(6-5), even if the speech signal voltage is small (i.e., the numerator), due to the extreme small quantization error (i.e., the denominator), the ratio is large enough.

$$\frac{S}{N} \;=\; 20 \times \log \frac{Small}{Very\ small} \;=\; Acceptable \tag{6-5}$$

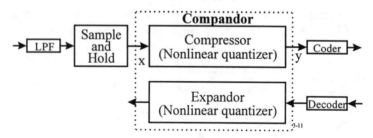

Figure 6-10   A **Comp**ressor and An Ex**pandor** (**Compandor**).

This section describes how the nonlinear quantization characteristic, shown in Figure 6-9, is implemented? Figure 6-10 shows the building blocks of a PCM system using a nonlinear quantizer, which is also known as a **compandor**. A compandor is the combination of a **comp**ressor, which amplifies small signals while extremely large signals are compressed (at the transmitter), and an ex**pandor** which performs the inverse function of a compressor (at the receiver). Before discussing its theoretical properties, the functions of a compandor will be examined by referring to Figure 6-11. The bottom left graph shows the positive half of the companding characteristic with "x" is the input signal and "y" is the output signal. Note that the negative half of the compressor characteristic is symmetrical to the origin.

The bottom right graph in Figure 6-11(B), highlighted with "small circles", is a portion of a speech signal. The "small circles" represent the speech sample values, and the vertical lines on the horizontal axis representing the sampling points at specific time intervals. When the speech signal is sampled, the sample value (*x*) is fed into the compressor. From the companding characteristic curve [Figure 6-11(B)], it can be seen that for an input signal with sample values "x", the corresponding output signal from the compressor is "y" which is highlighted with "small triangles". Note that the small speech signal sample (x) has been enlarged, and thus a better signal-to-noise ratio will be resulted. This is because the signal-to-noise ratio performance for large signals is superior, as previously described. Therefore, when a compressor with non-linear quantizing characteristics is used in a PCM system, the overall transmission quality (signal-to-noise ratio) is improved considerably.

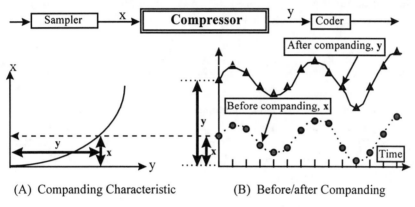

(A) Companding Characteristic   (B) Before/after Companding

Figure 6-11   Companding Characteristics.

Two different methods have been used in the telecommunications industry to implement the companding function. One is the μ-law compandor (widely used in North America), and the other is the A-law compandor (used in other countries).

- A-law Companding

In **A-law companding**, the output signal of the compressor (y), the input signal (x), and the A-law design parameter (A) related as shown in Eq.(6-6). It should be noted that the input signal is normalized to a value of "1" volt in Eq.(6-6). Therefore, the input signal is divided into two segments: (1) between 0 and 1/A, and (2) between 1/A and 1.

In actual PCM system implementations, the companding function given in Eq.(6-6) can not be realized by hardware, software, or firmware. Eq.(6-6) represents the continuous curve as shown in Figure 6-11(A), which is the companding characteristic. The curve can be represented by segmented pieces. In practical systems, a 13-segment (curve) approximation is applied to implement the companding function.

$$y = \text{sgn}(x) \frac{A\,|x|}{1 + \log_e(A)} \qquad \textit{for } 0 \le |x| \le \frac{1}{A}$$

$$= \text{sgn}(x) \frac{1 + \log_e |Ax|}{1 + \log_e(A)} \qquad \textit{for } \frac{1}{A} \le |x| \le 1 \qquad (6\text{-}6)$$

In Eq.(6-6), sgn(x) is "+" if the input signal (*x*) is positive, and "–" if the input signal is negative. The term "*A*" is the A-law design parameter, which typically has a value of 87.6 in modern digital networks. Since Eq.(6-6) contains the design parameter "*A*", which has been adjusted and refined with different values over the past several decades, this algorithm is called A-law companding.

- μ-law Companding

In μ-**law companding**, the relationship between the input signal (*x*), the output signal (*y*), and the μ-law design parameter (μ) is expressed by Eq.(6-7).

$$y = \text{sgn}(x) \frac{\log_e (1 + \mu |x|)}{\log_e (1 + \mu)} \tag{6-7}$$

The notation sgn(x) has the same meaning as sgn(x) in the A-law expression [see Eq.(6-6)]. The typical value for "μ" in modern North American digital networks is 255 (originally 100 until it was revised in 1970). Similar to A-law systems, the actual implementation of the μ-law companding algorithm given in Eq.(6-7) is accomplished by using a 15-segment approximation.

- Companding Improvement Factor

The performance of the companding technique for both A-law and μ-law is measured by a factor known as the "companding improvement factor, $\sqrt{C_A}$, or $\sqrt{C_\mu}$" (with respect to linear quantization), as follows:

$$\sqrt{C_A} = \frac{A}{1 + \log_e (A)} \tag{6-8}$$

$$\sqrt{C_\mu} = \frac{\mu}{\log_e (1 + \mu)} \tag{6-9}$$

For practical A-law systems (i.e., the 2.048 Mbps system known as the E1 or CEPT1 signal) has a companding improvement C_A = 24.1 dB when "A" = 87.6. Similarly, the North American μ-law T1 carrier system (i.e., 1.544 Mbps DS1 signal) has a companding improvement C_μ = 33.3 dB when "μ" = 255. Theoretically, the μ-law algorithm is better the A-law algorithm, but the transmission quality achieved by A-law already exceeds the quality level acceptable for practical applications. Hence the quality difference between A-law and μ-law is essentially unperceptable.

6.3.3 Coding (Sampled/Quantized) Speech Signal

When using companding, each speech sample is encoded into an 8 bit codeword (e.g., "00000000", "00000001", "00000010", ..., or "11111111") for an 8-bit PCM system. There are two common coding methods: (1) binary assignment, and (2) Gray's algorithm (see Table 6-3).

The *binary assignment* encoding method follows the binary rule "2^n"; where n = 0 to 7 for an 8-bit system; that is, 2, 4, 8, 16, 32, 64, ...". That is, the code assignment is implemented as shown in the 1st column (Binary assignment) of Table 6-3:

- The "most right-hand" bit (known as the least significant bit: LSB) alternates between **zero** ("0") and **one** ("1"). That is, the 2^0 bits follow an alternating pattern: "0, 1, 0, 1, 0, 1, 0, 1, 0, 1, ...", as indicated by the "LSB" column.

- The next column to the left of the LSB has bits alternating between **two** zeros ("00") and **two** ones ("11"). That is, the 2^1 bits follow an alternating pattern: "00, 11, 00, 11, 00, 11, ...".

- The third column to the left of the LSB has bits alternating between **four** zeros ("0000") and **four** ones or "1111" [where the "**four**" comes from 2^2].

- This pattern continues the remaining columns as shown in Table 6-3.

Table 6-3   Binary and Gray Algorithms.

Binary assignment	Codeword No.	Gray algorithm
MSB .... LSB		MSB .... LSB
0 0 0 0 0 0 0 0	1	0 0 0 0 0 0 0 0
0 0 0 0 0 0 0 1	2	0 0 0 0 0 0 0 1
0 0 0 0 0 0 1 0	3	0 0 0 0 0 0 1 1
0 0 0 0 0 0 1 1	4	0 0 0 0 0 0 1 0
0 0 0 0 0 1 0 0	5	0 0 0 0 0 1 1 0
0 0 0 0 0 1 0 1	6	0 0 0 0 0 1 1 1
0 0 0 0 0 1 1 0	7	0 0 0 0 0 1 0 1
0 0 0 0 0 1 1 1	8	0 0 0 0 0 1 0 0
0 0 0 0 1 0 0 0	9	0 0 0 0 1 1 0 0
0 0 0 0 1 0 0 1	10	0 0 0 0 1 1 0 1
0 0 0 0 1 0 1 0	11	0 0 0 0 1 1 1 1
0 0 0 0 1 0 1 1	12	0 0 0 0 1 1 1 0
0 0 0 0 1 1 0 0	13	0 0 0 0 1 0 1 0
0 0 0 0 1 1 0 1	14	0 0 0 0 1 0 0 1
.	.	.
.	.	.
.	.	.
1 1 1 1 0 1 0 0	245	1 0 0 0 1 1 1 0
1 1 1 1 0 1 0 1	246	1 0 0 0 1 1 1 1
1 1 1 1 0 1 1 0	247	1 0 0 0 1 1 0 1
1 1 1 1 0 1 1 1	248	1 0 0 0 1 1 0 0
1 1 1 1 1 0 0 0	249	1 0 0 0 0 1 0 0
1 1 1 1 1 0 0 1	250	1 0 0 0 0 1 0 1
1 1 1 1 1 0 1 0	251	1 0 0 0 0 1 1 1
1 1 1 1 1 0 1 1	252	1 0 0 0 0 1 1 0
1 1 1 1 1 1 0 0	253	1 0 0 0 0 0 1 0
1 1 1 1 1 1 0 1	254	1 0 0 0 0 0 1 1
1 1 1 1 1 1 1 0	255	1 0 0 0 0 0 0 1
1 1 1 1 1 1 1 1	256	1 0 0 0 0 0 0 0

This complete group of codewords is called the "*code space*" of the 8-bit binary code. The code space of the 8-bit Gray code (i.e., the second method for assigning 256 different codewords for an 8-bit PCM system; and shown in the column marked "Gray algorithm" in Table 6-3) is obtained as follows:

- The LSB column (2^0): Starts with "0", followed by two ones ("11"), followed by two zeros, and so on (i.e., 0, 11, 00, 11, 00, 11, 00, 11, 00, 11, ...).

- The second column to the left of LSB (2^1): Starts with two zeros ("00"), followed by four ones ("1111"), followed by four zeros ("0000"), and so on (i.e., 00,1111, 0000, 1111, 0000, 1111, 0000, 1111, 0000, ...).

- The third column to the left of LBS (2^2): Starts with four zeros ("0000"), followed by eight ones ("11111111"), followed by eight zeros ("00000000"), and so on (i.e., 0000, 11111111, 00000000, 11111111, 00000000, ...).

- The fourth column to the left of LBS (2^3): Starts with eight zeros ("00000000"), followed by sixteen ones ("11111111"), followed by sixteen zeros ("00000000"), and so on (i.e., 00000000, 1111111111111111, 0000000000000000, 1111111111111111, 0000000000000000, ...).

- This pattern continues for the remaining columns as shown in Table 6-3.

The code space obtained by applying the Gray algorithm has a unique characteristic: any two adjacent codewords in the code space (as shown in Table 6-3) have only one bit different between each sequential codeword. For example, the first and the second codewords, "0000000<u>0</u>" and "0000000<u>1</u>", are identical except for the last digit (the LSB; the 2^0 bit). The 2nd and the 3rd codewords, "000000<u>0</u>1" and "000000<u>1</u>1", are identical except the 2^1 bit.. This relationship is consistent for all adjacent codewords. Even the 256th and the 1st codewords, "<u>1</u>0000000" and "<u>0</u>0000000" only differ in the 2^7 bit. The error detection/correction advantages of applying Gray's algorithm have made it a popular coding algorithm in communications networks (described later in Example 6-4).

"*A table that contains all possible codewords is called the code space of a code*". For an 8-bit PCM system, there are 256 ($\equiv 2^8$) possible combinations. Thus, the code space of an 8-bit PCM system has a total of 256 different codewords available for assigning all the possible sample values. From the discussion in this section, it can be seen that a speech signal with a bandwidth of 4 kHz can be digitized using an 8-bit A-law or μ-law companding system. Therefore, the digitized speech (voice-frequency) signal has a speed (or rate) of :

$$Signal\ rate = 8\ \frac{bits}{sample}\ \times\ 8000\ \frac{samples}{sec} = 64\ \ kbps \qquad (6\text{-}10)$$

This 64 kbps signal is called an *E0* or a *CEPT-0* signal in the A-law digital hierarchy, and is a *DS0* signal in the μ-law digital hierarchy (both digital hierarchies will be described later in this chapter). In addition to DS0 and E0, the ISDN B-channel (information-<u>b</u>earing channel) also has a rate of 64 kbps.

Besides PCM voice digitization, there are various ways to perform the same function. Appendix 6A-2 (at the end of this chapter) describes Differential PCM (DPCM), Delta Modulation (DM), Adapted DPCM (ADPCM), Adapted DM (ADM), and speech encoding that have been applied for voice digitization in various applications.

Modern communications networks contain various signal rates (e.g., DS1, DS1C, DS3, , DS3C, DS4, E1, E2, E3, E4, STS-1, STS-3, STS-12, ..., STM-1, STM-4, etc.; see Figure 6-3 and Table 6-2), to insure efficient use of transmission facilities. To generate higher-rate signals two important communication technologies must be applied: (1) modulation, and (2) multiplexing. These techniques are described in the following section.

6.4 MODULATION AND MUTLTIPLEXING

The bandwidth available for a communication system (e.g., wire, wireless or optical fiber) has a finite limit. In addition, customer demand for additional/new services is constantly increasing. Therefore, bandwidth management has become an important area of research, and more efficient/effective methods are being studied and developed.

Bandwidth availability depends upon the specific system (i.e., media). However, typically the bandwidth available for carrying users is never adequate, especially since "Internet" services have become globally available. Therefore, it is important to manage the available bandwidth effectively. The simplified concept of bandwidth management is illustrated by Figure 6-12.

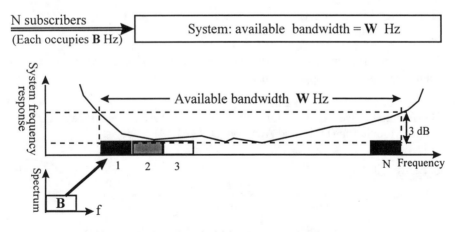

Figure 6-12   Bandwidth Management Concept.

A communication system (i.e., wire, radio, satellite, wireless/cellular, or fiber) has a finite bandwidth, W Hz. For example, a coaxial cable may have a bandwidth of 50 MHz, and an optical fiber, presently, may have a bandwidth of 100 GHz. The available bandwidth is used to transport N customers, and each customer occupies a bandwidth of B Hz (as shown in

Figure 6-12; for example, a voice customer has a bandwidth of 4 kHz). Managing the available bandwidth effectively depends upon two fundamental technologies:

(1) **Modulation**: Generally, the individual user occupies a frequency range that is not within the frequency range available for transmission. That is, the signal bandwidth and the system bandwidth are not coincident. Therefore, special methods must be applied to the individual signal so that it will occupy a frequency range that is within the system's available bandwidth. This technology is called "*modulation*". Modulation is required for performing bandwidth management in any communication system. The PCM (Pulse Code Modulation), described earlier is a common example of modulation.

(2) **Multiplexing**: Since the system bandwidth, W Hz, is always much larger than the bandwidth of one subscriber, B Hz, (i.e., $W \gg B$), multiplexing technology must be applied to utilize the system bandwidth efficiently. Therefore, multiplexing is also a technique required for managing system bandwidth.

6.4.1 Modulation Applications

From a signal transport viewpoint, there are three basic modulation technologies, as shown in Figure 6-13). They are briefly described as follows:

(1) ***Modulation for transporting an analog signal over an analog system***: The modulation technologies available are:

* Amplitude Modulation (AM)
* Frequency Modulation (FM)
* Phase Modulation (PM)

(2) ***Modulation for transporting an analog signal over a digital system***: The technologies available for this type of applications are:

* Pulse Code Modulation (PCM)
* Differential Pulse Code Modulation (DPCM), and Adaptive DPCM (ADPCM)
* Delta Modulation (DM), and Adaptive DM (ADM)

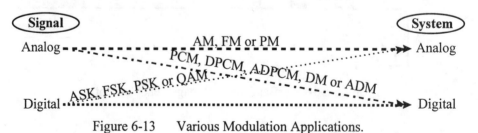

Figure 6-13   Various Modulation Applications.

(3) ***Modulation for transporting a digital signal over an analog system***: This technology is particularly important for wireless services. Commonly used modulation methods are:

* Amplitude Shift Keying (ASK)
* Frequency Shift Keying (FSK), and modified FSK
* Phase Shift Keying (PSK), and modified PSK
* Quadrature Amplitude Modulation (QAM), and modified QAM

6.4.2 The Need for Modulation

The need for modulation has been briefly discussed earlier. Two simple examples will be given here to illustrate the need for modulation. Example 6-1 describes methods that have been implemented for several decades. Example 6-2 describes modulation wireless application.

Example 6-1: An analog carrier system has an available bandwidth from 60 kHz to 108 kHz (i.e., a 48 kHz **bandwidth** as shown in Figure 6-14). Explain the need for modulation, and determine the maximum number of voice channels this system can carry.

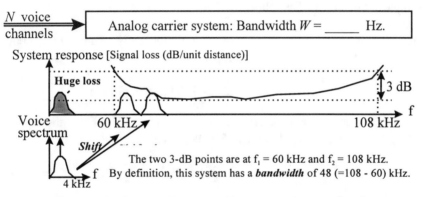

Figure 6-14   The System Frequency Response of an Analog System.

If a voice signal, which occupies the frequency range from 0 to 4 kHz, is not modulated, it will experience huge signal attenuation (loss; as shown in Figure 6-14) because of the system frequency response characteristics. Thus, this 4 kHz voice signal can not propagate far enough for practical applications. However, if the first voice signal is "shifted" to the frequency range of 60 kHz to 64 kHz; the second is "shifted" to the range of 64 kHz to 68 kHz; and so on, these "shifted" voice signals will experience very small signal attenuation (i.e., within the bandwidth from 60 kHz to 108 kHz). As a result, the voice signals can be carried over long distances for practical applications. The process of "shifting" a signal from its original frequency range to a higher frequency range is called "*modulation*". The unmodulated signal is called the **baseband signal**, and the modulated signal is called the **bandpass** (or **passband**) **signal**.

With modulation, since this system has a bandwidth of 48 kHz, the system has a capacity of **12 voice channels (or trunks)** derived as follows:

$$System\ capacity = \frac{System\ bandwidth}{Signal\ bandwidth} = \frac{48\ kHz}{4\ kHz} = 12\ \ voice\ channels\ (trunks)$$

Example 6-2: For a ***traditional*** analog radio transport system, the antenna length of a simple antenna is given by Eq.(6-11). Using this equation, determine the antenna length for: (1) a radio system that does not modulate voice signals; (2) an AM radio station used for voice transport; and, (3) an FM radio station used for voice transport.

$$L_{antenna} = (\frac{1}{10} \sim \frac{1}{20}) \times \lambda \tag{6-11}$$

where $L_{antenna}$ is the antenna length of a simple antenna, and λ is the wavelength of the signal to be transmitted. The relationship between the frequency and the wavelength of a signal is given by the following equation: $\lambda f = c$, *where c is the speed of light in vacuum, and has a value of* 3×10^8 *m/sec* (see Chapter 7 for the exact value of c). The antenna lengths for the three cases in this example are determined as follows:

(1) No modulation applied: The maximum voice frequency is assumed to be 4000 Hz. From the equation $\lambda f = c$; obtain $\lambda = 750$ km, and applying Eq.(6-11):

$$L_{antenna} = (1/20) \times 750\ km = 2.33\ miles$$

(2) AM applied: The maximum carrier frequency is 1600 kHz. From the equation $\lambda f = c$; obtain $\lambda = 187.5$ m, and applying Eq.(6-11):

$$L_{antenna} = (1/20) \times 187.5\ m = 31\ ft$$

(3) FM applied: The maximum carrier frequency is 108 MHz. From the equation $\lambda f = c$; obtain $\lambda = 2.78$ m, and applying Eq.(6-11):

$$L_{antenna} = (1/20) \times 2.78\ m = 5.5\ in$$

From this simplified example, it can be concluded that: (1) without modulation, an antenna length of 2.3 miles is impractical; (2) with amplitude modulation, an antenna length of 31 ft is acceptable; and (3) with frequency modulation, an antenna length of 5.4 inches is very attractive. Hence, the advantage of using modulation is obvious.

6.4.3 Amplitude (Frequency, or Phase) Modulation

To understand the various modulation methods (e.g., ASK, FSK, PSK, or QAM) used for advanced services, it is important to have a knowledge of amplitude modulation, frequency modulation, and phase modulation techniques.

Amplitude Modulation (AM) was originally developed in the late 1890's. The amplitude modulator at the transmitter is shown in Figure 6-15. From a frequency-domain viewpoint, a voice message occupies the frequency range from approximately 200 Hz to 3400 Hz. For simplicity, a voice signal is assumed to have a maximum frequency $f_m = 4$ kHz. When a voice signal is (amplitude) modulated by a carrier signal, with a carrier frequency of f_c, the amplitude modulated signal will occupy the frequency range from f_c to $f_c + f_m$ (as shown in the top right of Figure 6-15). The AM signal shown here is known as a "Single Side Band" AM (SSBAM) signal, rather than he second type of AM [Double-Side Band AM (DBSAM)].

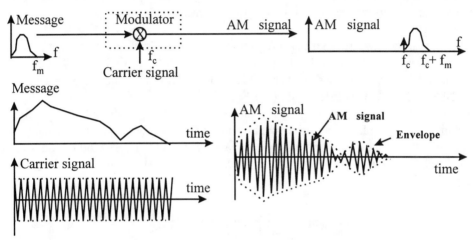

Figure 6-15   Amplitude Modulation.

The time domain signal for this AM system is illustrated by variation in message amplitude with respect to time. A carrier signal can be a sinusoidal wave or a saw-tooth wave. The carrier signal in Figure 6-15 has a constant amplitude and a fixed (carrier) frequency of f_c. In practical AM systems, the value of f_c is much higher than f_m (i.e., $f_c \geq 10 f_m$). The AM signal (in the time domain) consists of the carrier signal with varying amplitude (instead of a constant amplitude), as shown in Figure 6-15 (lower right graph).

It is important to note that the amplitude of the AM signal envelope tracks the amplitude of the original message. At the receiver, the AM signal will be demodulated (i.e., detected) using an envelope detector, that "traces" the envelope of the AM signal. This trace is the restored message signal. Using this simplified example, the definitions of amplitude modulation, frequency modulation, and phase modulation are described as follows.

__Amplitude Modulation Definition:__ The **amplitude** of the carrier signal is modulated (i.e., changed) in accordance with (the amplitude of) the message signal.

__Frequency Modulation Definition:__ The **frequency** of the carrier signal is modulated (i.e., changed) in accordance with (the amplitude of) the message signal.

Phase Modulation Definition: The **phase** of the carrier signal is modulated (i.e., changed) in accordance with (the amplitude of) the message signal.

6.4.4 Modulation Used to Carry Digital Signals over an Analog System

From Figure 6-13, it can be seen that Amplitude Modulation, Frequency Modulation, or Phase Modulation can be used to carry an analog signal over an analog system. This type of modulation technology is typically adopted for analog wireless services such as FDMA (see Chapter 8) besides the broadcast AM, FM radio systems.

In comparison, for transporting a digital signal over an analog system (e.g., the airway channel) different modulation technologies must be applied. From Figure 6-13, it can be seen that the modulation technologies used are ASK, FSK, PSK, QAM, and variations such as 16-FSK, 32-PSK, 64-QAM, DPSK, MSK (see Figure 6-16, and Table 6-4.) From Table 6-4, it can be concluded that ASK is the AM technique for modulating a digital signal; FSK is the FM technique for modulating a digital signal, and PSK is the PM technique for modulating a digital signal. Note that there is no analog equivalent for QAM. These modulation techniques (e.g., ASK, FSK, etc.) have also been adopted for data communication "modem" (i.e., modulation and demodulation) technologies.

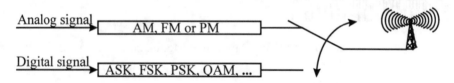

Figure 6-16   Modulation for Analog Systems.

Table 6-4   (AM, FM, PM) versus (ASK, FSK, PSK).

Analog signals	AM	FM	PM	Not applicable
Digital signals	ASK	FSK	PSK	QAM

The principle of Amplitude Shift Keying (ASK), Frequency Shift Keying (FSK), Phase Shift Keying (PSK), or Quadrature Amplitude Modulation (QAM) is to convert a digital signal into an analog signal so that it (the digital bit stream) can be effectively transported over an analog system (e.g., airlink of a cellular network, or analog local loop for a modem).

An analog signal can be represented by the sum of many sinusoidal signals by using the Fourier series principle. Therefore, a single-tone (single frequency) analog signal can be represented the following equation:

$$v(t) = A(t)\, cos\,(\omega\, t + \theta)$$ (6-12)

In Eq. (6-12), $v(t)$ is the analog signal (e.g., voltage or current), $A(t)$ is the amplitude of the signal, ω is the frequency of the signal, θ is the phase of the signal, and "cos" represents a sinusoidal wave function. The overview of using a sinusoidal wave for carrying digital bit stream is given as follows (additional descriptions of ASK, FSK, PSK, QAM, M-FSK, M-PSK, M-QAM, etc. are provided in subsequent sections in this chapter).

- **Amplitude Shift Keying (ASK):** If the amplitude, $A(t)$, in Eq. (6-12) changes, the signal characteristics will also change. This fact can be used to represent different digital signals using a scheme referred to as amplitude shift keying. For example, if the digital signal is a logical "1", then the amplitude $A(t)$ of the analog signal is "1-volt" [i.e., the signal transmitted over the airlink is $v(t) = cos\ (\omega\ t + \theta\ )$]. If the digital signal is a logical "0" then the amplitude $A(t)$ is "0-volt" [i.e., the signal transmitted over the airlink is $v(t) = 0$]. Additional details are provided in Section 6.4.4.1.

- **Frequency Shift Keying (FSK):** If the frequency, ω, in Eq. (6-12) changes, the signal characteristics will also change. This frequency change can be used to represent various digital signals by using frequency shift keying. The amplitude and the phase of the analog signal remains unchanged, but the frequency changes [i.e., the signal representing the digital bit stream has the following format: $v(t) = A \cdot cos\ \omega_i t$, where "$i$" can be any integer depending upon the digital stream]. Additional details of FSK and its variations are provided in Sections 6.4.4.1 and 6.4.4.2.

Table 6-5   Basic and Extended Modulations for Digital Signals.

	Amplitude	Frequency	Phase	Amplitude & Phase
Basic (Binary)	ASK	FSK	PSK	QAM
Extended (modified) Schemes		M-FSK 16-FSK 64-FSK MSK	M-PSK 16-PSK 64-PSK MSK DPSK QPSK	M-QAM 16-QAM 64-QAM QPSK

- **Phase Shift Keying (PSK):** If the phase, θ, in Eq. (6-11) changes, the signal characteristics also change. This phase change can be used to represent various digital signals by using phase shift keying. The amplitude and the frequency of the analog signal, Eq(6-12) remain unchanged. Additional details are provided in Sections 6.4.4.1 and 6.4.4.2.

- **Quadrature Amplitude Modulation (QAM):** If both the amplitude, A(t), and the phase, θ, are changed in Eq. (6-12) to represent different digital signals, the modulation technique is called quadrature amplitude modulation.

The modulation technologies listed above are summarized in Table 6-5, and details are discussed in detail in subsequent sections of this chapter.

6.4.4.1 Basic (Binary) ASK, FSK and PSK

Assume that a saw-tooth waveform is used as the carrier signal in the discussion of the three basic modulation methods: ASK, FSK and PSK (see Figure 6-17). In all three modulations, the digital bit stream is assumed to be "1011". The actual frequency of the analog signal is not important in the discussion of these three methods. The resulting signals are called ASK, BFSK (FSK), and BPSK (PSK), where the prefix "B" represents basic or binary.

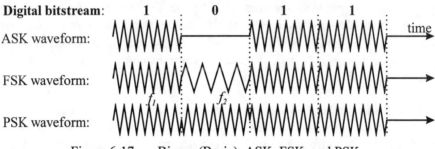

Figure 6-17   Binary (Basic) ASK, FSK, and PSK.

The ASK waveform (top waveform in Figure 6-17) has a carrier signal with an amplitude that changes in accordance with the message signal. That is, the amplitude of the carrier signal is maximum when the message value is a logical "1", and the amplitude of the carrier signal is zero when the message value is a logical "0". According to the definition of AM (see Section 6.4.3), an ASK waveform is an amplitude modulated signal. Likewise, ASK is considered an "amplitude modulation" technique for modulating a digital signal (Table 6-4).

The FSK (BFSK) waveform (middle waveform in Figure 6-17) contains two different frequencies. The frequency of the carrier signal is f_1 when the message value is a logical "1", and f_2 when the message value is logical "0". According to the definition of FM (see Section 6.4.3), an FSK waveform is a "frequency modulation" technique for modulating a digital signal (see Table 6-4).

The PSK (BPSK) waveform (bottom waveform in Figure 6-17) contains two different phases. The phase of the carrier signal is θ_1 when the message value is a logical "1", and the θ_2 when the message value is logical "0". According to the definition of PM (see Section 6.4.3), an PSK waveform is a "phase modulation" technique for modulating a digital signal.

6.4.4.2 M-ary FSK (or M-FSK)

A practical frequency shift keying method is M-FSK, where $M = 2^n$ (n = an integer). Note that M-FSK is pronounced as "M-m'ary FSK" as b'i'nary for binary. An FSK (BFSK) is a special M-FSK with an "n" of 2, and has been used in several earlier communication systems. For example, there are modems using 16-FSK, 32-FSK or 64-FSK technology. The modulation algorithm for a 16-FSK is provided in Example 6-3 to illustrate the basic principle of the M-FSK method. The same process, used in Example 6-3, can be used for any "M" value.

Example 6-3: Describe the modulation scheme for a 16-FSK system, assuming the digital data stream message as follows:

"0011 0010 0001 0110 1000 1110 … ".

An M-FSK modulation algorithm follows the three rules described below. In this example, the value for M is 16.

Rule No. 1: For a 16-FSK application, 16 different frequencies are used in the ***modulated*** signal. The 16 frequencies are equally spaced, as shown in Figure 6-18. Each of these frequencies is allocated a specific bandwidth on a per subscriber basis.

$$v(t) = A \cdot cos \ \omega_j \ t \qquad (j = 1, 2, 3, 4, …, 16) \tag{6-13}$$

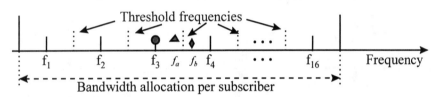

Figure 6-18   Frequency Allocation and Threshold Frequencies (16 FSK).

Note that "threshold frequencies" are applied at the receiver as shown in Figure 6-18. This is because the received frequency can be slightly different from the transmitted frequency. Therefore, threshold frequencies must be applied to determine the "most-likely" transmitted frequency. For example, assume f_3 is transmitted but f_a is received. In this case, a frequency of f_3 is restored because f_a is within the f_3 threshold range (see Figure 6-18). As a result, the recovered digital bit stream is correct. However, if f_3 is transmitted but f_b is received, then f_4 will be restored because f_b is outside the f_3 threshold, and the received (restored) signal will be decoded erroneously.

Rule No. 2: In a 16-FSK application M = 16, and since $16 = 2^4$, every 4 consecutive bits of the digital bit stream are carried by the individual sinusoidal signals given by Eq. (6-13). The specific frequency assigned are listed in Table 6-6. For example, the first four bits "0011" of the digital bit stream (in this example) are assigned the frequency of ω_3, and the analog signal transmitted over the airlink (analog local loop) is $A \cdot cos \ \omega_3 t$. The 2nd four-bit group "0010" is transmitted as $A \cdot cos \ \omega_4 t$. The 3rd four-bit group "0001" is transmitted as $A \cdot cos \ \omega_2 t$. This modulating scheme is continued for the remaining bit groups. Note that bit groups in Example 6-3 are in Gray code format (described next).

Rule No. 3: Applying the **Gray algorithm,** as shown in Table 6-6, to assign the 16 frequencies required for a 16-FSK application is typically used in modern communication systems. Table 6-6 lists: (1) the binary coding for the four-bit data groups, (2) the Gray coding for the four-bit data groups, (3) the 16 frequencies assigned for a 16-FSK modulation technique corresponding to either binary or Gray code bit groups, and (4) the 16 phases assigned for a 16-PSK modulation technique corresponding to either binary or Gray code bit groups. Gray coding is typically used because of error detection/correction advantages (see Example 6-4).

If the bit stream is not a multiple of four bits (in 16-FSK or 16-PSK applications), then any group of four bits, with the leading bit(s) matching the user's bit(s), can be chosen, and the decoded data stream will be correct. For example, if the 3 bits of a user's bit stream are "001", the frequency assigned to this group can be either ω_3 or ω_4 (assuming Gray code is used). If ω_3 is transmitted, then "0011" will be restored, and the first three bits (i.e., "001") will be restored correctly. Likewise, if ω_4 is transmitted then "0010" is restored, and the first three bits are still "001". Therefore, in either case the proper bits will be restored (i.e., "001").

Table 6-6　Frequency/Phase Assignment for 16-FSK/16-PSK.

Binary code	Gray code	Frequency/Phase
0000	0000	ω_1 / θ_1
0001	0001	ω_2 / θ_2
0010	0011	ω_3 / θ_3
0011	0010	ω_4 / θ_4
0100	0110	ω_5 / θ_5
0101	0111	ω_6 / θ_6
0110	0101	ω_7 / θ_7
0111	0100	ω_8 / θ_8
1000	1100	ω_9 / θ_9
1001	1101	$\omega_{10} / \theta_{10}$
1010	1111	$\omega_{11} / \theta_{11}$
1011	1110	$\omega_{12} / \theta_{12}$
1100	1010	$\omega_{13} / \theta_{13}$
1101	1011	$\omega_{14} / \theta_{14}$
1110	1001	$\omega_{15} / \theta_{15}$
1111	1000	$\omega_{16} / \theta_{16}$

Example 6-4: Explain why Gray's algorithm is applied when assigning the frequencies given in Table 6-6 for the 16-FSK modulation technology. Note that: (1) Gray's algorithm has been used in a PCM coding (described in Table 6-3), (2) Gray's algorithm can be used for an M-FSK, with M having a value other than 16, and (3) Gray's algorithm can also be used for a M-PSK, or QAM applications.

It can be shown that Gray's algorithm will produce better signal quality at the receiver. Assume that the transmitted frequency is ω_8, however, the received signal is incorrectly interpreted as the frequency ω_9 because of a system frequency shift. Two different cases are presented to show the advantage of Gray's algorithm.

Case 1: If the binary algorithm is used to assign the frequencies, and the transmitted bit group is assumed to be "0111", then the transmitted signal is represented by ω_8. Assuming there is a system frequency shift, and the frequency is incorrectly interpreted as ω_9, from Table 6-6, the restored binary bits for ω_9 are "1000". Thus, all four restored binary bits are in error: (i.e., "1000" instead of "0111"). The BER performance is definitely unacceptable.

Case 2: If Gray's algorithm is applied, and the transmitted frequency is assumed to be ω_8, that represents transmitted bit group of "0100". Assuming the system erroneously receives ω_9 instead of ω_8, the restored Gray's code information bits are "1100" (see Table 6-6). In this case there is only one bit error (i.e., "1100" instead of "0100"). Obviously, Gray's algorithm provides improved signal quality (i.e., a better bit error rate).

In summary, *"if code assignments are required for a series of values (e.g., coding analog samples or assigning the frequencies for an M-FSK system), and adjacent values have a close correlation (e.g., small differences), it is often advantageous to apply Gray's algorithm instead of binary coding."*

6.4.4.3 M-ary PSK (or M-PSK)

A practical phase shift keying method is the M-PSK or DPSK (differential PSK). M-PSK is similar to M-FSK, however the phase, rather than frequency is used to perform modulation. For a 16-PSK application, the 3rd column in Table 6-6 lists 16 possible phases. The three rules stated in Example 6-3 are modified for 16-PSK as follows:

Rule No. 1: For 16-PSK, 16 different phases are used in the ***modulated*** signal, represented by Eq.(6-14), the amplitude and frequency remain constant for all phases (θ_i).

$$v(t) = A \cos(\omega_c t + \theta_i) \qquad (j = 1, 2, 3, 4, ..., 16) \qquad (6-14)$$

Rule No. 2: For a 16-PSK application, M = 16 and since $16 = 2^4$, every 4 consecutive bits will be carried by the sinusoidal signal defined by Eq. (6-14), with a corresponding phase listed in Table 6-6 (note that 16 phases are also shown in the column containing the 16 frequencies).

Rule No. 3: Apply **Gray's algorithm** (see Table 6-6) to assign the 16 phases appropriately to represent the 4 information bits using 16-PSK modulation, in accordance with Eq. (6-14).

Figure 6-19 shows all 16 phase assignment for a 16-PSK modulation system. For example, the phase assigned for the information bit group "0000" is 0°, the phase assigned for the information bit group "0001" is 22.5°, the phase assigned for the information bit group

"0011" is 45°, the phase assigned for the information bit group "0010" is 67.6°, etc. This diagram is called the signal constellation for a 16-PSK modulation system. In this diagram, 16 threshold angles are also shown. The function of the threshold angles is the same as the threshold frequencies described in the 16-FSK application (Example 6-3).

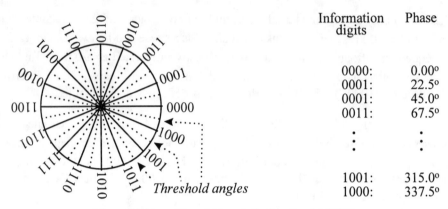

Information digits	Phase
0000:	0.00°
0001:	22.5°
0001:	45.0°
0011:	67.5°
⋮	⋮
1001:	315.0°
1000:	337.5°

Figure 6-19   Signal Constellation for 16-PSK.

Example 6-5: For a 16-PSK system, assuming the received signal is $A \cdot cos\ (\omega t + 98.5°)$, followed by $A \cdot cos\ (\omega t + 129.5°)$, $A \cdot cos\ (\omega t + 223.5°)$, $A \cdot cos\ (\omega t - 38.5°)$, and $A \cdot cos\ (\omega t - 298.5°)$, ... , determine the restored information signal bitstream.

By applying the signal constellation diagram of Figure 6-19, it can be seen that the phase of $98.5°$ is between the threshold angles of $78.75°$ and $101.25°$ *[derived from $(90° - 11.25°)$ and $(90° + 11.25°)$]*. Therefore, the restored phase is $90°$ and the corresponding information signal is "0110". Using the same approach, the remaining four message signals are restored as follows: "0101", "1111", "1001", "0010", ...

6.4.4.4 Quadrature Amplitude Modulation (QAM)

Three techniques for modulating a digital data stream so that the information bitstream can be transmitted over an analog system have been presented. They are ASK, FSK (and M-FSK), and PSK (and M-PSK). In each case, an analog signal is used to represent the digital data stream, and the analog signal has a changing amplitude, a changing frequency, or a changing phase. In Quadrature Amplitude Modulation (M-QAM), the analog signal that is used to represent a information bitstream can have different "*amplitude* and *phase*", as shown in Eq. (6-15).

$$v(t) = A_i \cdot cos\ (\omega_c t + \theta_j) \qquad (6-15)$$

The same three rules used in M-FSK and M-PSK are applicable to M-QAM, with the following exception: M-FSK uses Gray's algorithm to assign various frequencies, and M-PSK uses Gray's algorithm to assign various phases, but M-QAM uses Gray's algorithm to assign sets of (amplitude, phase) as shown in Figure 6-20. Each point in the signal constellation represents an analog signal with a paired set of (amplitude, phase) [i.e., (A_i, θ_i)].

Example 6-6: Draw the signal constellation of a 16-QAM system.

As shown in Figure 6-20, 16-QAM requires four points per quadrant ($4 \equiv 16/4$; where the number "4" is used as the divisor since there are four quadrants in the space). Each point (or star) in the signal constellation (Figure 6-20) represents a four-bit digital signal. For example, the point shown with amplitude of A_1 and a phase of θ_1 represents the digital signal bit stream "0000"; the point shown with amplitude of A_1 and a phase of $(\theta_1 + 90°)$ represents the digital signal bit stream "0001"; the point shown with amplitude of A_1 and a phase of $(\theta_1 + 180°)$ represents the digital signal bit stream "0011"; ..., etc.

The 16-QAM receiver requires two detection thresholds. One is the amplitude threshold detector that decides if the received signal has an amplitude of A_1, A_2, or A_3. The other is the phase threshold detector that determines if the receiver signal has a phase of θ_1, θ_2, θ_3, etc.

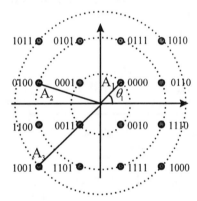

Figure 6-20   The Signal Constellation of a 16-QAM.

6.4.4.5 Variation of Modulation: 4-PSK, 4-QAM, and QPSK

A signal constellation can represent many different aspects of modulation. Figure 6-21 is used to illustrate this concept. A 4-PSK is a sinusoidal waveform used to carry a digital signal, that has four possible phases . These phases can be (0°, 90°, 180° and 270°), (45°, 135°, 225° and 315°), or others. That is, the assignment of four phases is not unique. Therefore, both Figure 6-21(A) and Figure 6-21(B) represent a 4-PSK signal constellation. In general, the assignments must be standardized on a regional or global basis.

Figure 6-21(B) also represents a 4-QAM signal constellation, therefore, it is reasonable to call the constellation in Figure 6-21(B) a QPSK modulation diagram. In addition, the phase of the star in the first quadrant of Figure 6-21(B) is 45° (π/4), hence, Figure 6-21(B) can also be called as a π/4-QPSK modulation constellation diagram.

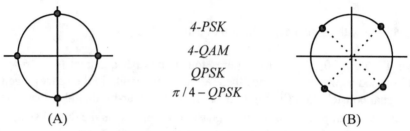

4-PSK

4-QAM

QPSK

$\pi / 4 - QPSK$

(A)   (B)

Figure 6-21   Variation of 4-PSK or 4-QAM.

6.4.4.6 Differential Phase Shift Keying (DPSK)

When the PSK method is used to modulate a digital signal, there is an issue that must be resolved. For wireless services, the signal phase varies while the conversation is in progress. The delay of the signal from the transmitter to the receiver varies, because the receiver is moving and multipath transmission exists.

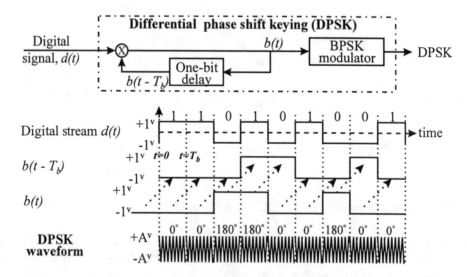

Figure 6-22   Implementation of DPSK Modulation.

By using the phase of an analog signal [e.g., $A \cos (\omega t + \theta_i)$] to represent a digital signal, the receiver must constantly estimate the phase relationship between the transmitter and the receiver. This activity is referred to as **phase estimation**. One of the most common phase estimation techniques is Differential Phase Shift Keying (DPSK), which is illustrated in Figure 6-22. The input digital data stream, $d(t)$, is fed to a differential phase shift keying modulator. The feedback signal, $b(t - T_b)$, is a one-bit delayed version of the signal $b(t)$, which is the product of the two signals, $d(t)$ and $b(t - T_b)$, that is:

$$b(t) = d(t) \times b(t - T_b) \tag{6-16}$$

In Figure 6-22 the digital data stream is assumed to be "110101001...". **The system assumes *the initial state with $b(t) = -1^v$* [for $- T_b \leq t \leq 0$]**, as shown in Figure 6-22. The formation of the combined signal, $b(t)$ is illustrated below:

Digital signal $d(t)$:	$+1^v$	$+1^v$	-1^v	$+1^v$	-1^v	$+1^v$	-1^v	-1^v	$+1^v$
$b(t - T_b)$:	-1^v	-1^v	-1^v	$+1^v$	$+1^v$	-1^v	-1^v	$+1^v$	-1^v
$b(t)$:	-1^v	-1^v	$+1^v$	$+1^v$	-1^v	-1^v	$+1^v$	-1^v	-1^v

Note that the products: $(+1^v)(+1^v) = +1^v$; $(+1^v)(-1^v) = -1^v$; $(-1^v)(+1^v) = -1^v$; and $(-1^v)(-1^v) = +1^v$ have been applied. It can be seen that the signal, $b(t)$, derived for the DPSK modulation is obtained as above [i.e., $b(t) = (-1^v, -1^v, +1^v, +1^v, -1^v, -1^v, +1^v, -1^v, -1^v]$.

After the signal $b(t)$ has been formed, BPSK (described in Section 6.4.4.1; Figure 6-17) is applied. The 1st, the 2nd, the 5th, the 6th, the 8th and the 9th bits of $b(t)$ are assigned a carrier signal with a phase θ_0 of 0° [since the waveform of $b(t)$, in Figure 6-22, is -1^v]. The 3rd, the 4th, and the 7th bits of b(t) are assigned a carrier signal with a phase θ_1 of 180° [since the waveform of b(t), in Figure 6-22, is $+1^v$]. The resulting signal is a DPSK signal, as shown on the bottom graph of Figure 6-22. That is, the modulated signals representing the bitstream of "110101001 ...", using DPSK modulation technique are:

A sin ωt; A sin ωt; A sin (ωt +180°); A sin (ωt +180°); A sin ωt; A sin ωt;
A sin (ωt +180°); A sin ωt; A sin ωt; ...

6.4.5 Comparison of Modulation Techniques

Two measurements used to compare various modulation techniques are: (1) fractional out-of-band power, and (2) probability of error for a given communication environment. The purpose of this section is to provide the reader with sufficient background to determine whether one modulation technique is "better" than another.

The power spectrum of any digital signal assumes the general shape illustrated in Figure 6-23. That is, the signal power spectrum density representing the digital signal frequency characteristics contains two parts: (1) one main lobe, and (2) many side lobes. It can be seen that the power of the main lobe is significantly stronger than side lobes. The parameter, T_b, is the bit interval of the digital signal that is modulated, and the main lobe typically contains 80% to 95% of the total signal energy (depending upon the modulation method).

   Two modulation techniques, QPSK and MSK, are used as examples for describing fractional out-of-band power. Any digital signal possesses an infinite bandwidth (i.e., one main lobe and an infinite number of side lobes as shown in Figure 6-23). However, the analog system used to transport the modulated digital data stream has a finite bandwidth. Therefore, only a fraction of the digital signal power can be transmitted over the analog system. The larger the fraction of digital signal power transmitted, the easier it is to restore the original digital signal. The portion of signal power that is not transmitted for a given system bandwidth (assigned to each customer) is defined as the "fractional out-of-band power". Thus, the fractional out-of-band power is a meaningful measurement of the effectiveness of a modulation technology.

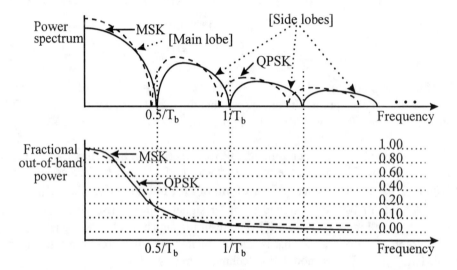

Figure 6-23   Power Spectrum & Fraction of Out-of-Band Power.

   The bottom graph of Figure 6-23 shows the fractional out-of-band power for MSK and QPSK. If the bandwidth allocated to carry the modulated signal is $0.5/T_b$, the MSK modulation technique has a fractional out-of-band power about 0.18 (18%), while the QPSK modulation technique has a fractional out-of-band power about 0.17 (17%). Therefore, QPSK modulation is slightly better than MSK, from a fractional out-of-band power viewpoint for this case. If the analog system has an available bandwidth of $1/T_b$, the fractional out-of-band power is about 0.05 (5%0 and 0.06 (6%) for QPSK and MSK, respectively. Therefore, in this case, MSK is slightly better than QPSK, with respect to fractional out-of-band power.

   Another performance measurement for comparing various modulation techniques is the probability of error for a sinusoidal signal that is transported over an analog system. Figure 6-24 shows the probability of error with respect to environmental condition (i.e., E_b/N_o) for four modulation schemes (QPSK, MSK, FSK, and DPSK). The condition of the communication environment is often expressed as the Signal-to-Noise ratio (S/N) for analog services. For digital services, the condition of the communication environment is expressed as the bit Energy to Noise ratio (E_b/N_o). Consider the case for a communication environment that has an E_b/N_o of 7.5 (see Figure 6-24). For the four modulation methods, Basic (binary) FSK (BFSK, FSK) modulation has an approximate probability of error (P_e) of 6×10^{-3}; Differential PSK (DPSK) has better performance than FSK with $P_e \approx 1 \times 10^{-3}$; and QPSK or MSK has the best performance with $P_e \approx 2 \times 10^{-4}$.

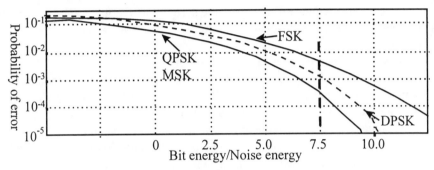

Figure 6-24   Probability of Error versus E_b/N_o.

6.4.6 Bandwidth Management using Multiplexing Techniques

Multiplexing is a technique that is simpler than modulation in concept, and can be used to manage system bandwidth effectively. A system can be analyzed from three different viewpoints: frequency, time, and wavelength. Therefore, various multiplexing methods have been developed accordingly: frequency, time and wavelength division multiplexing schemes.

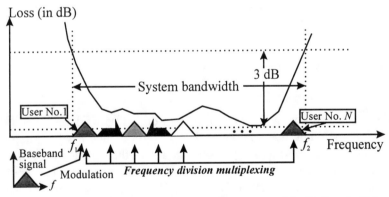

Figure 6-24   System Frequency Response and 3-dB Bandwidth.

- **Frequency Division Multiplexing** (FDM): All systems possess a finite bandwidth, as illustrated in Figure 6-24. There are several "standard" definitions of system bandwidth, however, a 3-dB bandwidth is commonly used in the communication field. The available system bandwidth is typically sufficient for carrying multiple users (channels), as shown in Figure 6-24. Therefore, the system bandwidth can be divided into "N" portions that can be assigned to "N" users. This method is called frequency division multiplexing (FDM). The FDM method can be considered a "*frequency-sharing*" technique, because the system bandwidth (i.e., a range of frequencies) is shared by N users. In addition to multiplexing, modulation is also used to implement FDM in practical systems. This is necessary because the baseband signal (e.g., voice signal, as indicated at the bottom left of Figure 6-24) is outside of the frequency range of the available system bandwidth. A modulation technique is used to "shift" the baseband signal to within the system bandwidth.

A schematic diagram of a FDM transmitter is shown in Figure 6-25. Each user is allocated a bandwidth of f_m, and is modulated by its individual carrier frequency f_{ci} ($i = 1$, N). The modulated signal is fed to a BandPass Filter (BPF) to reduce/eliminate crosstalk between adjacent channels. For example, user No. 1 with a bandwidth of f_m is modulated by a carrier signal with carrier frequency f_{c1}, and the modulated signal is fed to a bandpass filter with a passband of (f_{c1} to $f_{c1}+f_m$). User No. 2 with a bandwidth of f_m is modulated by a carrier signal with carrier frequency f_{c2}. The modulated signal is fed to a bandpass filter with a passband of (f_{c2} to $f_{c2}+f_m$). This procedure is applied to N users. In practical FDM systems, there is a guard band frequency used between two adjacent channels for further crosstalk reduction/elimination.

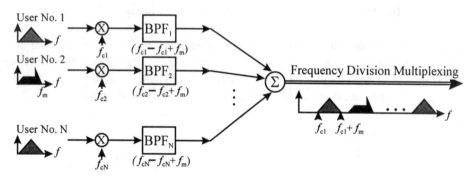

Figure 6-25   Modulation and Frequency Division Multiplexing.

- **Time Division Multiplexing** (TDM): In modern digital networking time is divided into "frames". A typical frame is a time interval of 125-µm, which has been derived from the Nyquist sampling rate [see Eq.(6-2)]. Each frame is further divided into "timeslots", also known as channels. Each time slot is assigned to an individual user. This method is called Time Division Multiplexing (TDM). Therefore, TDM is a "*time-sharing*" technique since a frame (a time interval of 125-µm) is shared by N users.

Figure 6-26 illustrates the concept of TDM. The heart of a TDM system consists of a selector, a clock, and a counter. During the first time slot (TS 1, shown in Figure 6-26), the selector "reads" 8 bits (11111111) from user No. 1. The counter performs the counting function. After 8 bits have been selected from user No. 1 and placed in TS 1, the selector then "reads" the information bits (10101010) from use No. 2 and "places" those bits in TS 2. This procedure is continued until the information bits (11001100) have been placed in TS N, and the multiplexing of one "frame" of information has been completed. The multiplexing process will then proceed to the next frame (i.e., repeats again starting with user No.1). More examples of practical TDM systems are provided in subsequent sections.

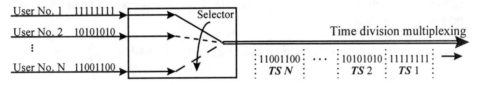

Figure 6-26   Time Division Multiplexing (TDM).

- **TDMA/FDMA**: In digital cellular systems, the multiplexing technique is a combination of "time-sharing" and "frequency-sharing". The technique is called Time Division Multiple Access plus Frequency Division Multiple Access (TDMA/FDMA). Figure 6-27 illustrates the concept of a TDMA/FDMA system. The available system bandwidth is divided into "M" increments, and a frame is divided into "N" time slots (TS 1, TS 2, TS 3, TS 3, ..., TS N). Hence, the capacity of the system is $M \times N$.

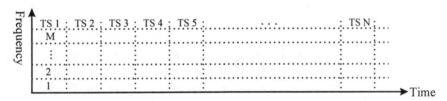

Figure 6-27   Concept of TDMA/FDMA.

- **Wavelength Division Multiplexing** (WDM): WDM is similar to the FDM technique illustrated in Figure 6-25. In an optical fiber system, besides FDM and TDM, WDM technology can also be applied. A WDM system typically applies WDM in addition to TDM. Similar to FDM and TDM, the WDM system applies a "wavelength-sharing" method. A detailed description of a WDM system is provided in Chapter 7.

6.5 THE μ-LAW DIGITAL HIERARCHY

The basic principles of modulation and multiplexing in a digital network have been presented. Now, the technique of combining lower-rate digital signals to form higher-rate signals is explained in the following sections. For example, many DS0 signals with a speed of 64 kbps (described in Section 6.3) can be time division multiplexed to generate a higher-speed DS1 signal. Likewise, several DS1 signals (each having a speed of 1.544 Mbps) can be multiplexed to form a DS1C signal or a DS2 signal. Table 6-7 lists all of the signal characteristics in the μ-law digital hierarchy. The first digital signal level, DS0, with a rate of 64 kbps can be either of the following signals:

- A digitized voice signal, obtained from an analog voice signal with a bandwidth of 4 kHz, sampled at the Nyquist rate of 8 kHz and coded by a μ-law compandor into 8-bit codewords (described in Section 6.3). The rate of 64 kbps is obtained from 8 bits/sample × 8000 samples/second.

- Data signals, with lower rates, can be combined to form a 64-kbps DS0 signal.

Table 6-7   The μ-law Digital Signal Hierarchy.

Signal	System	Service	Rate (Mbps)	Voice channels
DS0	–	–	0.064	1
DS1	T1	T1.5	1.544	24
DS1C	T1C	–	3.152	48
DS2	T2	–	6.312	96
DS3	T3	T45	44.736	672
DS3C	–	–	90.000	1,344
DS4	T4M	–	274.176	4,032

The Digital Signal-level 1 (DS1) was the first digital signal rate designed to be transported over long distances, typically up to 200 miles. The carrier system used to carry DS1 signals is called a T1 digital carrier system, or simply a T1 system. T1 systems are very common in modern digital networking. Regenerators (discussed later in this chapter) with a span of 6000 feet are used along the transmission path of T1 carrier systems. The service provided by a DS1 signal (T1 carrier) is called T1.5 service. Since a DS1 signal can carry 24 voice channels (or voice-equivalent channels) an application that occupies less than the full 24-channels is called a "fractional T1.5 service". A modified DS1 signal, called DS1C, was designed to 48 voice channels using twisted-pair wires. The DS1C signal is not as popular as DS1 signals, and has not been widely deployed (except in North Eastern parts of the US).

The Digital Signal-level 2 (DS2) signal rate was developed in the mid-1960s. At that time, it was intended to transport a Picturephone® signal (the pioneer of teleconference service) or carry four DS1 signals. Today, DS2 signals can only be found within switching offices, and is used for intra-office connections.

The Digital Signal-level 3 (DS3) signal rate is widely used in North American digital networks. DS3 can be carried by coaxial cables, digital radio systems, or optical fiber systems. The most popular optical fiber system, prior to SONET/SDH, was FT-G. The FT-G systems can carry 9, 36 or 72 DS3 signals (using a single fiber) at the corresponding speeds of 417

Mbps, 1.7 Gbps, and 3.4 Gbps (WDM is used for 3.4 Gbps). The service provided by a DS3 signal is known as "T45 service" (based on its speed of 44.736 Mbps). A modified DS3 signal, called DS3C was developed in the late-1970s for transport over optical fiber systems. DS3C contains two DS3 signals, and was commonly used when optical carrier systems were limited to speeds of approximately 90 Mbps.

The Digital Signal-level 4 (DS4) signal rate was developed in the late-1960s. Only a few of these systems were introduced in the New York City metropolitan area. They were known as T4M (M: metropolitan applications). However, optical fiber systems began to displace metallic wire digital carrier systems (e.g., T3 and T4), in the early 1980s. In modern digital networks, T3 and T4 coaxial digital carrier systems are no longer popular for long distance network applications.

Table 6-8   Speeds and Line Codes of the μ-law Signals.

Signal	Line rate (Mbps)	Tolerance (ppm)	Line code	Multiplexing
DS1	*1.544*	± 50	*B8ZS*	*24 channels*
DS2	*6.312*	± 30	*B6ZS*	*4 DS1s*
DS3	*44.736*	± 20	*B3ZS*	*7 DS2s*
DS4	*274.176*	± 10	–	*6 DS3s*

ppm: parts per million
BNZS: Bipolar N-zero substitution

ITU-T Rec. G.733 (DS1)
ITU-T Rec. G.743 (DS2)
ITU-T Rec. G.752 (DS3)

Table 6-8 lists the line rate (speed), tolerance, and line rate for each of the signals in the μ-law digital hierarchy. The line codes are discussed later in this chapter. The standard documentation that specifies DS1, DS2, and DS3 signals/performance are ITU-T Rec. G.733, G.743 and G.752, respectively.

6.5.1 DS1 Signals/T1 Digital Carrier Systems

It has been mentioned earlier that a T1 carrier system can transport 24 voice channels or "voice-equivalent" channels. The DS1 signal (transported by a T1 digital carrier system) format is illustrated in Figure 6-28. A DS1 signal starts with a framing bit, and is followed by 24 channels of payload consisting of : either voice, data, or video. Each channel has a capacity of 8 bits. Therefore, there are 193 bits within each 125-μs frame (i.e., 8000 frames per second). The line rate (speed) of the DS1 signal is derived as follows:

$$Speed = (1 + 24 \times 8) \; \frac{bits}{frame} \times 8000 \; \frac{frames}{sec} = 1.544 \; Mbps$$

For the case of DS1 voice-frequency signal transport, each voice signal is fed to a Pulse Code Modulator (PCM), which was previously described in Section 6.3 (see Figure 6-28). After the PCM processing, the voice signal is represented by an 8-bit codeword every 125 μs (i.e. once each **frame**). Since the voice frequency signal is sampled at the rate of 8 kHz, each voice signal has 8 bits of information transmitted over the T1 carrier system every 125 μs. Therefore, within one-frame interval (125-μs), a T1 digital carrier system is **time-shared** by 24 voice channels. Before transmitting the 24 voice channels, one framing bit is sent to indicate the beginning of each frame, hence there are a total of 193 bits in a DS1 signal (i.e., $193 = 1 + 24 \times 8$ per 125-μs frame interval).

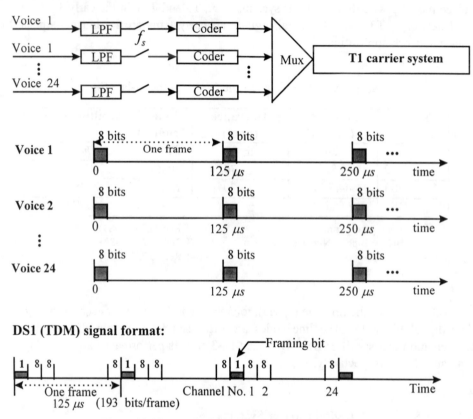

Figure 6-28   DS1 Signal/T1 Carrier System.

6.5.1.1 DS1 Signal Format and DS1 Signal Types

A T1 digital carrier system (i.e., a system carrying a DS1 signal) is based on time sharing. A frame is subdivided into time slots (channels), and each channel is occupied by one user (one voice channel or equivalent). The receiver in a T1 digital system demultiplexes and distributes the data stream to the appropriate users, since the received signal contains many timeslots (users). The receiver must also identify the starting point of each 125-μs frame. Therefore, frame synchronization is established by adding overhead bits to the transmitted data stream. Two different approaches have been developed for frame synchronization as shown in Figure 6-29: (1) SuperFrame (SF; of 12 frames in length) and (2) Extended SuperFrame (ESF; of 24 frames in length), which are described in the following sections.

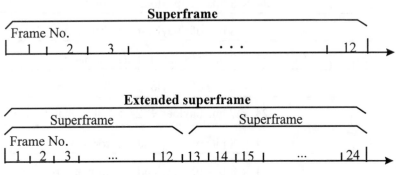

Figure 6-29   DS1 Signal Framing Formats.

6.5.1.2 DS1 Superframe (SF) Framing Format

The original DS1 signal (with a 24-channel capacity) requires 12 frames to establish frame synchronization. An interval of 12 frames (No. 1 to No. 12) is defined as a **SuperFrame (SF)**. This type of DS1 signal framing was introduced in 1960-1961. It is sometimes called "superframe DS1", or "rob-bit signaling DS1". This DS1 signal is different from Extended SuperFrame (ESF) DS1, which was a later enhancement developed for DS1 synchronization.

Each frame has one framing bit (or marker bit) that leads the frame. This framing bit is followed by 24 channels of payload and/or signaling information (see Figure 6-30). The frame structure (or format) of a SF DS1 signal is described as follows:

- Frames Nos. 1 to 5, and 7 to 11 have the following format:
  * One framing bit
  * 24 channels (8 bits each) of payload field

- Frame No. 6 has the following format:
  * One framing bit
  * 24 channels (7 bits each) payload field
  * 24 bits of signaling filed, known as the rob-bit signal scheme (also called the **A** channel signaling bits)

- Frame No. 12 has the following format:
  * One framing bit
  * 24 channels (7 bits each) payload field
  * 24 bits of signaling filed, known as the rob-bit signal scheme (also called the **B** channel signaling bits)

Frame No.	Framing bit	24 timeslots (Information payload field and/or signaling field)
1	1	24×8 bits of payload (information)
2	0	24×8 bits of payload (information)
3	0	24×8 bits of payload (information)
4	0	24×8 bits of payload (information)
5	1	24×8 bits of payload (information)
⑥	1	24×7 bits payload + 24×1 bit signaling
7	0	24×8 bits of payload (information)
8	1	24×8 bits of payload (information)
9	1	24×8 bits of payload (information)
10	1	24×8 bits of payload (information)
11	0	24×8 bits of payload (information)
⑫	0	24×7 bits payload + 24×1 bit signaling

Figure 6-30   Signal Format of a Superframe DS1 Signal.

The frame pattern, used to obtain synchronization for a DS1 signal in a T1 system, is formed by the 12 framing bits (i.e., the value of the first bit in each frame) over the superframe interval (1.5 ms interval = twelve 125-μs frames). The 12-bit frame pattern is given as follows:

$$1\ 0\ 0\ 0\ 1\ 1\ 0\ 1\ 1\ 1\ 0\ 0$$

which can be decomposed into subgroup patterns:

Terminal frame bits: 1 0 1 0 1 0   for **odd** frames   (1, 3, 5, 7, 9, and 11)
Signal framing bits: 0 0 1 1 1 0   for **even** frames (2, 4, 6, 8, 10, and 12)

Note that the *eighth* bit in each of the 24 channels in *Frame Nos. 6 and 12* (Figure 6-30) are "robbed" by the network, and used for control signal transport. Therefore, in each 1.5 ms superframe, 48 bits ($\equiv$ 1 bit/channel $\times$ 24 channels/frame $\times$ 2 frames/superframe) are designated for signaling purposes. Since these bit positions could have been used to carry user information, the SF DS1 format is also called "rob-bit" DS1 signaling.

Zero suppression (i.e., zero substitution, preventing the transmission of long strings of zeros) is required to insure the DS1 signal contains a sufficient level of energy (activity) for timing synchronization functions. The line code, applied to perform zero suppression, used for the SF DS1 signaling is bipolar [i.e., Alternate Mark Inversion (AMI)] code (described later in this chapter). Another line code, known as Bipolar 8 Zero Substitution (B8ZS) is also commonly used for DS1 signals (described later in this chapter).

6.5.1.3 Extended Superframe DS1 Signal Format

It is important to note that in the SF DS1 signaling format, the full eight (8) bit capacity of each time slot is not continuously available to carry user information. Because of rob-bit signaling in frame No. 6 and No.12, only seven out of eight bits are used for payload information during these intervals. For data transport applications prior to the 1980s, "rob-bit" signaling was acceptable. That is, the highest data rate prior to 1980 was:

$$7 \text{ (bits/sample)} \times 8000 \text{ (samples/sec)} \equiv \textbf{56 kbps}$$

In modern communications applications, the need for higher data rate transport has increased dramatically, and 56 kbps data rate is not always adequate.

A capacity of 64 kbps is more desirable not only because of the higher speed, but also because of better frame bit organization. Being able to carry 64 kbps data, defined as a ***clear (64 kbps) channel***, can be implemented by using the extended superframe DS1 signaling format. An Extended SuperFrame (ESF) DS1 signal uses a 24-frame interval (24 × 125-μs = 3 ms) as the basic transport time unit. Each 125-μs frame has a leading bit (called the framing bit) as in the superframe pattern. These bits are called **marker bits**, since not all 24 bits are used for framing. The ESF frame bit organization does not depend upon rob-bit signaling to carry control signaling. Therefore, ESF is called a "clear channel" DS1 signal.

The 24 marker bits spread over the 24 frame interval (one marker bit leading each frame), are divided into three classes (see Figure 6-31):

(1) Framing pattern field: This field consists of 6 bits (instead of 12 bits as in SF DS1 signal) used for frame synchronization purposes. The 6 framing bits are distributed over six frames: Nos. 4, 8, 12, 16, 20 and 24 as shown in Figure 6-31. The resulting frame pattern (of ESF DS1 signal) is given as follows:

$$\textbf{0 0 1 0 1 1}$$

Note that this bit stream has a decimal value of $11 \equiv (32 \times 0) + (16 \times 0) + (8 \times 1) + (4 \times 0) + (2 \times 1) + (1 \times 1)$; this may help the reader remember the frame pattern.

(2) Signaling field: This field is often referred to as the ***data link***. The signaling field replaces the rob-bit signaling strategy used in SF systems. A total of 12 bits are assigned for (control) signaling functions. They are distributed in frame Nos. 1, 3, 5, 7, 9, 11, 13, 15, 17, 19, 21 and 23 as shown in Figure 6-31.

(3) CRC (Cyclic Redundancy Check) field: The principles of CRC are discussed in detail in Chapter 13. The 6 CRC bits are used to detect/correct error(s) in the marker bits. This technique is known as a (24, 18) CRC code or a CRC-6 code. The CRC bits are distributed in frame Nos. 2, 6, 10, 14, 18, and 22 as shown in Figure 6-31.

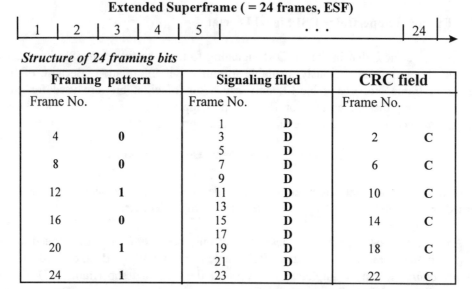

Extended Superframe (= 24 frames, ESF)

Structure of 24 framing bits

Framing pattern		Signaling filed		CRC field	
Frame No.		Frame No.		Frame No.	
		1	D		
4	0	3	D	2	C
		5	D		
8	0	7	D	6	C
		9	D		
12	1	11	D	10	C
		13	D		
16	0	15	D	14	C
		17	D		
20	1	19	D	18	C
		21	D		
24	1	23	D	22	C

Figure 6-31   Signal Format of an Extended Superframe DS1 Signal.

It should be noted that the ESF DS1 signal format is suitable for digital network having common-channel signaling capabilities (e.g., SS7 or CCS7 signaling systems). That is, with the aid of SS7 or CCS7 signaling capabilities, the functions provided by the "data link" in odd frames (Nos. 1, 3, 5, .., and 23), have proven to be adequate. However, some private networks, with the ESF DS1 24-frame structure still adopt a "rob-bit" signaling scheme. In this application, the 8th bit of each channel in frame Nos. 6, 12, 18, and 24 are assigned for signaling. They are referred to as the "A", "B", "C", and "D" signaling channels. This DS1 application can be considered a 3rd type of DS1 signal. Comparisons between SuperFrame (SF) and Extended SuperFrame (ESF) are given as follows:

- T1 digital carrier systems carrying SuperFrame (SF) DS1 signals are older, and some have been partially converted to the Extended SuperFrame (ESF) format. Eventually, ESF DS1 is expected to displace all the existing SF DS1 systems. The ESF DS1with "A", "B", "C", and "D" channel rob-bit signaling may serve as a transitional system.

- In SuperFrame (SF) DS1 signals, the 8th bit in the 6th and 12th frames of each superframe are "robbed" (designated) for signaling purposes. For voice applications, this rob-bit signaling has only a minor effect on transmission quality, and a DS0 voice channel effectively has a speed of:

8 (bit/sample) × 8000 (samples/frame) = **64 kbps**

However, for data applications, only 7 bits out of the 8 bits/channel can be used for data transport. Therefore, the data speed is limited to a maximum speed of:

7 (bit/sample) × 8000 (samples/frame) = **56 kbps**

- In Extended SuperFrame (ESF) DS1 signals, rob-bit signaling is not implemented. Therefore, a true 64 kbps bandwidth is available for each channel in every frame. The name "**clear (64 kbps) channel**" DS1 signal is given to this improved ESF DS1 signal. The line code for this type of DS1 signals is called B8ZS, which is applied to guarantee ones density used for timing synchronization.

Figure 6-32 is an alternative way (i.e., a two-dimensional structure) of viewing the signal format of an ESF DS1 signal. The advantage of this two-dimensional graph is that the overhead bits are shown separately from the payload information field. That is, the first column of this two-dimensional graph represents the overhead (or marker) bits, while the remaining columns are the payload (information) field.

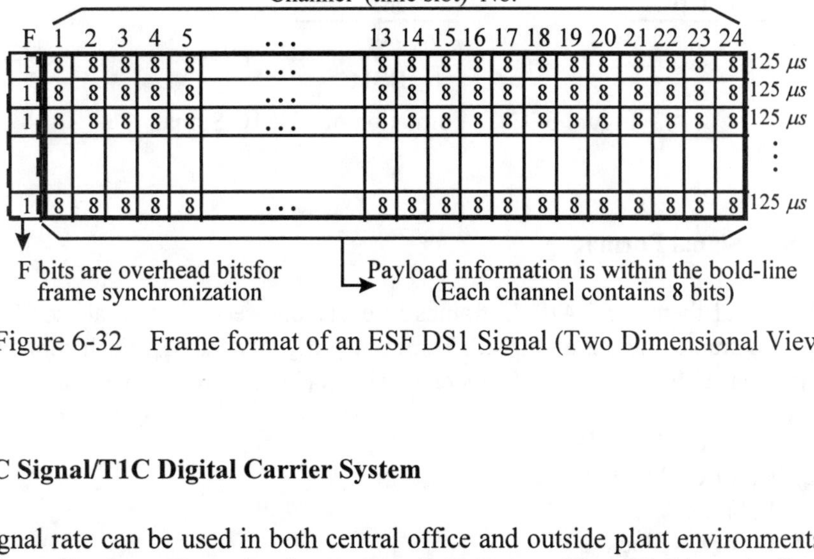

Figure 6-32   Frame format of an ESF DS1 Signal (Two Dimensional View).

6.5.2 DS1C Signal/T1C Digital Carrier System

The T1C signal rate can be used in both central office and outside plant environments, and is similar to T1 applications. However, for modern digital networking, T1C is typically found in the central office environment, where regenerator, powering, and maintenance functions are available. Although T1 facilities (transmission media) can also be used to carry T1C signals, the speed of DS1C is **3.152 Mbps ± 30 ppm** (i.e., twice the speed of DS1).

Two DS1 signals are multiplexed to form a DS1C signal (see Figure 6-33). The second DS1 signal bitstream is bit-inverted to improve signal statistics (i.e., improve the timing recovery capability). A one-bit scrambler is used to "exclusive OR" the present bit with the previous bit to guarantee sufficient one's density. In addition to scrambling, a 50% duty cycle bipolar format (discussed later in this chapter) is applied to the DS1C signal as a line encoding method. Note that "Exclusive ORing" two signals "A" and "B" [i.e. (A⊕B)] has the following results:

$$0 \oplus 0 = 0; \qquad 1 \oplus 0 = 1; \qquad 0 \oplus 1 = 1; \qquad \text{and} \qquad 1 \oplus 1 = 0$$

Pulse stuffing is always required for synchronization whenever low-rate signals are multiplexed to form a higher-rate signal. In this example, the DS1 signals are asynchronous, and two are multiplexed into a higher-rate DS1C signal. Control bits are also inserted in the information payload bitstream when pulse stuffing is used so the receiver can remove the appropriate stuffed bits from the received bitstream.

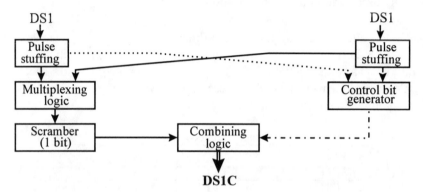

Figure 6-33   Formation of a DS1C Signal.

6.5.2.1 DS1C Signal Format

In a DS1C signal there are 2,478 *M* frames in every one-second interval (see Figure 6-34). Within each *M* frame, there are four **subframes**: M_1, M_2, M_3 and M_4. Each subframe contains 318 bits. In each subframe ($M_1 \sim M_4$) there are six blocks (Figure 6-34) as follows:

(1) M_i block ($i = 1, 4$)
(2) C_{11} or C_{21} block
(3) F_0 block
(4) C_{12} or C_{22} block
(5) C_{13} or C_{23} block
(6) F_1 block

There are three fields in each block:

(1) One overhead bit [M_i (i = 1, 4), C_{ij} (i = 1, 2; j = 1, 3) or F_i (i = 0, 1)]
(2) 26 bits from DS1 No. 1 (bit-interleaved multiplexed with DS1 No. 2)
(3) 26 bits from DS1 No. 2 (bit-interleaved multiplexed with DS1 No. 1)

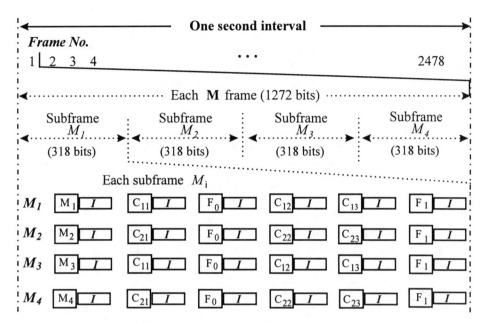

Figure 6-34   DS1C Signal Format.

The DS1C speed can be derived as follows:

$$2478 \times [4 \times 6 \times (1 + 26 \times 2)] \cong 3.152 \quad Mbps$$

where 2,478 is the number of M frames/second, 4 is the number of subframes ($M_1 \sim M_4$), 6 is the six blocks per subframe, 1 is the stuff bit position, 26 is the information bits from each DS1, and 2 is the number of DS1 signals (see Figure 6-34).

6.5.2.2 Control Signals (Overhead) for DS1C Signals

There are three types of control signals required to transport a DS1C signal: M bits, F bits, and C bits as shown in Figure 6-34. In the four subframes, the leading bit is the M control bit. These bits have the format of: (0 1 1 X); where X is the signaling bit. Two functions (frame synchronization and surveillance) performed by the M bits are described as follows:

- **Frame synchronization**: A framing pattern (011) is repeatedly sent by the transmitter. The framing pattern search procedure used by the receiver can be divided into three states:

  (1) "Hunt state": A receiver enters the "hunt state" during the system power-up/testing interval, or after the system has declared lost of frame while in the "sync state". The receiver continuously searches for the framing pattern "bit-by-bit" when in the "hunt state"

(2) "Pre-sync state": After a framing pattern is found, the receiver declares (enters) the "pre-sync state". While in the "pre-sync state", the receiver must detect six (the number "six" may vary with different system implementations) consecutive framing patterns before entering the "sync state". During the "pre-sync state", if a framing pattern is "missed" the system will return to the "hunt state" (shown by the dashed line in Figure 6-35).

(3) "Sync state": Once a system enters the "sync state", it can start transporting the digital signal. If frame synchronization is lost [e.g., the receiver has lost seven consecutive framing patterns (this number "seven" may vary with different system designs)], the system will revert to the "hunt state" (see Figure 6-35). But, if less than seven consecutive framing patterns are lost (in this example), the system will return to the "sync state" (shown by the dashed line in Figure 6-35). During this "hunting" period, only a test signal (e.g., the idle-channel signal) is transported over the digital system. That is, no user information is transported during the "hunting interval".

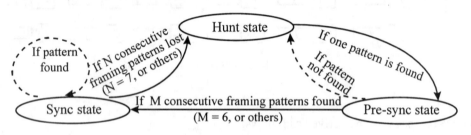

Figure 6-35   Frame Synchronization States.

- **Surveillance**: The second function of the M bits is surveillance. For $(M_1, M_2, M_3, M_4) =$ (011X), the "X" bit serves as the signaling bit. If X = "1", no alarm is present. If X = "0", this indicates there is an alarm condition in the system.

Another control signal is carried by the F bits. The two F bits (F_0 and F_1) alternate between "0" and "1" so that the receiver can identify (locate) the scrambled signals and the control bit time slots. That is, all the control bits ($M_1, M_2, M_3, M_4, F_0, F_1, C_{11}, C_{21}, C_{12}, C_{22}, C_{13}$, and C_{23}) must be properly identified (located).

The third control signal C_{ij} (i = 1, 2; j = 1, 3) is described in the following section, along with "pulse stuffing" used for a DS1C signal.

6.5.2.3 Pulse Stuffing for DS1C Signals

Two DS1 signals are bit-interleaved multiplexed (using Time Division Multiplexing: TDM) to form one DS1C signal (see Figure 6-36). The numbered boxes signify data bits from the 1st DS1 signal, and the numbered hexagons signify data bits from the 2nd DS1 signal.

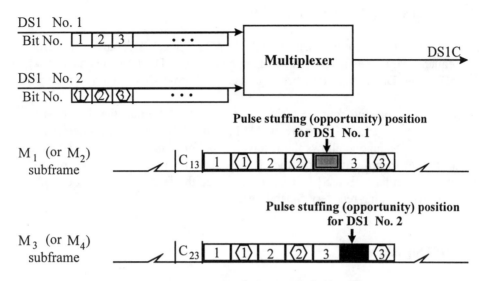

Figure 6-36   Generation of DS1C Signals, and Pulse Stuffing.

As previously described, there are six blocks in subframe M_1 through M_4 (see Figure 6-34). The line rate of a DS1C signal must be maintained at a constant speed, however there are instances when the DS1 No.1 and/or DS1 No.2 signals will have rates that are slower than expected. When this happens, "dummy bits" are stuffed into the DS1C data stream so that the line rate is held at a nearly constant value. This procedure is referred to as ***pulse stuffing***.

When necessary, stuff pulses are inserted at the 5th block of each M_i (i = 1, 4) subframe. That is, the block beginning with C_{i3} (i = 1, 2) is the position where pulses are stuffed when needed to maintain the DS1C "nominal" data stream rate.

- The 5th bit after C_{13}, during the M_1 or M_3 subframe, is the stuff bit position reserved for ***DS1 No.1*** (the first bit after C_{13} is the information bit from DS1 No. 1, the second bit is the information bit from DS1 No. 2, the third bit is the information bit from DS1 No. 1, etc.).

  * If $(C_{11}, C_{12}, C_{13}) = (111)$, stuffing is required.

  * If $(C_{11}, C_{12}, C_{13}) = (000)$, stuffing is not required.

- The 6th bit after C_{23}, during the M_2 or M_4 subframe, is the stuff bit position reserved for ***DS1 No.2*** (the first bit after C_{23} is the information bit from DS1 No. 1, the second bit is the information bit from DS1 No. 2, the third bit is the information bit from DS1 No. 1, etc.).

  * If $(C_{21}, C_{22}, C_{23}) = (111)$, stuffing is required.

  * If $(C_{21}, C_{22}, C_{23}) = (000)$, stuffing is not required.

6.5.3 DS2 Signals/T2 Digital Carrier Systems

There are two different environments in which a DS2 signal is deployed: (1) Digitizing 96 voice-frequency signals using a digital channel bank (e.g., a D4 digital channel bank), and, (2) Multiplexing four DS1 signals together (e.g., using a DDM-1000 digital data multiplexer).

Before multiplexing four DS1 signals into the DS2 format, the data bitstream (logic signals) of DS1 No.2 and No.4 are inverted to improve the overall signal characteristics of the DS2 signal. This technique allows the receiver to perform more accurate timing recovery. In addition to bitstream inversion, the B6ZS line code [Bipolar 6-Zero Substitution (Suppression); similar to B8ZS] is used to guarantee the ones density of the DS2 signal for further improving timing recovery capabilities.

6.5.3.1 DS2 Signal Format

In a DS2 signal there are 5,367 M frames in each one second interval. Within each M frame there are four **subframes**, called M_1, M_2, M_3 and M_4, which is identical to the DS1C digital signal format.

As shown in Figure 6-37, there are six blocks in each subframe ($M_1 \sim M_4$) as follows:

> (1) M_i block ($i = 1, 4$)
> (2) C_{11}, C_{21}, C_{31} or C_{41} block
> (3) F_0 block
> (4) C_{12}, C_{22}, C_{32} or C_{42} block
> (5) C_{13}, C_{23}, C_{33} or C_{43} block
> (6) F_1 block

In each block, there are five fields:

> (1) One overhead bit [M_i, (i = 1, 4), C_{ij} (i = 1, 4; j = 1, 3) or F_i (i = 0, 1)]
> (2) 12 bits from DS1 No. 1 (bit-interleaved multiplexed)
> (3) 12 bits from DS1 No. 2 (bit-interleaved multiplexed)
> (4) 12 bits from DS1 No. 3 (bit-interleaved multiplexed)
> (5) 12 bits from DS1 No. 4 (bit-interleaved multiplexed)

The DS1C speed can be derived as follows:

$$5367 \times [4 \times 6 \times (1 + 12 \times 4)] \cong 6.312 \quad Mbps$$

where 5,367 is the number of M frames/second, 4 is the number of subframes ($M_1 \sim M_4$), 6 is the number of blocks per subframe, 1 is the stuff bit position, 12 is the information bits from each DS1, and 4 is the number of DS1 signals.

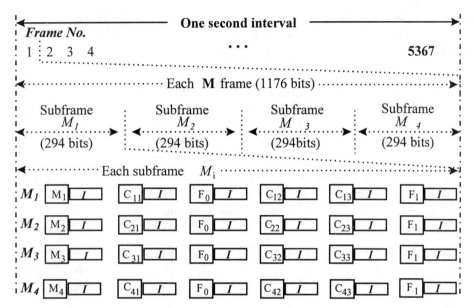

Figure 6-37   DS2 Signal Format.

6.5.3.2 Control Signals (Overhead) for DS2 Signals

There are three types of control signals used to transport a DS2 signal: (1) M bits, (2) F bits, and (3) C bits (Figure 6-37). In the four subframes, the leading bit is the M control bit. These bits take the format of: (0 1 1 X), where X is the signaling bit.

For frame synchronization, a framing pattern (011) is repeatedly sent by the transmitter. The receiver continuously searches for the framing pattern. This frame synchronization function/implementation is identical to that of a DS1C signal.

The "X" bit and the "F" bits perform the same synchronization, surveillance, and control functions as those in DS1C signals (refer to Section 6.5.2.2 for details).

6.5.3.3 Pulse Stuffing for DS2 Signals

A DS2 signal consists of several lower speed digital signals (e.g., DS1 signals). The DS1 signals within a DS2 signal have speeds that occasionally vary. When these signals are slower than expected, the system will stuff "dummy bits" into the outgoing signal (i.e., the DS2 signal) to maintain the line rate of 6.312 Mbps.

In a DS2 signal, the 6th block [i.e., the block led by an F_1 bit in each M_i (i = 1, 4) subframe] is reserved for pulse stuffing. Referring to Figure 6-38, within each M_i subframe, the 6th block is headed by the F_1 overhead bit. The first bit following the F_1 bit is the stuffing position for DS1 No.1, the second is for DS1 No. 2., the third is for DS1 No. 3, and the fourth is for DS1 No. 4. The relationship of the stuff bit assignments are shown as follows:

- For M_1 subframe, the 1st bit after the F_1 is the stuffing position for DS1 No.1.
- For M_2 subframe, the 2nd bit after the F_1 is the stuffing position for DS1 No.2.
- For M_3 subframe, the 3rd bit after the F_1 is the stuffing position for DS1 No.3.
- For M_4 subframe, the 4th bit after the F_1 is the stuffing position for DS1 No.4.

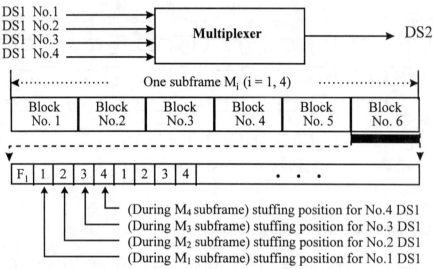

Figure 6-38   Stuffing Positions for Various DS1 Signals within a DS2.

The stuffing control (or indication) signals are assigned as follows:

- For No. 1 DS1

  * $(C_{11}, C_{12}, C_{13}) = (1\ 1\ 1)$, stuffing is required
  * $(C_{11}, C_{12}, C_{13}) = (0\ 0\ 0)$, stuffing is not required

- For No.2 DS1

  * $(C_{21}, C_{22}, C_{23}) = (1\ 1\ 1)$, stuffing is required
  * $(C_{21}, C_{22}, C_{23}) = (0\ 0\ 0)$, stuffing is not required

- For No. 3 DS1

  * $(C_{31}, C_{32}, C_{33}) = (1\ 1\ 1)$, stuffing is required
  * $(C_{31}, C_{32}, C_{33}) = (0\ 0\ 0)$, stuffing is not required

- For No. 4 DS1

  * $(C_{41}, C_{42}, C_{43}) = (1\ 1\ 1)$, stuffing is required
  * $(C_{41}, C_{42}, C_{43}) = (0\ 0\ 0)$, stuffing is not required

6.5.4 DS3 Signals/T3 Digital Carrier Systems

A DS3 signal can be carried by a coaxial cable or optical fiber. When a DS3 signal is carried by coaxial cable, the system is often referred to as a T3 digital carrier system. A common optical system for carrying DS3 signals is the FT-series G (or FT-G). There are various signal sources from which a DS3 signal can be generated:

- From DS1 signals: 28 DS1 signals can be time division multiplexed (TDM) to form one DS3 signal.

- From DS2 signals: seven DS2 signals can be multiplexed to form one DS3.

- From DS2, DS1C and/or DS1: any combination (e.g. one DS2, 20 DS1s and two DS1Cs) can be multiplexed to form one DS3.

For timing recovery/synchronization, a 50% duty cycle B3ZS line code is applied to a DS3 signal.

6.5.4.1 DS3 Signal Format

In a DS3 signal, there are (9,398~9,399) *M* frames in each one second interval. Within each *M* frame (4,760 bits), there are seven **subframes**, designated M_1, M_2, M_3, M_4, M_5, M_6 and M_7 subframes (each subframe contains 680 bits).

As shown in Figure 6-39 in each subframe, there are eight blocks:

(1) X, P, or M_i block (i = 1, 0)

(2) F_1 block

(3) C_{11}, C_{21}, C_{31}, ..., or C_{71} block

(4) F_0 block

(5) C_{12}, C_{22}, C_{12}, ..., or C_{72} block

(6) F_0 block

(7) C_{13}, C_{23}, C_{33}, ..., or C_{73} block

(8) F_1 block

In each block, there are two fields:

(1) One overhead bit field [X, P, M_i, (i = 1, 3), C_{ij} (i = 1, 7; j = 1, 3) or F_i (i = 0, 1)]

(2) Information bit field (e.g., for DS1 applications, there are 84 bits in this field for each of the seven DS1 signals).

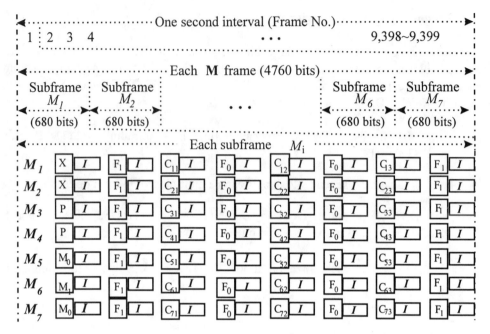

Figure 6-39   DS3 Signal Format.

The DS3 signal speed can be derived as follows:

$$(9{,}398 \sim 9{,}399) \times [7 \times 8 \times (1 + 84)] \cong 44.736 \quad Mbps$$

where $(9{,}398 \sim 9{,}399)$ is the number of M frames/second, 7 is the number of subframes ($M_1 \sim M_7$), 8 is the number of blocks per subframe, 1 is the stuff bit position, and 84 is the number of information bits contributed by each DS1 signal. Note that the identity of information bits will vary if different signal sources (e.g., DS2, DS1C, etc.) are used to form the DS3 signal.

6.5.4.2 Control Signals (Overhead) for DS3 Signals

Five types of control signals are used in a DS3 signal: X bits, P bits, M bits, F bits, and C bits. Their applications are briefly described as follows:

- X bits are used for alarm indication:

    * If (XX) = (00), there is no alarm in the system
    * If (XX) = (11), the system is in alarm conditions.

- P bits are used for parity-check function

    * If (PP) = (00), then Σ (1's in payload) = 0
    * If (PP) = (11), then Σ (1's in payload) = 1

- M bits are used to indicate the beginning of a multiframe interval

  * For multiframe alignment applications: $(M_0, M_1, M_0) = (0, 1, 0)$

- F bits are used for frame alignment

  * For frame alignment applications so the receiver can identify the control bit time slots: $(F_1, F_0, F_0, F_1) = (1, 0, 0, 1)$

- C bits are used for stuffing indication (or control)

  * $(C_{j1}, C_{j2}, C_{j3}) = (1, 1, 1)$ stuffing is required for DS2 No. j (j = 1, 7).

  * $(C_{j1}, C_{j2}, C_{j3}) = (0, 0, 0)$ stuffing is not required for DS2 No. j (j = 1, 7).

  * Note that bit stuffing is performed on a per subframe basis if necessary.

6.5.4.3 Pulse Stuffing for DS3 Signals

Pulse stuffing in a DS3 signal, carrying seven DS2 signals, is briefly discussed in this section. The seven DS2 signals are bit interleaved multiplexed (using TDM) to form one DS3 signal. The pulse stuffing bit is always inserted in the 8th block of each M_i (i = 1, 7) subframe (i.e., in the block which is lead by the 2nd overhead bit F1as shown in Figure 6-40) as follows:

- For M_1 subframe, the 1st bit after the F_1 is the stuffing position for DS2 No.1.
- For M_2 subframe, the 2nd bit after the F_1 is the stuffing position for DS2 No.2.
- For M_3 subframe, the 3rd bit after the F_1 is the stuffing position for DS2 No.3.
- For M_4 subframe, the 4th bit after the F_1 is the stuffing position for DS2 No.4.
- For M_5 subframe, the 2nd bit after the F_1 is the stuffing position for DS2 No.5.
- For M_6 subframe, the 3rd bit after the F_1 is the stuffing position for DS2 No.6.
- For M_7 subframe, the 4th bit after the F_1 is the stuffing position for DS2 No.7.

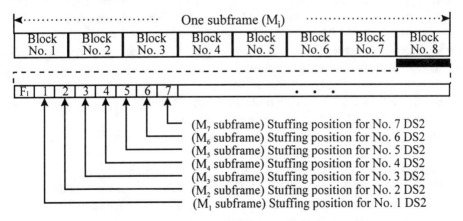

Figure 6-40   Stuffing Positions for a DS3 Signal.

Example 6-7 Referring to ITU-T G.703, ANSI T1.107, ANSI T1.107a, and Bellcore GR499, describe the Master-frame DS3 signal (see Figure 6-39) used to transport an Asynchronous Transfer mode (ATM) signal.

The master frame DS3 signal structure shown in Figure 6-39 is re-drawn in Figure 6-41 to illustrate ATM transport application.

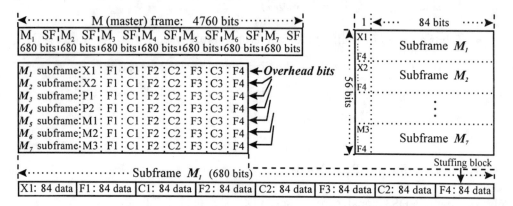

Figure 6-41   Mater-Frame DS3 Format for ATM Mapping.

As shown in Figure 6-41, the DS3 master frame contains 4760 bits. This structure is a block of 56 bits by 85 bits: where 56 = 7 (subframes) × 8 (blocks/subframe), and 85 = 1 (overhead bit) + 84 (data bits/block). For example, in the 1st subframe (M_1) the overhead bits leading the 8 blocks (of data bits) are (X1, F1, C1, F2, C2, F3, C3, and F4), as shown in the left side of Figure 6-41. As in the conventional DS3 structure, the DS3 for ATM applications has a total of 680 bits per subframe, and a total of 4760 bits per frame. Likewise, the 8th block in each subframe is used for bit stuffing to maintain the DS3 line rate of 44.736 Mbps.

The ATM cells are mapped into the 84-bit data field shown on the right side of Figure 6-41. One DS3 frame can carry approximately 11 ATM cells. The DS3 line rate is 44.736 Mbps while the ATM rate is 44.210 Mbps. Each ATM nibble (4 bits in length) is aligned with respect to the overhead bits. The overhead bits (i.e., C bits) are used to perform network management in ATM applications as follows:

- (Subframe M_1) C1: Application Identification Channel (AIC) function (required).
- (Subframe M_1) C2: network requirement bit (required).
- (Subframe M_1) C3: far end alarm and control channel (required).
- (Subframe M_2) C1-C3: reserved for future use.
- (Subframe M_3) C1-C3: parity-check bits (path parity indicator) (required).
- (Subframe M_4) C1-C3: Far End Block Error (FEBE) indication (required).
- (Subframe M_5) C1-C3: terminal-to-terminal path maintenance data link (optional)
- (Subframe M_7) C1-C3: reserved for future use.

Example 6-8: Describe the applications for DS1, DS1C, DS2, DS3, and DS3C digital signals in the North American digital network.

- **DS1 signal applications**: DS1 applications are shown in Figure 6-42. DS1 signals are found between two commercial business buildings, between a business building and a telephone Central Office (CO), and between two COs. In addition, DS1 signals are used as transport signals between a Mobile Switching Center (MSC) and its associated Base Stations (BSs). The T1 digital carrier system (consisting 4-wire twisted-pairs) is often used as the Subscriber Loop Carrier (SLC) system. The subscribers (which can be at rural, suburban area, or metropolitan areas) are connected to a Remote Terminal (RT), which is connected (via DS1) to the local exchange office (a detailed description of digital loop carrier is provided later in this chapter). Note that a toll office can serve as both a local exchange and a toll exchange.

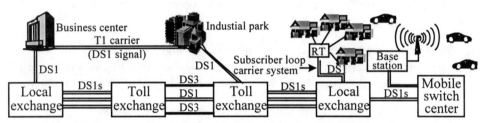

Figure 6-42   Applications of DS1 Signals (T1 Digital Carrier Systems).

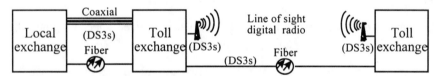

Figure 6-43   DS3 Signal Applications (T3 Digital Carrier System).

- **DS1C signal applications**: DS1C signal applications are similar to DS1 signal's. However, there are exception: (1) DS1C has twice the capacity of a DS1; and (2) it is less popular than DS1 signals (limited deployment in metropolitan areas).

- **DS2 signal applications**: DS2 was introduced as a higher capacity twisted-pair systems for transporting video signals (e.g., Picturephone® services). However, the applications was never realized, and as a result, DS2 signals have been used in the central offices.

- **DS3 signal applications**: DS3 signals have been widely used for long-haul transport. A DS3 signal can be carried by coaxial cable, digital radio (sometimes satellites), or optical fiber. The major applications are inter-office applications (Figure 6-43). For a SONET network, a DS3 signal can be carried by SONET OC-1, or STS-1signal.

- **DS3C signal applications**: DS3C has a rate twice the speed of DS3. DS3C signals are not deployed as wide spread as DS3 signals, and are typically carried by optical fibers.

Review Questions I for Chapter 6:

(1) Prior to SONET/SDH standards, there are three regional digital hierarchy in the PDH environment. They are: _____ , _____ and _____ standards.

(2) Since the modern telecommunications networks apply digital technology, and speech signals are analog in nature, therefore, it is necessary to _____ voice signals for distant transmission. A digital bitstream representing a voice signal typically has a rate of _____ .

(3) Name five primary functions of a Digital Channel Bank (DCB): _____ , _____ , _____ , _____ and _____ .

(4) Due the Nyquist sampling theorem, a speech signal must sample every _____-µs, this interval is an important time reference in modern communications networks, and is called a _____ . Therefore, there are _____ per second.

(5) There are basic two different techniques used to digitized speech signals. One used in North American networks, and is called the ___-law companding method. The other one used in the remaining parts of the world, and is called the ___-law companding method.

(6) Name four digital signals using the µ-law companding technique: _____ , _____ , _____ , _____ and _____ ; four digital signals using the A-law companding technique: _____ , _____ , _____ , and _____ . Several names have been used for E1, such as _____ , _____ , and _____ .

(7) There are two distinct ways to encode sampled/quantized speech signals: the "binary" coding, and the _____ coding algorithms.

(8) From transmission technology viewpoint, two important techniques used in managing system bandwidth for effective bandwidth usage are: _____ and _____ . Modulation methods used for analog signals are: AM, FM and PM, and digital signals are: ____ , _____ , _____ , and _____ . Multiplexing methods available for communications networks are FDM, _____ , and _____ .

(9) The µ-law digital hierarchy is known as a 24-timeslot system since a DS1 can carry ___ digitized voice signals, DS3 can carry _____ (= 28 × ____) digitized voice signals. Likewise, the A-law is known as a ___-timeslot system, but it is used to carry ___ digitized voice signals.

(10) To obtain frame synchronization at the receiver, the transmitter must insert periodically a standardized bitstream along with information bitstream. The bitstream is called a _____ _____ . For SuperFrame (SF) DS1, this pattern is derived from two bits groups: "101010", and "001110". Therefore, the pattern is _____ . However, in an Extended SF DS1, the pattern contains from six bits (i.e., __ __ __ __ __ __) and they are distributed over 24-frame interval in Frame Nos. ___ , ___ , ___ , ___ , ___ , and ___ . Three "states" are defined for frame synchronization implementations: the _____ state, the _____ state, and the _____ state.

6.6 THE A-LAW DIGITAL HIERARCHY

Table 6-9 lists all of the signals (E0, E1, E2, E3, and E4) used in the A-law digital hierarchy. This technology has been widely deployed globally (i.e., outside North America). Table 6-9 also shows the line rate, speed tolerance, line code used for zero suppression, and system capacity of each signal. The standards documentation for A-law PDH signals are: ITU-T Rec. G.732, G.742, and G.751.

Table 6-9   The A-law Digital Hierarchy.

Signal	Line rate (Mbps)	Tolerance (ppm)	Line code	Multiplexing
E-0	*0.064*	–	–	*1 voice channel*
E-1	*2.048*	± 50	*HDB3*	*32 timeslots/30 channels*
E-2	*8.448*	± 30	*HDB3*	*4 E-1s*
E-3	*34.368*	± 20	*HDB3*	*4 E-2s*
E-4	*139.264*	± 15	*CMI*	*4 E-3s*

ITU-T Rec. G.732 (E-1)   ITU-T Rec. G.742 (E-2)
ITU-T Rec. G.751 (E-3)   ITU-T Rec. G.751 (E-4)

The E-0 (also called PS-0 or CEPT-0), is a stand-alone signal, but a building block of the transport signals. The E-0 signal rate of 64 kbps can be obtained as follows:

- A digitized voice signal is obtained by sampling an analog voice signal with a bandwidth of 4 kHz, at the Nyquist rate of 8 kHz, and is then coded using an A-law compandor to form 8-bit codewords (as described in Section 6.3). The rate of 64 kbps is obtained from 8 bits/sample × 8000 samples/second.

- Data signals having lower rates, can be multiplexed together to form a 64-kbps E-0.

The E-1 (also called PS-1 or CEPT-1) signal is the first level digital signal adopted by the ITU-T for short-haul transport (similar to the DS1 signal used in T1 digital carrier systems for μ-law digital hierarchy). The carrier system used to carry an E-1 signal is called a 2-Mbps digital carrier because it has an average line of 2.048 Mbps. The E1 carrier system is popular for modern digital networking in countries other than North America (i.e., T1 carrier is used in North America). Occasionally, E1 is referred to as "European T1". The typical transmission distance for E1 applications is 400 km or less. Twisted pair wires, with regenerators having a span of 2 km are used in E1 transmission applications.

The second-level A-law digital signal, E-2, is also transported over twisted-pair wires. E-2 signals (usually found inside CO are multiplexed into E-3 or E-4 for distant transport.

The third-level A-law signal, E3, is carried by coaxial cables, digital radio, or optical fiber systems. The E-3 signals can interface directly to Synchronous Digital Hierarchy (SDH) networks (discussed later in this chapter).

The fourth-level signal in the A-law digital hierarchy, E-4, was developed in the late 1970s. E4 is the most popular Plesiochronous Digital Hierarchy (PDH) signal used in global digital networking. The E4 signal is typically carried by digital radio or optical fiber systems. For example, TAT-8 and TAT-9 (Trans-Atlantic Transport optical fiber systems) are designed to carry two and four E-4 signals, respectively.

6.6.1 E-1 (2 Mbps) Digital Carrier Systems

An E-1 signal can transport 32 voice channels or voice-equivalent channels. However, the actual number of voice channels carried by an E-1 are 30 or 31 (see Figure 6-44). The E-1 signal rate is derived as follows:

$$Speed = (32 \times 8) \frac{bits}{frame} \times 8000 \frac{frames}{sec} = 2.048 \ Mbps$$

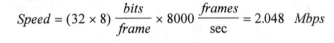

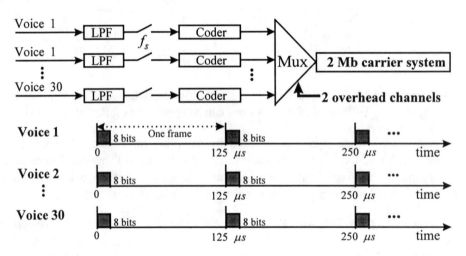

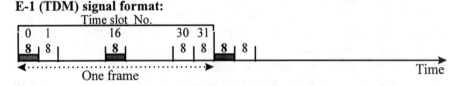

Figure 6-44   E-1 Signal/2 Mb Carrier System.

In the case of voice-frequency signal transport, each voice signal is fed to a PCM as previously described in Section 6.3. At the output of the PCM, each Voice Frequency (VF) signal is represented by an 8-bit codeword every 125 μs (i.e., **one frame**). That is, the voice frequency signal is sampled at the rate of 8 kHz, therefore each voice sample contains 8 bits of information that are transmitted (by the E-1 signal) every 125 μs. Therefore, the frame interval of an E-1 signal is **time-shared** by 30 voice channels (Channel Nos. 1 through 15, and 17 through 30 as shown in Figure 6-44). Note that before transmitting 30 the voice channels, two channels (Channel Nos. 0 and 16) are assigned for overhead functions. The overhead channels are discussed in the following sections. Note that the digital carrier system used to transport an E-1 signal is commonly referred to as a 2 Mbps system, or a "T1E system". This name implies that technology similar to T1 system is used for E-1 signal transport.

6.6.1.1 E-1 Signal Format and E-1 Signal Types

The first-level ITU-T digital signal, E-1, has a bandwidth (capacity) of 32 voice channels or time slots per 125 µs frame. Figure 6-44 shows the E-1 frame structure. An equivalent frame format for an E-1 signal can be represented by a two dimensional matrix as shown in Figures 6-45(A) and (B). Figures 6-45(A) and (B) indicate explicitly that the overhead bytes and payload field are totally separate functions. That is, in Figure 6-45(B), the first column time slots (No. 0, No.16) are dedicated for overhead functionality, and the other columns (No.1 through 15, and No.17 through 31) represent the payload envelope for transporting digitized voice or voice-equivalent signals.

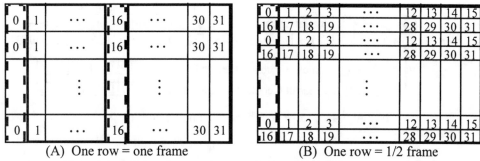

(A) One row = one frame   (B) One row = 1/2 frame

- Channel time slot numbers are shown; each time slot has 8 bits
- Overhead bytes: time slots 0 and 16   • Payload: time slots 1 to 15 and 17 to 31

Figure 6-45   E-1 Signal Format (Two Dimensional View).

Note that within the 32 channel capacity, two channels are assigned for overhead functionality. They are time slots Nos. 0 and 16. The first channel (time slot) is called channel 0 in each frame as shown in Figure 6-46. Channel 0 is designated as an overhead channel, followed by 15 payload time slots containing digitized voices or voice-equivalent signals (i.e., channels 1 through 15). Channel No. 16 is the second overhead channel, followed by channels 17 through 31 containing payload information.

A second type of E-1 signal is referred to as byte-synchronous 31 channel E-1 signal. In this application, the No. 16 time slot is not used for overhead applications, Hence it can carry payload information [i.e., increases the capacity from 30 to 31 voice or voice-equivalent (e.g., data signals) channels].

6.6.1.2 E-1 Overhead Channels

The timeslot 0 overhead channel is located at the beginning of each 125-μs frame (see Figure 6-46). Timeslot 0 performs two separate functions in alternate frames. That is, the first function is implemented (see Figure 6-46) using timeslot No. 0 of **odd** numbered frames, and the second function is implemented using timeslot 0 of **even** numbered frames. These two functions are described as follows:

Case (A) timeslot 0 sync function: Frame synchronization is implemented using timeslot 0 of odd numbered frames. During odd frames, timeslot 0 (8 bits) has the following format: "***X0011011***" [see Figure 6-46 case (A)]. The "X" bit position (value) is reserved for international use. The next seven bits "*0011011*" is known as the "framing pattern". This pattern is generated at the transmitter and is used by the receiver to achieve frame synchronization. Once frame synchronization is established, the receiver consistently identifies the position of timeslot 0 in each frame, along with the relative positions of the following 31 channels contained in each 125-μs frame. Thus, the voice/data information carried in the 30 payload channels can be delivered to the correct destinations.

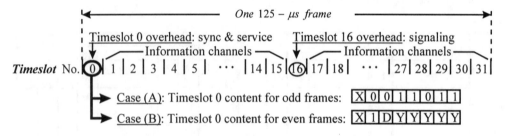

Figure 6-46   E-1 Overhead Channel: Timeslot 0.

Case (B) timeslot 0 service function: A "service word" is implemented using timeslot 0 of even numbered frames. That is, during even numbered frames, timeslot 0 (8 bits) has the following format: "**X1DYYYYY**" [see Figure 6-46 case (B)]. The "X" bit position (value) is reserved for international use, and the "Y" bit positions are reserved for national use. The "1" bit has no special function, however, it may improve the "ones density" of the data stream so the receiver can easily implement timing recovery. The "D" bit position is used as the Alarm Indication Signal (AIS) using the following conventions:

- D = 0 indicates no "urgent alarm" (no critical failures).
- D = 1 indicates an "urgent alarm" exists (a critical failure has occurred).

The timeslot 16 overhead channel is located in the middle of each 125-μs frame (see Figure 6-47). The use of timeslot 16 is recommended for the following applications:

(1) Common channel signaling applications: Timeslot 16 may be used for common channel signaling systems having a maximum rate up to 64 kbps. The method of obtaining signal alignment depends upon the particular common channel signaling specification.

(2) Channel Associated Signaling (CAS) applications: To perform the CAS overhead function using timeslot 16 requires information to be collected for 16 consecutive frames [i.e., an interval of 20 ms (16 × 125-μs) is needed to provide the required overhead capacity]. This 16-frame interval is called a ***multiframe***. The 16 frames are identified as Nos. 0, 1, 2, ..., 14, and 15. In Figure 6-47, the timeslot 16 content for frame No. 0 and frame No. *k* (*k* = 1 through 15) is shown.

Frame No. 0 has the following format "00001N_xD_x1"; and frame No. k (k = to 15) has the following format: "abcdabcd". The overhead functions performed by timeslot 16 are described as follows:

* ***Multiframe alignment signal***: The first nibble (4 bits) of timeslot 16 (during frame 0 of the 2-*ms* multiframe interval) has the pattern "0000", and is used to establish multiframe alignment (see Figure 6-47).

[***Note***: an individual bit position in any byte or channel is sometimes called a ***digital time slot*** to distinguish it from a ***channel time slot***, which refers to an 8-bit (byte) position.]

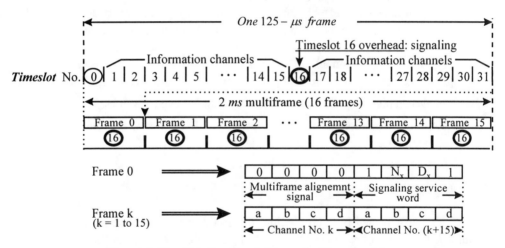

Figure 6-47　E-1 Overhead Channel: Timeslot 16.

* ***Alarm surveillance***: The second nibble (4 bits) of timeslot 16 in frame 0 of the 2-*ms* multiframe interval (see Figure 6-47) is used for alarm indication as follows:

  ◆ D_x = 0 and N_x = 1 indicates no alarm
  ◆ D_x = 1 (N_x = don't care) indicates an urgent alarm
  ◆ N_x = 0 (D_x = don't care) indicates non-urgent alarm

* ***Channel associated signaling***: The 64 kbps signaling overhead (contained in timeslot 16 of frames 1 to 15) is demultiplexed using the multiframe alignment signal as a reference. The information channels and overhead timeslot 16 are described as follows:

♦ The time slots in each frame are designated as Nos. 0, 1, 2, 3, 4, ..., and 31.

♦ Time slot Nos. 1 to 15, and Nos. 17 to 31 are used to carry information payload.

♦ Timeslot Nos. 0 and 16 are assigned as overhead channels.

♦ Information channels are designated as channels 1, 2, 3, 4, ..., and 30. For example, information channel 1 occupies timeslot No.1, information channel 2 occupies timeslot No. 2, ..., information channel 16 occupies timeslot No. 17, information channel 17 occupies timeslot No. 18, ..., and information channel 30 occupies timeslot No. 31.

♦ The bit allocation provides four 500 bps signaling channels, designated "a", "b", "c", and "d", for telephone or voice-equivalent services. For example, if the value of "k" (shown in Figure 6-47) is 1, the first four bits of timeslot No. 16 (the overhead channel) of frame No. 1 (k = 1) are assigned as the "abcd" bits for information channel 1 (k = 1). Simultaneously, the next four bits of timeslot No. 16 (the overhead channel) of frame No. 1 are assigned as the "abcd" bits for information channel 16 (since k+15 = 16 for k =1; see Figure 6-47). If the value of "k" (shown in Figure 6-47) is 2, then, the first four bits in timeslot No. 16 (the overhead channel) of frame No. 2 (k = 2) are assigned as the "abcd" bits for information channel 2 (since k = 2). Similarly, the next four bits in timeslot No. 16 (the overhead channel) of frame No. 2 are assigned as the "abcd" bits for information channel 17 (since k+15 = 17 for k =2, see Figure 6-47). The same assignment procedure is applied for all cases for the values of k from 1 to 15.

♦ When bits b, c, or d are not used they should have the values: b = 1, c = 0, and d = 1.

♦ It is recommended that the pattern 0000 for bits a, b, c and d not be used for signaling purposes in channels 1-15.

6.6.2 E-2 Signals

An E-2 signal is formed by multiplexing four E-1 signals. As in the case of the μ-law digital hierarchy, the speed of a higher rate A-law signals are *not* even multiples of the lower speed signal rates because of overhead bits:

$$\textit{Speed of E-2} = 4 \times \textit{speed of E-1 "+" overhead (OH) bits}$$

That is, the speed of an E-2 signal (8.448 Mbps) is not equal to four times the speed of an E-1 signal (2.048 Mbps), because additional overhead bits must be added to the information bitstream. This is applied for every stage of multiplexing as shown in Figure 6-48. Therefore, the following μ-law and A-law PDH signals rates have been standardized:

The *μ-law*:

DS3 speed = 44.736 Mbps	= 7 DS2 speed + overhead bits	= 7 × 6.312 Mbps + overhead
DS2 speed = 6.312 Mbps	= 4 DS1 speed + overhead bits	= 4 × 1.544 Mbps + overhead
DS1 speed = 1.544 Mbps	= 24 DS0 speed + overhead bits	= 24 × 64 kbps + overhead

The A-law:

E4 speed = 139.264 Mbps	= 4 E3 speed + overhead bits	= 4 × 34.368 Mbps + overhead
E3 speed = 34.368 Mbps	= 4 E2 speed + overhead bits	= 4 × 8.448 Mbps + overhead
E2 speed = 8.448 Mbps	= 4 E1 speed + overhead bits	= 4 × 2.048 Mbps + overhead
E1 speed = 2.048 Mbps	= 30 E0 speed + overhead bits	= 30 × 64 kbps + overhead

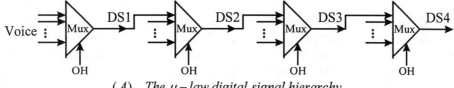

(A)   The μ – law digital signal hierarchy.

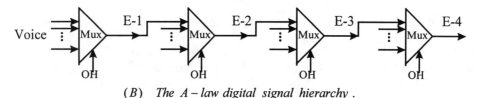

(B)   The A – law digital signal hierarchy .

Figure 6-48   Non-modularity of PDH Digital μ-law and A-law Signals.

6.6.2.1 E-2 Signal Format and Overhead

An 8.448 Mbps E-2 signal carries 132 channels. Each 64 kbps channel contains 8 bits, known as an octet (byte). There are 1,056 bits in each frame (125 μs). A frame carries 132 octet-interleaved 64 kbps channel time slots, numbered from 0 to 131. The assignment of the 132 time slots depends upon the specific application, as described in the following examples:

Case 1: Channel Associated Signaling (CAS) applications

Figure 6-49 shows the signal format of an E-2 signal using Channel Associated Signaling (CAS). That is, call-related control signaling information is carried by some of the 132 timeslots in an E-2 signal. The contents of the 125 μs frame interval are described as follows:

- Payload field: The information payload occupies 120 time slots (64 kbps per timeslot) for voice or voice-equivalent signals, distributed in four groups as follows:

    * E-2 channel time slots Nos.   5 to   32
    * E-2 channel time slots Nos.   34 to   65
    * E-2 channel time slots Nos.   71 to   98
    * E-2 channel time slots Nos. 100 to 131

These time slots are shown in Figure 6-49 within the bold-line boundaries. The mapping of voice-signal channels No. 1 to 120 into channel timeslot is shown in Table 6-10.

- Frame pattern field: E-2 timeslot 0 and the first six bits of E-2 timeslot 66 (a total of 14 overhead bits) contain framing pattern information.

- Service code: Bits 7 and 8 of E2 timeslot No. 66 are assigned for services.

- Signaling: E-2 timeslots 67 through 70 (32 overhead bits) are assigned for Channel Associated Signaling (CAS).

  A multiframe pattern consists of 16 consecutive frames (previously illustrated in Figure 6-47). The multiframe alignment signal, "0000", occupies digit timeslots 1 to 4 of the E-2 channel timeslots 67 through 70 in frame 0. The capacity of each of these four (67 ~ 70) 64 kbps channel timeslots is used to carry lower rate signaling channels (a, b, c, d) of the multiframe alignment signal.

- National use: E-2 timeslots 1 to 4 and 33 (40 bits) are reserved for national use.

- Channel timeslot 99: A CRC-6 procedure is used to monitor "end-to-end" transmission quality of the 8.448 Mbps E-2 signal. At the transmitter, the CRC is computed and inserted in bit positions C_1 to C_6 of channel timeslot No. 99 [see Figure 6-49(B)]. The CRC bit values are computed on a per frame basis. The CRC-6 block size is 132 octets (or 1,056 bits). In addition, the 7th bit (designated E) of channel time slot 99 is used to send the "transmitting direction" indication. This bit indicates whether or not the most recent CRC block arriving at the opposite end (i.e., the receiver) contains errors. The last bit ("S") of E-2 channel timeslot 99 is a stuffing bit used to maintain the 8.448 Mbps E-2 line rate.

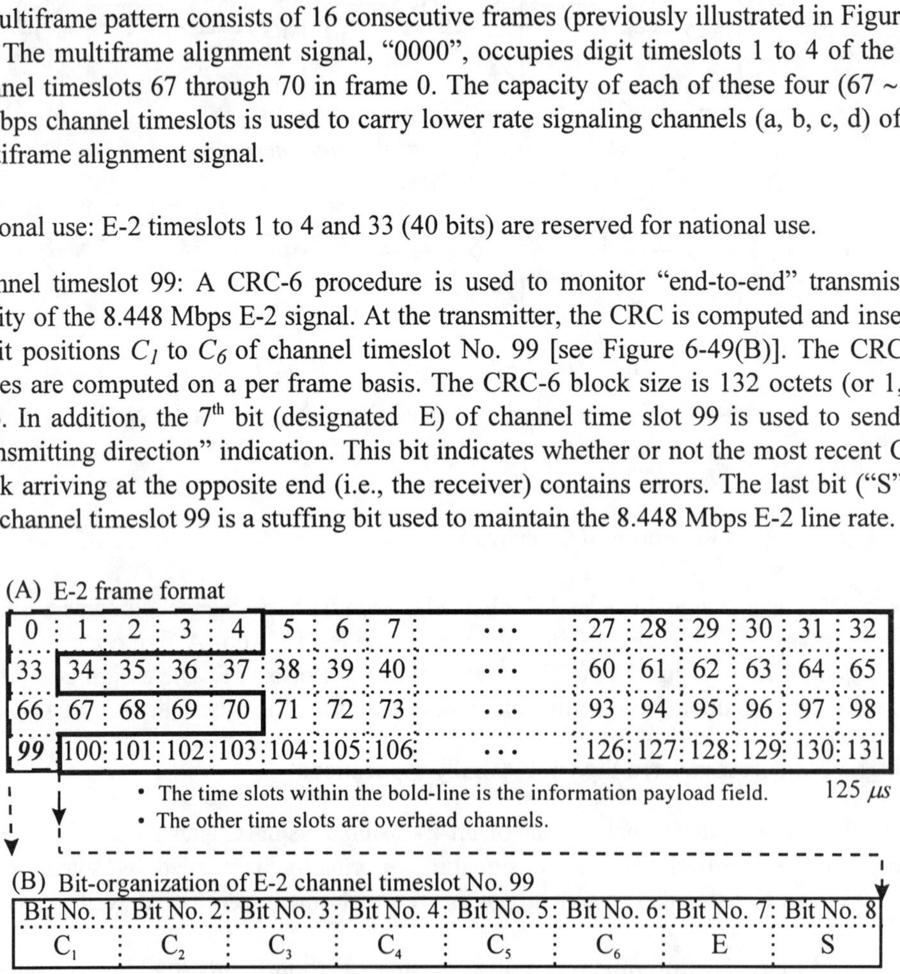

(A) E-2 frame format

0	1	2	3	4	5	6	7	• • •	27	28	29	30	31	32
33	34	35	36	37	38	39	40	• • •	60	61	62	63	64	65
66	67	68	69	70	71	72	73	• • •	93	94	95	96	97	98
99	100	101	102	103	104	105	106	• • •	126	127	128	129	130	131

- The time slots within the bold-line is the information payload field.   125 μs
- The other time slots are overhead channels.

(B) Bit-organization of E-2 channel timeslot No. 99

Bit No. 1	Bit No. 2	Bit No. 3	Bit No. 4	Bit No. 5	Bit No. 6	Bit No. 7	Bit No. 8
C_1	C_2	C_3	C_4	C_5	C_6	E	S

Figure 6-49   Frame Format for a Channel Associated Signaling E-2 Signal.

Table 6-10   E-2 Channel time slots versus voice channel numbers.
(for CAS Applications)

E-2 channel timeslot No.	5	. . .	32	34	. . .	65	71	. . .	98	100	. . .	131
Voice-signal channel No.	1	. . .	28	29	. . .	60	61	. . .	88	89	. . .	120

Case 2: Common Channel Signaling (CCS) applications

Figure 6-50 shows the format for an E-2 signal using Common Channel Signaling (CCS). In this application, a separate signaling network is used to transport call-related control signal information. There are 132 channels (0 ~ 131) per 125-μs frame in the E-2 format (see Figure 6-50). The contents of the 125-μs frame interval are described as follows.

0	1	2	3	4	5	6	7	$\cdots$	27	28	29	30	31	32
33	34	35	36	37	38	39	40	$\cdots$	60	61	62	63	64	65
66	67	68	69	70	71	72	73	$\cdots$	93	94	95	96	97	98
99	100	101	102	103	104	105	106	$\cdots$	126	127	128	129	130	131

- The time slots within the bold-line is the information payload field.   125 μs
- The other time slots are overhead channels.

Figure 6-50   Frame Format for a Common-Channel Signaling E-2 Signal.

- Payload field: The information payload occupies 127 timeslots (64 kbps per timeslot) for voice or voice-equivalent signals, distributed in four groups as follows:

  * E-2 channel time slots Nos.   2   to   32
  * E-2 channel time slots Nos.   34   to   65
  * E-2 channel time slots Nos.   67   to   98
  * E-2 channel time slots Nos. 100   to 131

These timeslots are shown in Figure 6-50 within the bold-line boundaries. The mapping of voice signal channels 1 to 127 into channel timeslots is shown in Table 6-11.

- Channel time slot 1: In accordance with CCS administration, this timeslot may be used for any of the following functions:

  * Carrying another voice signal
  * Carrying another service channel
  * Reserved for service applications within a digital exchange

- Frame pattern field: E-2 channel timeslot 0 and the first six bits of channel time slot No. 66 (a total of 14 overhead bits) contain framing pattern information.

Table 6-11   E-2 Channel time slots versus voice channel numbers.
(for CCS Application)

E-2 channel timeslot No.	2	. . .	32	34	. . .	65	71	. . .	98	100	. . .	131
Voice-signal channel No.	1	. . .	31	32	. . .	63	64	. . .	95	96	. . .	127

- Service code: Bits 7 and 8 of E-2 channel timeslot No. 66 are assigned for services.

- Signaling: E-2 channel timeslots 67 to 70 (32 overhead bits) are assigned for Common Channel Signaling (in descending order of priority) up to a rate of 64 kbps. The method of obtaining signal alignment depends upon the particular common channel signaling specification.

- National use: E-2 channel timeslot 33 (8 bits) is reserved for national use.

- Channel time slot 99: A CRC-6 procedure, identical to the CAS applications previously described for Case 1, is used for CCS channel timeslot 99.

6.6.3 E-3 Signals

An E-3 signal is formed by multiplexing four E-2 signals. However, because of overhead bits the signal rate of an E-3 signal is not four times the speed of an E-2 signal as shown in Figure 6-48(B). That is, the speed of an E-3 signal is given as follows:

Speed of E-3 = 4 × speed of E-2 "+" overhead (OH) bits

6.6.3.1 E-3 Signal Format

There are 537 octets (8-bit timeslots) in an E-3 125 μs frame. These timeslots are designed as: seven overhead bytes, and 530 payload bytes (see Figure 6-51). The speed of an E-3 signal is derived as follows:

(7 + 530) × 8 (bits/octet) × 8000 (octets/frame) = 34.368 Mbps

6.6.3.2 E-3 Overhead Channels

The seven E-3 overhead channel functions are described as follows:

- ***Frame synchronization bytes (frame alignment: FA1 and FA2)***: Frame synchronization bytes FA1 and FA2 have a pattern of "11110110 00101000" (F6 28 hex). This pattern is inserted at the transmitter at the beginning of each 125 μs frame interval. The receiver identifies this pattern to establish frame alignment so that all the other bytes, including the payload information, can be properly located and interpreted.

  It should be noted that this framing pattern, F628 hex is identical to the framing pattern used for SONET/SDH standards (discussed later in this chapter).

- ***Performance byte (Error Monitoring: EM)***: One byte of the E-3 frame is allocated for error monitoring. The even parity computation of this byte is done "bit-by-bit" for all the bytes in the ***previous*** 125 μs frame. Therefore, it is called a Bit-Interleaved Parity-check for 8-bit bytes (BIP-8). The receiver compares this received parity value with the parity computed by the receiver. The result of this comparison indicates the error performance of the connection between the transmitter and the receiver.

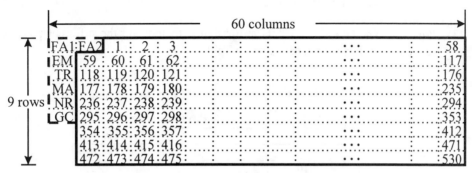

Figure 6-51   Frame Format for an E-3 Signal.

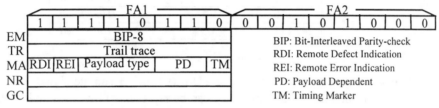

Figure 6-52   Overhead Bytes of an E-3 Signal.

- *Trail TRacing* byte (**TR**): This byte is used to continuously transmit a trail access point identifier so the receiver can verify that it is connected to the appropriate (intended) transmitter. A 16-byte frame (see Table 6-12) is defined for the transmission of the access point identifier. The first byte of the 16-byte frame is a "starter marker", with its value based on a CRC-7 [a generator polynomial of $x^7 + x^3 + 1$] calculation over the *previous frame*. The remaining 15 bytes are used to transport fifteen *T.50* characters (international T.50 reference version) required for accessing the point identifier function.

Table 6-12   Format of the 16-byte Trail Tracing (TR) frame.

Frame starter marker	1	C_1	C_2	C_3	C_4	C_5	C_6	C_7
Byte No. 2	0	X	X	X	X	X	X	X
⋮	⋮	⋮	⋮	⋮	⋮	⋮	⋮	⋮
Byte No. 16	0	X	X	X	X	X	X	X

- *Maintenance and Adaptation* byte (**MA**): This byte performs five different functions:

(1) *Remote Defect Indication (**RDI**)*: An alarm signal is sent back (upstream) to the network element indicating a failure in the opposite transmission direction. RDI is commonly called a "yellow alarm" in North America.

(2) *Remote Error Indication* (**REI**): This bit is set to "1" and sent to the remote trail termination if one or more errors are detected by the BIP-8 byte calculation. If there are no errors, this bit is set to "0".

(3) *Payload Type* (**PT**): Three bits are assigned to indicate the different payload types carried by an E-3 signal, as shown in Table 6-13:

Table 6-13   Payload Type Code Assignments.

Bits 3 to 5	Payload type
000	*Unequipped*
001	*Equipped, non-specified*
010	*ATM*
011	*SDH TU-12s*
100	*(Undefined)*
101	*(Undefined)*
110	*(Undefined)*
111	*(Undefined)*

(4) *Payload Dependent bits* (**PD**): The TU-n (i.e., SDH subrate signal for DS1, E1, or DS2 transport) multiframe indicator can be represented by bits 6 and 7 (PD field) of the "MA byte" (see Figure 6-52).

(5) *Timing Marker* bit (**TM**): If the timing source is traceable to a Primary Reference Clock [PRC; Primary Reference source (PRS)], this bit is set to "0", if not, it is set to "1".

- *Network Operator Byte* (**NOB, NR**): This byte is allocated for maintenance purposes specified by individual network operators. Its "transparency" is not guaranteed. That is, this byte may be modified at an intermediate point, and the "EM" byte parity value must be modified accordingly to ensure performance monitoring integrity. The NR byte can be used for tandem connection maintenance functions. In this application, bits 1 through 4 are used as an incoming error count, and bits 5 through 8 are used as a communications channel.

- *General Communication* byte (**GC**): It is reserved for general purpose communications functions (e.g., to provide a data/voice channel connection for maintenance purposes).

6.6.4 E-4 Signals

An E-4 signal is formed by multiplexing either four E-3 signals, sixteen E-2 signals, sixty-four E-1 signals, or combination. That is, an E-4 signal is 139.264 Mbps derived as follows:

$$\text{Speed of E-4} = 4 \times \text{speed of E-3 "+" overhead bits}$$
$$= 16 \times \text{speed of E-2 "+" overhead bits}$$
$$= 64 \times \text{speed of E-1 "+" overhead bits}$$

The speed of an E-4 signal 139.264 Mbps, is not an equal multiple of E-3, E-2, E1 signals, or combinations of them. Additional overhead bits [e.g., 1.792 ($\equiv$ 139.264 − 4 × 34.364) Mbps for E-3 multiplexing] are added to the information bitstream.

6.6.4.1 E-4 Signal Format

There are 2,176 octets (8-bit channels) in an E-4 signal during each 125 μs frame interval. The channels in an E-4 signal are designated as: nine defined overhead channels (FA1, FA2, EM, P1, TR, P2, MA, NR, GC), seven undefined overhead channels (highlighted as gray boxes in Figure 6-53), and 2,160 payload information channels. The speed of an E-4 signal can be derived as follows:

(16 + 2160) × 8 (bits/octet) × 8000 (octets/frame) = 139.264 Mbps

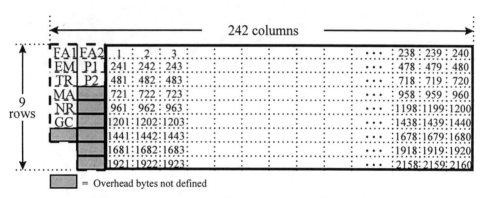

= Overhead bytes not defined

Figure 6-53   (Frame) Format for an E-4 Signal.

6.6.4.2 E-4 Overhead Channels

As previous indicated, 16 channels allocated for overhead use (see Figure 6-54). Seven overhead channels are not yet defined. Nine channels (FA1, FA2, EM, TR, MA, NR, GC, P1, and P2) have functions described as follows:

- *Frame synchronization bytes (frame alignment; FA1 and FA2)*: FA1 and FA2, the frame synchronization bytes, have a pattern of "11110110 00101000" (F628 hex). This pattern is inserted by the transmitter at the beginning of each 125 μs frame interval. The receiver identifies this pattern to establish frame alignment so that all the bytes, including the payload information, can be properly located and interpreted.

- *Performance byte (Error Monitoring, EM)*: One byte of the E-4 frame is allocated for error monitoring. The even parity computation of this byte is done "bit-by-bit" for all the bytes in the *previous* 125 μs frame. Therefore, it is called a Bit-Interleaved Parity-check for 8-bit bytes (BIP-8). The receiver compares this received parity information with the parity value computed by the receiver. The result indicates the error performance of the connection between the transmitter and receiver.

- *Trail Tracing*: The byte is used to continuously transmit a trial access point identifier so that the receiver can verify that it is connected to the appropriate (intended) transmitter. A 16-byte frame (see Table 6-12) is defined to transmit the access point identifier. The first byte of the 16-byte frame is a "starter marker", with its value based on a CRC-7 [a generator polynomial of $x^7 + x^3 + 1$] calculation over the ***previous frame***. The remaining 15 bytes are used to transport fifteen T.50 characters (international T.50 reference version) required for accessing the point identifier function.

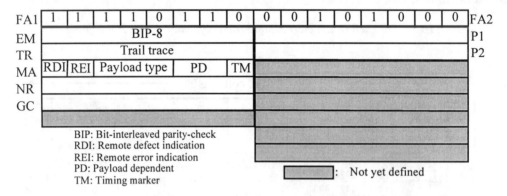

Figure 6-54   Overhead Bytes of an E-4 Signal.

- *Maintenance and Adaptation* byte (**MA**): This byte performs five different functions:

  (1) *Remote Defect Indication* (**RDI**): Refer to RDI bit description for an E-3 signal.

  (2) *Remote Error Indication* (**REI**): Refer to REI bit description for an E-3 signal

  (3) *Payload Type*: See Table 6-14.

  (4) *Payload Dependent bits* (**PD**): Refer to bits description for an E-3 signal.

  (5) *Timing Marker bit* (**TM**): Refer to TM bit description for an E-3 signal.

Table 6-14   E-4 Payload Type Code Assignment.

Bits 3 to 5	Payload type
000	*Unequipped*
001	*Equipped, non-specified*
010	*ATM*
011	*SDH TUG-2s*
100	*2 TUG-3s + 5 TUG-2s*
101	*(Undefined)*
110	*(Undefined)*
111	*(Undefined)*

TUG-2, TUG-3: discussed later.

- *Network Operator byte* (**NOB, NR**): This byte allocated for maintenance purposes specified by individual network operators. Its "transparency" is not guaranteed. That is, this byte may be modified at an intermediate point and the "EM" byte parity value must be modified accordingly to ensure performance monitoring integrity. The NR byte can be used for tandem connection maintenance functions. In this application, bits 1 through 4 are used as an incoming error count, and bits 5 through 8 are used as a communications channel.

- *General Communication* byte (**GC**): One "GC" byte is reserved for general purpose communications functions (e.g., to provide a data/voice channel connection for maintenance purposes).

- *Automatic Protection Switching* (**P1/P2**): These two channels support communication between the transmitter and receiver whenever protection switching is required for recovery from network failures. Their functions are similar to the "K1" and "K2" bytes (discussed later in this chapter) used in SONET and SDH networks.

Example 6-9: Describe applications for E-1, E-2, E-3 and E-4 digital signals in ITU-T digital networks.

- **E-1 signal applications**: E-1 applications are shown in Figure 6-55. E-1 facilities can be found between two business buildings, between a business building and a central office, and between two central offices. In addition, E-1 signals are typically the transport rate used between a Mobile Switching Center (MSC) and its associated Base Stations (BSs) in GSM cellular networks (see Chapter 8). The E-1 digital carrier system, a 4-wire (twisted-pair) system, is often used as the Integrated Digital Loop Carrier (IDLC) system for carrying voice, data, and fax signals within a concentrated subscriber areas. Subscribers located in rural, suburban, or metropolitan areas, are connected to a Remote Terminal (RT) that is connected (via a E-1 carrier system) to the local exchange office.

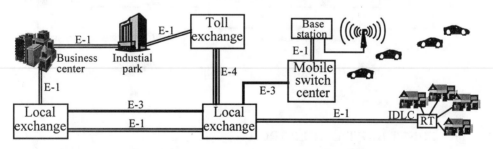

Figure 6-55   Applications of E-1 Signals (2 Mb Digital Carrier Systems).

- **E-2 signal applications**: E-2 was intended to be a high capacity twisted-pair system for "short-haul" services. However, E-2 was never widely deployed because higher speed signals (e.g., E-3) were introduced. Therefore, E-2 signals can only be found within central offices as intermediate signals (E-2 applications are similar to T-2 applications in the US).

- **E-3 signal applications**: E-3 signals have been widely deployed for long-haul transport in ITU-T digital networks. Similar to DS3 signals, an E-3 signal can be carried by a coaxial cable, digital radio (sometimes satellites), or optical fiber. The primary applications are inter-office transport (Figure 6-56). For example, a 560 Mbps optical fiber system can carry up to sixteen E-3 (or E-3 equivalent) signals. Each E-3 signal can carry 480 Voice-Frequency (VF) (or voice-equivalent) signals. In SDH networks, an STM-1 signal is designed to carrier three E-3 signals, and supports enhanced network management functions.

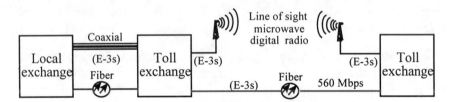

Figure 6-56   E-3 Signal Applications.

- **E-4 signal applications**: E-4 signals are the most common international transport digital carrier systems, with a capacity of 1920 voice or voice-equivalent channels. In undersea digital transmission applications, E-4 signals are carried by high-speed low-loss optical fibers. In the PDH domain, undersea fiber systems have a speed of 140 Mbps, 280 Mbps, or 560 Mbps (equivalent to one, two, and four E-4 signals, respectively). In trans-Atlantic or trans-Pacific undersea SDH networks, an STM-16 signal is designed to carry sixteen E-4 signals with enhanced network management functionality as shown in Figure 6-57.

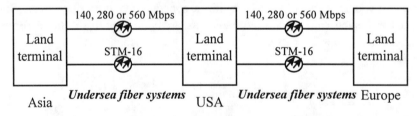

Figure 6-57   Undersea Transoceanic Optical Fiber Systems.

6.7 THE SONET DIGITAL HIERARCHY

The A-law and μ-law PDH digital signal hierarchies (see Figure 6-3) were described at the beginning of this chapter. This section is focused on the SONET digital hierarchy [i.e., SONET signals, formats, and overhead (network management) functionality].

The SONET hierarchy consists of five signals: STS-1, STS-3, STS-12, STS-48 and STS-192 (OC-1, OC-3, OC-12, OC-48 and OC-192 for optical fiber networking). The speed of an STS-1 signal is 51.84 Mbps. The speed of higher-rate SONET signals have the following ***modular*** relationship, which does not exist in the PDH signal hierarchy:

Speed of an STS-N signal = N × speed of an STS-1 signal

where N = 3, 12, 48 and 192.

The five SONET signals and their corresponding speeds are shown in Figure 6-58. The basic SONET signal is STS-1, which consists of three components: (1) the STS-1 *Synchronous Payload Envelope* (SPE), including the STS-1 Path Overhead (POH), having a capacity of 9 by 87 bytes per frame, (2) nine STS-1 section overhead bytes per frame, and (3) eighteen STS-1 line overhead bytes per frame. The STS-1 SPE can be formed from one DS3 signal or seven VTGs (***Virtual Tributary Groups***), plus "one column" of STS-1 path overhead bytes (i.e., nine bytes).

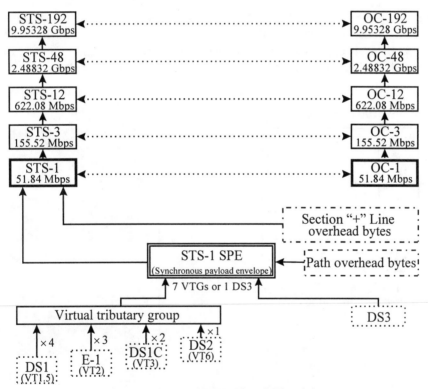

Figure 6-58   SONET Signal Hierarchy

A VTG, having a capacity of 9 by 12 bytes per frame, can contains: four VT1.5s, three VT2s, two VT3s, or one VT6. A VT-n (***Virtual Tributary*** signal; with n = 1.5, 2, 3 or 6) is also called SONET ***subrate*** signal. One VT1.5 signal, with a capacity of 9 by 3 bytes per frame, can carry one DS1 signal. One VT2 signal, with a capacity of 9 by 4 bytes per frame, can carry one E-1 signal. One VT3 signal, with a capacity of 9 by 6 bytes per frame, can carry one DS1C signal. Likewise, one VT6 signal, with a capacity of 9 by 12 bytes per frame, can carry one DS2 signal.

6.7.1 STS-1 Signal Format

The first level (basic) SONET signal is an STS-1 signal, which has a transmission rate of 51.84 Mbps, and a 125 μs frame structure arranged as 9 rows by 90 columns [see Figure 6-59(A)]. Each entry [indicated by "**B**" in Figure 6-59(A)] of this "9 × 90" frame is one byte (i.e., octet; 8 bits). The transmission sequence of the "9 × 90" frame is left-to-right and top-to-bottom. The most significant bit (MSB) of each byte is transmitted first, and the least significant bit (LSB) is transmitted last. Since ***one byte in any SONET signal is 64 kbps***, the speed of STS-1 can be obtained: 9 × 90 × 64 kbps = 51.84 Mbps (where 9 × 90 is the STS-1 frame structure).

The first three columns of each STS-1 frame are defined as the Transport Overhead bytes (which includes the section overhead and the line overhead). The remaining 87 columns (9 × 87 = 783 bytes) is called the STS-1 ***Synchronous Payload Envelope*** (SPE) [see Figure 6-59(A) and (B)]. It will be shown that an SPE does not always begin at column 4 of a STS-1.

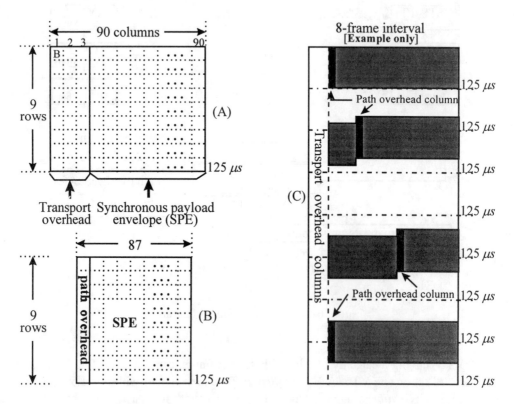

Figure 6-59   STS-1 Frame, SPE and Floating SPEs.

The first column of the STS-1 SPE is defined as the Path Overhead (POH) bytes [Figure 6-59(B)]. The payload field is in *"floating mode"*, therefore the SPE can start in any column between 4 and 90. Figure 6-59(C) represents an 8-frame interval that illustrates the possible locations [i.e., floating mode positions; "shaded areas" in Figure 6-59(C)] of an STS-1 SPE. An SPE can be located entirely within one 125 μs frame (a rectangular shape), or it can straddle two 125 μs frames. However, the STS-1 SPE capacity is always 783 bytes (derived from a structure of 9 by 87 bytes), and can be considered rectangular by "wrapping-around" and using the path overhead as the first column, or simply spread across two 125 μs frames.

An STS-1 SPE can carry PDH signals (e.g., one DS3 signal, seven DS2 signals, 14 DS1C signals, 21 E-1 signals, or 28 DS1 signals). A mixture of DS1s, DS1Cs, and DS2s can also be carried in a STS-1 SPE. Similarly, transport of an E-4 signal, E-3 signals, or a mixture of E1s and E3s is also possible. The transport of PDH signals with a data rate lower than DS3 rate, VTG (described in the next section) must be applied.

6.7.2 Virtual Tributaries and Virtual Tributary Groups

There are four types (or sizes) of virtual tributary signals defined in the SONET standards. They are designed to transport and switch "sub-STS-1 rate" signals. Table 6-15 lists the four virtual tributaries, their rates, and applications. The number of columns of the SPE required for each tributary carried by an STS-1 is also listed in Table 6.15, and shown in Figure 6-60.

Table 6-15   Virtual Tributaries, Speeds and Applications.

Virtual tributary	Rate (Mbps)	Columns required	Applications
VT1.5	*1.728*	*3*	*DS1 or equivalent*
VT2	*2.304*	*4*	*E-1 or equivalent*
VT3	*3.456*	*6*	*DS1C or equivalent*
VT6	*6.912*	*12*	*DS2 or equivalent*

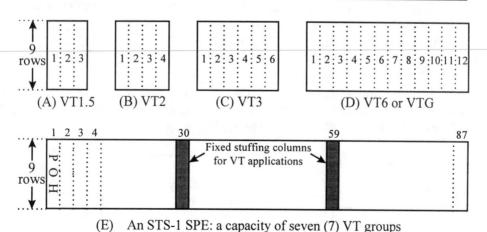

(A) VT1.5   (B) VT2   (C) VT3   (D) VT6 or VTG

(E)   An STS-1 SPE: a capacity of seven (7) VT groups

Figure 6-60   VT Types, VT Group, and SPE and VT Groups.

Table 6-16   VT Groups Column Assignments.

VTG No.	Columns occupied by each VTG within SPE											
1	*02*	*09*	*16*	*23*	*31*	*38*	*45*	*52*	*60*	*67*	*74*	*81*
2	*03*	*10*	*17*	*24*	*32*	*39*	*46*	*53*	*61*	*68*	*75*	*82*
3	*04*	*11*	*18*	*25*	*33*	*40*	*47*	*54*	*62*	*69*	*76*	*83*
4	*05*	*12*	*19*	*26*	*34*	*41*	*48*	*55*	*63*	*70*	*77*	*84*
5	*06*	*13*	*20*	*27*	*35*	*42*	*49*	*56*	*64*	*71*	*78*	*85*
6	*07*	*14*	*21*	*28*	*36*	*43*	*50*	*57*	*65*	*72*	*79*	*86*
7	*08*	*15*	*22*	*29*	*37*	*44*	*51*	*58*	*66*	*73*	*80*	*87*

A capacity of 12 SPE columns is defined as a **Virtual Tributary Group** (VTG). Therefore, a VT6 is equivalent to a VTG as shown in Figure 6-60(D). For VT applications (i.e., for transporting DS1, E-1, DS1C, and/or DS2 signals) an STS-1 SPE has a maximum capacity of seven (7) VTGs, as listed in Table 6-16. In addition to one column of path overhead, there are two ($\equiv 90 - 3 - 1 - 7 \times 12$) unused columns in the STS-1 SPE, as shown in Figure 6-60(E). These "unused columns" are referred to as "fixed stuffing columns", and assigned to column 30 and 59 of the STS-1 SPE. The seven VTGs are numbered 1 to 7. Table 6-16 shows the column assignments for all seven VTGs (note that there are 12 columns per VTG).

The applications for VTG is summarized as follows:

- *One VT group (e.g., VTG No. 1, 2, ..., or 7) can carry one, and only on, type of VT-n signal (where n = 1.5, 2, 3 or 6). No mixture of VT-n signals is allowed within a VTG.* However, a mixture of different types of VTGs can be carried in a STS-1 SPE.

- *One VT group* can be used to carry

  * **Four** (4)   VT1.5 signals (**DS1** signals; 4×1.728 Mbps = 6.192 Mbps)
  * **Three** (3)   VT2 signals (**E-1** signals; 3×2.304 Mbps = 6.192 Mbps)
  * **Two** (2)   VT3 signals (**DS1C** signals; 2×3.456 Mbps = 6.192 Mbps)
  * **One** (1)   VT6 signal   (**DS2** signal; 1×6.192 Mbps = 6.192 Mbps)

Example 6-10:   Determine the number of STS-1 signals (within an STS-N; N = 3, 12, 48 or 192) required to carry the services shown in Figure 6-61 (i.e., seven DS1 signals, five E-1 signals, 3 DS1C signals, one DS2 signal and two DS3 signals).

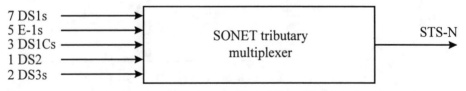

Figure 6-61   VT/VTG and STS-1 SPE Applications.

Table 6-17 shows a systematic approach for designing a solution to Example 6-10. The reasoning that supports these design choices is as follows:

The services involve both DS3 signals and sub-rate signals. It is important to realize that the DS3 signals must be transported separately, because one DS3 signal can be directly mapped into one STS-1 SPE. That is, VTG restrictions do not apply to the transport of DS3 signals. Therefore, a capacity of two STS-1 signals is required for the two DS3 signals.

All the remaining sub-rate signals must be transported on a per VTG basis. One VTG can carry 4 DS1 signals. Since seven DS1 signals need to be accommodated, two VTGs are required, and this arrangement will allow growth of one additional DS1 in the future ($2 \times 4 - 7 = 1$ spare). The VTG applications for the E1, DS1C, and DS2 signals are shown in Table 6-17. Note that in the case of E1 and DS1C, there is also spare capacity for the future growth. In accordance with Table 6-17, a total of seven VTGs are required to carry all the sub-rate signals. Since an STS-1 can contain a maximum of seven VTGs, one separate STS-1 is required to transport all sub-rate signals in this example.

In summary, to transport all the signals shown in Figure 6-61, a total of three separate STS-1 signals (or one STS-3) signal is required.

Table 6-17   VTGs and STS-1 SPEs Assignment.

Signals	VTG capacity	VTGs needed	Notes
7 DS1s	4	2	One spare DS1 for growth
5 E-1s	3	2	One spare E-1 for growth
3 DS1Cs	2	2	One spare DS1C for growth
1 DS2	1	1	No spare capacity
	Subtotal = 7 VTGs required to transport all VT signals; therefore, one STS-1 is required		
2 DS3s	Two DS3 signals require two STS-1 signals		
	Total capacity required = **three** STS-1 signals, or one STS-3 signal		

6.7.3 Path, Line, and Section

An end-to-end SONET connection consists of many network elements. Each element has a specific function and role with respect to network Operations, Administration, Maintenance, and Provisioning (OAM&P). The connection between two network elements can be classified as:

- A path (connecting two Path Terminating Equipment entities; PTEs)
- A line (connecting two Line Terminating Equipment entities; LTEs)
- A section (connecting two Section Terminating Equipment entities; STEs)

The concepts of path, line, and section are illustrated in Figure 6-62. A SONET end-to-end connection has a PDH interfaces on each end (i.e., a transmitter and receiver). Assume one end of the connection accepts 84 DS1 signals as its input. The DS1 signals are multiplexed, converted into a SONET signal format [at location "*A*"], and carried by a 2nd-level OC-3 SONET signal. For long distance transport, regenerators at locations "B", "C" and "D" are applied along the transmission path. Assume that some of the original 84 DS1 signals (i.e., one OC-1; 28 DS1s), are to be "dropped" at location "E" and carried to location "F" before being demultiplexed and converted back into PDH DS1 signals (see Figure 6-62).

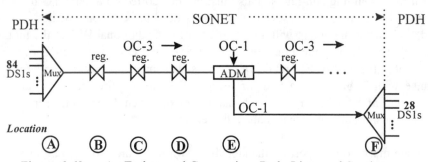

Figure 6-62   An End-to-end Connection: Path, Line, and Section.

- **_Path_**: The connection (link) between location "A" and location "F" is a SONET **path**. Therefore, the network elements at each end of a SONET path are called Path Terminating Equipment (PTE). A PTE can generate, modify, and interpret the path overhead bytes contained in a SONET signal.

- **_Line_**: The connection between two **adjacent** multiplexers is a SONET **line**. The network element at location "E" is an Add/Drop Multiplexer (ADM) and the network element at location "A" is a multiplexer. The connection between "A" and "E" is a SONET **_line_**. The network elements at "A" and "E" are called Line Terminating Equipment (LTE). Similarly, the connection between "E" and "F" is also a **_line_**. The equipment at locations "A", "E" and "F" are SONET LTE. A LTE can generate, modify, and interpret the line overhead bytes contained in a SONET signal.

- **_Section_**: The connection between two **adjacent** network elements is a SONET **section,** if one or both elements is a regenerator. The two elements of a SONET section are referred to as Section Terminating Equipment (STE). The connections "A" to "B", "B" to "C", "C" to "D", and "D" to "E" are SONET **_sections_**. Thus, equipment at location "A", "B", "C", "D" and "E" are SONET STEs that can generate, modify, and interpret the section overhead bytes contained in a SONET signal.

Caution: It should be understood that the link between "E" and "F" is a SONET **_section_** and a SONET **_line_**. The reason that "EF" is a "section" is because any digital receiver has a built-in regenerator. That is, the network element at location "F" consists of a regenerator prior to a demultiplexer.

In summary, the equipment at location "A" is classified as a SONET PTE, LTE, and STE. The equipment at locations "B", "C" and "D" are SONET LTEs only. The equipment at location "E" is classified as a SONET LTE and STE. The equipment at location "F" is classified a SONET PTE, LTE, and STE. Therefore, if a SONET network element is a PTE, it is also a LTE, and a STE. If a SONET network element is a LTE, it is also a STE.

6.7.4 SONET Overhead Channels (Bytes)

The SONET overhead channels assigned for performing SONET network operations, administration, maintenance, and provisioning consume about 4.5% of the total capacity of an STS-1 signal. The overhead bytes are distributed over the transport (section, line) overhead and the path overhead bytes. That is, SONET overhead bytes are classified into three groups:

1. Section overhead bytes: *A1, A2, J0, B1, E1, F1, D1, D2,* and *D3*.

2. Line overhead bytes: *H1, H2, H3, B2, K1, K2, D4, D5, D6, D7, D8, D9, D10, D11, D12, S1, M0,* and *E2*.

3. Path overhead bytes: *J1, B3, C2, G1, F2, H4, F3, K3,* and *N1*.

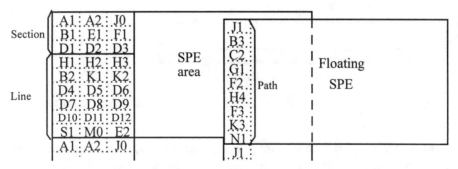

Figure 6-63   SONET Section, Line, and Path Overhead Bytes.

Figure 6-63 shows the position of the overhead channels (bytes). The functional description of each byte is discussed as follows:

- *Frame synchronization*: Every level of SONET signal is led by a framing pattern that is used by the receiver to establish frame synchronization. The framing pattern is carried by two bytes: "*A1*" and "*A2*", and has the pattern "1111 0110 0010 1000" (F628 hex). These are the first two bytes transmitted in a SONET STS-1 signal. For an STS-N signal, *N* "A1" bytes are followed by *N* "A2" bytes. That is, the transmitter will send out "*F6* hex" *N* times and "*28* hex" *N* times for an STS-N signal (where N = 3, 12, 48, or 192). SONET signals are scrambled to improve the signal statistics (i.e., ones density). However, the frame pattern is not scrambled because the descrambler at the receiver needs to be rest periodically, and the frame pattern is used to implement the descrambler reset sequence.

- ***Error (performance) monitoring***: There are three levels of error monitoring for a SONET end-to-end connection: section error monitoring (using the B1 byte), line error monitoring (using the B2 byte), and path error monitoring (using the B3 byte).

  Assume a SONET connection (path) consists of two lines as shown in Figure 6-64. Furthermore, the left-hand line in Figure 6-64 contains two sections. For example, the STE (denoted by *A*) at the "west end" of a SONET section sends the parity-check information byte, B1 (known as BIP-8: bit-interleaved parity check over 8 bit bytes) to the "east end" SONET STE (denoted by *B*). The "east end" STE compares the received B1 byte with the B1 value calculated at the "east end" STE. This is how error monitoring is performed. The same error monitoring procedure is applied to the line and the path connections using B2 and B3 bytes, respectively (i.e., B2 is used between LTEs A and C; and B3 is used between PTEs A and D).

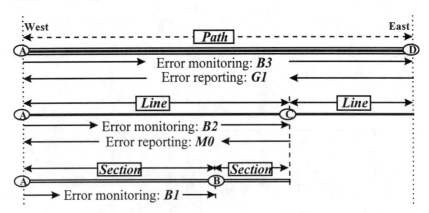

Figure 6-64   SONET Error Monitoring and Error Reporting.

- ***Error reporting***: For SONET line and path connections, in addition to error monitoring, error reporting is required. For example, if the "east end" SONET PTE (denoted by *D* in Figure 6-64) detects errors in the B3 information sent by the "west end" PTE (denoted by *A*), this condition is reported. That is, if the value of the received B3 byte (sent by "west end" PTE A) is different from the calculated B3 byte value (generated by "east end" PTE D), then error reporting information is sent back to the "west end" PTE using the G1 byte, as indicated in Figure 6-64.

  The same error monitoring/reporting procedure is applied to SONET "LTEs" using the B2 and M0 bytes as shown in Figure 6-64.

  Note: Error monitoring is commonly referred to as "forward error monitoring" and error reporting is frequently called "backward error reporting".

- ***Signal label***: One byte, C2, is assigned to indicate the signal type carried by each STS-N signal. For example, when C2 = 00010011, this indicates that an ATM signal is being carried the SONET signal.

- *Data communications channel*: There are two sets of Data Communications Channels (DCCs) allocated as part of the SONET overhead. The first DCC set consists of bytes D1, D2, and D3, which form a 192 kbps channel used by the STE for alarms, maintenance, control, monitoring, administration, and other communication between **section** terminating entities. The second DCC set consists of bytes D4 through D12, which form a 576 kbps channel that is used for the same functions (e.g., alarms, maintenance, control, monitoring, administration, etc.) between **line** terminating entities.

- *Orderwire*: The orderwire channel is a dedicated message (voice) channel used by maintenance personnel. The E1 byte is designated for use in the section connection, and the E2 byte is designated for use in the line connection.

- *User programmable bytes*: One byte, F1, is set aside for user applications within a SONET section. This byte must be passed from one section level entity to another, and it must be terminated by all section level equipment. Two additional bytes, F2 and F3, are provided for the same purpose at the path level.

- *Pointer and pointer action*: The pointer bytes, H1 and H2, are used to perform three functions. The first function is to serve as the "new data flag" to indicate the starting point of a new data/service. The second function is to indicate the exact location of the payload within the STS-1 SPE. The third function is to indicate the need for pulse stuffing. Another pointer byte, H3, is used to perform the actual stuffing function.

- *Tracing*: There is one section trace byte (J0), and one path trace byte (J1). They are used to continuously transmit two separate 64 kbps, fixed-length strings so the receiving terminal (in a path) can verify the connection to the intended transmitter. That is, the J0 and J1 bytes are used for connectivity checking.

- *Automatic protection*: When a failure occurs in an active transmission path, the network switches to a protection (spare) path automatically. Three bytes, K1, K2 and K3, are designated for communications between the transmitter and the receiver whenever automatic protection switching is required.

- *Tandem connection maintenance*: One byte, N1, is designated to perform the tandem connection maintenance function used to isolate the troubles in the transmission path.

- *Synchronization status*: One byte, S1, is designated to report network synchronization status messages.

6.7.5 SONET Multiplexing

SONET STS-1 signals are not used for long-haul transport applications. SONET STS-12 or STS-48 signals are typically used for long distance applications at the present time, and STS-

192 will be used in the future. These signals are typically found in optical fiber networks, and are designated as OC-12, OC-48, and OC-192, respectively. Figure 6-65(A) represents a SONET multiplexing point-to-point connection. Several PDH signals are multiplexed and converted into a SONET STS-1 signal. This STS-1 signal is multiplexed with two other STS-1 signals to form a higher rate STS-3signal used for a short-haul connection. For long distance transport, four STS-3 signals are multiplexed to form a STS-12 signal. Furthermore, four STS-12 (OC-12) signals can be multiplexed to form a STS-48 (OC-48) signal. Both STS-12 and STS-48 signals are usually carried by optical fibers (circle symbol with arrows). The optical carrier signals are called OC-12 or OC-48 [Figure 6-65(A)].

SONET networks often use a ring topology for "self-healing" applications [see Figure 6-65(B); ANSI T1.105.01]. ADM denotes add/drop multiplexer, and in a ring architecture the service (i.e., customer traffic) can be maintained even if a link between any ADMs fails by traveling the ring in either a clockwise (A, B, C, D) or counter-clockwise (D, A, B, C) direction to access the appropriate ADM element.

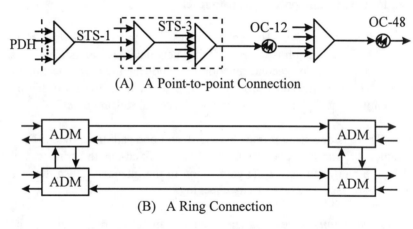

(A) A Point-to-point Connection

(B) A Ring Connection

Figure 6-65   SONET Signal Multiplexing.

6.8 THE SDH DIGITAL HIERARCHY

Technically, the SDH and SONET formats are consistent, but they have adopted different terminologies. Table 6-18 lists the SONET signals and the corresponding SDH signals.

Table 6-18   SDH Versus SONET Terminology.

SONET	STS-192	STS-48	STS-12	STS-3	STS-1
SDH	STM-64	STM-16	STM-4	STM-1	STM-0

SONET	–	VT6	VT3	VT2	VT1.5		Bytes & bits
SDH	TU-3	TU-2	–	TU-12	TU-11		Bytes & bits

As indicated in Table 6-18, SDH has adopted a terminology (i.e., naming convention) that is completely different from SONET. For example, the SDH Synchronous Transport Module (STM-1) signal used to transport a 139.264 Mbps signal (e.g. an E-4) is equivalent (but not identical) to a SONET STS-3 signal. Likewise, the SONET Synchronous Transport Signal (STS-1) signal used to carry a 44.736 Mbps (e.g. a DS3 signal) is equivalent (but not identical) to a SDH STM-0 signal. However, the SDH TU-3 signal used to carry an E-3 (34.368 Mbps) signal does not have a counterpart in the SONET digital hierarchy. Similarly, the SONET VT3 carrying a DS1C (6.312 Mbps) signal does not have a counterpart in the SDH hierarchy. The concept of "SONET is a subset of SDH, or vice versa" is inaccurate.

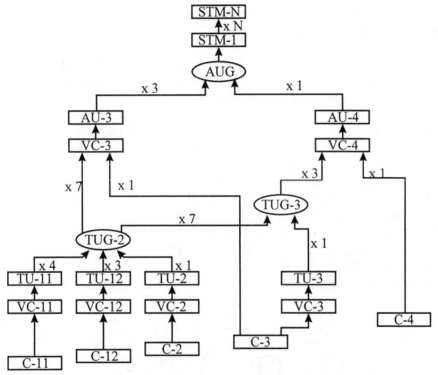

Figure 6-65   SDH Signal Hierarchy.

The SDH hierarchy consists of four signals: STM-1, STM-4, STM-16, and STM-64. Theoretically, there is a fifth signal, STM-0 that is identical to SONET STS-1 signal, but STM-0 is not generally used (the exception being specific applications in Japanese networks). For optical fiber networking, SDH signals are represented by STM-NO (O stands for optical signal; N = 4, 16, 64, etc.). The speed of an STM-1 signal is 155.52 Mbps, which is identical to the speed of a SONET STS-3 signal. The speed of higher-rate SDH signals have the following modular characteristic:

Speed of an STM-N signal = N × speed of an STM-1 signal

where N = 1, 4, 16 and 64.

The SDH signals and their associated sub-rate signals are shown in Figure 6-65. The details of the SDH signal hierarchy are beyond the scope of this book. A reference that describes these various terminologies is found in the book "Understanding SONET/SDH: Standards and Applications", ISBN 0-9650448-2-3, written by Ming-Chwan Chow.

In Figure 6-65, C-n (e.g. C-11) denotes Container-n, VC-n (e.g. VC-11) denotes Virtual Container-n, TU-n (e.g. TU-11) denotes Tributary Unit-n, TUG (TUG-2 and TUG-3) denotes Tributary Unit group-n (n =2 or 3), AU-n (e.g. AU-3) denotes Administrative Unit-n (n = 3 or 4), and AUG denotes Administrative Unit Group.

Since the SDH and SONET formats are technically consistent, only the differences between them will be highlighted:

- An STM-1 signal has a speed that is three times the speed of an STS-1 signal, and the transport overhead of an STM-1 has nine columns instead of three. The STM-1 transport overhead has the following organization (see Figure 6-66):

  * Row 1: A1, A1, A1, A2, A2, A2, J0, R1, R1 (R1s is reserved for national use).
  * Row 2: B1, Δ, Δ, E1, Δ, R2, F1, R1, R1 (Δ is media dependent; R2 reserved for future use; R1s for national use).
  * Row 3: D1, Δ, Δ, D2, Δ, R2, D3, R2, R2.
  * Row 4: H1, H1, H1, H2, H2, H2, H3, H3, H3.
  * Row 5: B2, B2, B2, K1, R2, R2, K2, R2, R2.
  * Row 6: D4, R2, R2, D5, R2, R2, D6, R2, R2.
  * Row 7: D7, R2, R2, D8, R2, R2, D9, R2, R2.
  * Row 8: D10, R2, R2, D11, R2, R2, D12, R2, R2.
  * Row 9: S1, Z1, Z1, Z2, Z2, M1, E2, R1, R1 (Z1s and Z2s for future growth).

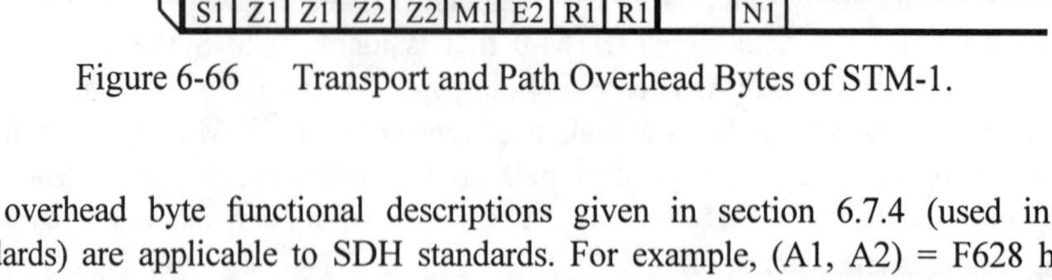

Reg. section overhead												
A1	A1	A1	A2	A2	A2	J0	R1	R1		J1		
B1	Δ	Δ	E1	Δ	R2	F1	R1	R1		B3		
D1	Δ	Δ	D2	Δ	R2	D3	R2	R2		C2		
H1	H1	H1	H2	H2	H2	H3	H3	H3		G1	Path overhead	
B2	B2	B2	K1	R2	R2	K2	R2	R2		F2		
D4	R2	R2	D5	R2	R2	D6	R2	R2		H4		
D7	R2	R2	D8	R2	R2	D9	R2	R2		F3		
D10	R2	R2	D11	R2	R2	D12	R2	R2		K3		
S1	Z1	Z1	Z2	Z2	M1	E2	R1	R1		N1		

(Rows 1-3: Reg. section overhead; Rows 4-9: Mux. section overhead)

Figure 6-66   Transport and Path Overhead Bytes of STM-1.

- The overhead byte functional descriptions given in section 6.7.4 (used in SONET standards) are applicable to SDH standards. For example, (A1, A2) = F628 hex is the framing pattern used by SDH signals to establish STM-1 signal frame synchronization, bytes B1, B2, and B3 are used to monitor transport errors in the transmission link, and D1 through D12 are used for alarms, maintenance, control, monitoring, administration, etc.

- The byte used for line error reporting (in response to B2 byte parity-check error monitoring) is called M0 in SONET STS-1 signals. In SDH it is called M1, which is the same as in the SONET STS-3, STS-12, STS-48, and STS-192 signals.

- A significant difference is the naming convention used for SDH versus SONET connections. Table 6-19 provides a summary of the SONET and corresponding SDH terminology (i.e., "line" is used in SONET while "mux section" is used in SDH).

Table 6-19   Terminology Differences between SDH and SONET Connections.

SONET	*Path*	*Line*	*Section*
SDH	*Path*	*Multiplexer section*	*Regenerator section*

- A SDH TUG-2 is equivalent (but not identical) to SONET VTG (VT group), occupying capacity of 12 columns.

- As shown in Table 6-18, TU-11 is identical to VT1.5, TU-12 is identical to VT2, and TU-2 is identical to VT6, but there is no SDH signal that corresponds to the SONET VT3 signal.

6.9 TRANSMISSION EQUIPMENT/TERMINALS

A typical telecommunications service involves Public Switched Telephone Network (PSTN; previously described in Chapter 2) and Internet (previously described in Chapter 2, further described in Chapter 11). In comparison, a typical wireless service involves both the cellular network and the PSTN, as shown in Figure 6-67. The cellular network for wireless services will be described further in Chapter 8. The signal formats used to carry information throughout the network (cellular and PSTN) have been presented in this chapter. This section describes access systems (provide user access to the PSTN), Network Elements (NEs connect the network), and regenerator equipment (used for long distance transport).A PSTN also requires switching machines (previously described in Chapter 5) and signaling network equipment (previously described in Chapter 4).

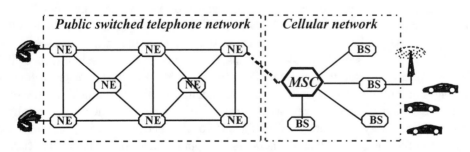

Figure 6-67   An End-to-End Connection for Wireless Service.

Five digital transmission network elements are discussed in this book, and are classified into three application groups as follows:

(1) Access applications:

        ＊ Digital channel banks
        ＊ Integrated digital loop carrier systems

(2) Network applications:

        ＊ Digital cross-connect systems
        ＊ Digital multiplexers

(3) Facility applications:

        ＊ Regenerator

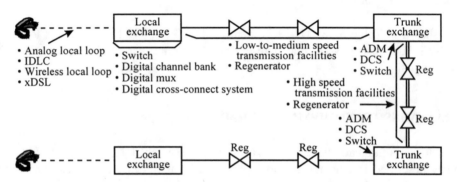

Figure 6-68   Network Elements in a PSTN End-to-End Connection.

These five network elements (besides switches; discussed in Chapter 5) are found in different parts of a PSTN. Figure 6-68 illustrates a typical end-to-end connection of a PSTN.

A customer can be connected to the PSTN via a traditional (analog) loop, an Integrated Loop Digital Carrier (IDLC) system, xDSL (HDSL, ADSL, or VDSL). or a wireless local loop (fixed wireless, or airloop). At the local exchange switching office, a voice signal is digitized and multiplexed with other signals (e.g., voice, data, or video) to form a low-speed digital signal (e.g., 1.544 Mbps DS1 signal, or 2.048 Mbps E-1 signal). A digital add/drop multiplexer may be used to combine several low speed signals into a medium-speed signal (e.g., 34.368 Mbps E-3 signal, or 44.736 Mbps DS3 signal). These medium speed digital signals are typically transported over coaxial cable facilities. For longer distance transport, various stages of multiplexers are used. In general, medium-to-high speed facilities use optical fiber systems, or digital radio systems when terrain does not permit optical fiber installation. Digital cross-connect systems are used in the network to perform routing, grooming, and reconfiguration functions. Note that regenerators are applied throughout the transmission path to insure signal strength/quality is maintained.

6.9.1 Digital Channel Banks

Digital channel banks have been used for networking applications since 1961 when DS1 signals were introduced to carry 24 voice-frequency signals. The name was derived from the channel banks (e.g., A5 and A6 channel banks) used in analog networking. Traditional digital channel banks perform analog-to-digital conversion [Pulse Code Modulation (PCM) using sampling, quantizing, and coding; described earlier in this chapter] and multiplexing of voice signals to form digital data streams [Figure 6-69(A)]. The most common digital channel banks used in North American networks are **D4** channel banks.

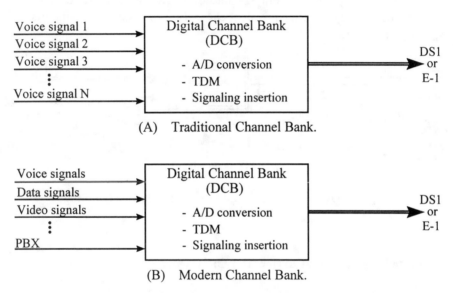

Figure 6-69   A Digital Channel Bank.

For modern applications, advanced digital multiplexing has been merged into traditional digital channel banks. As shown in Figure 6-69(B), a digital channel bank can accept a variety of analog signals that are converted into a digital bitstream. In addition to voice signals, modern channel banks can accept data signals (e.g., a groups of 2.4 kbps, 4.8 kbps, 9.6 kbps, 19.2 kbps, and/or 56 kbps) which are converted to a rate of 64 kbps (i.e., DS0 in North American networks, or E-0 in ITU-T digital networks). This 64 kbps signal occupies one timeslot (or channel) of a DS1 or E-1 signal. In the US, a 64 kbps digital signal representing a mixture of different rate data signals is called a DS0A or DS0B signal. Similarly, a digital channel bank can accept video signals or signals originating from a PBX.

Digital channel banks can be found in the following environments:

- Local exchange offices: The LEC (local exchange carrier) is the point in a network where voice-frequency signals are converted into digital signals for interfacing a Public Switched Telephone Network (PSTN).

- Business offices: Office buildings typically have a large number of voice-frequency, data, and other types of telecommunications signals. It is not economical to connect each individual signal to the switching office. Therefore, the signals are combined into a digital signal stream that is transported to the switching office (i.e., a central office).

Example 6-11 Describe the A-law access equipment (e.g., 8TR 641) used in telecommunications applications.

An 8TR-641 system is typically used in 32-timeslot A-law digital networks. This multiplexer can handle up to 31 voice or voice equivalent channels. The system also supports a variety of services, as shown in Figure 6-70.

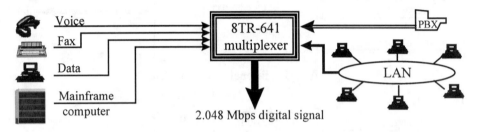

Figure 6-70   Access Multiplexer for A-law Networking.

6.9.2 Integrated Digital Loop Carrier (IDLC) Systems

The development of the Integrated Digital Loop Carrier (IDLC) system was originally intended to provide basic telephone services for rural areas. However, applications for IDLC are also frequently found in metropolitan areas.

As shown in Figure 6-71, the three major components of a traditional IDLC system are:

(1) *Central Office Terminal* **(COT):** The digital signals transmitted by a Remote Terminal (RT) may need to be delivered to different destinations. In general, voice signals are connected to the Public Switched Telephone Network (PSTN), which carries digitized voice signals via T1 carrier or optical fiber systems. However, data signals are often connected to private networks. Therefore, the Central Office Terminal (COT) provides separate interfaces (i.e., PSTN and private networks). *For modern transmission networks, the COT function has been integrated with other transmission equipment (e.g., digital multiplexers or digital cross-connect systems), and is no longer a "stand-alone" component in the switching office.*

(2) *Remote Terminal* **(RT):** A RT is located on the customer premises, where voice signals are digitized and multiplexed with data signals. For business applications, digitized voice

signals are multiplexed with data, and/or video signals. Therefore, a RT effectively performs the function of a digital channel bank (described earlier in this section).

(3) **Digital facility**: The digital facility is used to connect a RT and COT together, and this "long loop" is known as a digital carrier system. Since digital carrier systems have regenerators along the transmission path, the signal quality of an IDLC is much better than an analog loop system.

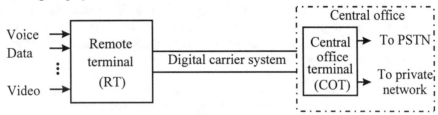

Figure 6-71   Major Components of an IDLC.

The reasons for deploying an IDLC in a network are described as follows:

* **Improve transmission quality**: Since digital carrier systems replaced traditional analog loop facilities, transmission quality has improved considerably. A comparison of analog and IDLC loop signal attenuation for voice band signal is shown in Figure 6-72.

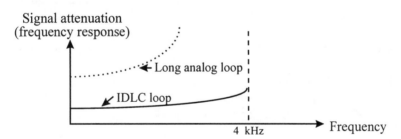

Figure 6-72   Transmission Quality Comparison.

* **Gain wire-pairs**: A typical digital carrier system in North America carries 24, 672, or even greater numbers of voice channels. The number of wires required to separately connect customers to the central office would be enormous. By using multiplexing techniques, an IDLC reduces wire connection density considerably, and is sometimes called a "pair gain system". Another common name for an IDLC is Subscriber Loop Carrier (SLC), because the system replaces traditional subscriber analog loops.

* **Relieve duct congestion**: When an IDLC is applied in metropolitan area, the advantage of "pair gain" can be used to reduce the number of wires in underground cable ducts.

* **Reduce communications costs**: All of the advantages previously described for an IDLC will result in lower "per user" costs.

Modern IDLC systems, such as the type used in the Next Generation Lightwave Networks (NGLNs), have the functional block diagram shown in Figure 6-73. The two major components in this block diagram are described as follows:

(1) Remote terminal: The modern remote terminal replaces the original remote terminal of an IDLC, and has the following capabilities:

* Access Resource Manager (ARM): The ARM is a digital multiplexer that can be used to multiplex digitized voice signals (e.g., POTS), data /video signals (e.g., ISDN), and DS1 signals to form a higher-speed signal (e.g., STS-3), as illustrated in Figure 6-73. In addition, the ARM has the ability to add and/or drop traffic like an ADM.

* Digital Channel Units (DCUs): A DCU performs the same functions as the digital channel bank previously described (see Figures 6-69 and 6-70) in this chapter.

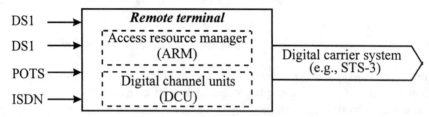

Figure 6-73   A Modern IDLC used for NGLN Applications.

(2) Digital carrier system: The digital facility is a long distance carrier system, and consists of:

* Multiple T1 carrier lines
* An STS-3 digital carrier system
* An OC-3 optical fiber carrier system

Modern IDLCs used in metropolitan areas are often referred to as Business Remote Terminals (BRTs). A BRT is a cabinet that contain various transmission equipment for use in either a SONET or SDH environment (e.g., NGLN applications).

6.9.3 Add/Drop Multiplexer (ADM)

A digital ADM can be configured in three different operational modes:

(1) *Terminal mode*: When a digital ADM is used as a multiplexer or a demultiplexer for point-to-point applications, it is operating in "terminal mode" [see Figure 6-74(A)]. In this mode, the equipment can multiplex many lower-rate tributaries (e.g., DS1, DS1C, DS2, and/or DS3) into a higher-rate digital signal (e.g., OC-3, OC-12, or OC-48). Simultaneously, this equipment can demultiplex a higher-rate digital signal to derive

many lower-rate digital signals. Traditional multiplexers have operated using this principle for many years. In the future, the lower-rate tributaries will be a combination of voice-frequency, voice-grade data, and/or video signals.

(2) ***Regenerator mode***: A "built-in" regenerator is part of the receiver in a digital ADM. Therefore, an ADM can be configured to operate in a regenerative mode [see Figure 6-74(B)]. Digital regenerators are discussed later in this chapter.

(3) ***Add/drop mode***: A modern digital multiplexer can be used as an ADM with an input port carrying a high-rate digital signal such as an OC-12. In the ADM mode, lower-rate tributaries (e.g., DS1 signals) can be "***dropped***" along the transmission path [see Figure 6-74(C)]. Likewise, the ADM can "***add***" (pick up) lower-rate tributaries and insert them into the higher-rate digital signal.

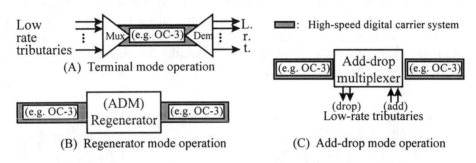

Figure 6-74   Different Modes of Operations for an ADM.

6.9.4 Digital Cross-connect System (DCS)

A Digital Cross-connect System (DCS) is often defined as: "***A DCS is a slow switching machine designed to perform cross-connect functions electronically. It functions like a digital switch, but has a blocking probability of '0' (i.e., it is a non-blocking slow switch)***". Some modern digital switches have a "built-in" DCS to perform both "slow" and "fast" switching functions. The external interfaces of a DCS are typically digital facilities (e.g., DS1, DS3, STS-1, OC-3, etc.). The major applications of a DCS are:

- ***Traffic consolidation***: There are occasions when several digital facilities are not carrying the maximum capacity load (traffic). In this case, it is economical to consolidate all the partially filled facilities into one (or more) common facility for transmission. For example, in Figure 6-75 the two input facilities, A_I and B_I, are partially filled. Since DCS performs the cross-connection function electronically, it can "***move***" (map) any channel of any input port to any channel of any output port. Thus, the DCS can "***move***" the channels from the A_I and B_I input ports (see Figure 6-75), to the single output port, A_O. This concentration function is particularly useful when the facilities have very high speeds (e.g., OC-3). Consolidating traffic to effectively produce an "extra" OC-3 facility results in a capacity increase of 155.52 Mbps (e.g., increasing capacity to carry an additional 2,016 voice channels, without physically adding more equipment). For a broadband DCS, that interfaces with higher rate signals (e.g., OC-12 or OC-48), the savings from consolidation is even more sigificant.

- ***Traffic segregation***: Digital facilities are designed to carry voice, data, and/or video signals. However, the service quality requirement for each type of signal is different. If all signal types are carried on the same facility, the facility must meet the most stringent quality requirement. However, if the signals are segregated (groomed), data signals can be carried on one facility and voices signals carried on a separate facility. For example, D_I and E_I inputs to the system in Figure 6-75, are both carrying a mixture of voice and data signals. Data signals typically require higher quality than voice signals so that the received data can be accurately restored. If a DCS was not applied, both output facilities would need to meet data signal quality requirements. However, by "grooming" voice from data (i.e., all the data traffic is routed to the output facility D_O, and all the voice traffic is routed to the output facility E_O), only one facility (i.e., D_O) is required to meet data quality standards. Since data quality facilities are expensive, traffic grooming obviously has economic advantages.

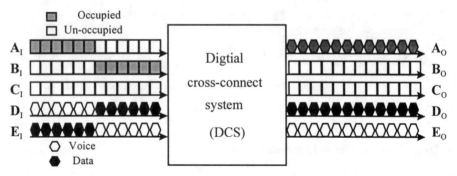

Figure 6-75   A Digital Cross-connect System (DCS).

- ***Test access***: A DCS can provide test signals to any desired facility for measuring and/or analyzing performance conditions. Figure 6-76 shows two facilities that need to be tested. A test signal (or test sequence) can be cross-connected to these facilities. The receiving ends of the facilities are monitored, and the results can be used to determine the condition of these facilities. This test access capability supports both "corrective" (e.g., responding to error reports) and "preventive" performance measurements (e.g., periodic testing).

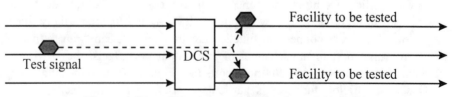

Figure 6-76   Test Access Application of the DCS.

- *Network restoration*: A DCS can perform the cross-connect function by moving any tributary from one facility to another facility. Therefore, a faulty facility can be isolated and removed from service as needed. Likewise, if a failure occurs in the "normal route", traffic can be re-routed (via DCSs) over an alternate path (see Figure 6-77) to maintain service.

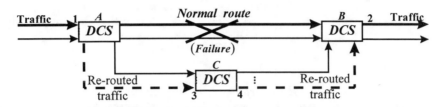

Figure 6-77   Network Restoration Using DCSs.

- *Centralized operation*: A DCS can "route" any signal from one port to another. In Figure 6-78, the network management information collected by Operations Systems (OSs) in various central offices (e.g., A, B, D, and E) can be routed via DCSs to a centralized OS located in central office C. Centralized processing reduce operational costs and increase efficiency.

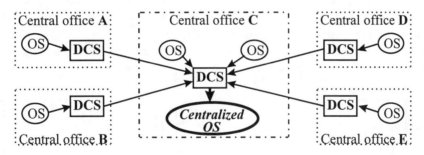

Figure 6-78   Centralized Operations Using DCSs.

6.9.5   Digital Regenerators

Long-distance transport systems use either repeaters or regenerators to maintain signal quality (see Figure 6-79).

- Repeaters [Figure 6-79(A)]

  * Application: analog networking
  * Function: amplification

- Regenerators [Figure 6-79(B)]

  * Application: digital networking
  * Functions:

  – Amplification/equalization (commonly called re-amplification)
  – Re-timing
  – Re-generating

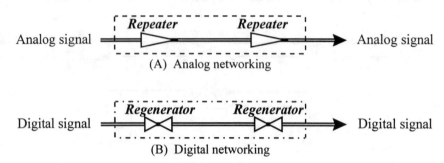

Figure 6-79   Long-distance Transport Systems with Repeaters/Regenerators.

Since modern networks use digital technology, only digital regenerators are described in this book. Assume a bipolar signal is originated at point "A", as shown in Figure 6-80. When this bipolar digital signal is transmitted over a long distance, the signal degrades (i.e., becomes smaller in amplitude and corrupted with noise) as shown at point "B" (Figure 6-80). To allow further transmission, this degraded signal must be "*regenerated*". With installation of an appropriate regenerator at this location, the degraded signal can be completely restored (i.e., regenerated without any erroneous bits).

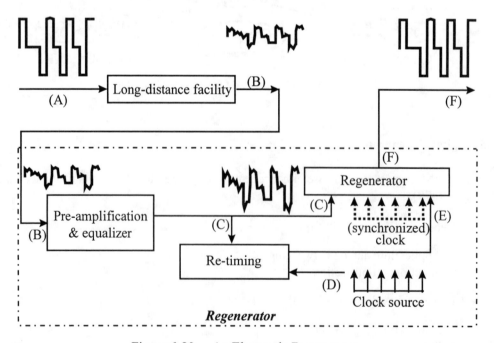

Figure 6-80   An Electronic Regenerator.

To understand the internal functions of a regenerator, it is necessary to know the digital signal restoration process used by the receiver of a digital link. Assuming a polar signal is used (i.e., a logical "1" is represented by 3 volts and a logical "0" is represented by −3 volts), the received signal is a degraded polar wave (see Figure 6-81). It requires a threshold voltage and a sampling clock (to provide decision making instances) at the receiver to interpret if the received voltage should be restored as a logical "1" or "0". If the received voltage at the sampling time is greater than the threshold voltage, the restored signal is a logical "1". Likewise, it the received voltage is smaller than the threshold voltage, the restored signal is a logical "0". In this example, two different clocks [with the **same clock rate** (i.e., the **clock accuracy**), but with different "starting point" (i.e., two clocks are **not in sync**)] are used to restore the degraded (received) signal. The restored signal using clock No.1 is "1010...", but the restored signal using clock No.2 is "0010...". It is clear that one of these two bitstreams contains erroneous bits. Therefore, it can be concluded that the errors are caused by improper clock synchronization. In summary, to restore a degraded digital signal at the receiver, the receiver clocks must meet the following two conditions:

(1) A specific accuracy (e.g., 10^{-8}, described in Chapter 14)
(2) Synchronized to a reference source (typically the incoming bit stream)

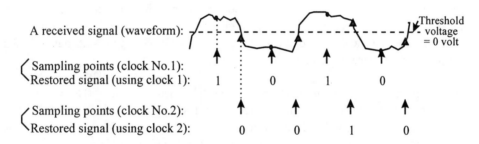

Figure 6-81   Signal Restoration: Clock and Clock Synchronization.

The internal functions of a digital regenerator are illustrated in the bottom half of Figure 6-80. The first step is to pre-amplify and equalize the signal [i.e., the degraded signal shown by the waveform (B)] to produce a string of clean transitions (from "1" to "0" and/or from "0" to "1"). This function is sometimes referred to as "re-amplification". The pre-amplified and equalized signal (C) is then used to synchronize the receiver clock. The clock at the receiver may have the required accuracy [e.g., 10^{-8} as shown in (D)], but it may not be synchronized to the incoming data stream (C), which serves as the reference clock. The synchronization (i.e., re-timing) function is very important for proper signal regeneration as illustrated in Figure 6-81. After the clocks are synchronized (E) with the incoming signal (C), they can be used to regenerate the pre-amplified/equalized signal. This is because the "*accurate and synchronized*" clock can provide "*precise decision making points*" to interpret if the incoming voltage should be restored as a logical "1" or "0" using threshold voltages. The output of the regenerator (F) is a "clean signal" that is equivalent to the original signal (produced by the far-end transmitter) provided channel noise did not cause any erroneous bits.

The electronic regenerator shown in Figure 6-80 is often referred to as a "3R" (Re-amplification, Re-timing, and Re-generating) "regenerative repeater". In digital optical fiber systems, a photodiode must be used to convert the optical signal into an electrical signal before it can be applied to an electronic regenerator. Similarly, after the electrical signal has been regenerated, it must be converted into an optical signal (via a laser) at the output of the regenerator. Other methods (i.e., Erbium doped fiber optical amplifiers, and photonic regenerators) of restoring signals in digital optical fiber systems are described in Chapter 7.

6.9.6 Transmission Equipment Applications in the Network

Figures 6-2 and 6-68 illustrate the high-level system configurations in a digital network. That is, (1) an Integrated Digital Loop Carrier (IDLC) system is used to connect subscribers (voice, data and video) to a local exchange office, (2) in a local exchange office, besides digital switches, Digital Channel Banks (DCBs) are used to convert speech signals into 64 kbps digital bit streams; digital multiplexers are used to combine several low rate signals into a higher rate signal for distant transport; and Digital Cross-connect Systems (DCSs) are used to "move" signals from incoming trunks to "intended" outgoing trunks for traffic control purposes, (3) regenerators are always used along the transmission "path" to "re-strength" signals for farther transport, (4) in the trunk exchange office, similar equipment used in the local exchange office are required, but with greater operation speed, for example, the multiplexed signal at the local exchange office are typically OC-3 and OC-12, but OC-12 and OC-48 are typical for the trunk exchange applications.

Figures 6-82 illustrates typical applications of Digital Channel Banks (DCBs), which is typically located in the local exchange office. The low-speed interfaces of a DCB are traditional analog voice loops, fax lines, voice-grade data using modems, business Private Branch eXchanges (PBXs), Hosts (mainframe computers), Local Area Networks (LANs), and fractional T1 services (N × 64 kbps; DS0 or E0). The high-speed interfaces of a DCB are digital lines (e.g., DS1, DS1C in North America, or E1 in other countries). The primary functions of a DCB are: (1) Pulse Code Modulation (PCM) for analog-to-digital conversion for voice applications, (2) Time Division Multiplexing (TDM) for combining all digital bitstreams (including digitized voice) into a higher rate bit stream for further processing (e.g., multiplexing or cross-connection), and (3) control signals insertion.

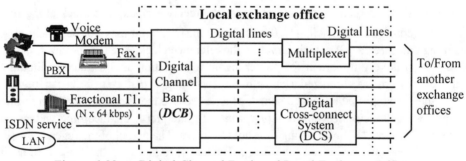

Figure 6-82   Digital Channel Bank and Local Exchange Office.

The digital lines of connecting a DCB are: (1) connected to multiplexers, that combine the digital lines from different DCBs, and digital lines from other digital equipment in the office (or from another offices) into a higher digital lines for farther transmission, (2) connected to Digital Cross-connect Systems (DCSs) in the same office for grooming and consolidating traffic preparing for effective distant transport, and (3) carried by digital carrier systems to another exchange office (i.e., another local exchange or a trunk exchange).

Besides the traditional access applications illustrated in Figure 6-82, there are many other access networks available in modern telecommunications networks. Examples are: (1) remote voice subscribers can be served by using Remote Terminal (RT), which is connected to local exchange office using digital lines (typically, T1 digital carrier trunks). Small business customer (with the need for voice and data transport) can be served by the same remote terminal, (2) Industry areas subscribers are connected to a digital multiplexer for effectively transport, and apartment complexes are first connected to a RT, which is then connected the digital multiplexer. The output interface of this multiplexer can be electrical or optical. Fibers are typically used in modern networks, (3) In the industry park or metropolitan business areas, modern Business Remote Terminals (BRTs) are often applied to handle heavy and mixture of voice, data, fax, and video signals. For traditional connections, digital coaxial cables are used to connect a BRT to a local exchange office. In contrast, modern access applications, OC-3 and OC-12 fiber feeders are used to connect BRT to local exchange office.

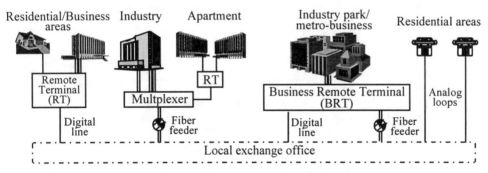

Figure 6-82   Access Network.

Besides the equipment shown in Figure 6-82, a typical local exchange office contains many other equipment as shown in Figure 6-83. Direct lines, that may be digital metallic lines, optical lines from business areas can be connected to the local exchange switch. Digital radio systems (typically operating in the microwave frequency range) can terminate at the switch. From access network, Business Remote Terminals (BRTs), Subscriber Loop Carrier (SLC) systems, and Integrated Digital Loop Carrier (IDLC) systems are connected local switch using a Central Office Terminal (COT). Special services trunks [e.g., Foreign eXchange (FX) and PBX tie trunks) are typically connected to Digital Cross-connect Systems (DCSs). The external interface speed of a DCS can be low-speed, medium-speed and high-speed digital lines (e.g., T1, T3, E1, E3, FT-G, STS-1, STS-3, OC-1, OC-3, STM-1, etc.). Various types of digital multiplexers (i.e., low-, medium-, and high-speed) are the essential equipment in a local and trunk exchange office. They can be configured in three modes: (1) a terminal mode for multiplexing/demultiplexing, (2) regenerator mode for regenerating signals so that they can be transported farther, and (3) add-drop mode for adding and/or dropping traffic as needed. In modern offices, fiber optical equipment are required to convert electrical signal to optical signal (and vice versa), multiplex optical signals, and regenerating signals.

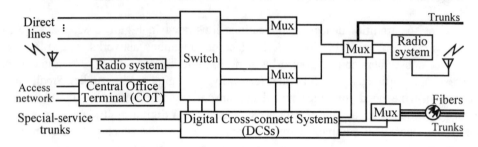

Figure 6-83   Typical Equipment in a Local Exchange Office.

Transmission equipment used for interconnecting: (1) local exchange offices, (2) local and trunk exchange offices, or (3) trunk exchange offices are described as follows:

- In a PDH environment: Figure 6-84 illustrates the applications of transmission equipment for connecting two local exchange offices.

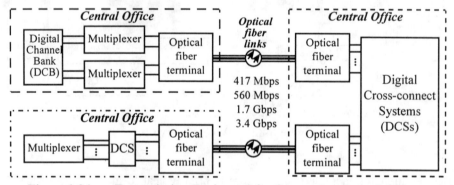

Figure 6-84   Transmission Equipment for Connecting Central Offices.

In this example, three Central Offices (COs) are used to illustrate the required transmission equipment connecting COs. The upper-left CO is a typical local exchange where Digital Channel Banks (DCBs), digital multiplexers, and optical fiber terminals are required. The transmission facilities used to connect this CO to another CO are typically optical fiber links (with speeds of 417 Mbps, 560 Mbps, 1.7 Gbps or 3.4 Gbps, depending on the required system capacity). The lower-left CO may be a trunk exchange office or a tandem exchange office. The transmission equipment used in this type of central offices are digital multiplexers, Digital Cross-connect Systems (DCSs), and optical fiber terminals. The transmission facilities used to connect this CO to another CO are optical fiber links. The right-hand side CO is a trunk exchange office, which is dominated by DCSs and optical fiber terminals (however, digital multiplexers are also needed).

• In SONET/SDH environment: Interconnecting central offices in SONET/SDH environment uses two different architectures: (1) linear (i.e., point-to-point), and (2) ring. Figure 6-85 illustrates the transmission equipment required for linear connection between two central offices. They are digital multiplexers, Digital Cross-connect Systems (DCSs), and optical fiber terminals. The facilities are OC-3 or OC-12 for intra-office connections, and OC-48 (which may be upgraded to OC-192 in the future) for inter-office connection.

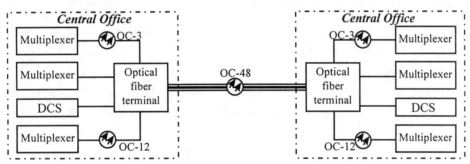

Figure 6-85   Point-to-Point Connection (CO to CO).

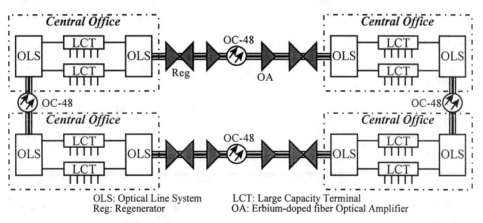

OLS: Optical Line System   LCT: Large Capacity Terminal
Reg: Regenerator   OA: Erbium-doped fiber Optical Amplifier

Figure 6-86   Inter-Office Connecting Using SONET Ring.

The most commonly used inter-office connection in the SONET/SDH environment is "ring". Figure 6-86 illustrates four central offices interconnected by a SONET OC-48 ring. Large Capacity Terminal (LCT) is an Add-Drop Mux (ADM) interfacing between SONET and PDH. Optical Line System is SONET optical terminal that performs functions required for optical transport besides multiplexing. Erbium-doped fiber Optical Amplifier (OA) and regenerators may have to be used along the transmission path.

Review Questions II for Chapter 6:

(11) The line codes for performing zero-suppression function used for each of the following signals are: _____ or _____ (for DS1), _____ (for DS2), _____ (for DS3), _____ (for E1, E2, and E3), and _____ (for E4). But, SONET/SDH applies _____ for zero suppression, the generator polynomial used is _____.

(12) For international connections, prior to SDH, the commonly used digital signals are the ____ signals, which can carry _____ voice or voice-equivalent signal simultaneously.

(13) A typical E1 signal has a capacity of ____ voice trunks, and _____ kbps overhead capacity for performing call-related, and network signaling functionality.

(14) Prior to SDH networks, the undersea optical fiber systems carry signals with a rate of 140, 280, and 560 Mbps. These systems are designed to carry __, __, and __ E4 signals.

(15) A SONET OC-192 can carry ____ DS3s, _____ DS2s, _____ DS1Cs, or ____ DS1s simultaneously. If digitized voice signals (carried by DS1, etc.) are finally transported over OC-192, _____ voice signals can be carried by OC-192 single fiber. In comparison, a T1 metallic digital carrier systems can only carry ___ voice signals.

(16) A SDH STM-64 can carry ____ E4s, _____ E3s, or ____ E11s simultaneously. If digitized voice signals (carried by E1, etc.) are finally transported over STM-64, _____ voice signals can be carried by STM-64 single fiber. In comparison, a T1 metallic digital carrier systems can only carry ___ voice signals.

(17) SONET/SDH standards can be modeled as three-layer model. These three layers are the _____ layer, _____ layer (SONET), or _____ layer (SDH), and _____ layer (SONET) or _____ layer (SDH).

(18) There are four sub-STS-1 signals in SONET standards, and known as Virtual Tributary (VT) signals. To carry DS1, ____ is used, to carry E1, _____ is used, to carry DS1C, ____ is used, and to carry DS2, ____ is used. Likewise, there are four sub-STM-1 signals in SDH standards, and known as Tributary Unit (TU) signals. To carry DS1, ____ is used, to carry E1, ____ is used, to carry DS2, ____ is used, and to carry E3, ____ is used.

(19) When VT or TU signals, they are always grouped together to form a special logical signal called _____ (i.e., _____ in SONET), and _____ and _____ (i.e., _____ and _____ in SDH).

(20) The group of overhead bytes performing framing synchronization, error monitoring, ordering, etc. is called _____ overhead bytes. The group of overhead bytes performing new data flag indication, frequency justification indication, automatic protection switch, error monitoring, error reporting, etc. is called _____ overhead (in SONET) or _____ overhead (in SDH). The third group overhead bytes is called _____ (in SONET) or _____ overhead (in SDH), and is used to perform connectivity check, error monitoring, error reporting, tandem connection maintenance, etc.

Appendix 6A-1

Extended Superframe for Channel Associated Signaling

Figure 6-31 shows the application of a DS1 signal with Extended SuperFrame (ESF) Format when the signal is used with a Common Channel Signaling network (e.g., CCS7, SS7). A ESF DS1 can also be applied in a network without Common-Channel Signaling. That is, a ESF DS1 can be used in a Channel Associated Signaling (CAS) network. The principle is the same as used in the SuperFrame (SF) DS1 signal applications (see Figure 6-30). The 8th bit in each time slot of frame Nos. 6, 12, 18 and 24 are used for rob-bit signaling. The frame pattern is the same as the one shown in Figure 6-31. That is, everything in Figure 6-31 is applicable for the channel associated signaling algorithm, with the addition of the rob-bit signaling shown in Table 6A-1.1. The "robbed-bits" from frame No. 6 are called the "*A*" channel signaling bits; the "robbed-bits" from frame No. 12 are called the "*B*" channel signaling bits; the "robbed-bits" from frame No. 18 are called the "*C*" channel signaling bits; and, the "robbed-bits" from frame No. 24 are called the "*D*" channel signaling bits.

Table 6A-1.1   Additional Rob-bit Signaling Bits for a ESF DS1 Signal.

Frame No.	Information bits per time slot	Signaling bit (rob-bit)	Designation (channel)
1 to 5	8	0	-
6	7	1	A
7 to 11	8	0	-
12	7	1	B
13 to 17	8	0	-
18	7	1	C
19 to 23	8	0	-
24	7	1	D

In summary, there are three types of DS1 signals (see Figure 6A-1.1) used in the μ-law digital networks: (1) SuperFrame DS1 with "robbed bits" in Frame Nos. 6 and 12, (2) Extended SuperFrame DS1 for Channel Associated Signaling (CAS) networks, with "A", "B", "C", and "D" signaling channels, and (3) Extended SuperFrame DS1, with clear channel capability and no "robbed bits" signaling, for common-channel signaling applications.

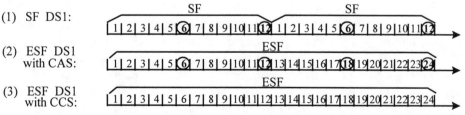

Numbers represent Frame Nos. in a SF or ESF   "Circled" numbered frame: Robbed bit positions

Figure 6A-1.1   Three Types of DS1 Signals.

Appendix 6A-2

Bit Compression Techniques for Voice Applications

There are several pulse code modulation techniques that used different waveform encoding algorithm such as Differential Pulse Code Modulation (DPCM), Adaptive DPCM (ADPCM), Delta Modulation (DM), and Adaptive DM (ADM). The details of these bit compression techniques are discussed in the following sections.

6A.2.1 DPCM and ADPCM

Figure 6A-2.1 illustrates the differences between a conventional pulse code modulation (PCM) system and a differential PCM (DPCM) system. The two major differences between a PCM system and a DPCM system are:

(1) Instead of applying the sample speech signal to a "coder" [a compandor plus an 8-bit encoder] for encoding each sample into an 8-bit codeword, a DPCM system applies the *difference* of the current speech sample and the previous speech sample to a coder.

(2) Second difference is that a DPCM system has a feedback circuit known as a predictor or an integrator, which is used to provide (predict) the sample value of the previous speech sample.

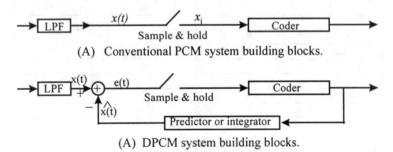

(A) Conventional PCM system building blocks.

(A) DPCM system building blocks.

Figure 6A-2.1   Building Blocks of a PCM and DPCM System.

Because the DPCM system applies the **differences** in a speech sample to the coder, the name **differential PCM** is used to describe this technique. Figure 6A-2.2 illustrates the compression function of a DPCM system compared to a conventional PCM system.

A conventional PCM system accepts the speech signal, x(t), as its input. It samples the speech at a rate of 8000 samples per second. As shown in Figure 6A-2.1(A), the speech sample signal, x_i, is fed to a compressor (the transmitting part of a compandor) and digital encoder (a compandor and a encoder is simply referred to as a coder). As previously described an 8-bit companding PCM system requires 256 quantization levels to encode all possible speech sample sizes (see the upper half of Figure 6A-2.2). This is necessary to achieve an acceptable speech quality at the receiver, after the digital samples are restored to an analog signal. This quality is equivalent to a 13-bit PCM system if a compandor is not used.

In comparison, a DPCM system feeds the difference signal, *e(t)*, instead of the original speech signal, *x(t)*, to the coder to generate an 8-bit codeword (see Figure 6A-2.1). The difference signal is obtained from the following equation:

$$e(t) = x(t) - \hat{x}(t) \qquad (6A\text{-}2.1)$$

where *x(t)* is the sample value of the original speech signal and $\hat{x}(t)$ is the predicted speech signal [$\bar{x}(t)$ represents the predicted speech sample value of the previous sample]. Thus, the signal e(t) represents the difference between the present speech sample value and the previous speech sample value.

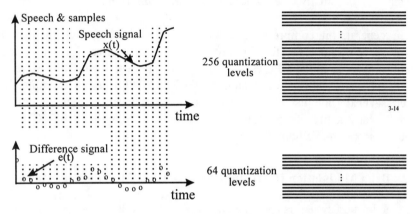

Figure 6A-2.2   Speech Signal, Sampled Signal and Difference Signal.

The compression capability of a DPCM system is shown in the bottom half of Figure 6A-2.2. The difference signal, *e(t)*, has an amplitude value range which is much smaller than the original speech signal. In other words, the value of *e(t)* spreads over (in the vertical direction) a much smaller scale. As shown in the upper half of Figure 6A-2.2, a conventional PCM system requires 256 levels ($\equiv 2^8$, for an 8-bit companding coder) to encode all original speech samples, while the DPCM difference signal, *e(t)*, needs only 64 quantization levels to encode all possible samples. This is summarized in Table 6A-2.2.

Table 6A-2.2   Comparison Between PCM and DPCM.

System	Required levels	Bits/sample	Signal rate (kbps)
PCM	256	8	64
DPCM	64	6	48

How can a DPCM system achieve a compression savings of 16 kbps ($\equiv 64 - 48$) for transporting speech signals? The speech signal is sampled at a rate of 8000 samples per second (i.e., every 125 µs) for conversion into digital code. The nature of speech signals is that volume (signal amplitude) does not change within a short interval (e.g., 125 µs). In other words, the difference signal, *e(t)*, in Figure 6A-2.1 or 6A-2.2 is quite small. Therefore, the difference signal, *e(t)*, does not spread far from the original sample signal. Hence, for a typical voice signal, 64 quantization levels, instead of 256 levels, results in acceptable quality.

$$e(t) = x(t) - \hat{x}(t) \approx 0 \qquad or \qquad x(t) \approx \hat{x}(t) \qquad (6A-2.2)$$

It is known that the sample value $\hat{x}(t)$ for the next sample of $x(t)$ is very close to the previous sample. That is, there is a ***high occurrence of redundancy*** between any two adjacent samples. By encoding the difference signal, the signal redundancy between two adjacent samples is effectively not transmitted. This, in turn, saves system bandwidth. Therefore, a DPCM system is actually a ***redundancy removal technology***.

In general, bit compression techniques take advantages of the correlation between adjacent speech samples. The simplest method is to utilize the correlation between two adjacent samples as in the DPCM example previously described. Other more methods applied in a DPCM system are called Adaptive DPCM (ADPCM). For example, if the correction between **several** (not just two) adjacent samples is used, further bit compression can be achieved. Many sophisticated methods have been proposed for ADPCM systems that are beyond the scope of this book.

6A.2.2   Delta Modulation (DM) and Adaptive Delta Modulation (ADM)

A delta modulation system is a special DPCM application. It is a one-bit DPCM system, in which every sample is encoded into a single bit. Figure 6A-2.1(B) can represent a DM system with a simple coder. This coder is effectively a threshold detector, that performs the following functions:

$$If \quad e(t) \geq 0, \quad then \quad a \quad "1" \text{ is transmitted}$$
$$If \quad e(t) < 0, \quad then \quad a \quad "0" \text{ is transmitted}$$

It can be seen in Figure 6A-2.3, that a logical "1" is transmitted for an "+A" volt signal, and a logical "0" is transmitted for a "−A" volt signal. The most significant advantage of using delta modulation for converting an analog speech signal into a digital signal is that the coder is simple and easy to implement (it is a threshold detector). However, in a delta modulation system a severe slope overload can occur. To understand this, requires familiarity with the concept of quantization noise (error). Two types of quantization impairments result from modulation (PCM, DPCM, adaptive DPCM, DM or adaptive DM):

(1) **Slope overload**: Because of the steep slope of the original analog signal, it cannot be accurately restored from the modulated signal. Figure 6A-2.3 shows the original analog signal (dashed curve), the modulated signal of a DM system (stepped curve), and the restored analog signal (solid curve created by smoothing the stepped signal with a filter). There is a substantial difference between the original speech signal and the restored signal, called the slope overload quantization noise (error). The speech quality during the slope overload period is typically unacceptable.

(2) **Granular noise**: Since the original analog signal is a continuous curve, the restored analog signal will never be an exact duplicate of the original signal, even if the slope is almost flat.

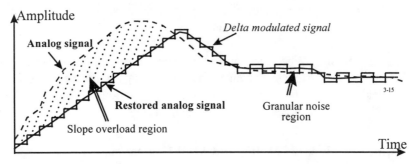

Figure 6A-2.3   Analog Signal and Delta Modulation

Due to the severe slope overload problem in a delta modulation system, the speech quality is unacceptable for practical applications. Therefore, an adaptive scheme must be applied to achieve acceptable speech quality. There are many ways to implement adaptive delta modulation. The following two methods (separate or combined) have been recommended for practical applications:

(1) **Oversampling the analog signal** [Figure 6A-2.4(A)]: For example, instead of sampling at the Nyquist rate of 8 kHz, four times the sampling rate (32 kHz), eight times oversampling (64 kHz), or sixteen times oversampling (128 kHz) can be used to improve signal quality. Four times the sampling rate (32 kHz) is shown in Figure 6A-2.4(A).

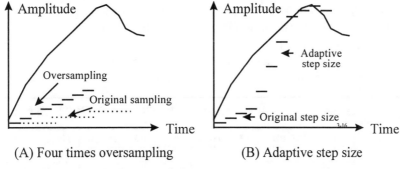

(A) Four times oversampling   (B) Adaptive step size

Figure 6A-2.4   ADM: Oversampling and Adaptive Step Size.

Table 4-3   PCM, DM and ADM comparison.

Modulation	PCM	DM	ADM		
Sampling rate (kHz)	8	8	32	64	128
Bits/sample	8	1	1	1	1
Signal rate (kbps)	64	8	32	64	128

Table 4-3 compares an 8-bit PCM, a DM, and an ADM system with 4 times, 8 times and 16 times oversampling.

In general, delta modulation and four times oversampling ADM will not provide an acceptable speech quality. However, eight times or sixteen times oversampling will produce in good quality restored signal.

(2) **Changing step size**: When the transmitter detects a severe slope load [as in Figure 6A-2.3(A)], the ADM system can change the step size (e.g., four times the original step size), and a "flag" is sent to the receiver. After the slope overload condition no longer existing, the step size can be resumed to its original size.

6A.2.3 PCM, DPCM, ADPCM, DM and ADM Comparison

As previously mentioned, transmission of speech signals over telephone lines, radio channels, and satellite channels constitutes the largest type of daily communications. Therefore, it is understandable that more research has been performed on speech encoding than any other information-bearing signal. It is also appropriate to compare the efficiency of waveform encoding methods (e.g., PCM, DPCM, ADPCM, DM and ADM) for telephone quality speech with respect to the bit rate required to transmit speech signals.

For the purpose of comparison, the speech signal is assumed to be band-limited to a frequency range of 200 to 3200 Hz and sampled at a nominal rate of 8000 samples per second for all waveform encoders except delta modulation (DM). In the case of DM, the sampling rate (f_s) is identical to the bit rate (provided the signal is over-sampled eight times). The simplest form of waveform coding is linear pulse-code modulation, in which an analog signal is linear (uniformly) quantized as described earlier in this chapter. This approach is widely used for analog-to-digital conversion. Two common variations of PCM coding for telephony, μ-law and A-law, are based on nonlinear (nonuniform) quantization of the signal amplitude using a logarithmic scale (rather than a linear scale). These PCM coders utilize the static characteristics of amplitude nonstationarity of human speech to achieve acceptable signal quality at bit rates of 56 and 64 kbps, as seen in Table 6A-2.4.

Table 6A-2.4   Comparison of PCM, DPCM, ADPCM, DM, and ADM

Modulation	Quantization	Bits/sample	Rate (kbps)
PCM	*Linear*	*12-13*	*96-104*
PCM	*Companding*	*7-8*	*56-64*
DPCM	*Companding*	*4-6*	*32-48*
ADPCM	*Adaptive*	*3-4*	*24-32*
DM	*Binary*	*1*	*64*
ADM	*Adaptive*	*1*	*16-32*

In differential coding, efficiency is obtained by quantizing sample-to-sample differences in the speech signal rather than the signal itself. Because the average sample-to-sample correlation of speech is high, the root-mean-square (rms) level of the difference signal is lower than the rms level of the actual signal. Thus, smaller quantizer step-sizes can be used, which leads to a lower overall quantization noise. Used in combination with adaptive step-size methods, this approach can be extended to Adaptive Differential Pulse-Code Modulation (ADPCM) and Adaptive Delta Modulation (ADM).

The quality of the speech signal synthesized at the receiver, assuming an error-free binary sequence, provides telephone (toll) quality speech for all methods listed in Table 6A-2.4. That is, a listener would have difficulty discerning the difference between the digitized speech and the analog speech waveform. It should be noted that ADPCM and ADM are particularly efficient waveform encoding techniques.

6A.2.4 Speech Encoder

The goal of a speech encoding is to reduce the number of bits needed for transmitting a speech signal, thereby lowering the bandwidth required for the transmission equipment/facility. For example, a traditional PCM system requires a rate of 64 kbps to represent a speech signal. For wireless applications, this 64 kbps rate would need a bandwidth that is impossible for the airway to transport. By using a speech encoder, the 64 kbps rate can be reduced, which allows more users to simultaneously share the same transport channel. The final result is reduced communications costs for customers, with a speech quality that meets or exceeds the end-user's expectation level.

Today's cellular telephone system is based on repeated use of 832 ($\equiv 2 \times 416$ RF channels) Frequency-Modulation (FM) radio channels within the radio spectrum allocated by the Federal Communications Commission. To meet the anticipated growth of cellular telephone service (2 million subscribers to 18 million), a new generation of digital cellular equipment is needed. This equipment will utilize digital signal processing techniques for low-bit-rate speech coding, spectrally efficient modems, and fast adaptive equalization. High-quality speech coders that operate at the lowest possible bit rate will play a key role in a spectrally efficient digital cellular system. A 7.95-kbps speech coding algorithm has been selected for use in digital cellular equipment.

The Telecommunications Industry Association (TIA) has agreed to use Time-Division Multiple-Access (TDMA) technology for the new system, which is based on the 7.95-kbps speech coding algorithm. Channel bandwidth will remain at 30 kHz to ease the transition from analog to digital technology. Thus, an existing analog channel unit serving one conversation could be replaced by a digital unit serving three customers in the same 30-kHz channel spectrum without affecting adjacent analog channels. A 30-kHz channel could support data rates up to 48.6 kilobytes per second (i.e., 16.2 kbps per conversion). However, taking overhead into account, the upper limit for the speech coding rate is 13 kbps.

In speech coders with low data rates, loss of information can cause serious degradation of speech quality. A plot of speech quality as a function of bit rate for several classes of speech coders is shown in Figure 6A-2.4. A PCM coder used in the public network, yields high quality, but requires a high bit rate (64 kbps). However, even with a high bit rate, a PCM system may yield a poor quality because of other impairments. A high-complexity hybrid coder, such as the Code-Exited Linear Predictive (CELP) coder, yields high quality for wireless applications, even at relatively low bit rates. As shown in Figure 6A-2.4, a high-complexity waveform coder (e.g., ADPCM), and low-complexity waveform coder (e.g., DM) can provide good quality provided a high bit rate is used. However, if the bit rate is low, they cannot provide good quality. It can also be seen that a low bit rate high-complexity Linear Predictive Coder (LPC) does not provide acceptable quality.

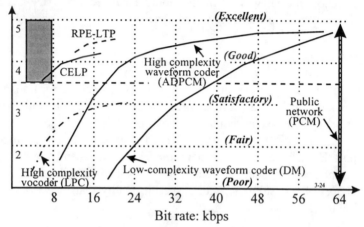

Figure 6A-2.5   Speech Encoding Quality Comparison.

Current research to produce high quality at bit rates below 10 kbps (shown by the shaded area in Figure 6A-2.5) is focused on coders of the high-complexity hybrid type. A list of **representative hybrid coders** are shown in Table 6A-2.5.

Table 6A-2.5   Three Major Hybrid Speech Coders.

Multi-pulse linear predictive (MP-LPC)	Used by British Telecom in Skyphone™ (a trademark of British Telecom) service
Regular Pulse Excited Long Term Prediction coder (RPE-LTP)	A variation of LPC selected for use in the Pan-European digital cellular system, GSM.
Code-Excited Linear Predictive (CELP)	A variation of CELP selected for use in the North American digital cellular system.

Table 6A-2.6   Complexity Comparison for Different Coders.

Coder	Bit rate (kbps)	Complexity (MIPS)
PCM	64	< 1
ADPCM	32	< 1
RPE-LTP	13	6
MP-LPC	10	10
CELP	8	15~30

Table 6A-2.6 represents speech coder complexity in terms of the number of "Millions of Instructions Per Second" (MIPS) required for typical speech coders. As shown in Table 6A-2.6, the MIPS required for CELP coders is at least an order of magnitude greater than required by a PCM coder. As a result, speech coder technology is being driven by advances in Digital Signal Processing (DSP) and Very Large Scale Integration (VLSI) devices.

Example 6A-2.1: Speech encoder performance comparison using Mean Opinion Score (MOS).

The speech quality comparison for PCM, low complexity DM, ADPCM, CELP and RPE-LTP speech encoding methods has been previously discussed (see Figure 6A-2.5) where the Grade Of Service (GOS) was applied. A similar parameter, the Mean Opinion Score (MOS) has been used for comparison purpose. Similar to the GOS, the MOS has a scale of 5; with a "5" as the best speech quality and a "1" as the worst speech quality.

There are two conventions used to compare the speech quality using MOS: (1) MOS versus channel impairment, and (2) MOS versus frame error rate. Table 6A-2.7 lists the MOS ratings for several speech encoder algorithms. For instance, the Low (Data) rate CELP (LD-CELP) coder provides a speech rate of 16 kbps, and the Extended Variable-Rate Coder (EVRC) provides a speech rate of 8 kbps. However, the EVRC applies the statistics of the speech signal using variable-length (also called the run-length) code to achieve a high-degree of bit compression. An 8 kbps EVRC speech encoder can provide speech quality equivalent to the 13 kbps RPE-LTP speech encoder as shown in Table 6A-2.7.

Table 6A-2.7   Speech Mean Opinion Scale (MOS) Comparison.

Speech encoder		Mean Opinion Scale (MOS)
The μ-law PCM	(64 kbps)	*4.3*
The ADPCM	(32 kbps)	*4.1*
The LD-CELP	(16 kbps)	*4.15*
The CDMA-CELP	(13 kbps)	*4.15*
The EVRC	(8 kbps)	*4.1*
The IS-54 VSELP	(8 kbps)	*3.64*
The GSM RPE-LTP	(13 kbps)	*3.64*

Example 4-5: MOS versus percentage frame error rate.

Figure 6A-2.6 shows the MOS ratings for three speech encoder algorithms (proposed for CDMA applications) versus the percentage frame error rate.

The MOS versus the percentage frame error rate graph (Figure 6A-2.6) is one of several practical performance measurements which wireless customers use to judge whether wireless services have met the service specifications. Typical wireless system performance is expected to have a frame error rate between 1% and 2%. Likewise, an MOS of 3.0 or better is an unofficial industrial acceptable speech quality level which is actually lower than the required level of MOS = 4.0 for landline services (i.e., trunk quality).

The percentage frame error rate of 1% implies that (statistically) only one frame of information received will experience one error, assuming 100 frames of information have been received.

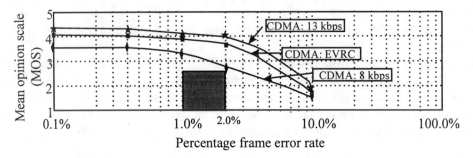

Figure 6A-2.6   MOS versus Percentage Frame Error Rate.

Three different speech encoders for CDMA applications are shown in Figure 6A-2.6: an 8-kbps speech encoder, a 13-kbps speech encoder, and an Extended Variable Rate speech enCoder (EVRC). It can be seen that:

- If system performance is at a level of "1% frame error rate" or better, all three speech encoders provide acceptable speech quality (see Figure 6A-2.6).

- If system performance is at a level of "2% frame error rate", then both the EVRC speech encoder and the 13-kbps speech encoder still provide an acceptable level of speech quality. However, the 8-kbps speech encoder performs slightly below a MOS of 3.0, and is considered to have an unacceptable speech quality.

- From Table 6A-2.7, it can be seen that the EVRC is an 8-kbps speech encoder. But, its performance (MOS) is almost as good as the 13-kbps speech encoder (see Figure 6A-2.6). Therefore, the EVRC speech encoder can be applied in future CDMA wireless systems, including cellular and PCS.

CHAPTER 7

Optical Fiber Communications

Chapter Objectives

Upon the completion of this chapter, you should be able to:

- Describe the relationship between frequency and spectral wavelength for several well-established communications systems, such as digital radio and optical fibers.

- Describe optical fundamentals: speed of light in vacuum, in glass medium, Snell's law, total internal reflection, lenses [including quarter-pitch Graded Index (GRIN) rod lenses] and lens properties (e.g., light focusing and light collimating).

- Describe the three components of an optical fiber link: (1) the light source (types of light sources, operating principles, characteristics including spectral width, rise time, central wavelength, radiation power, and signal modulation); (2) the optical fiber (materials, structure, operational modes, types, characteristics including dispersion, attenuation, and operating windows); and (3) the photodiodes (types, operating principles, and characteristics including responsivity, sensitivity, and rise time).

- Discuss the procedures for designing/analyzing an optical fiber link (selecting the proper light source, fiber, and photodiode) to achieve a specific system performance. Illustrate network applications using optical fiber links.

- Describe optical networking technique [all-optical amplifiers, regenerators, wavelength add/drop multiplexers, routers, digital cross-connection, switch, Wavelength Division Multiplexing (WDM) systems] and associated equipment.

7.1 INTRODUCTION

Various types of media can be used as transport facilities for telecommunications networks. The signal hierarchy (e.g., DS1, DS1C, DS2, DS3, E1, E2, E3 and E4) discussed in Chapter 6 can be transported by one or more media shown in Table 7-1. A brief summary of media applications is given as follows:

- Twisted-pair wires: typically used for DS1 and E-1 signal transport.

- Coaxial cables: typically used to transport DS3 and E-3 signals, sometimes used to carry SONET STS-1 or STS-3 signals in short-haul applications.

- Waveguides: typically used for microwave frequency radio transmission.

- Airway: typically used for digital radio, wireless, cellular, and PCS applications.

- Fibers: typically used for high-capacity and low bit-error rate applications.

Table 7-1   Comparison of Various Media.

Medium	Voice capacity	Applications/distance
Twisted pairs	*Tens*	*T1/short-medium*
Coaxial cables	*Hundreds*	*T3/medium-long*
Waveguides	*Hundreds*	*Radio*/short-medium*
Airway	*Thousands*	*Radio, satellite/short-long*
Optical fibers	*Tens of thousands*	*OC-N/short-long*

* Electrical signal to radio: waveguide current applications

From Table 7-1, it can be seen that optical fibers are suitable for both short and long-haul transport. For example, fibers can be used for "fiber-to-the-home" and "loop" applications. Optical fiber is found in trunk applications, and practically all undersea international "long-haul" facilities.

As a foundation for understanding optical fiber communication systems, the concepts of frequency and wavelength will be briefly reviewed. The frequency (wavelength) spectrum of common communications applications are shown in Figure 7-1. Note that the relationship between frequency and wavelength is given as follows:

$$\lambda f = c \tag{7-1}$$

where λ is the wavelength, f is the frequency (of a signal), and c is the speed of light in vacuum (a value of approximately 3×10^8 m/s). It can be seen that if frequency increases, wavelength decreases (and vice versa). The applications listed in Figure 7-1 are not intended to be complete, but are provided for the purpose of illustrating the relative position of the optical fiber spectrum used in communications systems. The following examples are common "reference points" in the telecommunications field.

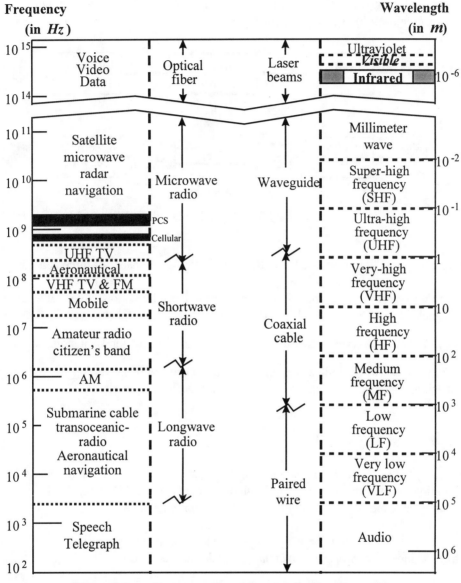

Figure 7-1   Frequency Spectrum Application Examples.

- AM/FM radio (upper frequency is approximately 1600 kHz or 108 MHz).

- Cellular communications technology (allocated to the 900 MHz spectrum range).

- PCS technology (occupies the 1900 MHz spectrum range).

- Optical fiber communications (discussed in this chapter) has the frequency spectrum in the range called infrared (i.e. beyond the wavelength of 700 nm or 0.7 μm). There are three popular wavelengths used in optical fiber communications: **820 nm**, **1310**

nm, and **1550 nm**. Note that in optical fiber communication the unit used is the *wavelength* of the signal (not the frequency).

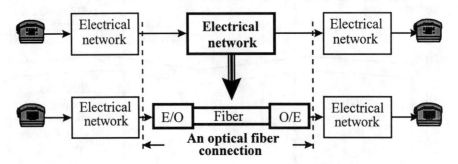

Figure 7-2   An Optical Fiber Link Replacing an Electrical Link.

A typical optical fiber link (see the bottom of Figure 7-2) consists of three components:

- *Light source:* serves as the "signal" for transmitting information. Note that an electrical signal is converted into a light signal (i.e., Electrical to Optical converter: E/O).

- *Optical fibers:* serves as the transmission medium for carrying light signals.

- *Photodiode* (or photodetector): serves as the receiver for converting the light signals back into an electrical signal (i.e., Optical to Electrical converter: O/E).

7.2 OPTICAL FUNDAMENTALS

Light propagation (incident, reflection, refraction, and Snell's law), total internal reflection, absorption, emission, radiation, stimulation, focusing, collimating, and lenses are fundamental technologies used in optical fiber communications.

Before introducing these optical fundamentals, an important term known as the refraction index, or "*index of refraction*", *n* is defined as follows:

$$n = \frac{c}{v} \tag{7-2}$$

where *c* is the speed of light in a vacuum (having a value of approximately 3×10^8 m/s), and *v* is the speed of light for a given material having an index of refraction of *n*.

7.2.1 Incident, Reflection, Refraction and Total Internal Reflection

When a light ray (in material 1) enters the boundary region between two different types of material, a portion of light energy is reflected back (known as a reflection ray) into material 1, and another portion of the light energy is transmitted (refracted) across the boundary (known as a refraction or transmission ray) into material 2 (see Figure 7-3).

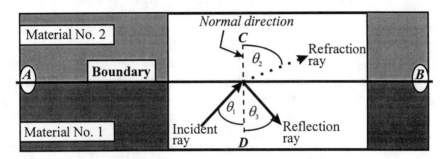

Figure 7-3   Incident, Reflection and Refraction.

Material 1 (with an index of refraction of n_1) meets material 2 (with an index of refraction of n_2) and forms a boundary <u>AB</u>. The direction, <u>CD</u>, which is perpendicular to boundary <u>AB</u> is known as the "normal direction". Three angles, θ_1, θ_2, and θ_3 are known as the incident, the refraction, and the reflection angles, respectively. They are all measured with respect to the "normal direction <u>CD</u>", and have the following relationships:

$$\theta_1 = \theta_3 \tag{7-3}$$

and,

$$n_1 \bullet \sin \theta_1 = n_2 \bullet \sin \theta_2 \tag{7-4}$$

Eq. (7-4) is called Snell's law, and its application in optical fiber communications is illustrated in Examples 7-1 and 7-2.

Example 7-1 Describe *total internal reflection and fiber structure*: This example is used to illustrate (1) the application of Snell's law; (2) the theory of total internal reflection; and (3) fiber structure using total internal reflection to achieve better light transmission efficiency (see Figure 7-4, which shows two fiber layers: core and cladding).

In Figure 7-4 two cases of light-ray (path) transmission (incident, reflection, and refraction) are illustrated. Case one is represented by "dashed" rays, and case two is shown by "solid" rays.

<u>Case 1</u>: When the light ray (marked "2") enters the fiber core it hits the "core/cladding" boundary, and generates reflection ray "3" that stays within the fiber core, thereby carrying an information signal to its destination. Simultaneously, the incident ray, "2", generates refraction ray "4" within the fiber cladding. Ray "4" remains in the cladding region and can not reach the destination since only those rays traveling within the fiber

core can be detected by the photodetector at the receiver. Thus, this refraction ray "4" contributes to signal loss. This phenomenon is a result of Snell's law as follows:

$$n_1 \cdot \sin \theta_1 = n_2 \cdot \sin \theta_2; \quad \text{where} \quad n_1 \, (= 1.48) > n_2 \, (= .46)$$

$$\Rightarrow \theta_2 > \theta_1$$

Therefore, light rays that enter the core (from the light source) with an angle larger than θ_a (e.g., ray "5") will have attenuated signal strength due to refraction, as indicated by ray "4" (see Figure 7-4). By applying Snell's law, this undesirable condition can be improved by the principle of "total internal reflection", as described in Case 2.

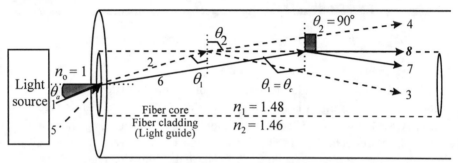

Figure 7-4   Fiber Structure: Core and Cladding Layers.

Case 2: When a light ray enters the fiber core (marked ray "6") it hits the "core/cladding" boundary with angle known as the "critical angle" (θ_c) and refraction ray "8" remains within the fiber core. Therefore, the information carried by ray "8" will reach its destination. This condition increases the signal power at the receiver, and is called "total internal reflection". Snell's law [Eq.(7-4)] is applied to the "core/cladding" boundary as follows:

$$1.48 \cdot \sin \theta_c = 1.46 \cdot \sin 90^\circ$$

Using this relationship the critical angle (θ_c) which allows "total internal reflection" to occur for the best signal transmission within the fiber core, can be calculated as follows:

$$\theta_c \text{ (the critical angle)} = 80.57^\circ$$

Then, applying Snell's law at the "air/fiber" boundary as follows:

$$1 \cdot \sin \theta_a = 1.48 \cdot \sin 80.57^\circ$$

The maximum acceptance angle (θ_a) for the fiber can be obtained as follows:

$$\theta_\alpha \text{ (the maximum acceptance angle)} = 14^\circ$$

Hence, light rays from the light source that enter the fiber within the maximum acceptance angle (θ_a) will experience total internal reflection, which is a desirable operating condition. The total internal reflection was discovered by John Tyndall.

Example 7-2: Describe the need for using "index matching gel" to improve transmission quality when fibers are "spliced".

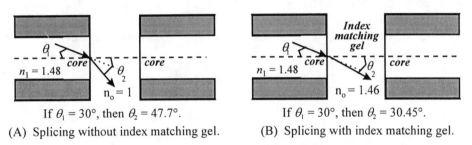

| If $\theta_1 = 30°$, then $\theta_2 = 47.7°$. | If $\theta_1 = 30°$, then $\theta_2 = 30.45°$. |
| (A) Splicing without index matching gel. | (B) Splicing with index matching gel. |

Figure 7-5   Splicing without and with Index Matching Gel.

Figure 7-5(A) shows the space that exists between two fibers that have been "spliced" (i.e., physically joined together). The space in Figure 7-5(A) is exaggerated to illustrate the cause of splicing loss. The index of refraction of the fiber core typically has a value of 1.48, and the index of refraction of air typically has a value of 1.0. A light ray entering the splicing space is assumed to have an angle of 30° (see Figure 7-5). By applying Snell's law, the refraction angle (θ_2) can be found to be 47.7°. It can be seen that after the light ray enters the splicing space, it bends away from the fiber core. This contributes to signal attenuation, and is known as "splicing loss", which results in degraded performance.

In Figure 7-5(B), "index matching gel" is used to fill the splicing space, and increases the index of refraction from 1.0 to 1.46, which is the index of refraction for typical index matching gels. By applying Snell's law, the refraction angle (θ_2) can be found to be 30.45°. That is, the refraction ray and the incident ray are approximately a straight line. Therefore, the signal attenuation is considerably reduced, and performance is improved.

"Index matching gels" have been widely applied by "array splicing" technique that connect ribbon fibers (of 12 fibers one each side) together.

7.2.2 Lenses and Lens Properties

An optical fiber communication system often requires lenses to improve the efficiency of light coupling. The general requirements for these lenses are: they must be thin, exhibit small absorption, have low reflection loss, and introduce minimum aberration. Two types of lenses used for optical fiber communication systems are: (1) spherical, and (2) GRIN (**gr**aded **in**dex) ¼-pitch rod lenses. Both types of lens possess a set of characteristics (properties) that can applied in optical fiber systems. Three primary properties of optical lenses have been applied to optical fiber applications, and are briefly described as follows:

- **Focusing**: A spherical lens has two spherical surfaces. The boundary of the these two surfaces and the lens axis join at the lens center "O", as shown in Figure 7-6. There are two focal points, "P" and "Q" on the lens axis (the plane passing through the focal point and perpendicular to the lens axis is called the focal plane). If the radii of the two spheres forming the lens are R_1 and R_2, and the index of refraction of the lens is "n", then, the focal length "f" of the lens is given by the following equation:

$$\frac{1}{f} = (n-1)(\frac{1}{R_1} + \frac{1}{R_2}) \tag{7-5}$$

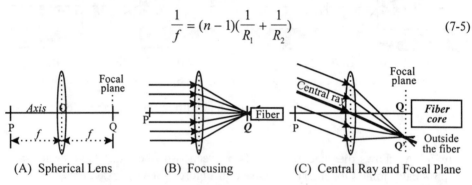

(A) Spherical Lens   (B) Focusing   (C) Central Ray and Focal Plane

Figure 7-6   A Spherical Lens, and its Focusing Property.

When light rays enter the lens in parallel with the lens axis, they will focus at the lens focal point "Q" as shown in Figure 7-5 (B). This property can be applied in an optical link to increase the light coupling efficiency, thereby reducing the coupling loss between a light source (the transmitter of the optical link) and the optical fiber core region. Figure 7-6(B) illustrates that almost all light rays enter the fiber.

When light rays enter the lens in parallel with one another, but not parallel to the lens axis, they will not converge at the focal point "Q". Instead, they will focus at the point Q', which is the intersection of the central ray and the focal plane of the lens. The *central ray* is the light ray that passes through the center ["O"; Figure 7-6(A)] of a lens without changing its propagation direction. This focusing property confirms that if a lens is not properly aligned with the fiber, there will be signal losses due to the light rays staying outside of the fiber [see Figure 7-6(C)].

- **Collimating**: The "collimating" property, shown in Figure 7-7, can be considered the inverse of the lens focusing property. This can be seen by comparing Figure 7-6(B) and Figure 7-7(A), and comparing Figure 7-6(C) and Figure 7-7(B). That is, if light rays originate at the lens focal point ("P"; which is one of the two focal points of a spherical lens, and "Q" is the second focal point) and enter the lens, they exit the lens in parallel with the lens axis. The collimating property can be applied to increase light coupling efficiency (described in Example 7-3). If the light rays originate at another point in the focal plane, for example at point P', they will exit the lens in parallel with the central ray (not with the lens axis) as shown in Figure 7-7(B).

(A) Collimating from a focal point   (B) Collimating from a point on focal plane

Figure 7-7   The Collimating Property of a Spherical Lens.

- **Imaging**: An object (indicated as $\underline{AB}$ in Figure 7-8) placed in front of a lens will form an image, which can be a "real" or a "virtual" image (indicated as $\underline{A'B'}$ in Figure 7-8). If the object is placed beyond the focal point "P" of the lens, the image is "real". A "virtual" image is formed if the object is placed between the focal point "P" and the center of the lens ("O"). The "image" of a point on an "object" is formed by two light rays. One is the central ray and the other is the "parallel ray" (with respect to the lens axis) passing through the lens focal point.

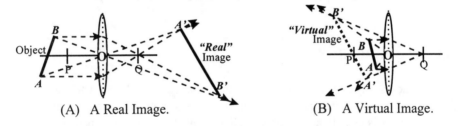

(A) A Real Image.   (B) A Virtual Image.

Figure 7-8   An Object and Its Real or Virtual Image.

Example 7-3 Describe a quarter-pitch GRIN (GRaded Index) rod lens, and its focusing and collimating properties.

The index profile for this type of lens is a graded index, which has a maximum index at the lens axis and decreases gradually towards the edge of the lens.

When the index of refraction is "properly graded", any light ray entering at the axis of the rod shaped lens will have a projection that is a "sine" curve. The period of a sinusoidal wave is also called the pitch (P). The slope at the P/4 (quarter-pitch of the sine wave), 3 × (P/4), 5 × (P/4), 7 × (P/4), ..., points has a value of "0". That is, the propagation direction of a light ray "*leaving*" any of these points (e.g., quarter-pitch P/4-point, etc.) is horizontal. A quarter-pitch GRIN rod lens is manufactured based on this principle. A quarter-pitch rod lens can have a length of P/4 (quarter-pitch), $3 \times$ (P/4), $5 \times$ (P/4), $7 \times$ (P/4), etc. By using a quarter-pitch GRIN rod lens, as shown in Figure 7-8, advantages of "focusing" and "collimating" properties can be easily illustrated.

For practical optical fiber links, the collimating property is implemented by applying a quarter-pitch GRIN rod lens to improve light coupling efficiency at the transmitter. As shown in Figure 7-8(B), when the light rays leave the light source and enter the quarter-pitch GRIN rod lens at the lens axis, their travelling projectors are sinusoidal shaped. These light rays leave the lens in parallel with the lens axis. If the optical fiber is placed in-line with this lens, the light rays will enter the fiber core in parallel. This will increase the coupling efficiency, and will also reduce delay dispersion (described later in this chapter). Thus, the transmission quality is improved.

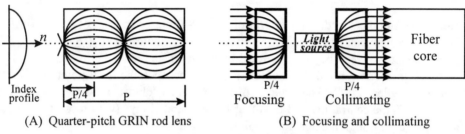

(A) Quarter-pitch GRIN rod lens   (B) Focusing and collimating

Figure 7-8   Quarter-Pitch GRIN Rod Lens and Its Properties.

7.3 LIGHT SOURCES

Figure 7-9 shows a typical transmitter used in an optical fiber link. The signal source is connected to a coder, which is then connected to driver circuitry, and finally to a light source. The coder is usually an analog-to-digital converter, a channel encoder (described in Chapter 13) and/or a line encoder (described in Chapter 6). A driver is usually provided because the light source requires an electrical signal at a lower level than the coder normally generates. Two common designs are Transistor-To-Transistor Logic (TTL) and prebiased drivers. A light source must be provided for converting an electrical signal into an optical signal. Two types of light sources used in modern optical fiber communications networks are:

(1)   Light emitting diode: It is used for low to medium speed optical fiber systems.

(2)   Laser (laser diode): It is used for high speed optical fiber system. Lasers are further divided into four categories:

    *   Multi-Longitudinal Mode (MLM) lasers: A MLM laser is typically used in an optical fiber system with data rate higher than the systems using LED.

    *   Single Longitudinal Mode (SLM) lasers: A SLM laser is used in high-speed optical fiber systems.

* Single Frequency (SF) lasers: A SF laser is used in super-high-speed systems; optical fiber systems using SF lasers have very small dispersion, thus SF lasers are suitable for super high speed transport systems.

* Tunable lasers: They are used for optical fiber systems that employ Wavelength Division Multiplexing (WDM); the wavelengths used in a WDM system are typically separated by 5 nm or less, therefore, tunable lasers are used so that the operating wavelengths can be set appropriately.

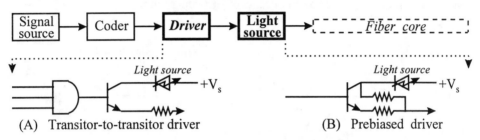

(A) Transitor-to-transitor driver   (B) Prebiased driver

Figure 7-9   A Typical Transmitter of An Optical Link.

When designing or analyzing a lightwave optical fiber system, the following parameters of a light source must be considered:

* Optical output radiated power: The light source output power directly influences the power-budget design of an optical link [described later in Eq.(7-63)].

* Power spectral density: A light source radiates light with various light intensities over a range of wavelength (described later; Figure 7-11). This characteristic is used for bandwidth (rise-time, system speed) design of an optical link. There are many light source characteristics associated with the power spectral density including: the central (nominal) wavelength, the minimum and the maximum wavelengths, and the spectral (line) width of the light source.

* Rise time: The light source rise time is another parameter used in bandwidth design.

* Mode stability: Mode stability is critical in system performance (i.e., data rate), but it only exists with laser diodes.

* Transfer characteristics: They provide the relationship between the electrical signal and optical signal, and used to modulate the signal for transmission.

* Temperature characteristics: The output power of a light source changes as the operating temperature varies; and an important factor in system operations.

The details of these parameters (i.e., radiation power, power spectrum, rise time, spectral width, etc.) are defined in the following sections for Light Emitting Diode (LED) and laser (Light Amplification by Stimulated Emission of Radiation) light sources.

7.3.1 Light Emitting Diode (LED)

A light emitting diode is a p-n junction semiconductor device that is implemented using either homojunction and heterojunction technology. A homojunction LED is simple in structure and operating principle, but has poor light-coupling characteristics that are not suitable for practical applications in modern optical networks. A heterojunction LED is complex, but has excellent light-coupling properties. Although heterojunction LEDs are commonly used in optical networks, the simplicity of homojunction technology is convenient for defining the basic concepts of LED operation.

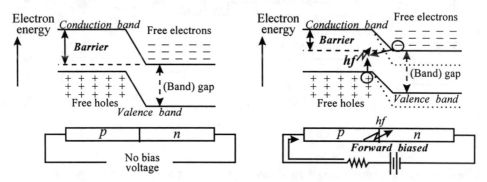

Figure 7-10   Homojunction LED: Valence and Conduction Bands.

Figure 7-10 shows the (electron) energy distribution of a p-n junction (homojunction) device. Free electrons are distributed in the "*n*" (negative; since free electrons possess negative charges) region (i.e., conduction band) while free holes are distributed in the "*p*" (positive) region (i.e., valence band). If no bias voltage is applied to this p-n junction device, the band gap forms a "barrier" which is too large (wide) for electrons and holes to perform any interaction. However, if a forward bias voltage is applied to the device, as shown in Figure 7-10 (right side), the "barrier" will be reduced to a level that allows the free electrons to interact with free holes. This process generates "photons" (i.e., light) for transmission, and is known as "*recombination*".

The relationship the radiated wavelength and the band gap energy given by the following equation while the energy of a photon is given as $W = hf$ (*f* is the frequency).

$$\lambda = \frac{hc}{W_g} \qquad [W_g \text{ in Joules; and } \lambda \text{ in } \mu m] \qquad (7\text{-}6)$$

$$= \frac{1.24}{W_g} \qquad [W_g \text{ in eV; and } \lambda \text{ in } \mu m] \qquad (7\text{-}7)$$

where h is Planck's constant ($= 6.626 \times 10^{-34}$ Joules • sec), c is the speed of light in a vacuum ($\approx 3 \times 10^8$ m/s), and W_g is the band gap energy (material dependent: Table 7-2).

Table 7-2   (LED) Radiation Wavelength and Band Gap Energy.

Material	Band gap, W_g, energy (eV)	Radiated wavelength (λ in μm)
GaAs	1.4	0.8857
AlGaAs	1.4~1.55	0.8000~0.8857
InGaAs	0.95~1.24	1.0000~1.3052
InGaAsP	0.73~1.35	0.9185~1.6986

Ga: Gallium   As: Arsenide   Al: Aluminum   In: Indium   P: Phosphor

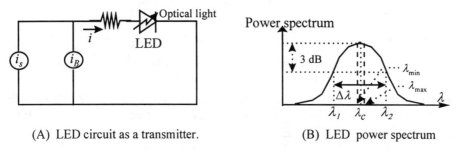

(A) LED circuit as a transmitter.   (B) LED power spectrum

Figure 7-11   An LED with its Power Spectrum Density.

An LED is driven by an electrical current (i) which contains two parts: the signal current (i_s) and the biased current (i_B). The light generated by the LED has different light intensities with respect to the light wavelength. This characteristic is called the power spectrum, which is one of the most important parameters of a light source [see Figure 7-11(B)] of an LED. It can be seen that an LED is a continuous device since it has a continuous spectrum across a wide range of wavelengths. Assuming λ_1 and λ_2 are the wavelengths of the two 3-dB points on the spectrum curve [Figure 7-11(B)], the 3-dB *spectral (line) width*, $\Delta\lambda$, of the LED is given as follows:

$$\Delta\lambda = \lambda_2 - \lambda_1 \tag{7-8}$$

There are various definitions of spectral width (e.g., 3-dB, 10-dB, 15-dB or 20-dB spectral width) for LEDs used in different applications. Generally, higher system speed applications have larger dB spectral width parameters. For example, for low-speed and short-haul transport systems, a 3-dB spectral width is used to define the light source characteristics, and a 15-dB spectral width must be used to specify the light source characteristics of a high-speed and long-haul system. Later in this chapter it is proven that there is a direct correlation between spectral width and system operating speed.

To describe the operating wavelength of an optical fiber system, the nominal (central) wavelength (λ_c) of the light source is used. λ_c is defined as follows (where λ and λ are the two 3-dB points on the power spectrum curve):

$$\lambda_c = \frac{\lambda_1 + \lambda_2}{2} \tag{7-9}$$

When a light source is specified by its central wavelength (e.g., 0.82 μm, 1.31 μm, 1.55 μm), a range of tolerance must also be specified. Tolerance represents the minimum and the maximum wavelengths of the central wavelength of the light source. For example, if the central wavelength of an optical system is required to be 1310 *nm*, then the minimum and the maximum wavelengths may be specified as 1305 *nm*, 1315 *nm*, respectively, with a tolerance of ±5 *nm*.

The radiated optical power of an LED is given by the following equation:

$$P = \frac{\eta W_g}{e} \cdot i \qquad [W_g \text{ in Joules}]$$

$$= \eta W_g \cdot i \qquad [W_g \text{ in eV}] \qquad (7\text{-}10)$$

where P is the radiated power of the LED, i is the driven current [previously described in Figure 7-11(A)], W_g is the band gap of the LED, e is the electronic charge, and η is the recombination efficiency of the LED.

The relationship between the input current (i) of the LED and the output radiated optical power (P) is plotted as in Figure 7-12(B). The curve, "P versus I", is called the transfer characteristic of the LED. Since this curve is approximately linear, an LED is considered to be a linear device (different from a laser, which is a nonlinear device, and described later in this chapter).

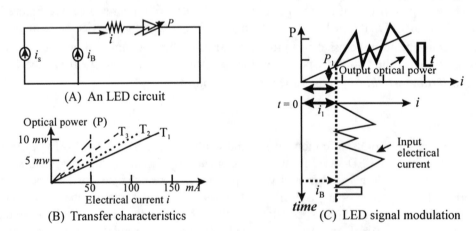

(A) An LED circuit

(B) Transfer characteristics

(C) LED signal modulation

Figure 7-12   Transfer Characteristic of an LED.

The transfer characteristics is used to analyze the modulation technique of an LED. The signal modulation of an LED is illustrated in Figure 7-12(C). The input electrical current used to drive the LED is shown by the "thin" curve (marked "Input electrical current"). For example, at $t = 0$, an input current i_1 generates an optical power P_1 in accordance with the transfer characteristics. By applying the same process for all input currents, the output optical power curve can be derived. The resulting power curve is called the modulated optical power (signal), and indicated by the "bold" curve. Note that a bias current is often required so that the net electrical current input to the LED is always positive since an LED can only operate if the driving current is positive.

The system (LED) performance is a function of system operating temperature. Figure 7-12 (B) also shows the output optical power changes as the device operating temperature changes. For example, if the input electrical current is constant [e.g., 50 mA; Figure 7-12(B)], the LED output power is not constant if the operating temperature changes (e.g., from T_3 to T_2, or T_1; $T_1 > T_2 > T_3$). The output is lower for an LED to operate at T_1 than operating at T_3. That is, a typical LED radiates a lower power, P, as its operating temperature increases, T, as follows:

$$P = P_o [1 - \alpha(T - T_o)] \tag{7-11}$$

where P_o is the power radiated by the LED at an operating temperature of T_o, and is known as the temperature coefficient of the LED.

Besides operating within minimum radiated power requirements, a light source must also meet the required rise time as specified for a particular optical fiber system. The rise time is defined as: "***the time required for the light source to respond to the input signal***". The common specification for rise time is: "***the time delay between 10% to 90% of the light source's final steady state response***".

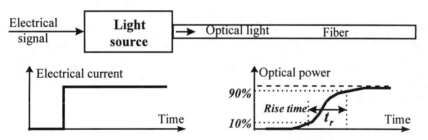

Figure 7-13   A Light Source and Its Rise Time.

It is assumed the electrical signal input (to the light source) is a "step function", however, the light source can never response immediately. That is, the optical power output of an LED can never be a "step function", but it will eventually reach a steady state value. Thus, the time for the LED to reach 90% of the steady value, measured by starting from its 10% point, is the "***rise time***" of the light source. The rise time of the light source (like spectral width) also effects the system speed of an optical fiber system (described later in this chapter).

Another parameter has been proposed to replace the rise time of a light source, and is called the Full Duration Half Maximum (FDHM), which is defined the time interval between 0% and 50%, instead of between 10% and 90%. Presently, FDHM has not been favorably adopted as the rise time.

7.3.2 Laser Diodes (LDs)

A laser diode can be considered a high-frequency oscillator, which requires three elements for proper operation: (1) amplification medium, (2) feedback (mirror), and (3) frequency determination (tuning mechanism).

Almost all the parameters used to specify an LED are also applicable to laser diodes. For example, power spectral density, rise time, spectral width, central wavelength, minimum wavelength, maximum wavelength, and transfer characteristics.

However, the major difference between a laser diode and an LED is that a laser diode is a "discrete" instead of a continuous device (compare Figure 7-14 with Figure 7-11). That is, the power spectrum of a laser is not continuous as the light wavelength varies. The power spectrum of a laser may be zero in a range of wavelengths, as the wavelength increases (or decreases), the power spectrum is no longer zero. This is illustrated by the "*spikes*" in Figure 7-14. Each of these spikes is called the "*longitudinal mode*" of the laser. The definitions of spectral width, central (nominal) wavelength, minimum and maximum wavelengths of a laser is the same as the definition of an LED.

There are three different types of laser diodes shown in Figure 7-14:

(1) Multi-Longitudinal Mode (MLM) laser: This device has a power spectrum shown in Figure 7-14(A), and contains more than one longitudinal mode within the spectral width ($\Delta\lambda$), i.e., within the two 3-dB (10-dB, 15-dB, etc.) points. It is typically used in medium to high-speed optical fiber system. A MLM laser can operate at higher speeds than an LED.

(2) Single Longitudinal Mode (SLM) laser: The power spectrum for a SLM laser is shown in Figure 7-14(B), which contains one and only one longitudinal mode within the spectral width. A SLM laser is typically used for high-speed optical fiber systems, and is also called a "*narrowband*" laser.

(3) Single Frequency (SF) laser: A SF laser is similar to a SLM laser, but it has an extremely small spectral width. A SF laser is suitable for very high-speed optical fiber system applications.

A SF laser diode's central wavelength may be tunable for special applications, such as Wavelength Division Multiplexing (WDM) optical fiber systems (described later in this chapter). It should be noted that it is important to maintain the central (nominal) wavelength of a light source(especially a laser diode). This is known as the mode stability requirement of a light source.

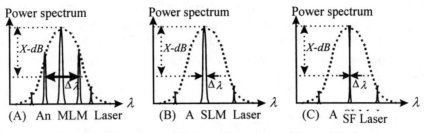

Figure 7-14   Power Spectral Densities of Laser Diodes.

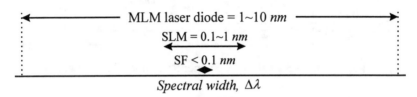

Figure 7-15   Spectral Width Improvement of Laser Diodes.

There is no clear difference between a SLM and a SF laser, other than spectral width. Figure 7-15 shows the relative spectral width boundaries for MLM, SLM, and SF laser diodes. It is clear SF lasers have the shortest spectral width.

Example 7-4 Define the *side-mode suppression ratio* of a laser:

This parameter is used to measure the spectral width variation of a laser diode. Referring to Figure 7-14, if the optical power of the dominant mode of a laser diode is M_1, and M_2 is the optical power of the most significant side-mode, then the Side-mode Suppression Ratio (SSR) of a laser is defined as follows:

$$SSR = 10 \times \log \frac{M_1}{M_2} \tag{7-12}$$

The larger the value of SSR, the higher the system speed. That is, to operate a high speed optical fiber system requires, the laser diode must have a SSR greater than 10 dB.

Example 7-5 Define the terms *rise time, fall time, on time, overshoot,* and *undershoot*:

In an optical fiber system, a digital signal of "1" or "0" is converted from its electrical from (voltage or current) into an optical signal. When a logic "1" is transmitted, the corresponding optical pulse shape is shown in Figure 7-16. Ideally, a logical level "1" implies that optical light is emitted by the light source; while a logical level "0" implies no emission of light. However, in an actual system, a logical "0" still allows a small amount of light to be emitted by the light source. In this example, the optical power level for a logical "1" is assumed to be "b_1", and the power level for a logical "0" is "b_0" (but not absolute 0). The difference between these two power levels is Δb ($\equiv b_1 - b_0$), and is called the pulse height.

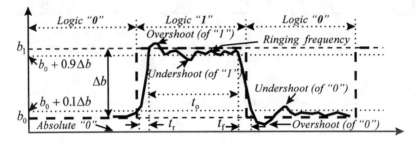

Figure 7-16   Optical Signal Pulse Shape: Rise time, Fall Time, and so on.

The "***rise time***" has been previously defined (see Figure 7-13), is shown as t_r in Figure 7-16. The "***fall time***" (t_f) is the time interval during which the pulse "falls" from 90% to 10% of pulse height. Note that in defining the rise time and the fall time, the terms "the first time" and "the last time" must be applied. That is, the time interval between the "first time" and the "last time" of 90% pulse height is called the "***on time***" (t_o). Larger "on time" values imply smaller "rise time and/or fall time", which translates into higher optical fiber system speed.

The "overshoot of logical 1", "undershoot of logical 0", "overshoot of logical 0", and "undershoot of logical 0" all have a direct impact on system noise performance. Too much overshoot and/or undershoot will degrade system signal-to-noise ratio performance, and will result in an unacceptable Bit Error Rate (BER) for a digital optical fiber system. The nominal optical power level of a logical "1" is b_1, therefore if the signal pulse goes beyond (is larger than) this level, the signal is said to ***overshoot*** its "1"s level. Likewise, if the power level is smaller than b_1, it is ***undershoot*** "1"s level signal. Similarly, the signal can either overshoot or undershoot the logical "0" level. Note that "overshoot of logic 0" actually occurs when the signal power goes below (not over) the nominal power level of a logical "1" (i.e., less than b_0).

Another important parameter is the "***extinction ratio***" (r_e) of an optical signal. This parameter is useful when designing an optical fiber system from a power budget viewpoint (i.e., S/N performance). It is defined by using the following equation:

$$r_e = 10 \times \log \frac{A}{B} \tag{7-13}$$

where "A" is the average optical energy in a logical "1" level, and "B" is the average optical energy in a logical "0" level.

Example 7-6: Identify various structures used for Distributed-FeedBack (*DFB*) laser diodes, Continued-Wave (*CW*) laser diodes, and *pulse* type laser diodes.

For high-bandwidth "direct detection" and "coherent" optical fiber systems, stable single-mode operation of laser diodes are required. To meet this need, special laser diodes have been developed that suppress multi-mode performance. One type of diode commonly used is the Distributed-FeedBack (DFB) laser diodes, which applies a *buried-heterostructure double-channel* form. In comparison, a CW laser diode utilizes a *gain-guided double-heterostructure* form. The detailed description of these structures is beyond the scope of this book

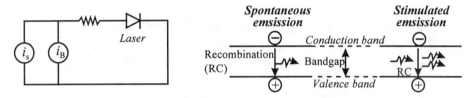

Figure 7-17   Laser Diode Spontaneous and Stimulated Emission.

Two mechanisms are involved in the light generation process of a laser diode: "spontaneous emission" and "stimulated emission" (Figure 7-17). Spontaneous emission of light is a result of the recombination of excited electrons (negative charges) in the conduction band with holes in the valence band (Figures 7-10 and 7-17) (i.e., similar to the process in an LED). However, light photons in a laser diode are predominantly generated by stimulated (induced) emission. That is, "*photons trigger the generation of more photons, by stimulating additional recombination of electrons and holes*". This is effectively a "chain reaction" process. In general, all the photons produced by stimulated emission have the same wavelength and phase. The stimulated photons are thus called "*coherent*", which results in a narrow spectrum. The *single-heterostructure* from, which is simpler than the double-heterostructure and double channel structure, is used in pulse laser diodes. This type of laser diode is used in high-power pulse applications [e.g., designing an Optical Time-Domain Reflectometer (OTDR), which is described later in this chapter].

Example 7-7 Describe signal modulation of a laser diode, and its pumping threshold.

A major difference between a laser diode and an LED is the transfer characteristics, as shown in Figures 7-12 and 7-18. That is, a laser is a nonlinear device because the transfer characteristic curve (i.e., *P* versus *i* curve; *P* the optical power and *i* the electrical driving current of a laser) is nonlinear. The region below the "pumping threshold" (threshold current; I_{th}) is not suitable for signal modulation because a very small increase in optical output power will be produced even if the input electrical current increase is considerably large. This small modulated signal will be easily corrupted by channel noise, and cannot be restored at the receiver. However, in the region beyond the pumping threshold, a laser diode operates like an LED. That is, it operates linearly. Therefore, for any applications using a laser diode as its light source, a bias (current) must be used to guarantee the laser diode always operates in its linear region. For example, if the signal to be modulated (transmitted) has a range of -12 mA to 12 mA, and the pumping threshold of the laser diode in the optical system is 8 mA, then the minimum bias current (i_B) required is 20 mA [$= 8 - (-12)$]. Generally, the required biased current of a laser is given as follows:

$$i_B = I_{th} - i_{smin} \tag{7-14}$$

where i_B is the required bias current for the laser to operate in the linear region, I_{th} is the pumping threshold of the laser and i_{smin} is the minimum signal current.

It should be noted that the signal modulation illustrated in Figures 7-12 and 7-18 is called an "***intensity modulation***". This is because the intensity of the light (i.e., power; P) increases in accordance with the increase of the LED (or laser) driving current (i).

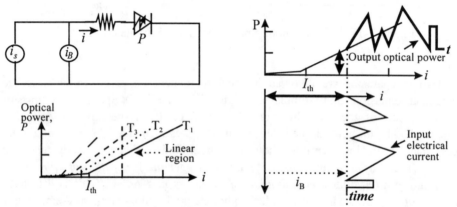

Figure 7-18   Transfer Characteristic and Signal Modulation of a Laser.

It is important to note that the transfer characteristic curve of a laser diode also changes as the operating temperature changes (like an LED). In addition, the pumping threshold of a laser diode also increases as the operating temperature increases. Since the bias current is a function of the pumping threshold, the bias current will also increase as the operating temperature increases. Hence, a laser diode is much more temperature sensitive than an LED. Therefore, it is important to maintain the operating temperature of a laser diode within specified range to insure proper performance.

7.4 OPTICAL FIBERS

Optical fibers are commonly used for back-haul (long-haul) transmission facilities in modern digital networks. There are many advantages of using fibers for transmission facilities. Two important advantages of using optical fiber systems are its high capacity (high speed) and excellent transmission quality. Fiber cables have practically no crosstalk interference. In addition, fiber size is extremely small, which is important for cable duct congestion control. It is easy to handle fiber cable due to its light weight and its flexibility. Based on its large capacity and relatively low cost material, optical fiber systems present an economical method for transport. In addition, optical fiber systems are more secure than other systems using different types of transmission media. Eavesdropping is practically impossible in fiber applications.

Optical fibers can be applied in many area as transmission media: customer loops, intra-office trunks, inter-office trunks, broadband-ISDN, CATV, HDTV, computers, data links, and local area networks. Other applications such as surveillance, remote monitoring, automobile wiring, ship wiring, aircraft wiring, gyroscope, position sensors, temperature sensors, etc. also utilize fiber optic technology.

To design and analyze an optical fiber system, many parameters and terminologies of optical fibers must be considered, including:

- Fiber structure: fiber core, cladding, and protection layers.
- Fiber dimension: fiber core and cladding.
- Index of refraction: fiber core and cladding; Step-Index (SI) and GRaded INdex (GRIN) structure.
- Fiber materials: fiber core and cladding.
- Fiber modes: single mode and multi-mode operation.
- Fiber signal attenuation: material, wavelength, splices, connectors.
- Fiber dispersion: (delay spread) material, wavelength, inter-modal distortion.
- Fiber bandwidth: power spectrum, numerical aperture.

The details of these parameters and terminologies are described in the following sections.

7.4.1 Fiber Structure

Optical fibers have a "waveguide-type" structure consisting of many layers. The inner most layer is the fiber **core**, which is the transmission medium that carries the optical light signals. From signal transport viewpoint, core layer is the most important layer. This layer and its surrounding layer, the **cladding**, perform the waveguide or *lightguide* function. Fibers must be properly designed so that the light energy carrying the information signals will be contained within the fiber core over long distances. A phenomena known as the **total internal reflection** (previously described in this chapter) is the foundation of the lightguide function.

Fibers typically have two or more layers, besides the fiber core and cladding layers, such as an inner protection layer and an outer protection layer. These additional layers are used to protect the fiber cladding and the fiber core, while also increasing the strength of the fiber cable. Figure 7-19 shows a typical four-layer fiber structure.

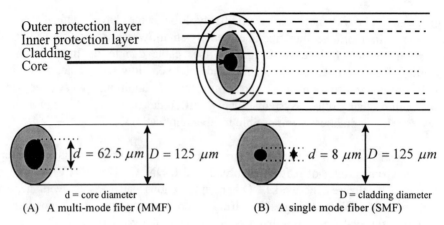

Outer protection layer
Inner protection layer
Cladding
Core

$d = 62.5 \ \mu m$ $D = 125 \ \mu m$ $d = 8 \ \mu m$ $D = 125 \ \mu m$

d = core diameter   D = cladding diameter
(A) A multi-mode fiber (MMF)   (B) A single mode fiber (SMF)

Figure 7-19   Typical Four-layer Structure/Dimensions of an Optical Fiber.

7.4.1.1 Fiber Dimensions

The diameter of the fiber cladding is typically 125 µm, which is slightly thicker than a human hair (100 to 110 µm). Therefore, one single fiber is often referred to as a "hair". The core diameter is much smaller than the cladding diameter. As shown in Figure 7-19, there are two common sizes, for commercial optical fiber communications systems:

(1) Core diameter = 62.5 µm: for Multi-Mode Fibers (MMF); sometimes referred to as enhanced fibers; a diameter of 50 µm has been adopted by some vendors).

(2) Core diameter = 8 ~ 10 µm: for Single Mode Fibers (SMF).

7.4.1.2 Index of Refraction

All materials have a specific index of refraction. This characteristic makes it possible for optical fibers to serve as light guide systems. That is, two materials with two different indices of refraction can perform the light guide function. as described in Example 7-1 previously presented in this chapter, described the principle of incident, reflection, refraction, and total internal reflection. The index of refraction (n), of a material is defined as follows:

$$n = \frac{c}{v} \tag{7-15}$$

where c is the speed of light in a vacuum and v is the speed of light in the material (which has the index of refraction n). The index of refraction plays an important role in fiber characteristics, and there are several ways to construct the fiber core and cladding to obtain a specific level of performance. Two common index profiles are: (1) Step-Index profile (SI fibers), and (2) GRaded-INdex profile (GRIN fibers).

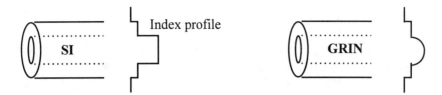

Figure 7-20   Index Profiles, SI and GRIN Fibers.

For multi-mode operation, GRIN fibers have different optical characteristics compared to SI fibers. For example, GRIN fibers have better dispersion properties, and as a result can carry more data signals. Single mode fibers are typically modeled as SI fibers because of the small core sizes, which result in small variation in index of refraction.

7.4.1.3 Fiber Materials

Three types of material are used to manufacture fiber optic cables: (1) glass, (2) PCS (Plastic Cladding Silica), and (3) plastic. Glass fibers (with the fiber core and fiber cladding made of glass) are used for the highest speed systems, and plastic fibers (with fiber core and fiber cladding made of plastic) are used for lower speed systems.

Table 7-3   Glass, PCS and Plastic Fibers.

Type	Core material	Cladding material
Glass	*Glass*	*Glass*
PCS	*Glass*	*Plastic*
Plastic	*Plastic*	*Plastic*

Based on the fiber dimension and index profile of the cladding/core, glass fibers can be grouped into four categories (as shown in Table 7-4). Single mode GRIN fibers are not practical because the core diameter is small and it supports only a few modes (e.g. 2 or 3 modes, compared to multi-mode fiber that typically supports 1500 to 2000 modes). The advantages of the GRIN structure can not be realized for single mode fibers, therefore all single mode fibers are SI. From a transmission speed viewpoint, single fibers offers the best performance (compared to GRIN and SI multi-mode fibers). As indicated in Table 7-4, GRIN multi-mode fibers have a higher speed capability than SI multi-mode fibers. In general, SI multi-mode fibers are used in Local Area Networks (LANs), Fiber-To-The-Home (FTTH), and other access networks.

Table 7-4   Fiber Types and Speeds.

	Single mode	Multi-mode
GRIN	–	Medium-high speed
SI	High speed	Low-medium speed

7.4.2 Fiber Signal Attenuation

As a light signal propagates through a fiber, the signal is attenuated along the transmission path as in the case of other media (e.g. twisted pair wires, coaxial cable, and airway). Fiber attenuation is usually expressed in dB/km. The actual attenuation is dependent upon the fiber material dependent and wavelength, as shown in Figure 7-21. Three commonly-used "windows" are in the ranges of 820 *nm*, 1310 *nm* and 1550 *nm*, which are in the infrared region. Note that the wavelength range between 400 and 700 *nm* is known as the (human) visible light, where 700 *nm* is the wavelength of a "red" light.

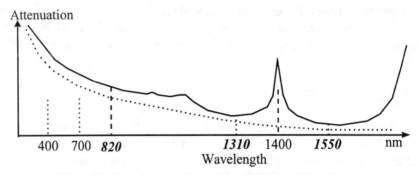

Figure 7-21   Attenuation versus Wavelength.

The 820 *nm* window was developed for optical fiber communications systems in the late 1970s and early 1980s (noted that low-loss 820 *nm* fiber optic technology was proposed in the mid 1960s). Low-speed and/or short haul systems [e.g. Local Area Networks (LANs)] often apply the 820 *nm* window. The 1310 *nm* window was developed in early 1980s for field deployment. Systems using 1310 *nm* are found in some moderate to high speed systems. The 1550 *nm* window (technology; was first developed in the mid 1980s) is the most recent high speed/super high speed optical fiber technology, and is referred to as *long wavelength technology*. Wavelength Division Multiplexing (WDM) technology (described later in this chapter) is used to increase system capacity and also operates in the 1550 *nm* region.

7.4.2.1 Signal Losses Along the Transmission Path

Figure 7-22 shows the source of signal losses along the transmission path of a typical optical fiber. The light signal originates from the light source (i.e. a LED or laser diode), however, a portion of this light signal will not enter the fiber core. These losses are referred to as coupling and connector losses. The portion of light signal that entered the fiber core will be gradually attenuated as it propagates long the fiber core. Splicing two fibers together (for long distance applications) introduces splicing and connector losses. Finally, at the receiver end, another signal loss is introduced by the connector to the photo-detector (known as connector loss, which also occurs at the transmitter: Figure 7-22).

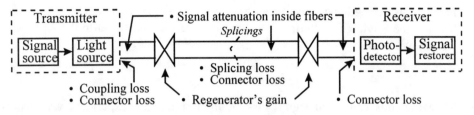

Figure 7-22   Signal Loss along the Transmission Path of an Optical Fiber Link.

If the signal power (strength) is too low for a receiver to detect (a photodetector or photodiode is always used), regenerators with specific gains must be applied at proper locations along the transmission path. In an optical fiber system, optical amplifiers (discussed later in this chapter) and regenerators are used simultaneously to compensate for signal losses along the transmission path.

7.4.2.2 Signal Loss within Optical Fibers

The light signal traveling inside a fiber core is attenuated as it propagates along the length of fiber cable. These signal losses may be categorized into three types:

(1) **Glass absorption loss**: Glass absorbs light intrinsically.

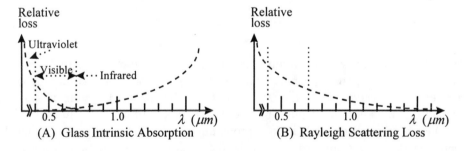

Figure 7-23   Glass Intrinsic Absorption and Rayleigh Scattering Loss.

As shown in Figure 7-23(A), the absorption losses in the ultraviolet and in the infrared regions are significant, but the loss is diminished in the visible region.

(2) **Rayleigh scattering loss**: This loss is caused by spatial variations in the optical density of glass, and small bubbles or crystals embedded in the glass. Since the non-homogeneous properties of glass that scatter light are typically smaller than the wavelength (λ) of the light signal, the scattering loss of a fiber follow (approximately) the Rayleigh scattering law. That is, the losses are very high for small wavelengths, and decrease with increasing wavelength [see Figure 7-23(B)].

For long wavelength systems (e.g., 1.55-μm systems) scattering losses are negligible because the Rayleigh scattering losses are inversely proportional to the 4th power of λ as shown below:

$$\text{Rayleigh losses} \propto \lambda^{-4} \tag{7-16}$$

(3) **Miscellaneous causes of signal loss**: A light signal can experience signal power losses due to:

- Microbending caused by surface distortion
- Diameter variations (causing connector loss)
- Bubbles penetrating the fiber core
- Core ovality and eccentricity (causing splicing loss)
- Index profile errors (causing signal distortion)
- Material absorption of impurities such as hydroxide (OH^-), etc.

Combining all the signal losses described above, the "net light signal loss" is represented by the attenuation curve shown in Figure 7-21. Note that the peak at 1400 nm corresponds to the natural oscillation frequencies (or harmonics) of specific impurities in glass (typically OH^-).

Attenuation is an important factor that determines the transmission distance of an optical fiber link. An optical fiber system must also satisfy the system power budget requirement, that is,

$$P_r = P_t - L_c - L_s - L_f - L_m - D_m \tag{7-17}$$

where P_r is the receiver sensitivity (i.e., the minimum power required), P_t is the transmitted power, L_c are the connector losses (one connector at the transmitting end and the other at the receiving end), L_s is the sum of all splice losses, L_m is the miscellaneous loss, and D_m is the design margin. The fiber loss (L_f) is the fiber attenuation (in dB/km) multiplied by the fiber length in km. From Eq. (7-17), it can be seen that the fiber attenuation (L_f) is a factor that influences the maximum transmission distance of an optical fiber system. An example of a power budget design for an optical fiber link is presented later in this chapter.

7.4.3 Fiber Dispersion

A digital pulse will spread out as it propagates through an optical fiber, as illustrated in Figure 7-24. This spreading characteristic is called "delay spread", "signal distortion", or "dispersion". The effect of dispersion is a reduction in transmission speed. As dispersion increases, the interval in which a logical "0" resides can have enough energy level to be misinterpreted as a logical "1". This phenomena is known as Inter-Symbol Interference (ISI). For example, three bits (101) have been transmitted over long distance (Figure 7-

24). The "tail end" of the first "1" is extended to the time interval of the "0" bit. Likewise, the "head end" of the second "1" is also extended to the time interval of the "0". Both "extended energies" to the logical "0" result in ISI. When the ISI is severe enough, the restored bit is "1" instead of "0". The transmission Bit Error Rate (BER) is then increased. To maintain an acceptable BER, the transmission rate must be reduced in order to reduce the ISI, then better BER can be obtained. Methods for reducing dispersion include: linear dispersion compensation, dispersion-shift (discussed later in this chapter), and non-linear dispersion compensation schemes.

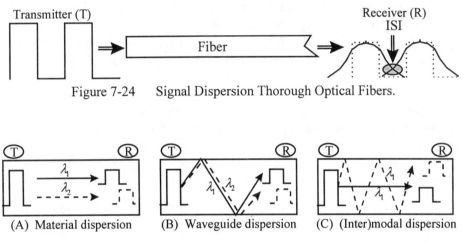

Figure 7-24   Signal Dispersion Thorough Optical Fibers.

(A) Material dispersion   (B) Waveguide dispersion   (C) (Inter)modal dispersion

Figure 7-25   Three Causes of Signal Dispersion.

There are three types of signal dispersion: (1) material dispersion, (2) waveguide dispersion, and (3) (inter)modal distortion (delay distortion, or dispersion). Sometimes, the combination of material and waveguide dispersion is referred to as "chromatic dispersion". For single mode fibers, chromatic dispersion dominants the signal dispersion characteristics. This is because single mode fiber operates in only a few modes (discussed later in this chapter), hence modal distortion is negligible. However, modal distortion is a major factor for multi-mode fibers because a typical multi-mode fiber operates in thousands of modes (described later in this chapter). Each of these signal dispersion characteristics is described as a separate topic in the following sections.

7.4.3.1 Material Dispersion

The basic material used in making glass fibers is SiO_2 (silicon dioxide; commonly called sand). Specific "impurities" are deliberately added to achieve the desired indices of refraction in the fiber core and cladding regions. The optical fibers used for communication systems are "Germanium (Ge) doped silica". Based on the properties of the fiber material, different wavelengths of light travel at different speeds. Thus, these signals arrive at the end of the fiber at different instances in time as shown in Figure 7-25(A) (Note that this figure only shows the fiber core, not the fiber cladding). The difference between the first and the last arrival significant wavelength arrival times is known as material dispersion (M). A mathematical expression of material dispersion is given as follows:

$$\Delta\tau = -M \times \Delta\lambda \times L \qquad\qquad (7\text{-}18)$$

$$\Delta\tau = t_{\lambda_2} - t_{\lambda_1} \qquad (\lambda_2 > \lambda_1) \qquad\qquad (7\text{-}19)$$

where $\Delta\tau$ (in *ps*) is the traveling time difference between two wavelength λ_2 and λ_1 (i.e., t_{λ_2} is the traveling time for the light with a wavelength of λ_2, and t_{λ_1} is the traveling time for the light with a wavelength of λ_1), M is the material dispersion [in ps/(nm × km)], $\Delta\lambda$ is the spectral width of the light source, and L is the length of the optical fiber link.

Example 7-8: Describe the relative speed for signals having various wavelengths transported by a Ge-doped silica fiber that has a material dispersion (M) as shown in Figure 7-26(A).

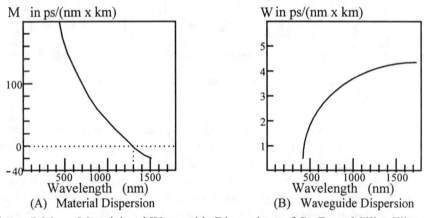

Figure 7-26   Material and Waveguide Dispersions of Ge-Doped Silica Fibers.

From Figure 7-26(A) it can be seen that if the wavelengths of the light signals (λ_2 and λ_1; assuming $\lambda_2 > \lambda_1$) are smaller than 1300 *nm* (i.e., $\lambda_1 < 1300$ nm, $\lambda_2 < 1300$ nm) the value of M is positive. The length of the fiber (L) and the spectral width of the light source ($\Delta\lambda$) are always positive, therefore by applying Eq.(7-18) with M having a positive value, the derived $\Delta\tau$ is negative. From Eq. (7-19), with a negative $\Delta\tau$, it can be obtained:

$$t_{\lambda_1} > t_{\lambda_2}$$

Therefore, a light with a longer wavelength (< 1300 *nm*) will travel faster in the fiber. That is, for the light with a wavelength ≤ 1300 *nm*, the 1300 *nm* light travels the fastest.

In contrast, for signals with wavelengths larger than 1300 *nm*, the shorter wavelength travels faster than the longer wavelength. The result is obtained as follows. From Figure 7-26, it can be seen that M has a negative value (L and $\Delta\lambda$ are both positive), from Eq.(7-18), the value of $\Delta\tau$ is derived as "positive". That is, from Eq. (7-19), the following is true:

$$t_{\lambda_1} < t_{\lambda_2}$$

Therefore, the traveling time of a shorter wavelength signal (t_{λ_1}; $\lambda_1 > 1300$ nm) is always less than the traveling time of a longer wavelength signal (t_{λ_2}; $\lambda_2 > 1300$ nm). That is, for the light with a wavelength ≥ 1300 *nm*, the 1300 *nm* light travels the fastest.

7.4.3.2 Waveguide Dispersion

The second type of dispersion that exists in a single mode fiber is waveguide dispersion [as illustrated in Figure 7-25(B)]. The fiber core and the fiber cladding form a waveguide for light signals. Therefore signals with different wavelength will not behave the same at the core-cladding boundary. That is, even if two wavelengths (λ_2 and λ_1) enter the boundary at the same instance in time, they will "bounce" at different angles, and will arrive at the destination in different time instances. This characteristic is known as waveguide dispersion (W), which is expressed as follows:

$$\Delta\tau = -W \times \Delta\lambda \times L \tag{7-20}$$

$$\Delta\tau = t_{\lambda_2} - t_{\lambda_1} \qquad (\lambda_2 > \lambda_1) \tag{7-21}$$

where $\Delta\tau$ (in *ps*) is the traveling time difference between two wavelength λ_2 and λ_1 (i.e., t_{λ_2} is the traveling time for the light with a wavelength of λ_2, and t_{λ_1} is the traveling time for the light with a wavelength of λ_1), W is the waveguide dispersion [in ps/(nm × km)], $\Delta\lambda$ is the spectral width of the light source, and L is the length of the optical fiber link.

Eqs.(7-20) and (7-21) are almost identical to the equations representing material dispersion [i.e., Eqs.(7-18) and (7-19)]. The waveguide dispersion characteristic for commercial grade Ge-doped silica fibers is shown in Figure 7-26(B). It can be seen that the waveguide dispersion has a positive value for all wavelengths (i.e., W > 0).

For conventional single mode fibers, the total dispersion is equal to the sum of the material dispersion (M) and the waveguide dispersion (W), as shown in Figure 7-27(A). Therefore, the total dispersion of a conventional single-mode fiber is given as follows:

$$\Delta\tau = |(M + W)| \times \Delta\lambda \times L \tag{7-22}$$

Example 7-9: Compare the characteristics of Dispersion-Shifted Single Mode Fiber (DS-SMF), and Dispersion-Flattened Single Mode Fiber (DF-SMF) using Figure 7-27.

Eq.(7-22) is the most important equation that can be used to describe the dispersion-shift and dispersion-flattened techniques. From Figure 7-26, it can be seen that commercial Ge-doped silica fibers have negative values for M and positive value for W if the system operating wavelength is the long wavelength range (around 1550 *nm*). That is,

(1) The material dispersion (M) is practical equal to zero at 1310 *nm*, and has negative values for wavelengths larger than 1310 *nm* [see Figure 7-26(A)]. At 1550 *nm*, the material dispersion is approximately –20 *ps/(nm* x *km*).

(2) The waveguide dispersion (W) always has a positive value in the range from 1310 *nm* to 1600 *nm*. At 1550 nm, the waveguide dispersion is 4.5 *ps/(nm* x *km*) [see Figure 7-26(B) or Figure 7-27)].

Therefore, in the range of (1310 *nm*, 1600 *nm*) cancellation between material dispersion (i.e., "+" values) and waveguide dispersion (i.e., "–" values) is possible. For example,

$$
\begin{aligned}
M+W \quad &= |-20 + 4.5| \quad = 15.5 \\
&= |-16 + 6.5| \quad = 9.5 \\
&= |-12 + 9.0| \quad = 3.0 \\
&= |-10 + 9.5| \quad = 0.5
\end{aligned}
$$

That is, by applying dispersion shift technique [shifting M from –20 *ps/(nm* x *km*) to –10 *ps/(nm* x *km*), and shifting W from 4.5 *ps/(nm* x *km*) to 9.5 *ps/(nm* x *km*)], it is possible to have an "almost zero" total (net) dispersion, by substituting (M+W) = 0.5 into Eq.(7-23) and obtain $\Delta\tau \approx 0$. This type of special single mode fiber is called Dispersion-Shifted Single Mode Fiber (DS-SMF), as shown in Figure 7-27(B). The technique for changing material dispersion (M) is to add proper impurities into the fiber core. Likewise, the technique for changing waveguide dispersion (W) is to design different index of refraction profile of the fiber cladding and core.

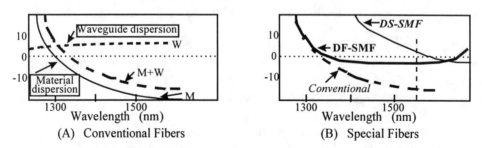

Figure 7-27   Conventional, Dispersion-Shifted & Dispersion-Flattened Fibers.

In general, a single mode fiber is suitable for operating at a specific wavelength (e.g., 1310 *nm*, or 1550 *nm*). The DS-SMF is suitable for this type of operation because it has a nearly zero dispersion at the operating wavelength, but a considerably large dispersion elsewhere. This characteristic may not be acceptable in Wavelength Division Multiplexed (WDM) systems because a range of wavelengths is required for the systems. For this type of application, a very special fiber type known as Dispersion Flattened Single Mode Fiber (DF-SMF) is needed. This type of fiber has a "very small" and "flat" net (total) dispersion in the range of 1320 *nm* to 1570 *nm*. Therefore, it can be efficiently operated at any wavelength in that range [see Figure 7-27(B)].

7.4.3.3 (Inter)-Modal Distortion

The third type of fiber dispersion [shown in Figure 7-25(C)] is called (inter)-modal distortion and is the dominant dispersion condition in multi-mode fibers. When the same wavelength light signals enter at the fiber core at different angles, the propagation paths are different. The light taking a straight path will arrive at "the end of the fiber" first. The light signal taking a "zig-zag" path will arrive at a later instance in time. The difference in traveling time between the fastest path and the slowest path is called (inter)-mode distortion. The theoretical modal distortion is given by Eqs.(7-23) and (7-24). However, the practical modal distortion (described later in this chapter) is significantly different from the following expressions.

(A) *SI multi-mode fibers*: The (inter)-modal distortion is calculated as follows.

$$\Delta\tau = \frac{n_1(n_1 - n_2)}{cn_2} \times L \tag{7-23}$$

$$\approx \frac{n_1}{c}\Delta \quad if \ n_1 \approx n_2 \tag{7-24}$$

where $\Delta\tau$ is the (inter)-modal distortion, L is the length of the fiber, c is the speed of light in a vacuum (= 3×10^8 m/s), the n_1 is the index of refraction of the fiber core, n_2 is the index of refraction of the fiber cladding, and $\Delta = $ *fraction index* $= (n_1 - n_2)/n_1$.

Example 7-10: Compute the estimated data rate over a 100 km SI multi-mode fibers with $n_1 = 1.48$, and $n_2 = 1.46$.

By substituting $n_1 = 1.48$, and $n_2 = 1.46$ into Eq.(7-23), one can obtain

$$\Delta\tau_{SI} = \frac{1.48(1.48 - 1.46)}{3 \times 10^8 \ (m/s)(1.46)} \times (100 \ km) = 6.76\mu s$$

The "rule of thumb" relationship between the data rate and the dispersion is given by the following equation. For this example, it is assumed that the (inter)-modal distortion dominants the total fiber dispersion.

$$R = Data \ rate = 1/(2\Delta\tau) = 1/(2\times 6.76 \ \mu s) \approx 73 \ kbps \tag{7-25}$$

The estimated data rate for this example is calculated to be R = 73 kbps, which is a relatively slow speed for optical fiber applications.

(B) *GRIN multi-mode fibers*: The (inter)-modal distortion of GRIN MMF is given by

$$\Delta \tau_{GRIN} = \frac{n_1}{c} \cdot \frac{\Delta^2}{2} \cdot L = (\frac{\Delta}{2}) \cdot (\Delta \tau)_{SI} \qquad (7\text{-}26)$$

It can be seen if $n_1 = 1.48$, and $n_2 = 1.46$, then $\Delta = fraction\ index = (n_1 - n_2) / n_1 = 0.00135$. Substitute these values into Eq.(7-26) yields the following result:

$$(\Delta \tau)_{GRIN} = 0.68\% \times (\Delta \tau)_{SI} = 4.6\ ns$$

$$R = Data\ rate = 1 / (2 \Delta \tau) = 1 / (2 \times 4.6\ ns)\ \approx 10.735\ Mbps$$

Therefore, GRIN multi-mode fiber has better performance than SI multi-mode fiber, with respect to data rate. That is, a GRIN multi-mode fiber has lower (inter)-modal distortion than an SI multi-mode fiber, which allows GRIN multi-mode fibers to support higher data rates than SI multi-mode fibers.

Eqs.(7-23) and (7-26) represent the theoretical (inter)-modal distortion of SI multi-mode and GRIN multi-mode fibers, respectively. However, these formulas provide the theoretical values, which are a result of the assumption that an optical fiber link is a "straight line". In this model, the fastest ray (mode) will always reach the end of the fiber first, and the slowest ray will always reach the end of the fiber end last.

In practical applications an optical fiber link has many "macrobends" along the transmission path. At the beginning of the path, one ray may be ahead of the others. However, due to fiber bending, the fastest ray may take a longer path and may not be the first ray that reaches the end of the fiber. Likewise, the slowest ray at the "head end" of the fiber link (due to fiber bending) may be the first ray that reaches the end of the fiber Therefore, the actual modal distortion is found to be much smaller than the theoretical value, and can be expressed by the following equations:

$$\Delta \tau_d = (\frac{\Delta \tau}{L})_{theoretical} \times L \qquad \text{if } L \le L_e \qquad (7\text{-}27)$$

$$= (\frac{\Delta \tau}{L})_{theoretical} \times \sqrt{LL_e} \qquad \text{if } L > L_e \qquad (7\text{-}28)$$

where $(\Delta \tau / L)_{theoretical}$ is the theoretical modal distortion [calculated from Eq. (7-23) or (7-26)], and for typical Ge-doped silica fibers the equilibrium length, L_e, is approximately 1 km. That is, the (inter)-modal distortion is directly proportional to the fiber length for a fiber less than L_e. The distortion increases rapidly as the length increases. However, for fiber with a length longer than L_e, the distortion is proportional the square root of the fiber length. Therefore, the distortion will not increase as fast as before.

The total signal dispersion (i.e. the sum of material dispersion, waveguide dispersion and modal distortion) of a multi-mode fiber is given as follows:

$$(\Delta\tau)^2_{total} = [(M + W) \times \Delta\lambda \times L]^2 + (\Delta\tau_d)^2 \qquad (7\text{-}29)$$

where M and W are the material and waveguide dispersions, $\Delta\lambda$ is the spectral width of the light source, L is the length of the optical fiber link, and $\Delta\tau_d$ is the practical (inter)-modal distortion given by Eq.(7-28) when $L > L_e$.

7.4.4 Fiber Modes

Engineering applications (rather than an extensive theoretical definition) of fiber modes are presented in this book. Fiber modes can be viewed as light rays (light paths) propagating through the fiber core (see Figure 7-28). The terminology and characteristics of fiber modes is described in Examples 7-11, 7-12 and 7-13.

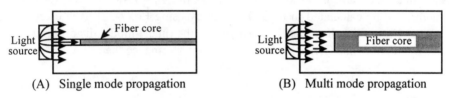

    (A)   Single mode propagation   (B)   Multi mode propagation

Figure 7-28   Single/Multi-mode Fiber Propagation.

Example 7-11: Describe the operational characteristics of a single mode fiber.

A (true) theoretical single-mode light propagation must meet the following condition:

$$d \leq 2.405\lambda / (\pi \times \sqrt{n_1^2 - n_2^2}) \qquad (7\text{-}30)$$

where d is the fiber core diameter, λ is the operating (nominal) wavelength, and n_1 and n_2 are the indices of refraction of the fiber core and cladding, respectively. For an optical fiber with a central wavelength of 1550 nm, $n_1 = 1.48$, and $n_2 = 1.46$, the fiber core must have core diameter $d \leq 4.89$ μm. A fiber with such as small core size has not been successfully commercially produced.

Example 7-12: Describe the relationship between the theoretical and practical operation of a single mod fiber.

A commercial single mode fiber with a core diameter $d = 8$ μm and $\lambda = 1.55$ μm has been theoretically calculated to have six modes (not one mode). However, for an actual transmission distance between 25 miles and 45 miles, 2 to 3 modes have been measured.

Example 7-13: Describe and apply the formula [Eqs.(7-31) and (7-32)] used to calculate the number of fiber modes (N_m) for SI and GRIN multi-mode fibers.

Eq.(7-31) provides the formula for calculating the number of fiber modes (N_m) for SI multi-mode fibers. In Eq.(7-31), d is the fiber core diameter, λ is the operating wavelength, n_1 is the index of refraction of fiber core, and n_2 is the index of refraction of fiber cladding. The parameter V is defined as the normalized frequency.

$$N_m = \frac{V^2}{2}; \quad V = \frac{\pi d}{\lambda} \times \sqrt{n_1^2 - n_2^2} \qquad \text{for SI -MMF} \qquad (7\text{-}31)$$

$$N_m = \frac{V^2}{4} \qquad\qquad\qquad \text{for GRIN MMF} \qquad (7\text{-}32)$$

An SI multi-mode fiber with a core diameter $d = 62.5$ μm, $\lambda = 0.82$ μm, $n_1 = 1.48$, and $n_2 = 1.46$, the normalized frequency V, and the number of modes can be calculated to be $V = 58.062$ and $N_m = 1686$ modes by using Eq.(7-31). Likewise, a GRIN multi-mode fiber with a core diameter $d = 62.5$ μm, $\lambda = 1.31$ μm, $n_1 = 1.48$, and $n_2 = 1.46$, the number of modes can be calculated to be 330 modes by using Eq.(7-32). It is clear that GRIN-MMF has less modes than SI-MMF. Therefore, the (inter)-modal distortion is smaller for GRIN-MMF, and thus GRIN-MMF has a better performance than SI-MMF.

7.4.5 Numerical Aperture

The maximum acceptance angle (θ_α) for coupling a light source to the fiber core was previously defined in Section 7.2.1 (Example 7-1). Another parameter, Numerical Aperture (NA) is related to the maximum acceptance angle as follows:

(1) For SI fibers, the Numerical Aperture (NA) and the maximum acceptance angle relationship is given by the following equation:

$$\text{NA} = sin\ \theta_\alpha = \sqrt{n_1^2 - n_2^2} \qquad (7\text{-}33)$$

For an SI fiber with $n_1 = 1.48$, and $n_2 = 1.46$, the maximum acceptance angle was calculated to be 14° (Example 7-1). Therefore, the Numerical Aperture (NA) can be calculated from to be 0.24 using Eq.(7-33).

(2) For parabolic GRIN fibers, which have an index (of refraction) profile of a parabola, the numerical aperture and the maximum acceptance angle are related as follows:

$$NA = n_1\ (2\Delta)^{1/2}\ \sqrt{1 - (r/d)^2} \qquad (7\text{-}34)$$

where Δ is the fraction index (previously defined), r is the distance from the core axis to the entry point of the light source, and d is the core diameter.

If the light enters at the core axis (i.e. the center of the core), the numerical aperture can be simplified as Eq.(7-35). For typical fibers with $n_1 = 1.48$ and $n_2 = 1.46$, the numerical aperture and the maximum acceptance angle can be calculated as 0.2433 and 14.08°, which are similar to those of SI multi-mode fibers.

$$NA = n_1 \sqrt{2\Delta} \qquad (7-35)$$

7.4.6 Fiber Specifications

Optical fiber systems can be used to carry either analog or digital signals. Hence, one parameter (***3-dB bandwidth***) is very useful in specifying a fiber.

Example 7-14: Determine the maximum number of voice channels (4 kHz each) if the transmission distance is (a) 10 km, and (b) 5 km using fibers with a 32 MHz•km bandwidth (3-dB bandwidth), no guard band between channels is used.

(a) For the 10 km applications: By definition, it is given that

$$f_{3\text{-dB}} \times L = 32 \text{ MHz•km}$$

If $L = 10$ km, then $f_{3\text{-dB}} = 32$ MHz (= 32 MHz•km/10 km) is the 3-dB bandwidth. Thus, the maximum number of voice channels, $N_c = (32 \text{ MHz})/(4 \text{ kHz}) = 8,000$.

(b) For 5 km applications: $L = 5$ km, $f_{3\text{-dB}} = 64$ MHz (=32 MHz•km /5 km), and $N_c = (64$ MHz)/(4 kHz) = 16,000 voice channels.

Example 7-15: Technical specifications of 8.2-μm single mode (SI) fibers.

Fiber parameter	Specifications	
Core diameter		*8.2 μm (nominal)*
Cladding diameter		*125.0 μm*
Coating diameter		*250 μm*
Mode field diameter		*8.8 ± 0.5 μm*
Core eccentricity		*≤ 1.0 μm*
Attenuation range	*(TPD):*	*0.38 to 0.50 dB/km @ 1310 nm*
		0.25 to 0.45 dB/km @ 1550 nm
	(MIFL):	*0.40 to 0.60 dB/km @ 1310 nm*
		0.30 to 0.50 dB/km @ 1550 nm
Zero dispersion wavelength		*1310 ± 10 nm*
Maximum dispersion range		*3.2 ps/(nm × km) [1285 to 1330 nm]*
		19 ps/(nm × km) [1550 nm]

TPD: Transmission performance designators;   MIFL: Maximum individual fiber loss

Example 7-15 illustrates typical fiber specifications of 8.2-μm (SI) single mode fiber. Besides physical specifications, attenuation and dispersion are two primary parameters that must be specified so that they can be verified for their performance in the field. Since single mode fibers are typically used for high-speed systems, dispersion requirement is practically the most important system specification.

Examples 7-16 and 7-17 provide similar specifications for 62.5 μm multi mode fibers (known as enhanced fibers), and 50 μm multi mode fibers (known as standard fibers). Bandwidth requirement is an important specification for multi mode fiber applications.

Example 7-16: Technical specifications of 62.5-μm multi mode fibers.

Fiber parameter	Specifications		
Core diameter	*62.5 μm (nominal)*		
Cladding diameter	*125.0 μm*		
Coating diameter	*250 μm*		
Core ovality	*20% maximum (typical value 4%)*		
Core eccentricity	*7.5% maximum (typical value 1.5%)*		
Attenuation range	*(TPD):*	*3.45 to 3.8 dB/km @ 850 nm*	
		0.85 to 2.0 dB/km @ 1300 nm	
	(MIFL):	*4.15 to 4.3 dB/km @ 850 nm*	
		1.0 to 2.0 dB/km @ 1330 nm	
Refraction index delta	*2%*		
Bandwidth range	*300 to 1,000 MHz-km @ 850 nm*		
	150 to 300 MHz-km @ 1300 nm		

TPD: Transmission performance designators;   MIFL: Maximum individual fiber loss

Example 7-17: Technical specifications of 50-μm multi mode fibers.

Fiber parameter	Specifications		
Core diameter	*50 μm (nominal)*		
Cladding diameter	*125.0 μm*		
Coating diameter	*250 ± 15μm*		
Core ovality	*20% maximum (typical value 4%)*		
Core eccentricity	*8.5% maximum (typical value 1.5%)*		
Attenuation range	*(TPD):*	*2.9 to 3.6 dB/km @ 850 nm*	
		0.85 to 2.0 dB/km @ 1300 nm	
	(MIFL):	*3.5 to 4.2 dB/km @ 850 nm*	
		1.0 to 2.4 dB/km @ 1330 nm	
Refraction index delta	*1.3%*		
Bandwidth range	*300 to 900 MHz-km @ 850 nm*		
	300 to 1,400 MHz-km @ 1300 nm		

TPD: Transmission performance designators;   MIFL: Maximum individual fiber loss

7.4.7 Fiber Cables

A generic fiber cable [Figure 7-29(A)] consists of three major layers: (1) a cable core, which is a tube containing many fibers; (2) a cable sheath, which is a protective layer having different designs based on environmental conditions, and (3) cable oversheath, which is an external protective layer for unusually harsh environment. For example, a fiber cable can be designed to have an oversheath for tensile strength, crush resistance, protection from excessive bending, abrasion resistance, vibration isolation, moisture and chemical protection, radiation resistance, extreme temperatures (flame retarded), etc. Commercial sheath and oversheath types are listed as follows:

- Sheath types: Commonly used sheaths are: steel reinforced crossply, non-metallic crossply, rodent-lightning protected crossply, PVC riser crossply, Lightguide Express Entry (LXE), dielectric sheath, metallic sheath, and fiber-glass reinforced sheath. LXE consists of two steel strength members running longitudinally along a coated carbon steel armor, and is an economical cable that is excellent for mid-span applications. Table 7-5 lists recommended cable sheaths for specific applications. In areas exposed to lightning, fiber-glass reinforced sheath can be used (instead of steel reinforced cable) with additional rodent-lightning protection. The oversheath for this type of application is known as "B-oversheath", which is typically required if rodent protection is needed. In aerial plant applications, where high lightning exposure is a consideration, the low strength fiber-glass reinforced sheath is also used.

Table 7-5   Fiber Cable Applications and Sheath Type.

Cable application	Environmental hazard	Sheath/oversheath
Aerial	*None*	*LXE-ME*
	Rodent/lightning	*LXE-RL*
Buried PVC pipe	*Rodent/lightning*	*LXE-ME*
Directed buried,	*Lightning*	*LXE-ME*
trenched or plowed	*Rodent/lightning*	*LXE-RL*
Underground plant and buried PVC pipe	*None*	*LXE-DE*
Indoor	*Fire*	*PVC crossply*
Lake/river crossings	*None*	*C-oversheath*
	Bottom currents/snagging	*Wire armored oversheath*

LXE: Lightguide express entry;   DE: Dielectric;   ME: Metallic;   RL: Rodent lightning

- Oversheath types: Three commonly used types of oversheath are: B-oversheath for rodent protection, C-oversheath for water-proofing, and wire-armored oversheath for extra strength applications.

There are several commercial fiber cable types: ribbon cables, Lightpack® cables, building cables, lightweight cables, armored cables, fish-bite cables, etc. For outside plant applications, ribbon or Lightpack® cable core is often used with metallic, rodent/lightning or dielectric sheaths. For metro-high risers, either ribbon or Lightpack® cable core is used with a fire retardant sheath. Figure 7-29(B) illustrates a commonly used commercial ribbon cable. One ribbon consists of 12 single coated fibers (color coded for easy identification), which is surrounded by a protective adhesive material, and polyester tape. Ribbon cable is available within a variety of configurations (2, 6, 12, 18 or 24 ribbons per cable). The ribbons are contained in a core tube with air or an inert filling material [see Figure 7-29(B)].

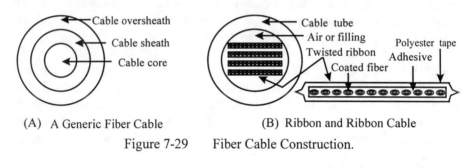

(A) A Generic Fiber Cable   (B) Ribbon and Ribbon Cable

Figure 7-29   Fiber Cable Construction.

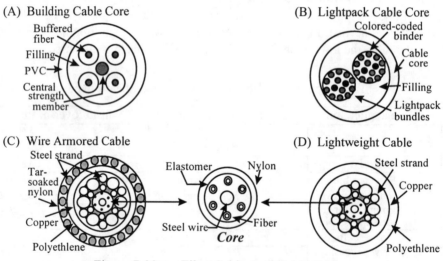

Figure 7-30   Fiber Cables and Cable Cores.

Figure 7-30 shows the structures of four different fiber cable types: (A) building cables, (B) Lightpack® cables, (C) wire armored cables, and (D) lightweight cables. Building cables contain 2, 4, 6 or 12 fibers per cable. The buffered fibers are color coded, and contained within PVC jackets. The central strength member is metallic. Lightpack® cable contains 12 fibers (color coded) per bundle within spiral wrapped, color coded binders. A cable can contain up to 8 bundles. The cable core is filled with water-blocking compound. Wire armored cables built with extra protection strength. Lightweight cables are built for easy handling. Both types have a central core with six fibers.

Rodent protection is available for several cable types. Two spiral overlapping layers of stainless steel are placed directly over the standard lightguide cable sheath and covered with an additional layer of high density polyethylene. Rodent protected cables are also designed to be used in areas exposed to lightning or where rodent infestation has caused damage to buried cable, aerial telephone wires, power, or cable TV facilities.

7.4.8 Passive Components

Connectors are a common example of the passive components used in optical fiber links. They are used to connect the transmitter to the fiber, the fiber to the receiver, and for splicing. Connectors are usually temporary, and introduce 0.2 to 0.5 dB loss per connection (at the transmitter and receiver). Splices are usually permanent and introduce 0.05 to 0.1 dB per splice. In designing/selecting connectors and splicing technologies, it is important to consider "insertion loss", "loss repeatability", "reflections", "environmental stability", and "ease of assembly". Three important steps are required in connectorization and splicing: fiber end preparation, alignment of fibers, and retention of fibers in the aligned position. There are many types of connectors commercially available: biconic connectors, ST (Straight Tip) connectors, SC connectors, FC connectors, etc. The commonly used splicing methods are rotary splice, rapid ribbon splice, and array splice.

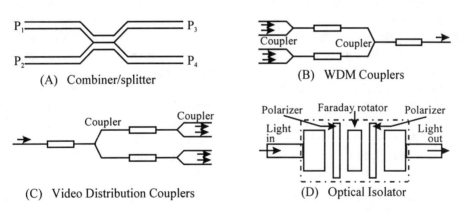

(A) Combiner/splitter

(B) WDM Couplers

(C) Video Distribution Couplers

(D) Optical Isolator

Figure 7-31   Combiners/Splitters, Couplers and Isolators.

A second category of passive components are combiners and splitters (couplers). Important parameters to be considered are insertion loss, and the coupling (splitting) ratio. These components are used in Wavelength Division Multiplex (WDM) equipment, and video distribution systems, etc. [see Figure 7-31(A) to 7-31(C)].

Another passive component is the optical isolator, using polarizer and Faraday rotators to isolate light propagation to a desired direction [see Figure 7-31(D)]. They are usually applied to single mode fibers and biconic connectors.

Passive attenuators are often used at the receiving end of an optical fiber link to provide the needed loss when the signal is "too hot", or as a Line Built-Out (LBO) network. Attenuators can be used for "single-mode to single-mode", and "single-mode to multi-mode" fiber connections, and typically have a return loss greater than 30 dB.

7.5 LIGHT DETECTORS

A light detector (photodetector, or photodiode) is the receiving component in an optical fiber link. Two commercial light detector types, used as receivers for optical fiber links, have been developed for telecommunications applications: PIN (**P**-type, **I**ntrinsic-type, and **N**-type semiconductor) detectors, and APD (Avalanche Photo-Detector or Photo Diode) detectors. PIN receivers are typically used for short optical fiber links, and APDs are used for longer links. Several optical properties of a photodiode must be specified for an optical link: rise time, responsivity, sensitivity, quantum efficiency, APD internal gain, VI (Voltage/Current) characteristics, and transfer characteristics. These parameters are described in the following sections.

7.5.1 The Receiver Front-end Circuit for an Optical Link

Figure 7-32 shows the major components of the front-end circuit for a typical optical fiber link. The receiver is connected to a photodiode, which accepts light signals from the fiber core. The photodiode must be reverse biased, that is, the negative port of the biased voltage (V_B) must be connected to the positive port of the photodiode. The output of the diode is connected to a load resistor (R_L) which provides the input voltage to a preamplifier. An AC coupling is used to connect the diode to the preamplifier. One or more post-amplifiers may be applied after the preamplifier to achieve acceptable signal level to meet Signal-to-Noise ratio (S/N) and/or Bit Error Rate (BER) requirement. This is because the typical received optical power is too weak (−25 to −40 dBm) to be fed to the conventional electrical components connected the end of an optical link.

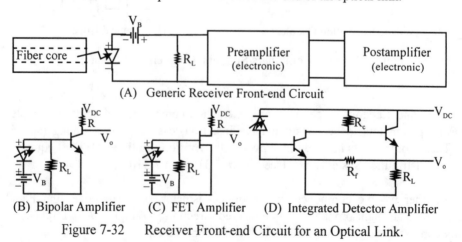

(A) Generic Receiver Front-end Circuit

(B) Bipolar Amplifier   (C) FET Amplifier   (D) Integrated Detector Amplifier

Figure 7-32   Receiver Front-end Circuit for an Optical Link.

7.5.2 Photodiode Parameters

Major photodiode parameters used in analyzing and designing an optical fiber link are: (1) rise time, (2) quantum efficiency, (3) responsivity, (4) APD internal gain, (5) sensitivity, and (6) VI characteristics. They are described in the following sections.

7.5.2.1 Rise Time

The rise time of a photodiode has the same definition as the rise time of a light source. That is, in Figure 7-13, the light source is replaced by a photodiode. Similarly, the graph for electrical current and optical power are exchanged (see Figure 7-13). However, the photodetector rise time can be decomposed into two parts: (1) transit time, and (2) circuit rise time. The transit time (t_{tr}) is device dependent. It is influenced by the depletion layer width of the device, and the electronic speed (v) in the device. The circuit rise time (t_{ct}) is dependent upon the junction capacitance (C_d) of the diode, and the value of the load resistor (R_L). Therefore, the (total) photodetector's rise time (t_{pd}) is given by the following equation:

$$t_{pd} = (t_{tr}^2 + t_{ct}^2)^{1/2} = [(\frac{Depletion\ layer\ width}{v})^2 + (2.19 \times R_L\ C_d)^2]^{1/2} \tag{7-36}$$

In Eq.(7-36), the transit time (t_{tr}), and the device junction capacitance (C_d) of a photodiode are both specified by the manufacturer. The load resistor (R_L) is a design parameter.

7.5.2.2 Quantum Efficiency

The quantum efficiency (η) is used to express the quality of a photodiode (i.e., its ability to effectively convert an optical signal into an electrical signal). If the number of incident photons from the fiber entering the photodetector is N_C, and the number of the emitted electrons generated by the photodetector is N_e, then the quantum efficiency of the photodetector (η) is defined as follows:

$$\eta = \frac{N_e}{N_c} \tag{7-37}$$

A typical photodiode has a quantum efficiency of 0.8 to 0.95.

7.5.2.3 Responsivity

Photodiode responsivity (this word is derived from "response"), which is very important is optical fiber link design, and defined as follows:

$$Responsivity\ = \rho = \frac{Received\ optical\ power}{Converted\ electrical\ current} \tag{7-38}$$

For example, if a receiver (photodetector) has a responsivity of $\rho = 0.8$ w/A, the receiver will generate (convert a photonic signal into electrical signal) 0.8 μA of electrical current for transmission, if the received optical power is 1μw. Table 7-6 lists three PIN photodetector types and their associated corresponding range, peak response wavelength, and peak responsivity. The three commercial photodiodes are:

(1) Silicon (Si) photodiode: Figure 7-33(A) shows the relative responsivity for the wavelength range listed in Table 7-6. It is most suitable for 800 *nm* applications.

(2) Germanium (Ge) photodiode: Figure 7-33(B) illustrates this device is the most suitable for 1550 *nm* applications. However, it can also be used for 1310 *nm* applications.

(3) Indium-Gallium-Arsenide (InGaAs) photodiode: Figure 7-33(C) indicates this device is most suitable for 1550 *nm* applications.

Table 7-6   Commercial PIN Photodiode (PIN) Types.

Material	*Wavelength range (μm)*	*Peak response wavelength (μm)*	*Peak responsivity (A/W)*
Silicon (Si)	0.3 ~ 1.1	0.80	0.5
Germanium (Ge)	0.5 ~ 1.8	1.55	0.7
InGaAs	1.0 ~ 1.7	1.70	1.1

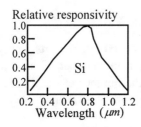 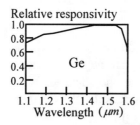 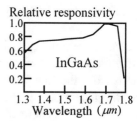

Figure 7-33   Relative Responsivity of Various PIN Photodiodes.

The current generated by the photodetector is given by the following equation:

$$i = \frac{\eta e P}{hf} = \frac{\eta e \lambda P}{hc}$$

(7-39)

where η is the quantum efficiency of the photodetector, P is the received optical power, *e* is the electronic charge (= 1.6×10^{-19}), *h* is Planck's constant (6.626×10^{-34} Joules•sec), *c* is the speed light in a vacuum ($\approx 3 \times 10^8$ m/s), λ is the wavelength of the light signal, and *f* is its associated frequency ($\lambda f = c$).

7.5.2.4 APD Internal Gain

Figure 7-34 illustrates the differences between a PIN and an APD photodetector. There are three layers in "PIN": Positive (P, possessing excessive positive charges called "holes"), Intrinsic (I, possessing equal amount of positive and negative charges), and Negative (N, possessing excessive negative charges called "electrons"). In contrast, an APD contains four layers: P+ (equivalent to "P" in PIN, possessing excessive positive charges), π (equivalent to "I" in PIN), "P" (possessing excessive positive charges, but not as much as in the P+ region), and "N+" (equivalent to "N" in PIN, possessing excessive negative charges). Note the "I" and ""are the regions where "recombination" takes place.

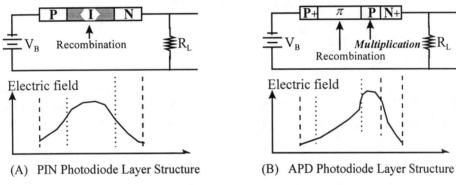

(A) PIN Photodiode Layer Structure   (B) APD Photodiode Layer Structure

Figure 7-34   Difference between a PIN and an APD.

Due to the layer structure of an APD (besides the recombination process) the electric field in the "P" region is very strong and allows an additional multiplication process. Therefore, the photo-generated current produced by an APD photodiode is much stronger than the current generated by a PIN (assuming the same amount of photonic energy is applied to both photodiodes). Therefore, there is an internal gain (M) in the APD photodiode process of converting a light signal into an electrical signal. The photo-generated current, produced by an APD, current is given by the following equations:

$$i = M \left(\frac{\eta e P}{hf}\right) = M \left(\frac{\eta e \lambda P}{hc}\right) \qquad \text{[refer to Eq.(7-39) for PIN]} \quad (7\text{-}40)$$

$$M = \text{APD internal gain} = \frac{1}{1 - (V_B / V_{BR})^n} \qquad (7\text{-}41)$$

where V_B and V_{BR} (typically between 20 ~ 500 volts) are the diode's reverse bias and reverse breakdown voltages, respectively, and n (< 1) is a device dependent parameter which is an empirical data. For commercial APDs, the typical internal gain has a value between 50 ~ 200. The responsivity defined for a photodiode has been previously described in Eq.(7-38) for PIN devices. The same definition is applied to APD devices. However, some manufacturers include the APD gain in specifying the device responsivity. Therefore, the APD values of responsivity are: 30 to 170 A/W.

7.5.2.5 Sensitivity

A photodiode receiver sensitivity is the minimum required received power for an optical fiber link to meet a specific operational (performance) requirements. For example, an optical fiber link may need a receiver with a sensitivity of −32 dBm if the required S/N of the link is 21 dB. Likewise, a receiver with a sensitivity of −36 dBm may be acceptable if the required S/N is 18 dB. Figure 7-35 shows the sensitivity ranges for a typical PIN and APD receiver. The receiver sensitivity (P_S, which is the required minimum received power) are expressed by the following equations:

$$P_S = \frac{1}{\rho} \sqrt{4kT(\Delta f / R_L) \cdot (S/N)} \qquad \text{(without preamplifier)} \qquad (7\text{-}42)$$

$$= \frac{1}{\rho} \sqrt{4kTF(\Delta f / R_L) \cdot (S/N)} \qquad \text{(with preamplifier)} \qquad (7\text{-}43)$$

where ρ is the photodiode responsivity (see Section 7.5.2.3), k is the Boltzmann constant (= 1.38 × 10^{-23} J/°K), T is the receiver ambient temperature in °K, F is the noise figure (noise factor) of the preamplifier, Δf is the system bandwidth, R_L is the load resistance, and S/N is the required system signal-to-noise ratio.

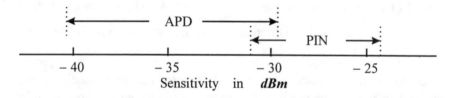

Figure 7-35   Receiver Sensitivity Ranges for a Typical PIN and APD.

Example 7-18: Determine the receiver sensitivity (P_S) for an optical fiber system that requires an S/N of 21 dB (125.9), assuming the receiver responsivity (ρ) is 0.8 A/W, the receiver ambient temperature (T) is 27°C, and the preamplifier has a noise figure (F) of 3 dB.

Applying Eq.(7-43), with the assumption that the system bandwidth (Δf) is 400 MHz, and the load resistance (R_L) 600 Ω, a typical value of receiver sensitivity = −26.8 dBm is derived as follows:

$$P_S = \frac{1}{\rho} \sqrt{4kTF(\Delta f / R_L) \cdot (S/N)}$$

$$= \frac{1}{0.8} \sqrt{4(1.38 \times 10^{-23})(273 + 27)(2)(400 \times 10^6 / 600) \times (125.9)} = 2.089 \ \mu w$$

$$= -26.8 \ \text{dBm}$$

7.5.2.6 VI Characteristics

An important process in designing the receiver for an optical fiber link is to determine the load resistance (R_L) that provides the proper voltage to drive the preamplifier [Figure 7-32(A)], and the reverse bias voltage (V_B) required to operate the photodiode. The VI characteristic (voltage/current) as shown in Figure 7-36(B) can be used to calculate R_L and V_B. Figure 7-36(A) shows the front-end circuit of the receiver, where the voltage across the photodiode is V_d, and the diode current is i_d (the diode current is assumed to flow from "+" polarity to "−" polarity through the diode). Therefore, in a photodiode, the diode current (i_d) always has negative values, as shown in Figure 7-36(B).

Figure 7-36(B) represents the VI characteristic of a photodiode. The graph is divided into two regions: (1) the photovoltaic, and (2) the photoconductive region. The photoconductive region is the normal diode operating mode, with the horizontal axis representing the photodiode voltage (V_d) and the vertical axis representing the photodiode current (i_d). Due to conventional notation of a electronic diode, both V_d and i_d have negative values [see Figure 7-36(B)].

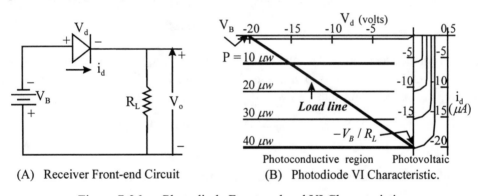

(A) Receiver Front-end Circuit	(B) Photodiode VI Characteristic.

Figure 7-36   Photodiode Front-end and VI Characteristic.

By applying the Kirchhoff (loop) voltage law to the circuit (loop) shown in Figure 7-36(A), the following relationship is derived:

$$V_B + i_d R_L + V_d = 0 \tag{7-44}$$

$$V_o = i_d R_L \tag{7-45}$$

By setting $i_d = 0$ and $V_d = 0$ in Eq.(7-44), respectively, it can be obtained the following:

$$V_d = -V_B \qquad \text{(with } i_d = 0\text{)} \tag{7-46}$$

$$i_d = -V_B/R_L \qquad \text{(with } V_d = 0\text{)} \tag{7-47}$$

This results in two points, $(-V_B, 0)$ and $(0, -V_B/R_L)$, on the VI characteristic.

Connecting these two points, $(-V_B, 0)$ and $(0, -V_B/R_L)$, on the VI characteristic graph in Figure 7-36(B) forms the "*load line*" for photodiode operation. The slope of the load line is given by the following relationship:

$$\text{Slope of load line} = -1/R_L \qquad (7\text{-}48)$$

Example 7-19: Determine the required receiver bias voltage (V_B) and load resistor (R_L) if the required output voltage of a photodiode at the receiver is in the range of $(-5$ to -20 volts), that is applied the pre-amplifier following the photodiode [see Figure 7-32(A)].

Assume the photodiode has the VI characteristic shown in Figure 7-36(B), and the received optical power from the photodiode is $(-20$ to -14 dBm). First, convert the optical power levels from dBm into μw (10 μw, 40 μw). These power levels correspond to $i_d = -5 \ \mu A$, and $i_d = -20 \ \mu A$ [the two bold lines in Figure 7-36(B)]. Next, substitute these values into Eq.(7-45) as follows:

$$V_o \ (\equiv -5 \ \text{volts}) = i_d \ (\equiv -5 \ \mu A) \times R_L$$

$$V_o \ (\equiv -20 \ \text{volts}) = i_d \ (\equiv -20 \ \mu A) \times R_L$$

From these equations, the value of the load resistor (R_L) is determined to be 1 MΩ. Next, by substituting $i_d = -20 \ \mu A$, and $R_L = 1$ MΩ into Eq.(7-45), the reverse bias voltage, V_B, can be obtained as follows:

$$V_B = -(i_d \times R_L) = -(-20 \ \mu A)(1 \ \text{M}\Omega) = 20 \ \text{volts}$$

Example 7-20: Summarize the components of a typical optical fiber link.

Figure 7-37 lists the major components of an optical fiber link. The listing is organized so that the top component of each category (e.g. light sources, optical fibers, and photodiodes, as shown in Figure 7-37 and the table below) is for low speed/short distance applications, while the bottom component is for high speed and/or long distance applications.

Light sources	*Fibers*	*Photodiodes*
• LED	• MMF	• PIN
• Laser	* SI MMF	* Bipolar
* MLM laser	* GRIN MMF	* FET
* SLM laser	• SMF	* IDT
* SF laser	* 1310 nm SMF	• APD
* Tunable laser	* 1550 nm SMF	
	* DS-SMF	
	* DF-SMF	
	* Special SMF	

Referring to Figure 7-37, (1) if an optical fiber link uses an LED as its light source, GRIN multi-mode fibers as its transmission medium, and PIN as its photodiode, then this link is suitable for low-to-medium speed or short-to-medium distance applications, and (2) if the link uses a Single Frequency (SF) laser diode as its light source, Dispersion-Shifted Single Mode Fibers (DS-SMF) as its transmission medium, and an APD as its photodiode, the link can carry very high speed signals over long distances.

In contrast, the combination of LED (low-speed device), single mode fibers (high-speed device) and PIN (low-speed device) for link components is not recommended.

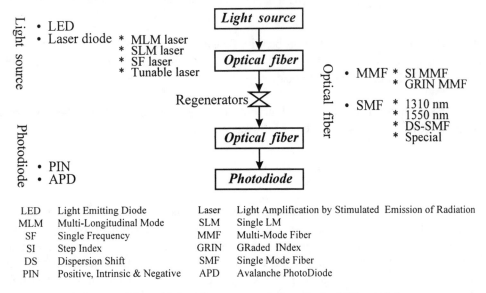

LED	Light Emitting Diode	Laser	Light Amplification by Stimulated Emission of Radiation
MLM	Multi-Longitudinal Mode	SLM	Single LM
SF	Single Frequency	MMF	Multi-Mode Fiber
SI	Step Index	GRIN	GRaded INdex
DS	Dispersion Shift	SMF	Single Mode Fiber
PIN	Positive, Intrinsic & Negative	APD	Avalanche PhotoDiode

Figure 7-37   Major Components of an Optical Fiber Link.

7.6 SYSTEM PERFORMANCE MEASUREMENTS

Like other communication systems, an optical fiber link is modeled as consisting of an input signal [$x(t)$], an additive channel noise [$n(t)$], and a received signal [$y(t)$]. The channel noise is assumed to be "white" and Gaussian (a white noise is not always Gaussian, or vice versa). Figure 7-38 represents the model for a typical optical fiber system for analyzing and designing of an optical link.

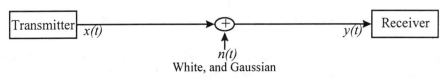

Figure 7-38   System Model of an Optical Fiber Link.

7.6.1 Noise Analysis

There are numerous noise sources in a communication system: thermal, shot, amplifier, laser, and background noises, etc. The distribution characteristics of each of these noise sources can be Gaussian, uniform, Poisson, Laplace, etc. However, in accordance with the Central Limit Theorem, if the number of noise sources (n_i, i = 1, K) is large enough (i.e., K = large) and the sources are independent, then the sum (n) of all the noises is a Gaussian noise, as shown in the following equation:

$$n = \sum_{1}^{K} n_i \qquad (n_i\text{: independent; } K\text{: large}); \qquad n \sim N(\mu, \sigma^2) \qquad (7\text{-}49)$$

where $N(\mu, \sigma^2)$ indicates that the probability density function is Gaussian, μ is the mean (i.e., the expectation value), σ is the standard deviation, and σ^2 is the variance of the random variable (i.e., the system noise) n.

The probability density function (pdf) [$f(n)$, as shown in Figure 7-36(A)] of a "practical" Gaussian noise is given as follows:

$$f(n) = \frac{1}{\sigma\sqrt{2\pi}} e^{-n^2/(2\sigma^2)}; \qquad n \sim N(0, \sigma^2) \qquad (7\text{-}50)$$

That is, a practical Gaussian noise has an average noise voltage of 0 volt, a standard deviation of σ, and a variance of σ^2. Another important characteristic of this Gaussian noise is "whiteness". A noise possessing constant power across the entire frequency spectrum is called a "white noise". This is shown in Figure 7-39(B), where both the "one-side" and the "two-side" spectrum [$S_n(f)$; N_0 = power density] are shown. It is much easier to analyze system noise performance if the noise source is white and Gaussian.

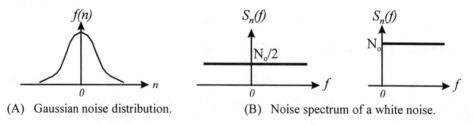

(A) Gaussian noise distribution.   (B) Noise spectrum of a white noise.

Figure 7-39   Gaussian Noise and White Noise.

7.6.2 Background for Noise Analysis

Two special functions are important for analyzing system noise performance: (1) the error function of "a" [erf(a)] which is defined by Eq.(7-51), and (2) the complementary error function of "a" [erfc(a)] which is defined by Eq.(7-52). Both functions are described in

Appendix 7-1 at the end of this chapter. The graphical representation (meanings) of these functions are illustrated in Figure 7-40.

$$erf\,(a) = \frac{2}{\sqrt{\pi}} \int_0^a e^{-x^2}\,dx \qquad \text{(error function)} \qquad (7\text{-}51)$$

$$erfc\,(a) = 1 - erf\,(a) \qquad \text{(complementary error function)} \qquad (7\text{-}52)$$

- The error function of "a", erf(a), represents the area, between $x = a$ and $x = -a$, shown in Figure 7-40(A), where random variable x is Gaussian, $x \sim N(0, \frac{1}{2})$. Note that this Gaussian distribution, $N(0, \frac{1}{2})$, has a "zero" mean and a variance of $\frac{1}{2}$.

- The complementary error function of "a", erfc(a), represent the sum of the two areas (a, ∞) and $(-a, -\infty)$. That is, the sum of the two "tails" shown in Figure 7-40(B).

- In general, for a Gaussian, $n \sim N(\mu, \sigma^2)$, where μ is the mean and σ^2 is the variance, the error function is related to the probability of a Gaussian distribution by the following equation. That is, the probability that the random variable, x, has a value between $(\mu - k\sigma)$ and $(\mu + k\sigma)$ is equal to the functional value of an error function with an augment of $k/\sqrt{2}$. The applications of Eq.(7-53) are important in analyzing and designing a telecommunications system, and examples are given later.

$$erf\,(\frac{k}{\sqrt{2}}) = probability[\mu - k\sigma \le x \le \mu + k\sigma] \qquad (7\text{-}53)$$

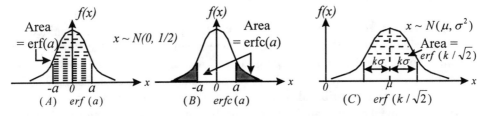

Figure 7-40   Graphical Representations of erf(a) and erfc(a).

Example7-21: Illustrate the probability of error (P_e) for transmitting digital signal over a digital communication system takes the following format:

$$P_e \propto erfc\,(K) \qquad (7\text{-}54)$$

Assume that a polar signal is transmitted as shown in Figure 7-41. That is, a logical "1" is transmitted by a value of "A volts", while a logical "0" is transmitted by "–A volts". In addition, assume that the probabilities for transmitting a logical "1" and a logical "0" are identical (e.g., the source generates the same number of "1s" and "0s"), that is,

$$P_1 = P_0 = 1/2 \qquad (7\text{-}55)$$

The notation $P_{1/0}$ denotes the probability that a logical "1" is received when a logical "0" has actually been transmitted (i.e., the probability of a misinterpreting a "1" as a "0"). Similarly, $P_{0/1}$ denotes the probability that a logical "0" is received when a logical "1" has actually been transmitted (i.e., the probability of a misinterpreting a "0" as a "1").

The probability that a received bit is in error (P_e) is given as follows:

$$P_e = P_0 \times P_{1/0} + P_1 \times P_{0/1} = (1/2) P_{1/0} + (1/2) P_{0/1}$$

$$= (1/2) (P_{1/0} + P_{0/1}) \tag{7-56}$$

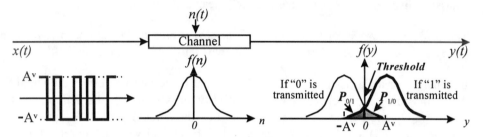

Figure 7-41   A Polar Signal Transmission over Gaussian Channel.

Referring to Figure 7-41, observe that the transmitted voltage is either "A" or "−A" volts, and the channel noise is assumed to be a Gaussian distribution with a "0" mean. Therefore, the two conditions (transmitting a logical "1" and a logical "0") are analyzed as follows, assuming a polar signal (waveform) is used in the system:

(1)   If a logical "1" (i.e., a voltage of "A" volts; a polar signal) is transmitted, the received signal, y(t), will be "A" volts if there is noise added to the signal (i.e., no channel noise, and the noise voltage is "0"). However, the noise is Gaussian with a possible noise voltage from −∞ volts to ∞ volts, and the received signal may be "A" volts ± the voltage contributed by noise. Hence, the probability density function of y(t) will assume the Gaussian form as shown by the "highlighted" curve in Figure 7-41. ***Since the transmitted signal is either "A" or "−A" volts, the receiver will adopt the "0" volt as the detection threshold.*** Therefore, if the received signal is less than "0" volts, it will be "restored" as a logical "0", which is an error. This probability (i.e., a "0" is restored when a "1" has actually been transmitted) is $P_{0/1}$, as shown by the "shaded" area marked as "$P_{0/1}$" in the received signal, y(t), in Figure 7-41.

(2)   Similarly, if a logical "0" is transmitted, the received signal, y(t), will be "−A" volts if the noise is "0", or −A ± the voltage contributed by noise. The probability density function of y(t) assumes the Gaussian as shown by the "thin-line" curve in Figure 7-41. If the received signal is greater than "0" volts, it will be "restored" as a logical "1", which is an error. This probability (i.e., "1" is restored when a "1" has actually been transmitted) is $P_{1/0}$ as shown in Figure 7-41.

Based on these results (i.e., one "shaded area" in y(t) shown in Figure 7-41 representing $P_{1/0}$, and the other "shaded area" representing $P_{0/1}$), Eq.(7-56) can be rewritten as follows:

$$P_e = (1/2)\,(P_{1/0} + P_{0/1}) = (1/2)\text{ (Two "shaded areas" in Figure 7-41)} \quad (7\text{-}56a)$$

But, from Figure 7-40(B), it is clear that the two shaded areas of a Gaussian distribution represents a complementary error function, erfc(K), K is an arbitrary number that is a function of "A" of the transmitted signal. Eq.(7-56a) becomes

$$P_e = (1/2)\;erfc\;(K) \quad (7\text{-}56b)$$

As indicated by Eq.(7-54) (for a polar digital communication system) the probability of error that a binary digit has been erroneously restored (transmitting a logical "1", but restoring a logical "0", $P_{0/1}$; or transmitting a logical "0", but restoring a logical "1" , $P_{1/0}$) can be expressed as *erfc(K).*

7.6.3 Noise Power in an Optical Fiber Link

An optical fiber link can be dominated by two types of noise: (1) thermal noise, and (2) shot noise. The average noise power is given by the following equations:

$$P_T = 4kT\Delta f \qquad\qquad \text{(Thermal noise)} \qquad (7\text{-}57)$$

$$P_S = 2\Delta f R_L \bullet e\,(\frac{\eta e P}{hf} + I_d) \qquad\qquad \text{(Shot noise)} \qquad (7\text{-}58)$$

where
- k = Boltzmann constant = 1.38×10^{-23} J/°K
- T = System temperature in °K
- Δf = system bandwidth
- R_L = Load resistance
- e = electron charge = 1.6×10^{-19} C
- η = receiver quantum efficiency
- P = received optical power
- h = Planck constant = 6.626×10^{-34} J sec
- f = Signal frequency = c/λ
- I_d = Receiver photodiode dark current

From Eqs.(7-40), (7-57), and (7-58), the Signal-to-Noise ratio (S/N) of an optical fiber link can be derived as follows:

$$\frac{S}{N} = \frac{(\dfrac{\eta e P M}{hf})^2 R_L}{4kT\Delta f + 2R_L e\Delta f\,(\dfrac{\eta e P}{hf} + I_d)M^n} \quad (7\text{-}59)$$

where *n* is an empirical parameter, with a typical value of 1~2, *M* is the APD gain.

Example 7-22: Determine the system Signal-to-Noise (S/N) performance for the optical fiber link shown in Figure 7-42.

(1) To calculate the received signal power, the received optical power must be calculated first as follows: $p_r = 10$ mw $\times [10^{-(40\ dB)/10}] = 1$ µw because

$$P_r \text{ (the received power)} = P_t \text{ (transmitted power)} \times 10^{(-\text{Loss in dB})/10}$$

With a receiver responsivity ($\rho = 0.5$ A/w) and received optical power ($p_r = 1$ µw) the photodetector will generate 0.5 µA. Therefore, the signal power of the received to be used for S/N calculation can be calculated as follows:

$$p_R = (0.5 \times 10^{-6})^2 \times 50 = 1.25 \times 10^{-11} \quad \text{w}$$

(2) To calculate the thermal noise power, Eq.(7-57) is applied as follows:

$$P_T = 4kT\Delta f = 4(1.38 \times 10^{-23})(273 + 27)(10^7) = 1.66 \times 10^{-13} \quad \text{w}$$

(3) To calculate shot noise power, Eq.(7-58) is applied as follows:

$$P_S = 2\Delta f R_L \cdot e(\frac{\eta e P}{hf} + I_d) = 2(1.6 \times 10^{-19})(10^7)(50)(3.2 \times 10^{-11} + 2 \times 10^{-9})$$

$$= 3.25 \times 10^{-17} \quad \text{w}$$

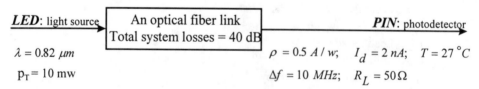

LED: light source | An optical fiber link — Total system losses = 40 dB | **PIN**: photodetector

$\lambda = 0.82$ *µm*

$p_T = 10$ mw

$\rho = 0.5$ *A*/*w*; $I_d = 2$ *nA*; $T = 27\ °C$

$\Delta f = 10$ *MHz*; $R_L = 50\Omega$

Figure 7-42   An Example Optical Fiber Link.

By substituting these powers, P_R, P_T, and P_s, into Eq.(7-59), the system signal-to-noise performance can be derived as follows:

$$\frac{S}{N} = \frac{1.25 \times 10^{-11}}{1.66 \times 10^{-13} + 3.25 \times 10^{-19}} = 75.3; \quad or,\ 18.8\ dB$$

This type of system is known as a thermal (noise) limited optical fiber system (see definition below) rather than a quantum limited system.

Definition 7-1: An optical fiber system can be classified as one of two types:

- A thermal (noise) limited system if $P_T \gg P_s$
- A shot (noise) or quantum limited system if $P_T \ll P_s$

7.7 OPTICAL LINK DESIGN

As a generalization, all of the commercial optical fiber systems presently deployed can be considered thermal noise limited systems. The system Signal-to-Noise (S/N) performance in Eq.(7-59) can be simplified as follows:

$$\frac{S}{N} = \frac{(\frac{\eta e PM}{hf})^2 R_L}{4kT\Delta f} = (\frac{\eta e PM}{hf})^2 \times \frac{R_L}{4kT\Delta f} \tag{7-60}$$

The corresponding probability of error (Pe) is derived as follows:

$$P_e = 0.5 \times erfc \,(0.354 \times \sqrt{S/N}) \tag{7-61}$$

The following approximation of erfc(a) is useful in performing this computation instead of utilizing several mathematical tables:

$$erfc \,(a) = \frac{1}{a\sqrt{\pi}} \times e^{-a^2} \qquad if \quad a > 1 \tag{7-62}$$

Example 7-23: Derive the probability of error (Pe) for the optical fiber link described in Example 7-22, which has a signal-to-noise ratio of 75.3 (18.8 dB).

$$P_e = 0.5 \times erfc \,(0.354 \times \sqrt{S/N}) = 0.5 \times erfc \,(0.354 \times \sqrt{75.3}) = 0.5 \, erfc \,(3.07186)$$

$$= 0.5 \times \frac{1}{3.07185\sqrt{\pi}} \times e^{-3.07186^2} = 7.3 \times 10^{-6}$$

A Pe = 7.3 × 10⁻⁶ is an acceptable quality for voice applications and some voice-grade data services, that require a minimum Pe = 10⁻⁴.

7.7.1 System Specifications

This book only describes digital applications over optical fiber. There are many parameters (characteristics) associated with digital applications, but only the following parameters are considered in optical link design/analysis:

- Transmission rate (system speed)
- Signal type (non-return-to-zero or return-to-zero unipolar signals)
- Transmission path length (link distance)
- Regenerator requirements
- Connector and splicing requirements
- System operating temperature
- P_e (probability of error) requirements

The goal of designing an optical fiber link is to select the proper link components (e.g., light source, fiber type, and photodetector) so that a required Bit Error Rate (BER) performance specification can be met. The design of an optical link can be processed in two steps: (1) power budget design, and (2) bandwidth (rise time) budget design.

7.7.2 Power Budget Design

The major components of a typical optical fiber link is shown in Figure 7-42, and the sources of transmission loss are also indicated.

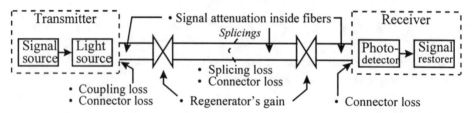

Figure 7-42   Major Components and Losses of an Optical Link.

The power budget design of an optical fiber link is specified as follows:

$$P_T - (L_{cp} + L_{ct} + L_{sp} + L_{fb}) - M \geq R_s \qquad (7\text{-}63)$$

where
P_T = Power transmitted (by the light source) in *dBm*
L_{cp} = Coupling losses between the light source and the fiber core, in *dB*
L_{ct} = Connector losses (including: light source and photodiode), in *dB*
L_{sp} = Splicing losses, in *dB*
L_{fb} = Fiber attenuation (loss), in *dB*
M = System loss margin requirement (specification), in *dB*
R_S = Receiver sensitivity requirement (specification), in *dBm*

7.7.3 Bandwidth (Rise-time) Budget Design

The proper design of an optical fiber link starts with the selection of the light source, fiber, and photodiode by analyzing the radiation optical power, the total attenuation along the transmission path, and the required minimum optical power at the photodiode. In addition, the coupling loss, connector loss and splicing loss must be considered by applying Eq.(7-63). After the power budget has been met, the designed (selected) components must be checked to verify that they meet the system rise time requirements. This procedure is called the rise time or bandwidth budget design, which requires extensive analysis and calculation. The appropriate equations have been presented in different parts of this book. For convenience, these equations are summarized here.

- **System transmission rate and rise time**: The relationship between the optical fiber system transmission rate (R_{NRZ}, R_{RZ}) and system rise time (t_s) is given by Eq.(7-64).

$$R_{NRZ} \leq \frac{0.7}{t_s} \;(NRZ\; signal\; rate); \quad or, \quad R_{RZ} \leq \frac{0.35}{t_s} \;(RZ\; signal\; rate) \qquad (7\text{-}64)$$

If a non return-to-zero (NRZ) unipolar signal is transmitted, 0.7 is used in calculating system transmission rate, R_{NRZ}. In contrast, if a return-to-zero unipolar signal is transmitted, 0.35 is used in calculating system transmission rate, R_{RZ}.

- **Composition of system rise time**: The system rise time (t_s) of an optical fiber link is composed of many components. Three of these components are significant in the design of the system: (1) light rise time (t_l), (2) fiber rise time (t_f), and (3) photodiode rise time (t_p). These parameters are related as follows:

$$t^2_{\;s} = t^2_l + t^2_f + t^2_p \qquad (7\text{-}65)$$

- **Fiber rise time**: The fiber rise time (fiber dispersion) has three components: (1) material dispersion, (2) waveguide dispersion, and (3) (inter)-modal distortion. Therefore, three cases need to be considered:

(1) For single mode fibers, only the material and waveguide dispersions are significant in computing fiber rise time. That is, the fiber rise time (t_f) is given by:

$$t_f = t_M = |(M + W)| \cdot \Delta\lambda \cdot L \qquad (7\text{-}66)$$

where M and W [in $ps/(nm{\cdot}km)$] are the material and waveguide dispersions, $\Delta\lambda$ (in nm) is the spectral width (line width) of the light source, and L (in km) is the physical fiber transmission path length. Note that t_M is the sum of the material and waveguide dispersions.

(2) For GRIN multi-mode fibers, in addition to the material and waveguide dispersions, the (inter)-modal distortion must be considered. The total fiber rise time is given by Eq.(7-67), where t_M is the sum of material dispersion and waveguide dispersion [given by Eq.(7-66)], and t_D is the (inter)-modal distortion [given by Eq.(7-68)]. In Eq.(7-68), the parameters n_1, n_2, and c are the index of the fiber

core, the index of the fiber cladding, and the speed of light in a vacuum, respectively. The parameter L is the fiber transmission path length.

$$t_f^2 = t_M^2 + t_D^2 \tag{7-67}$$

$$t_D = (n_1 - n_2)^2/(2cn_1) \bullet L \qquad \text{for GRIN MMF} \tag{7-68}$$

(3) For SI multi-mode fibers (similar to GRIN fibers), in addition to the material and waveguide dispersions, the (inter)-modal distortion must be considered. For this application, t_D is the (inter)-modal distortion as given by Eq.(7-69) [instead of Eq.(7-68)].

$$t_D = n_1 \bullet (n_1 - n_2)/(cn_2) \bullet L \qquad \text{for SI MMF} \tag{7-69}$$

In both cases (GRIN and SI multi-mode fibers), the distortion calculated using Eq.(7-68) or (7-69) yielding theoretical values. Hence, these calculations must be modified by Eq.(7-28) or (7-29) to derive the actual values. Note that t_D and $\Delta\tau$ are used interexchangeable in this book.

$$\Delta\tau_d = (\frac{\Delta\tau}{L})_{theoretical} \times L \qquad \text{if } L \leq L_e \tag{7-27}$$

$$= (\frac{\Delta\tau}{L})_{theoretical} \times \sqrt{LL_e} \qquad \text{if } L > L_e \tag{7-28}$$

- **Photodiode rise time**: It consists of two components (1) transit rise time and (2) circuit rise time. The transit rise time is device dependent, and is always specified by device manufactures. The circuit rise time is a function of the device parameters and the circuit load resistor, and given by Eq.(7-36). Note that t_p and t_{pd} are used interexchangeable in this book.

$$t_{pd} = (t_{tr}^2 + t_{ct}^2)^{1/2} = [(\frac{Depletion \ layer \ width}{v})^2 + (2.19 \times R_L \ C_d)^2]^{1/2} \tag{7-36}$$

7.7.4 System Performance Measurements

After the power budget and the rise time (bandwidth) design calculations have been completed, the system performance requirements (for thermal noise limited systems), Signal-to-Noise ratio (S/N), and Bit Error Rate (BER) must be verified by applying the following two performance equations.

$$\frac{S}{N} = (\frac{\eta e P M}{hf})^2 \times \frac{R_L}{4kT\Delta f} = \frac{P_S}{P_T} \tag{7-70}$$

$$P_e = 0.5 \times erfc \ (0.354 \times \sqrt{S/N}) \tag{7-71}$$

7.7.5 Digital Optical Fiber Link Design Example

Several fiber link design examples are discussed in this section. Example 7-24 assumes that the link components (including their parameters) are specified, and the exercise is to calculate the minimum/maximum system speeds, and the system performance. Example 7-25 assumes the system specifications are given, and the exercise is to analyze various methods to select appropriate link components using in the transmission path.

Example 7-24: An optical fiber link, carrying NRZ unipolar signals, is used for local area network applications. The specifications of the link are shown below, determine: (1) the minimum/maximum speeds, (2) receiver sensitivity requirement, and (3) the probability of error (bit error rate).

LED: light source | Fiber | *PIN*: photodetector

LED: light source
$P_T = 0.01$ *mw*
$\lambda = 0.82$ *μm*
$t_l = 10 \sim 50 ns$

Fiber
Coupling/connector losses = 1 dB
Splicing losses = 1 dB
Total fiber losses = 10 dB
Total fiber rise time = 20 ns

PIN: photodetector
Load resistor = 50 ohms
Temperature = 27°C
Rise time = 10 ns
Bandwidth = 10 MHz

(1) First, determining the power budget (assuming no design margin is required), the transmitted power (P_T) is calculated to be 10 log (0.01 mw) = −20 dBm. The total system loss (connectors, splice, fiber) is 12 dB (= 1 + 1 + 10). Therefore, from Eq.(7-63), the required receiver sensitivity is derived to be −32 dBm:

$$-20 \text{ dBm} - (1 + 1 + 10) \text{ dB} = -32 \text{ dBm}$$

(2) Rise time (bandwidth) budget: Assuming a rise time (t_t) of 50 *ns* for the light source, then the system rise time (t_s) can be calculated, using Eq.(7-65), as follows:

$$t_s = \sqrt{t_l^2 + t_f^2 + t_p^2} = \sqrt{50^2 + 20^2 + 10^2} = 54.77 \text{ ns}$$

From Eq.(7-64), the system speed can be obtained as 12.78 Mbps:

$$R_{NRZ} = \frac{0.7}{t_s} = \frac{0.7}{54.77 \text{ ns}} = 12.78 \text{ Mbps}$$

If a rise time of 10 *ns* (instead of 50 *ns*) for the light source is used, then the system rise time can be calculated as follows:

$$t_s = \sqrt{t_l^2 + t_f^2 + t_p^2} = \sqrt{20^2 + 20^2 + 10^2} = 30 \text{ ns}$$

From Eq.(7-64), the system speed can be obtained as 23.33 Mbps:

$$R_{NRZ} = \frac{0.7}{t_s} = \frac{0.7}{30 \text{ ns}} = 23.33 \text{ Mbps}$$

Therefore, the minimum and maximum system speeds of this optical fiber link are 12.78 Mbps and 23.33 Mbps, respectively.

(3) Performance measurement verification: To calculate system performance, the receiver quantum efficiency and the responsivity are both assumed to be 0.8. An internal gain does not exist for a PIN, but it must be considered for an APD receiver. Thus, M should be set to 1 in Eq.(7-70). The first step is to calculate the received optical power as follows:

$$P_R = P_T \times 10^{-\text{Loss}/10} \qquad\qquad (7\text{-}63a)$$

$$= 0.01 \text{ mw} \times 10^{-22/10} = 6.3096 \times 10^{-7}$$

From Eq.(7-39), the current generated by the received optical power is obtained as:

$$i_s = \rho P_R = 0.8 \times 6.3096 \times 10^{-7} = 5.0476 \times 10^{-7}$$

$$P_S = i_s^2 R_L = (5.0476 \times 10^{-7})^2 (50) = 1.274 \times 10^{-11} \text{ w}$$

$$p_T = 4kT\Delta f = 4\,(1.38 \times 10^{-23})(273 + 27)(10^7) = 1.656 \times 10^{-13} \text{ w}$$

$$\frac{S}{N} = \frac{Signal\ power}{Thermal\ noise\ power} = \frac{1.274 \times 10^{-11}}{1.656 \times 10^{-13}} = 76.93; \quad or, \quad 18.9\ dB$$

$$P_e = 0.5 \times erfc(0.354 \times \sqrt{76.93})$$

$$= 0.5 \times erfc(3.1049) = 0.5 \times \frac{e^{-3.1049^2}}{3.1049 \times \sqrt{\pi}}$$

$$= 5.9 \times 10^{-6}$$

Example 7-25: This example assumes that the components have not been selected. Therefore, various components (e.g., LEDs, lasers, various fiber types, PINs, APDs, etc.) having individual characteristics are available for use in the actual optical fiber link implementation. The link specifications are given in Figure 7-43.

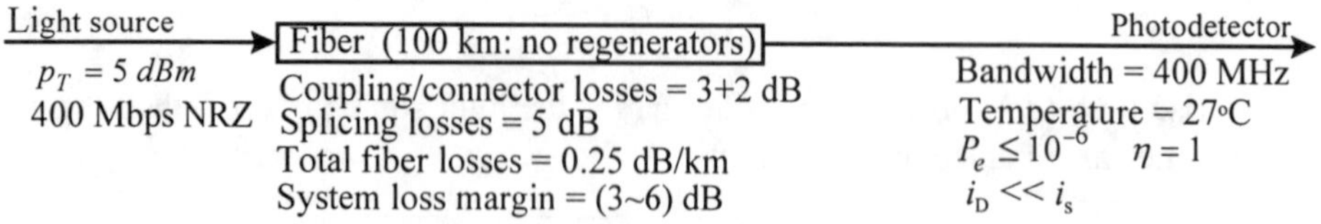

Light source	Fiber (100 km: no regenerators)	Photodetector
$p_T = 5\ dBm$	Coupling/connector losses = 3+2 dB	Bandwidth = 400 MHz
400 Mbps NRZ	Splicing losses = 5 dB	Temperature = 27°C
	Total fiber losses = 0.25 dB/km	$P_e \leq 10^{-6}$ $\eta = 1$
	System loss margin = (3~6) dB	$i_D \ll i_s$

Figure 7-43   System Specifications of an Optical Fiber Link.

The typical characteristics for various components are listed in Table 7-7. As indicated, a range of values are given for this design exercise.

Table 7-7   Characteristics for Some Available Components.

Component	Characteristics (Value range)	
LEDs	Rise time: 2 ~ 50 ns	Spectral width: 10 ~ 100 nm
Lasers	Rise time: 0.1 ~ 1 ns	Spectral width: 0.1 ~ 1 nm
Fibers	Rise time: 0.1 ~ 10 ns M: -20 ps/(nm×km)	Losses: 0.2 ~ 4 dB/km W:4.5 ps/(nm×km)
PINs	Rise time: 0.5 ~ 10 ns Transit time: 0.1 ~ 1 ns	Sensitivity: -20 ~ -32 dBm Junction capacity: 1 pf
APDs	Rise time: 0.1 ~ 1 ns Transit time: 0.1 ~ 1 ns	Sensitivity: -25 ~ -40 dBm Junction capacity: 1 pf

Since no particular devices or components have been specified in the link, the design traditionally starts with the rise time budget. Applying Eqs.(7-64) and (7-65), with a 400 Mbps NRZ signal, the system rise time allocation can be calculated as follows:

$$t_s = 0.7/(R_{NRZ}) = 0.7/(400 \times 10^6) = 1.75 \ ns$$

$$t_s^2 = t_t^2 + t_f^2 + t_p^2 = (1.75 \times 10^{-9})^2 \tag{7-72}$$

With three "unknown" variables (i.e., $t_t, t_f, and \ t_p$) in one equation [i.e., Eq.(7-72)], it is first necessary to "select" two values for two unknowns using the characteristics in Table 7-7. In general, a light source and fiber type are specified first: (1) an SLM 1.55 μm laser diode with a rise time of $t_l = 1 \ ns$, and a spectral width of 0.15 nm (a typical commercial component); and (2) a fiber with $M = -20$ ps/(nm×km), $W = 4.5$ ps/(nm×km), and a nominal wavelength of 1.55 μm. The fiber dispersion (t_f), which is approximately equal to the design rise time, is calculated using Eq.(7-66) as follows:

$$t_f = |M + W| \times \Delta\lambda \times L = |-20 + 4.5| \times 0.15 \times 100 = 0.23 \ ns$$

By substituting $t_l = 1 \ ns$, and $t_f = 0.23 \ ns$ into Eq.(7-72), the photodiode rise time (t_p) is calculated to be 1.4 ns., which consists of transit time (t_t) and circuit rise time (t_c) as follows:

$$t_p^2 = t_t^2 + t_c^2 = t_t^2 + (2.19 R_L C_d)^2 \tag{7-73}$$

Substituting $t_p = 1.4 \ ns$, $C_d = 1$ pf, $t_t = 0.5 \ ns$ into Eq.(7-73), the circuit rise time (t_c), and the load resistor (R_L) can be calculated to be 1.3 ns, and 600 Ω, respectively. This calculation completes the "first round of rise time budget" design. As a result of these calculations, the selected optical fiber link consists of the following components:

- A light source, operating at 1.55 μm, with a rise time of $t_l = 1 \ ns$, and a spectral width of 0.15 nm.

- A single fiber, operating at 1.55 μm, with a rise time of $t_f = 0.23 \ ns$.

- A PIN photodiode, having a cutoff wavelength (assumed) of 1.65 μm, a transit time $t_t =$ 0.5 ns, a junction capacity of $C_d = 1$ pf, a circuit rise $t_c = 1.3$ ns, and a load resistor of 600 Ω. This photodiode is assumed to have a receiver sensitivity of (−20~−32) dBm.

*<u>Note</u>: **The final goal of the "rise time (bandwidth)" budget design is to calculate the load resistor, which is used to determine the voltage for driving the preamplifier connected to the photodiode. The load resistor value also influences the system performance requirement (i.e., the signal-to-noise ratio).***

After the "first round" rise time (bandwidth) budget design is complete, the next step is to perform system power budget design by applying Eq.(7-63) as follows:

$$P_T - (L_{cp} + L_{ct} + L_{sp} + L_{fb}) - M \geq R_s \qquad (7\text{-}63)$$

$$5 \text{ dBm} - (3 + 2 + 5 + 100 \times 0.25) \text{ dB} - (3\text{~}6) \text{ dB} = (-33 \sim -36) \text{ dBm} \qquad (7\text{-}74)$$

Therefore, the receiver selected earlier, based on the rise time budget design, must have a receiver sensitivity of (−33 ~ −36) dBm or better. Obviously this requirement can not be satisfied because the photodiode selected has a receiver sensitivity of (−20 ~ −32) dBm. For practical system design, it is advised that another type of photodiode (which can meet the receiver sensitivity requirement) be selected. For example, an APD photodiode with a receiver sensitivity of (−25 ~ −42) dBm (shown in Table 7-7) may be used. However, for the purpose of design flexibility, it is assumed that PIN photodiode will be used. To accomplish this goal, one can reduce the system loss margin can be reduced to 1 dB, instead of (3~6) dB.

After the "first round" of power budget design is complete, the system S/N must be calculated so that the system bit error rate requirement can be verified. By substituting the following parameters in Eq.(7-70), the S/N ratio is computed to be 88.8 (19.5 dB). The parameters used are: the photodiode internal gain, M =1 (PIN, instead of APD is used), the receiver quantum efficiency $\eta = 1$ (for practical devices, this value should be 0.5~0.8), T = 27°C, $\Delta f = 400$ MHz, and P = −31 dBm [assuming the system loss margin is reduced from (3~6) dB to 1 dB, the result is −31 dBm or 0.794 μw].

The final step of the system design is to verify the bit error rate by applying Eq.(7-71). From Eq.(7-71), the probability of error (P_e) (assumed to be the system bit error rate) is calculated to be 1.24×10^{-6}, which does not meet the system requirement of 10^{-6}. Therefore, the first round system design is not satisfactory, and re-design is required. The generalized re-design guideline to be considered are listed as follows:

- Operate the system at larger wavelength (e.g., from 820 to 1310, to 1550 nm).

- Select a better light source (e.g., a smaller rise time, a smaller spectral width, and a stronger radiated output power, etc.).

- Select better fibers [e.g., a smaller dispersion (rise time), and lower attenuation].

- Select a better photodiode [e.g., PIN to APD, with better responsivity, sensitivity, lower transit time, lower junction capacity, lower rise time, etc.].

- Change the load resistor value (generally not recommended as the course of action).

Any combination of the actions listed above can be implemented to derive a satisfactory design.

For the purpose of illustration, the design is revised by selecting fibers with parameters (characteristics) that are the same as the one used in the first round design. However, the fibers have a lower loss (0.24 dB/km instead of 0.25 dB/km). Therefore, in a transmission path of 100 km, the signal loss along the link is reduced by 1 dB. That is, the received power is increased by one dB (from −31 dBm to −30 dBm).

Rule of thumb: For a thermal noise limited optical fiber link, if the received optical power is increased by "X dB", the system S/N ratio will be increased by "2X dB".

The revised system S/N value is 20.9 dB or 123.03. By substituting this value into Eq.(7-71), the probability of error is calculated to be 1.45×10^{-8}, which meets the system requirement of 10^{-6}. Thus, the system design procedure is complete.

Example 7-26: This example illustrates the applications of two important units used in a communication system: dB and dBm, and extends the concept of power budgeting an optical fiber link. The system specifications of the link are as shown in Figure 7-44, and a system loss margin of 3 dB is required. Determine the maximum allowable fiber attenuation in dB/km for this optical fiber link.

The transmitted power of 1 *mw* is equal to having a power level of "0 dBm". Therefore, apply power budget equation [Eq.(7-63)] as follows:

$$P_T - (L_{cp} + L_{ct} + L_{sp} + L_{fb} + M) \geq R_s \qquad (7\text{-}63)$$

$$0 \text{ dBm} - (1 + 10 \times 0.1 + 100 \times L + 1 + 3) \text{ dB} \geq -38 \text{ dBm}$$

The value of L is found to be **0.32** dB/km [the maximum allowable attenuation (loss) of the fiber used in this link], which produces a 3 dB system loss margin and −38 dBm receiver sensitivity as specified.

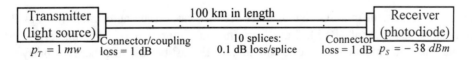

Figure 7-44   An Optical Fiber Link: Power and Losses.

Example 7-27: For practical applications, assume that fiber with 0.3 dB/km attenuation (instead of 0.32 dB/km as calculated in Example 7-23) are used. In this exercise, the receiver sensitivity is assumed to be −38 dBm rather than a range (e.g., −38 dBm ∼ −42 dBm), and the system loss margin requirement is still 3 dB. Determine the interface needed at the receiving end of this optical fiber link.

From the power budget equation [Eq.(7-63)], it can be calculated that the receiving end of the link terminates a power level of −36 dBm [0 dBm − (1 + 10 0.1 + 100 × 0.3 + 1 + 3) dB]. The receiver sensitivity is −38 dBm, which is 2 dB [−36 dBm − (−38 dBm)] lower than the −36 dBm signal power. Therefore, at the receiving end of this optical fiber link, an attenuator (line built-out network) must be used to reduce the level of the received signal so that it is compatible with the receiver sensitivity.

Example 7-28: A laser diode is used as the light source for the optical fiber link described in Example 7-24. This laser diode radiates 1 mw of optical power when initially installed as shown in Figure 7-44. However, it radiates 5% less power (with respect to its original power) each year. Assume the system is designed to operate for 10 years. Determine the fiber attenuation (in dB/km), and required action at the time of installation.

The radiated optical power at installation is 1 mw (0 dBm). The radiated power at the beginning of the 2nd year, 3rd year, 4th year, ..., and the 10th year are: 0.95, 0.9, 0.85, ..., and 0.5 mw (−3 dBm), respectively. Therefore, there is a 3 dB loss between the time of installation and the end of the 10th year caused by laser aging. Hence, the power budget design must be based on the power and loss values expected at the end of the 10th year. That is, applying Eq.(7-63) and obtain the following:

$$-3 \text{ dBm} - (1 + 10 \times 0.1 + \ 100 \times L + 1 + 3) \text{ dB} \geq -38 \text{ dBm}$$

The fiber attenuation is calculated to be 0.27 dB/km. Therefore, at the time of installation, the attenuator used at the receiver is set for 3 dB loss, and then gradually reduced to 0 dB loss by the 10th year.

Example 7-29: Assume that an existing FT-G optical fiber link is to be reused as an SONET optical fiber link to carry an OC-12 (622.08 Mbps) signal, and the power budget does not require any modification. Describe, from a rise time budget view, what needs to be done to insure this link will work properly.

Any network element in a SONET or SDH network can serve one or more roles as Path Terminating Equipment (PTE), Line Terminating Equipment (LTE), and/or Section Terminating Equipment (STE). For example, a PTE must be able to generate/modify path overhead bytes, read/interpret path overhead bytes, and implement these overhead bytes.

Similarly, an LTE must be able to generate/modify line overhead bytes, read/interpret line overhead bytes, and implement these overhead bytes; and an STE must be able to generate/modify section overhead bytes, read/interpret section overhead bytes, and implement these overhead bytes. However, FT-G equipment is not equipped with these capabilities. Therefore, the existing transmitter, receiver, and regenerators must be replaced with SONET transmitter, receiver, and regenerator equipment.

Example 7-30: Assume that an existing FT-G optical fiber link is to be reused as an SONET optical fiber link to carry an OC-48 NRZ unipolar signal with a signal speed of 2.44832 Gbps. For simplicity, assume that power budget does not require any modification. Describe, from rise time budget view, what needs to be done to insure this link will work properly as SONET optical fiber link.

The nominal rise time of the light source used in FT-G is $t_t = 0.2$ ns. The nominal rise time of the fibers (used in FT-G) is $t_f = 0.195$ ns, and the nominal rise time of the photodiode (used in FT-G) is $t_p = 0.3$ ns. These parameters are related to system rise time, (t_s) as follows [from Eq.(7-65)]:

$$0.4099^2 = 0.2^2 + 0.195^2 + 0.3^2$$

The system rise time, t_s, is related to the speed of the system [Eq.(7-64)] as follows:

$$t_s = 0.7/\text{Rate} = 0.7/(1.7 \text{ Gbps}) = 0.4099 \text{ ns}$$

Hence, this system must be modified for OC-48 applications, assuming the original fiber is used. That is, FT-G fiber will remain in service. To increase the speed (i.e., data rate) of the system from 1.7 Gbps to 2.44832 Gbps, the system rise time must be reduced [in accordance with Eq.(7-74)] as follows:

$$t_s = 0.7/\text{Rate} = 0.7/(2.44832 \text{ Gbps}) = 0.281 \text{ ns}$$

Applying Eq.(7-65), the rise time budget calculation starts with:

$$0.281^2 = t_t^2 + 0.195^2 + t_p^2 \tag{7-75}$$

A new light source with a rise time $t_t = 0.15$ ns (reduced from 0.2 ns), and a new photodiode with a rise time $t_p = 0.1336$ ns (reduced from 0.3 ns) are recommended as replacements for the existing FT-G components. These faster rise times meet the requirement in Eq.(7-75). Therefore, it is possible to upgrade the existing FT-G optical fiber link to carry 2.44832 Gbps unipolar signal by installing a new light source, photodiode, and regenerators. Besides meeting the rise times requirements, the new components must be SONET compatible. That is, they must be able to perform all the functions that SONET PTE, LTE and/or STE perform. Finally, the new optical link (i.e., the new light source, regenerators, and photodiodes) must have the BER verified.

7.7.6 Optical Fiber Measurements and OTDR

The three components, the light source, fibers, and the photodiode, in a fiber link have practically equal levels of importance. However, "fiber measurements" are of most important when specifying the system. These optical fiber measurements includes:

- Fiber diameter
- Fiber numerical aperture
- Fiber refractive index profile
- Fiber attenuation
- Fiber dispersion
- Fiber field performance

A common method for obtaining these measurements is Optical Time Domain Reflectometry (OTDR), both in the laboratory and in the field. The OTDR technique is based on the fiber attenuation measurements. The attenuation measurement is the averaged signal loss (in dB/km) over the entire fiber length. However, since signal loss along the optical link path varies, the OTDR measurement is more exact. For example, OTDR allows splicing loss and connector loss to be evaluated. In addition, the ability to locate faults (e.g., breaks and signal attenuation non-uniformity) along the fiber path is provided by OTDR technology.

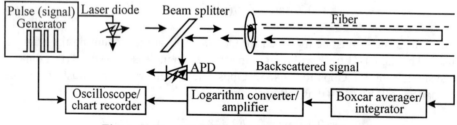

Figure 7-44   Building Blocks of an OTDR.

OTDR, which is also known as a *"backscatter measurement"* technique, is the measurement and analysis of the fraction of light that is reflected back within the numerical aperture of the fiber. This technique was first described by Barnoski and Jensen in 1976. The backscatter method does not require cutting the fiber, (i.e., it is nondestructive) to access the output end of the optical fiber end.

Figure 7-44 illustrates the OTDR principle, and the major building blocks of an OTDR measurement system. As shown, the input signal to a laser diode is a pulse (signal) generator, that couples an optical signal (with a power level of 10 dBm or higher) into a laser diode. The pulse signal repetition rate must be carefully chosen so that the signals returning from the fiber core *do not overlap*. The returning signal (from the fiber) is separated from the incident signal by using a directional coupler, which may be a twisted-pair coupler or a polarizing beam splitter. In general, the pulse widths may range from several *ns* to several *μs* for the following repetition rates:

- For long fiber lengths, a repetition rate of 1 kHz is typically used.
- For short fiber lengths, a repetition rate up to 20 kHz is typically used.

The returning/reflected signal is detected by an APD (rather than a PIN photodiode since an APD has a better "sensitivity" and thus results in a more accurate measurement than using a PIN). The APD photodiode converts the optical signal into an electrical signal, which is fed to a "boxcar averager/integrator", which improves the signal-to-noise ratio performance. Since the display of an OTDR uses logarithmic scales, a logarithmic converter/amplifier is required. An OTDR displays the "one-way" fiber attenuation on the vertical scale (axis), and the "one-way" fiber length on the horizontal scale using either an oscilloscope or chart recorder for displaying the measurement results.

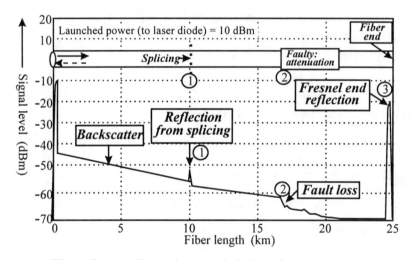

Figure 7-45   Example Recorded Chart from an OTDR.

Example 7-31: Figure 7-45 illustrates the OTDR display results derived by testing a single-mode fiber. The light power incident (launched) into this step-index fiber is assumed to be 10 dBm (i.e., 10 mw). Describe how this chart is interpreted

In this example, the length of the fiber is assumed to be 25 km, and the launched power into the laser diode (Figures 7-44 and 7-45) is 10 dBm. From the backscatter curve (which indicates the backscattered signal power level) it can be seen: (1) there is reflection power at 10 km from the light launching point, which is a splice joint at the 10 km point; (2) there is a loss at the 17 km point, which is a faulty region with light attenuation starting at the 17-km point; and (3) there is Fresnel end reflection at 25 km point, indicating the fiber ends at the 25 km point from the light launching point.

Review Questions I for Chapter 7:

(1) Rank the following media (twisted-pair, coaxial cable, fiber) from data rate viewpoint: _____ higher than _____, which is higher than _____.

(2) At the state of the art (from wavelength viewpoint), all present optical fiber systems apply the _____ technology.

(3) John Tyndall's _____ theory established fiber core/cladding structure principle.

(4) List four major parameters of a light source (LED or laser diode) in designing an optical fiber link: _____, _____, _____, and _____.

(5) (True, False) The spectral width (linewidth) of a light source can be defined as the wavelength range of the two 3-dB, 10-dB, or 15-dB points of light source spectral density function.

(6) In the US, typical multi-mode fiber has a core diameter about _____ μm, and the core diameter of a single mode fiber is about _____ μm. A single mode fiber can carry more data than a multi-mode fiber due to its smaller _____.

(7) List three major parameters of optical fiber in designing an optical fiber link: _____, _____ and _____. The three (central) operating wavelengths are _____nm, _____nm and _____nm.

(8) (True, False) For telecommunications networks, glass fibers are used.

(9) The primary signal loss along a transmission path includes: _____ loss, _____ loss, _____ fiber, and _____ loss. When the received signal is too "hot", an _____ is required at the receiver.

(10) Three type of signal dispersion when transporting over optical fiber: _____, _____ dispersion, and _____ distortion. For single mode operation, _____ is negligible.

(11) (True, False) Typical PIN receiver has a responsivity of 0.6 to 0.8 A/w, for an APD an internal gain of 50 to 200 is included.

(12) List the major parameters of a optical receiver in designing an optical link: _____, _____, and _____.

(13) Receiver sensitivity range for a typical _____ is (−24 dBm to −32 dBm), and for a typical _____ is (−28 dBm to −40 dBm).

(14) System noise is modeled as a Gaussian noise by applying _____.

7.8 LIGHTWAVE NETWORK AND APPLICATIONS

Several types (e.g., twisted-pair, coaxial cables, airway, waveguide, and fibers; see Table 7-1) of transmission media can be used along the transmission path of an end-to-end connection (see Figure 7-46). This end-to-end communication connection is divided into two parts: (1) the access connections between the end users and the network; and (2) the inter-office connections between various nodes within a network.

In general, the access facilities transport low to medium speed signals, and the inter-office facilities transport medium to high speed signals. An optical fiber link can be designed for low, medium, or high speed applications. Therefore, optical fiber links can be used for both access connections and inter-office connections. These applications are described in the following sections.

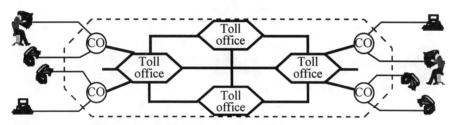

Figure 7-46   An End-to-End Connection: Access and Inter-Office.

7.8.1 Access Connection/Networks

The access connection to a "back-haul" network (used to connect the inter-office nodes) can be considered a small network that serves residential customers, small business customers, and large business customers. An access network can be composed of:

(1) Wireline facilities for connecting customers from rural areas, suburban areas, apartment complexes, and industrial/business centers to the local exchange switching center.

(2) Wireless facilities (e.g., cellular, mobile, fixed wireless, and wireless PBX) for connecting customers from remote or terrain-restricted areas (e.g., mountains, etc.).

(3) Fiber facilities for connecting customers from apartment complexes, business center and industrial parks, as shown in Figure 7-47.

Several commonly used access systems used in the PDH, SONET, and SDH environments are described in the following sections. These systems include: DDM-1000 (a PDH multiplexer), SLC-2000 (with RT-2000 located at the customer premises, in the SOENT environment), BRT-2000 (primary usage in metropolitan business buildings; in the SONET/SDH environment), Optimux (PDH multiplexer), etc.

7.8.1.1 DDM-1000

One of the most common PDH transmission elements used in North American digital networks for access applications is the DDM-1000 multiplexer. This equipment is normally configured for "terminal mode" operation. That is, at one end of an "end-to-end" connection, the DDM-1000 serves as a digital multiplexer, which combines several low-rate signals (either same or different speeds) together for transmission over the network. At the other end of the "end-to-end" connection, the DDM-1000 separates (demultiplexes) the signals for distribution.

The low-speed side of this mulitplexer can be 1.544 Mbps DS1 digital signals, 3.152 Mbps DS1C signals, 6.312 Mbps DS2 signals, 44.736 Mbps DS3 signals, or a combination of these signals (see Figure 7-48). The high speed side of the multiplexer can be DS3 or DS3C (90 Mbps or 180 Mbps). The DS3 output signal can be carried over a coaxial cable, digital radio, or an optical fiber link.

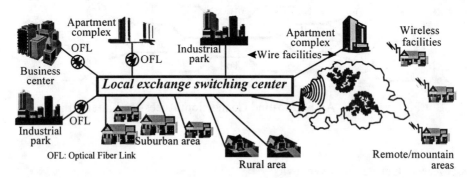

Figure 7-47   Access Network.

The DS3C (90 Mbps and 180 Mbps) output signals of a DDM-1000 are usually carried by *low-speed* optical fiber links. Therefore, an LED (instead of a laser diode) is typically used as the electrical-to-optical converter in conjunction with multi-mode fibers (instead of single mode fibers). If the distance is relatively short, Step Index (SI) multi-mode fibers are adequate. For example, most campus applications for local area networks use SI multi mode fibers. In comparison, for long distance applications GRIN multi-mode fibers are typically used. For high speed access applications, GRIN multi mode fibers may also be used. At the receiver, PIN photodiodes (instead of APD photodiodes) are adequate for optical-to-electrical conversion. For access network applications regenerators (normally required for connection distances of 25 miles or longer) are usually not required.

Figure 7-48 shows a configuration using DDM-1000 to access a DS3 network in the PDH environment. (Note that the DDM-1000 multiplexer is used as a multiplexer at the transmitting end, and a demultiplexer at the receiving end.) This configuration is commonly called *"fiber in the feeder plant"* arrangement.

    Another important point about the application shown in Figure 7-48 is the use of a
SLC-96 system. Besides accepting various digital signals, a DDM-1000 can be connected
directly to a Subscriber Loop Carrier system [e.g., SLC-96 or SLC-5, which is referred to
as integrated digital loop carrier (IDLC) system described in Chapter 6], to serve voice
customers, audio customers, fax customers, and video customers.

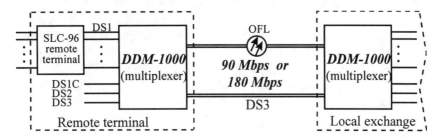

Figure 7-48   DDM-1000 Multiplexer as Feeder Plant.

7.8.1.2 SLC-2000/RT-2000

Fiber To The Home (FTTH) or fiber To The Curb (FTTC; Figure 7-49) is another
network access application using optical fiber links as transmission facilities. The RT-
2000 access system (a Remote Terminal of SLC-2000) consists of a SONET add-drop
multiplexer (e.g., DDM-2000) and a digital remote terminal.

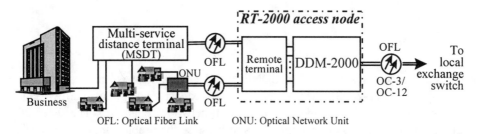

Figure 7-49   Fiber to the Home/Curb (FTTH/FTTC).

    The access facilities connected to the local exchange switching office are
terminated by a DDM-2000 multiplexer (similar to DDM-1000 but used in the SONET
environment; described in Chapter 6), at the access node. Business and residential
customers at remote sites are connected to the access node via a Multi Service Distance
Terminal (MSDT) and an Optical Fiber Link (OFL), terminated by the Remote Terminal
(RT-2000). An Optical Network Unit (ONU) can be used as a distribution node for fiber
to the home or fiber to the curb applications. The optical fiber links terminated by the
remote terminal are similar to those used in the fiber feeder plant (Figure 7-48). That is,
LEDs, multi-mode fibers, and PIN photodiodes are used as the link components.

7.8.1.3 Integrated Narrowband/Broadband Access system

Figure 7-50 illustrates an optical fiber links application used for Switched Digital Video (SDV) services. In addition to video services (e.g., broadcast video, high definition television, 2-way video, video dialtone, and interactive multimedia services), full-service telephony and high capacity digital services are also provided. These services can be provided by an integrated narrowband/broadband access system. On the access side, this system consists of three components: (1) Host Digital Terminal (HDT), (2) Optical Fiber Link (OFL), and (3) integrated Optical Network Unit (ONU). The local exchange access side utilizes digital carrier systems (e.g., T1 digital carrier), SONET/SDH optical fiber links (OC-1, OC-3, etc.) terminated by SONET DDM-2000, or SONET/SDH rings (OC-3, OC-12, or OC-48 terminating at an add-drop node).

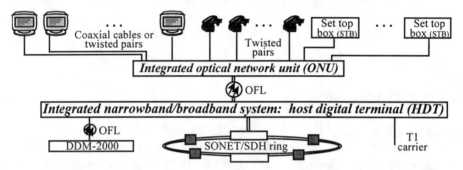

Figure 7-50   Integrated Narrowband/Broadband System: Video & Audio.

7.8.1.4 BRT-2000

An important access system used in SONET/SDH environment is the Business Remote Terminal-2000 (BRT-2000), which supports an extensive range of narrowband and broadband services, while providing improved network reliability. The BRT-2000's modularity allows network service providers to increase the usable bandwidth available to business customers without adding new fiber facilities. The BRT-2000 offers SONET/SDH functionality, therefore it can enhance network Operations, Administration, and Maintenance (OA&M) while reducing operational costs. An important feature is that BRT-2000 supports POTS, special services, ISDN, Centrex, DS1, DS3, etc. BRT-2000 offers full service protection, including self-healing fabric for DS1 (T1.5 Mbps) and DS3 (T45 Mbps) services.

The BRT-2000 access system is mounted in special cabinets (97A or 97B) for compatibility with SLC-2000, SLC-5, DDM-1000, and DDM-2000 in the SONET/PDH environment, or OLC-2000, Optimux, Combimux, ISM-2000 in the SDH/PDH environment. As shown in Figure 7-51, the optical fiber links can have speeds of 90 or 180 Mbps for low-speed applications. These links utilize LED, multi mode fibers, and

PIN photodiodes as link components. However, when applications require the speed of an OC-3 or OC-12, medium-speed optical fiber links may be used, depending upon the facility length. For longer paths, the system may require single mode fibers.

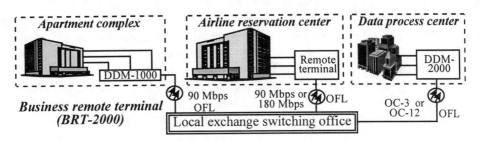

Figure 7-51   Business Remote Terminal (BRT-2000) Applications.

7.8.1.5 Optimux (PDH: 8TR661)

Optimux is an enhanced Network Access System (NSA) used with the existing NAS SLC-120, which can group any combination of voice and data services for ITU-T 30-channel hierarchy applications. The SLC-120 performs all the functions of a SLC-5 (used in the μ-law digital hierarchy. Besides processing digital facilities, SLC-120 contains two additional components. One component is the Customer Service Multiplexer [CSM; called the Remote Terminal (RT) in the μ-law digital hierarchy] located near the customers. The other component is the EXchange Multiplexer [EXM; equivalent to the Center Office Terminal (COT) in the μ-law digital hierarchy] located in the local exchange office. The output of SLC-120 is a 2.048 Mbps metallic facility that meets ITU-T recommendations.

The SLC-120 offers the flexibility of an open-ended architecture supporting a range of customer service applications. The SLC-120 service capabilities include a a variety of two-wire and four-wire data services (up to 19.2 kbps), switched services, PABX lines, coin, and 64 kbps digital data. It also supports Subscriber Pulse Metering (SPM) at 12 Hz and 16 Hz for individual lines (without separate applique equipment), and prepares the local loop plant to offer enhanced services in the future. This access system is widely deployed in digital networks in Australia, Asia (China), and Europe (e.g., England, Ireland, Belgium, France, Germany, Poland).

The SLC-120 system has been enhanced to handle up to 240 channels on 8 pairs of E1 metallic lines, with a hardened CSM for uncontrolled environments. It provides ISDN and remote OA&M capabilities. The option of 1+1 automatic protection switching is also available. This enhanced system is called SLC-240.

Optimux can support a capacity of up to 16 bidirectional 2.048 Mbps facilities. It is an integrated optical fiber link and digital multiplexing/demultiplexing system. Optimux is typically used in local, junction, and metropolitan area networks. The multiplexing and demultiplexing speeds are 2.048 Mbps, 8.448 Mbps, and 34.368 Mbps. That is, it can multiplex 2.048 Mbps signals into 8.448 Mbps signal and vice versa, and it can multiplex 8.448 Mbps signals into 34.368 Mbps signal and vice versa. The optical links can be 2.048 Mbps, 8.448 Mbps or 34.368 Mbps, depending the customers' needs. Since these optical fiber links are used for access networking, they are low speed links (LEDs, multi mode fibers, and PIN photodiodes). A 1+ 1 automatic protection switching option is available for reliability. This product has been implemented for access networks in China, Thailand, England, and France. It is typically used in areas where SLC-120 access systems have been deployed. The leading vendors manufacturing access networks are Lucent, Alcatel, Siemens, NKT, STC, NTI, DSC, Seicor, NEC, Fujitsu, and Ericsson.

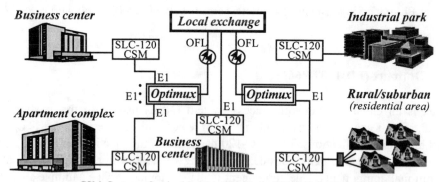

Figure 7-52   SLC-120 and Optimux Access System Applications.

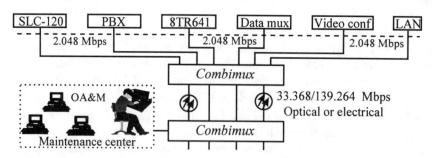

Figure 7-53   Combimux Access System Applications.

7.8.1.6 Combimux (PDH: 8TR671)

The Combimux access system is an enhanced version of Optimux that can be applied to inter-office networking. The major difference is the speed of the system. The interface speeds for Combimux are from 2.048 Mbps to 34.368 Mbps, and from 34.368 Mbps to

139.264 Mbps, and vice versa. The optical fiber links are either 34.368 Mbps or 139.264 Mbps facilities. These speeds are still low enough to use LED, multi mode fiber, and PIN photodiode technology. Another difference is Combimux has a "built-in" 1+1 Automatic Protection Switching (APS) feature. Areas implementing SLC-120, SLC-240, or Optimux are expected to use Combimux in their access or inter-office networks.

7.8.1.7 ISM-2000 Access System

The Intelligent Synchronous Multiplexer (ISM)-2000 access system supports optical interfaces with STM-1 or STM-4 rates, and is typically used in SDH/PDH environments (i.e., the PDH interfaces are E1, E3, and/or E4). This access system performs the same functions of DDM-1000 or DDM-2000 for different rates and formats.

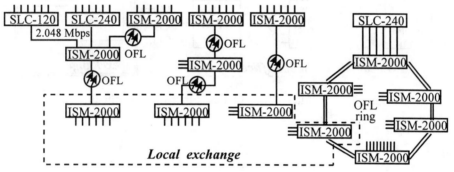

Figure 7-54   ISM-2000 Access System Applications.

7.8.2 Inter-office Connection

Inter-office networks are the transport facilities that tie local and toll exchanges (switching offices) together, that form the "telecommunications network". Inter-office transmission systems using optical fiber links are (1) DDM-1000, FT-G, 8TR694, 8TR695, Optimux, and Combimux in the PDH environment, (2) DDM-2000, FT-2000, LCT-2000 in the SONET environment, and (3) ISM-2000, Phase-FNS, and SLM-2000 in the SDH environment. These systems are briefly described in the following sections.

7.8.2.1 FT Series G

The FT Series G lightwave (PDH) system (see Figure 7-55) is the most commonly used asynchronous high speed optical fiber systems in digital networks. This system provides highly integrated equipment based on a flexible modular architecture that is suitable for all applications - long and short haul, high and low density traffic. The original system operated at 417 Mbps using single mode fibers applying 1310 nm, with a system capacity of 6,048 two-way voice (voice-equivalent) channels, or nine DS3 digital signals. The upgraded configuration operates at 1.7 Gbps and serves 36 DS3 signals [i.e., 24,192 voice or (voice-equivalent) channels]. An improved configuration applies two-wavelengths (1310 nm and 1550 nm) in Wavelength Division Multiplexing arrangement (WDM; discussed later in this chapter), to provide a capacity of 72 DS3 signals [i.e., 48,384 voice (voice-equivalent) channels], with reduced start-up and operational costs.

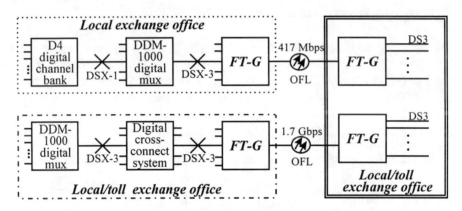

Figure 7-55   Inter-Office Optical Fiber Link Applications.

7.8.2.2 FT-2000

The FT-2000 lightwave system supports the highest speed optical links used in digital networks at the present time. These optical fiber links operate at 1310 or 1550 nm, depending upon the transmission path length. As shown in Figure 7-56 this system integrates the functions of an add-drop multiplexer and an optical fiber terminating with OC-48 as its high-speed port. It uses an open system architecture that embraces SONET, ISDN, ANSI, UL, Bellcore and ITU-T standards. The responsiveness, fault detection, enhanced OA&M capabilities, and network restoration capabilities make FT-2000 a "benchmark" in the lightwave industry. FT-2000 offers a solid foundation from which present asynchronous networks can gracefully evolve into advanced synchronous networks of the future.

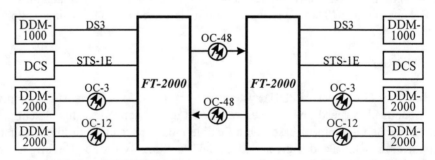

Figure 7-56   FT-2000 Lightwave System for Inter-Office Applications.

FT-2000 with one high-speed OC-48 optical fiber link is typically used for "2-fiber" SONET bidirectional line switched ring networking applications (see Figure 7-56). FT-2000 LCT (LCT-2000) is an upgraded system, with two high-speed OC-48 optical fiber links (instead of one link) in each direction of transmission in 4-fiber SONET bidirectional line switched ring network applications. In LCT-2000 the add-drop multiplexer is replaced by the Optical Line System (OLS-2000). The LCT-2000 operates only at the wavelength of 1550 nm (not supporting the 1310 nm wavelength). In addition, optical amplifiers are applied along the transmission path.

7.8.2.3 PDH 8TR694 and 8TR695

The PDH 8TR694 and 8TR695 optical fiber systems were the most common lightwave systems prior to SDH optical fiber systems. They perform the following functions: multiplexing and optical transport (see Figure 7-57). The optical fiber links support both 1310 nm and 1550 nm operations. The major difference is the system speed. The 8TR694 has two 139.264 Mbps low-speed ports, two 139.264 Mbps " high-speed" ports, and applies multi mode or single mode fibers depending on the path length. The 8TR695 has four 139.264 Mbps low-speed ports, two 565 Mbps high-speed ports, and always uses single fibers because of the 565 Mbps high-speed requirement.

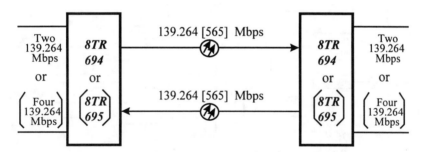

Figure 7-57   Applications of PDH 8TR694 and 8TR695.

7.8.2.4 SLM-2000

The Synchronous Line Multiplexer (SLM)-2000 lightwave system (see Figure 7-58) performs the same functions that the SONET FT-2000 system does, and is used in the SDH environment. There are two configurations are available: it can be either a SLM-4 (i.e., SLM-2000/STM-4; equivalent to SONET DDM-2000/OC-12 add-drop multiplexer) or SLM-16 (i.e., SLM-2000/STM-16; equivalent to SONET FT-2000 transport system). Several similar products appear in SDH environment are: Phase 4/4, Phase 16/4 (Lucent), ROC48, ROC192 (Alcatel), FLM-2400 (Fujitsu), ITS-2400 (NEC), etc. Note that all the access systems described in this section are used in SONET and SDH environments, and can perform add-drop, multiplexing and demultiplexing functions.

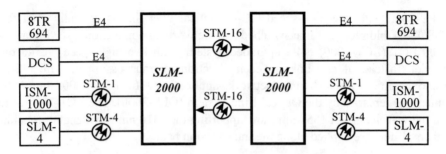

Figure 7-58   Applications of SDH SLM-2000.

7.8.3 Bandwidth Management

A transmission product that has been widely used for network bandwidth management is Digital Cross-connect System (DCS) (shown in Figure 7-59). There are two categories of DCSs used for bandwidth management: (1) PDH DCS, and (2) SONET/SDH DCS systems, and both categories can be either wideband or broadband designs.

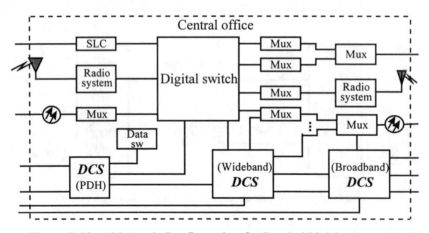

Figure 7-59   Network Configuration for Bandwidth Management.

A PDH DCS offers voice/data transport, and fully integrated network access capabilities. A DCS provides non-blocking digital channel cross-connection and test access. Lucent's DACS II (Digital Access Cross-connect System Version 2), deployed in North American digital and international gateway digital networks, is a multi-processor controlled system that performs DS0 cross-connection. In addition, 2.048 Mbps services can also be cross-connected from one terminating facility to any other facility. This capability is further extended to include subrate channels at 2.4 kbps, 4.8 kbps and 9.6 kbps signal rates. There is a further need for DACS II to interface with SONET signals (e.g., OC-3). This capability may soon be available.

In addition to cross-connection and test access, DACS II supports facility performance monitoring, equipment maintenance, and digital signal processing (e.g., multipoint bridging, subrate processing, etc.). The system assures efficient digital carrier system transmission by segregating (grooming) voice service channels from special service channels, permitting direct cabling to staging areas, eliminating the need for intermediate distribution frames to consolidate channels, and reducing the number of transmission facilities and associated transmission terminals. When used as an international gateway, DACS II provides A-law/µ-law signal format conversion, signaling translation, and rearranging DS0 channels within the primary signals. In the gateway application all the DACS II testing, monitoring, service protection, and bandwidth management functions are retained. Therefore, DACS II offers an integrated solution for all digital networking applications.

Figure 7-60 illustrates the network configuration for bandwidth management using WideBand (WB) and BroadBand (BB) DCSs (WB-DCSs and BB-DCSs) in a SONET environment. In SONET applications, a DCS is referred to as a WB-DCS if it interfaces with facilities having signal rates equal to or less than STS-1 or DS3 (e.g., signal rates ≤ 51.85 Mbps). If the interface signal rates are equal to or higher than STS-1 or DS3 signals (e.g., signal rates ≥ 51.85 Mbps), it is classified as a BB-DCS. A wideband DCS provides clear-channel switching at both the DS3 (44.736 Mbps) and STS-1 (51.84 Mbps) signal rates, and a BB-DCS provides cross-connection of DS1 or VT1.5 signal rates.

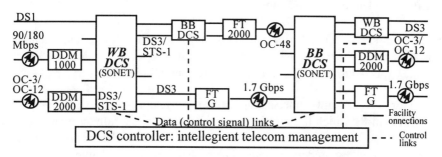

Figure 7-60   SONET Wideband and Broadband DCS for Bandwidth Management.

Both WB and BB-DCS systems provide dynamic network restoration capabilities. The provisioning, maintenance, and restoration functions of WB-DCS and BB-DCS can be automated via a centralized operating system [e.g., Intelligent Telecommunications Management (ITM)], serving as the DCS controller. Since WB-DCS and BB-DCS provide built-in test access capabilities, the need for manually testing facilities is eliminated. In the SONET environment, facility cable cuts, loss of signal frame, signal loss, pointer loss, equipment failure, or critical failures on high-capacity circuits can be disastrous. To handle these potential disasters, WB-DCS and BB-DCS offer reliable methods for quick restoring service. The automatic detection of Alarm Indication Signal (AIS) in-band message formats allows rapid re-routing of failed circuits to alternate traffic paths (i.e., automatic circuit restoration; self-healing networks).

Figure 7-61 shows a similar DCS arrangement in the SDH environment. Instead of DS3 (STS-1) signals being the basic rate for classifying WB-DCS and BB-DCS, the E-4 (STM-1) rate is used in SDH network. The switching fabrics are VC4 (E4), VC12 (E1), VC-3 (E3), or AU-3 (DS3) [in SONET the switching fabrics are VT1.5 (DS1) and DS3; switching fabric is referred to the unit of traffic can be cross-connected]. The DCS functions and network management capabilities are similar in both SONET and SDH environments. Note that in Figures 7-60 and 7-61, facility connections (customer traffic) are represented by solid lines and element control links (OAM&P) are shown as dotted lines.

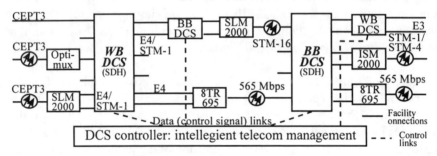

Figure 7-61   SDH Wideband and Broadband DCS for Bandwidth Management.

7.8.4 Undersea Lightwave Systems

Undersea high-speed optical fiber systems have replaced traditional metallic cable systems, especially for trans-pacific and trans-Atlantic telecommunications. These fiber systems are expected to dominate international telecommunications services.

The first significant undersea optical fiber system was TAT-8 (Trans-Atlantic Transport system-8), which has termination points at the Tuckerton terminal station (New Jersey, USA), Penmarch terminal station (France), and Widemouth terminal station (England) provided by undersea branching repeater in the Atlantic Ocean. The link applies 1.3 µm Multiple Longitudinal Mode (MLM) laser diodes as light sources, conventional (standard) Single Mode Fibers (SMF) operating at 1.3 µm for transmission media, electronic regenerators spaced at 70 km along the transmission path, and PIN photodiodes as receivers. The optical data rate is 280 Mbps (two CEPT4 signals), carrying 20,000 voice channels per fiber pair.

The second generation undersea optical fiber system is known as the SL-560 system (Submarine Lightwave 560 Mbps system), which has been deployed across both the Atlantic and Pacific oceans. Several systems carry the bulk of international traffic, (e.g., TAT-9, TAT-10, TAT-11, MAT-2, Columbus II and Americas 1) across the Atlantic ocean, and (e.g., TPC-4, HAW-5, PacRimEast, and PacRimWest) across the Pacific ocean. Single Longitudinal Mode (SLM) laser diodes (also known as narrow band lasers), conventional Single Mode Fibers (SMF), and APDs are components used in these systems, with (electronic) regenerators spaced 140 km along the transmission path.

The third generation undersea optical fiber systems apply SDH protocols (see Figure 7-62). Presently the SL-2000 systems dominate third generation Atlantic and Pacific ocean applications. The systems are: TAT-12, TAT-13, Columbus-B, and Americas-N for the Atlantic ocean, TPC-5, and APCN (Asia Pacific Cable Network) for the Pacific ocean, and FLAG (Fiber Link Around the Globe) originating in the Far-east and crossing the Indian ocean. Single Longitudinal Mode (SLM) laser diodes, Dispersion-Shifted Single Mode Fibers (DS-SMF), and APDs are the components used in these systems. In addition, optical amplifiers (spaced at 40-to-70 km) are used to reduce the number of electronic regenerators. The optical fiber links operate at 1.55 μm, with a system speed of 5 Gbps carrying 300,000 voice channels per fiber pair. A SDH path-switched 4-fiber bidirectional ring architecture is used for TAT-12 and TAT-13. Table 7-8 lists the technical specifications of the optical amplifiers used in SL-2000 undersea systems.

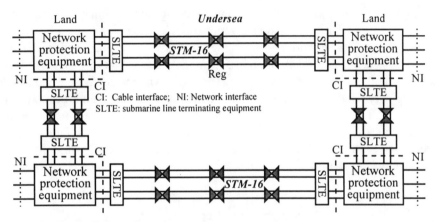

Figure 7-62   An Undersea Lightwave System Using SDH Ring.

Table 7-8 Technical Specifications of SL-2000 Optical Amplifiers.

Specification	Values
Repeater voltage drop	*7.5 volts (average) per amplifier pair* *30.2 volts maximum*
Line current	*0.92 amperes*
Power consumption	*7.4 w (average) per amplifier pair*
Temperature rise (ambient)	*5 ℃ to chassis*
Maximum IC temperature	*+50 ℃ (in shallow water)* *+40 ℃ (in deep water)*
Nominal input signal power	*−13 dBm*
Nominal amplifier gain	*17 dB*
Nominal output signal power	*+4 dBm*
Nominal amplifier noise figure	*6 dB*

7.8.5 Local Area Network Applications

A common local area network using optical fiber links is known as the FDDI ring (Fiber Distributed Data Interface). The data speed (of the workstation) is 100 Mbps, and the line rate is 125 Mbps because a special line code (4B/5B) has been adopted as the zero suppression method used to guarantee ones density in the data stream (see Chapter 6). That is, 100 Mbps × 5/4 = 125 Mbps, as every 4 consecutive bits (4B) must be converted into a pre-assigned 5-bit (5B) word to insure ones density.

The SONET OC-3c transmission facilities are typically used to carry FDDI signals. In this application the optical speed is 155.52 Mbps. The optical fiber links use 1300 nm LEDs as light sources, with multimode fibers, and PIN photodiodes.

A FDDI network consists of two counter-rotating rings (primary ring shown by the bold lines, and secondary ring shown by thin lines in Figure 7-63) with a maximum perimeter of 1 km. The network nodes in an FDDI network are classified into two types: (1) Dual Attached Stations (DASs), and (2) Single Attached Stations (SASs). A DAS is connected to both the primary and secondary rings, while a SAS is connected to the primary ring only. Under normal operational conditions (when no fault occurs) customer traffic is carried by both rings. However, each station will select data from the primary ring, while the traffic in the secondary will serve as "backup data" for faulty conditions. Thus, the primary ring is often referred to as the "working channel" while the secondary ring is called the "protection channel".

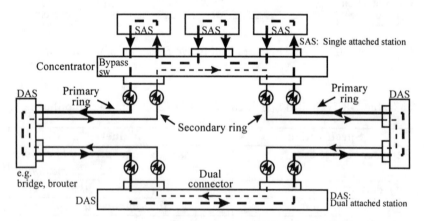

Figure 7-63   FDDI Ring Architecture: Primary and Secondary Rings.

The two ring topology offers network reliability by allowing operation to continue during link breaks (e.g., fiber cuts), station failures, or power failures. When a link breaks, the network decomposes into a single ring as shown in Figure 7-64. For a single break, each station in the ring is still connected for service, but, the ring length has been doubled and longer facility delays are experienced. If a second failure occurs, the network will be reconfigured into two smaller networks, and some stations may be isolated from other stations. Continuous monitoring of failure conditions allows the network to restore operation automatically after a repair is completed.

Each station is connected to (i.e., served by) a "bypass switch", which also serves as a "concentrator". If a station is malfunctioning or is "powered down", its "bypass switch" will route traffic around the station. In Figure 7-64 the upper-left SAS station is assumed to be malfunctioning or powered-down. As a result, its associated "bypass switch" re-routes traffic so that the "faulty" station is isolated from the network.

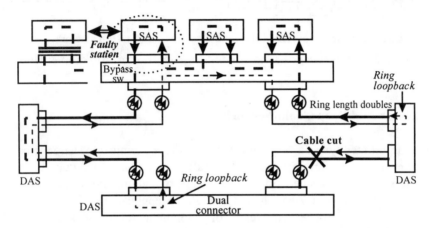

Figure 7-64   FDDI Ring Reconfiguration: Loopingbacks.

7.8.6 Various Network Layouts

The central offices using lightwave systems for inter-office connection can adopt one or combination of the following five network layouts: (1) point-to-point, (2) linear add/drop chain, (3) hubbing, (4) DCS-based mesh networks, and/or (5) ring add/drop.

7.8.6.1 Point-to-Point Networks

A point-to-point network [Figure 7-65(A)] is typically a dedicated network for private applications. The basic function is to perform multiplexing and demultiplexing. The optical fiber links for point-to-point applications may be operated at low, medium, or high speed. The choice of link components is summarized in Figure 7-37. For example, if the link speed is low (e.g. OC-1, or OC-3) and/or link length is short (e.g. 25 miles or less), the preferred link components are LEDs, multimode fibers, and PINs. If the link length is long, and the speed is OC-12 or OC-48, then laser diodes, single mode or DS-SMF (Dispersion-Shifted Single Mode) fibers, and APD are appropriate link components. If automatic protection feature is required, either the 1+1 or 1:n (1 by n) can be applied.

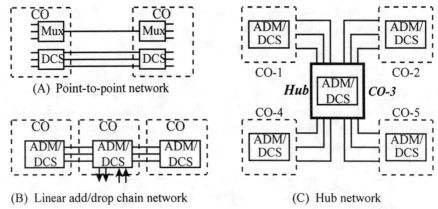

(A) Point-to-point network

(B) Linear add/drop chain network   (C) Hub network

Figure 7-65   Network Layouts: Point-to-point, Linear Add/drop & Hub.

7.8.6.2 Linear Add/Drop Chain Networks

When a transport facility has a very high capacity, it is economical to share the bandwidth among many users. Performing "add/drop" traffic operation is a required function in modern telecommunications networks. The principle of add/drop networking is illustrated in Figure 7-65(B).   Figure 7-65 illustrates add/drop applications. The linear add/drop function is analogous to the "enter/exit" of automobile traffic on a major highway. Add/Drop Multiplexers (ADMs) eliminate the need of using "back-to-back" multiplexing/demultiplexing equipment. In the SONET/SDH environment (see Figure 7-66), most ADM equipment can "add and drop" either electrical and/or optical signals as required.   Note that Digital Cross-connect Systems (DCSs) are used for signaling grooming conversion in this arrangement.

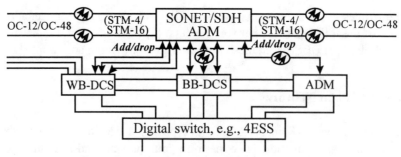

Figure 7-66   Linear Add/drop Applications.

7.8.6.3 Hub Networks

The purpose of using "hubbing" is to reduce the number of physical facilities needed to interconnect the nodes in a network. The number of facility connections (N_c) for connecting *"n"* network nodes to each other without and with a "hub" are given in Eqs.(7-76) and (7-77), respectively.

$$N_c = \frac{n(n-1)}{2} \qquad \text{(without a "hub")} \qquad (7\text{-}76)$$

$$N_c = n \qquad \text{(with a "hub")} \qquad (7\text{-}77)$$

Figure 7-65(C) shows that the network node identified as "CO-3" is a hub node, which is a "stepping stone" for any node to be connected to another node. For example, a user served by "CO-1", who intends to talk to a customer served by "CO-5", will first be connected to the "hub" (i.e., CO-3), and then connected from the hub to "CO-5".

7.8.6.4 DCS-Based Mesh Networks

Another network layout is a DCS-based mesh network as shown in Figure 7-67(A). Mesh networks are often connected to a "network controller" (not shown), which acts as the "brain" of the network. This controller is connected to any node (not a DCS) in the network, and to an Operations Systems (OS) that administers the network. A DCS-based network requires many facility connections, in accordance with Eq.(7-76). Therefore, only networks with a small or medium number of DCSs should be connected using the DCS-meshed technique. This configuration is typically used wherever fast network restoration (due to critical network failure conditions) is required.

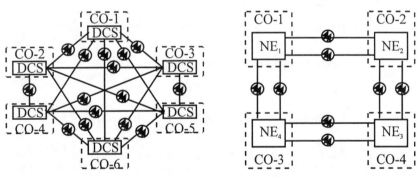

   (A) DCS-Based Mesh Configuration   (B) Ring Add/drop Configuration

    Figure 7-67   Network Layouts: DCS-Based Mesh and Ring Add/drop.

It should be noted that a network can use combined topologies: for example, a "hub" and DCS-based mesh to reduce the number facility connections (see Figure 7-68). In this example, assuming "hub" is used, the number of facilities required to interconnect six network nodes would have been 6(6 − 1)/2 = 15 [in accordance with Eq.(7-76)] instead of 14 (a larger savings in required number of facilities is expected for a larger network with more nodes). The mesh connections may be used for connecting "high priority" traffic. Any "low-priority" or overflowing "high-priority" traffic can be routed via the "hub" node. In Figure 7-68, the high-priority traffic is carried by the "bold" lines, that is often referred to as direct trunks. In contrast, the low priority or overflowing high priority traffic is routed via the hub (shown by thin lines in Figure 7-68).

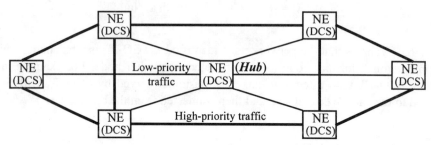

Figure 7-68   DCS Meshed and Hub Network.

7.8.6.5 Ring Add/Drop Networks

A common network layout is the ring add/drop configuration [see Figure 7-67(B)], which provides reliable self-healing capabilities under most failure conditions. Ring architectures (shown below) and operations have been previously discussed in Chapter 6 (see ANSI standards T1.105.01).

(1) *Line switched rings*

    • Unidirectional Line Swwitched Rings (ULSR) ⎰ * 2-fiber rings
                                              ⎱ * 4-fiber rings

    • Bidirectional Line Swwitched Rings (BLSR) ⎰ * 2-fiber rings
                                              ⎱ * 4-fiber rings

(2) *Path switched rings*

    • Unidirectional Path Swwitched Rings (UPSR) ⎰ * 2-fiber rings
                                              ⎱ * 4-fiber rings

    • Bidirectional Path Swwitched Rings (BPSR) ⎰ * 2-fiber rings
                                              ⎱ * 4-fiber rings

7.9 THE TREND: ALL-OPTICAL NETWORKING

The proposal for using low-loss optical fibers for telecommunications transport systems was originated in 1965. However, it was not until the early 1980s, that low-loss optical fiber systems were first made commercially available for long-haul transmission applications. In three decades, optical fiber communications technology progressed rapidly, and will continue to grow for decades to come. Among all the new technologies, Erbium-Doped Fiber Optical Amplifiers (EDFOA, EDFA, EOA, or OA) and Dense

Wavelength Division Multiplexing (DWDM) are considered to be "break-through" technologies. DWDM and several other technologies are briefly described in this section (e.g., all-optical regenerators, routing/switching, cross-connecting, wavelength add/drop, polarization, chromatic dispersion compensation, soliton, etc.).

7.9.1 Optical Networking

The development of Wavelength Division Multiplexing (WDM) systems opened a whole new dimension for designing high-speed communications networks. All WDM systems were originally used for point-to-point high capacity transport. New optical networking techniques offer a wide range of features and functions. The most important and promising functions are: (1) Wavelength routing, (2) Wavelength switching, (3) Wavelength cross-connection, (4) Wavelength add-drop, (5) Wavelength conversion (i.e., being able to perform all-optical transfer of information from one wavelength of light to another) to overcome possible wavelength contention at network nodes. Studies have indicated that wavelength conversion can also benefit network management and link restoration schemes. Many wavelength conversion technologies are promising, the most well-known technologies are cross-gain and cross-phase modulation, four wave mixing, and optical loop mirror.

7.9.2 Electronic Regenerators and Optical Amplifiers

As previously described in Chapter 6, a digital signal degrades along its transmission path because of signal attenuation and noise corruption. Therefore, regenerators must be applied along the path to maintain transmission quality. Electronic regenerators have be used for digital facilities, twisted-pair, coaxial cable, and optical fiber. For optical fiber applications several components are added to the electronic regenerator as shown in Figure 7-69. These components are briefly described as follows:

- A photodiode (either PIN or APD) is used as Optical-to-Electrical converter (O/E).

- (Re-)amplification/equalization, re-timing, and re-generating circuits are used to amplify and equalize the degraded signal so that the re-timing circuit can perform timing synchronization. The output of the "re-timing" circuit is a synchronized clock, used by the "re-generating" circuit to sample the amplified/equalized signal. This allows regenerated digital data stream to be prepared for further transmission.

- A light source, which is either an LED or laser diode, is used as electrical-to-optical conversion. These two components (O/E and E/O converters) are required since the conventional regenerator must be implemented in electronic domain.

In general, electronic regenerators are "bit rate sensitive". That is, the re-timing circuits are designed to operate at a specific speed. Therefore, systems with different speeds will require different regenerator design. In addition, regenerators require individual power supplies. For undersea applications, the cost of power supplies is considerably high. A large number of active components in a regenerator can result in lower reliability, especially for undersea applications. Therefore, in modern undersea optical fiber systems, Erbium-doped optical amplifiers are typically used because they do not contains many active components as in regenerators.

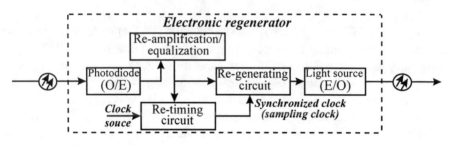

Figure 7-69   An Electronic Regenerator.

The use of optical amplifiers increases optical transparency (i.e., reduces the need for conversion) of a network, and will improve overall reliability. Eliminating the need to perform optical-to-electrical, and electrical-to-optical conversion reduces the costs of the system, improves system processing time and fewer regenerators are required. Overall system costs are less, the attenuation limits of the optical link are extended, and the network design is simplified.

Many types of optical amplifiers have been developed, and some are commercially available for practical applications. Three types appear to be promising for optical fiber link applications, and are listed as follows:

- Optical fiber amplifiers (the most mature technology)
- Semiconductor amplifiers
- Raman fiber amplifiers

Table 7-9   Various Optical Fiber Amplifier Technologies

Glass host	Active medium	Spectral region
• Silica glass fiber	• Erbium	• 800-to-900 *nm*
		• 1240-to-1340 *nm*
• ZBLAN glass	• Praseodymium	• 1520-to-1600 *nm*

ZBLAN: Zirconium, Barium, Lanthanum, Aluminum, and Sodium Fluorides

However, researchers and developers around the world will continue to search for other technologies to develop higher power, greater gain, and more bandwidth. This may be accomplishing using hybrid configurations of existing methods, and searching for different host fiber (glass host). Table 7-9 lists the spectral application, active medium, and glass host for several optical amplifiers.

The principle of all optical amplifiers is "amplification through stimulated emission". Figure 7-70(A) shows a simplified block diagram of an optical amplifier. Figure 7-70(B) illustrates the components of an optical amplifier: two optical isolators, one WDM, and erbium-doped fiber. The optical isolator at the input of an optical amplifier is used to eliminate possible disturbances caused by backward-traveling amplified spontaneous emission on the upstream span. The optical isolator at the output protects the device against possible reflected power from the downstream line. Pumping an optical amplifier is done by using a wavelength of either 980 nm or 1480 nm, preferably by means of semiconductor laser diodes. The optical signal input is fed to the active erbium-doped fiber along with the pump radiation through a Wavelength Division Multiplexing (WDM) coupler that minimizes the power loss of both input beams.

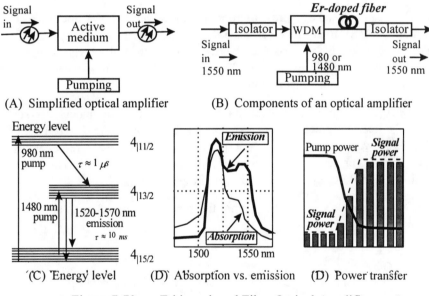

(A) Simplified optical amplifier   (B) Components of an optical amplifier

(C) Energy level   (D) Absorption vs. emission   (D) Power transfer

Figure 7-70   Erbium-doped Fiber Optical Amplifier.

The energy level of a typical erbium-doped fiber optical amplifier is shown in Figure 7-70(C). It can be seen that this type of Erbium-Doped Fiber Optical Amplifier (EDFA) can operate at either at 980 nm or 1480 nm. As shown in Figure 7-70(C), the 980 nm laser diode pumps from the "$4_{|15/2}$" state to the "$4_{|13/2}$" state, then back to the "$4_{|15/2}$" state via 1502 nm-1570 nm emission. Likewise, in the 1480 nm application, it starts from the "$4_{|15/2}$" state, pumps to the "$4_{|13/2}$" state, and then back to the "$4_{|15/2}$" state.

Figure 7-70(D) shows the relative absorption and emission spectra of an EDFA. The emission spectrum characteristics contribute to the optical signal amplification process. The concept of pumping can be visualized as "conservation of energy".

Figure 7-70(E) illustrates the transfer of optical power from the pumping laser to optical signal. The optical signal is attenuated as it propagates along the fiber link. When the signal power level becomes too small to continue, the pumping laser gives up its energy with the help of the active medium (erbium-doped fiber in this case), thereby transferring energy to the weak optical signal. If signal amplification is not done at the proper time, the signal will be corrupted by channel noise and becomes unrecognizable due to a low signal-to-noise ratio.

Example 7-32: Describe four applications for EDFA an optical amplifier.

(1) An EDFA can be used as a power amplifier at the transmitter end of a fiber link as shown in Figure 7-71(A). If the "launching power" from a light source is not strong enough to meet power budget requirements, the light source can be connected to an EDFA to provide the necessary power gain.

(2) An EDFA can be used as a pre-amplifier at the receiver end of a fiber link as shown in Figure 7-71(B). Every receiver has a specific sensitivity level (e.g., −32 dBm). If the received power is too weak (i.e., below receiver sensitivity level), then an EDFA can be used to provide extra gain. This is a reverse situation when the received power level is "too hot", that requires an attenuator.

(3) An EDFA can be used as in-line amplifiers (especially for undersea systems) to reduce the need for electronic regenerators as shown in Figure 7-71(C), and Figure 7-71(E). However, EDFA does not perform "re-timing" or "re-generating" the optical signals, presently it is necessary to use a regenerator (to perform these two functions) after (5 to 8) EDFAs as shown in Figure 7-71(E).

(4) An EDFA can be used as a splitter amplifier for various wavelength applications, as shown in Figure 7-71(D).

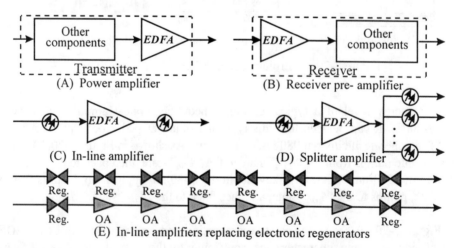

(A) Power amplifier

(B) Receiver pre- amplifier

(C) In-line amplifier

(D) Splitter amplifier

(E) In-line amplifiers replacing electronic regenerators

Figure 7-71   Various Applications of Erbium-doped Fiber Optical Amplifiers.

As the data traffic generated by Internet use increases, the demand for higher capacity terrestrial and undersea optical lightwave communications systems will also increases. As a result, the number of optical channels using Wavelength Division Multiplexing (WDM) must be increased accordingly. "***Widening the gain-bandwidth of an optical amplifier***" has become the most important technology used to increase the number of channels carried by optical fiber systems, thereby increasing system capacity. Presently, the commercial EDFAs provide an optical bandwidth of 35 nm (or less) for transmission systems that span more than 1,000 km. An essential aspect of increasing optical fiber communication system bandwidth is the ability to broaden the amplifier gain-bandwidth while maintaining high optical out power.

The conventional EDFAs deployed in many commercial optical fiber systems (especially undersea lightwave systems) have several advantages (e.g., high gain, low noise figure, etc.) over other optical amplifier technologies. However, EDFAs also have a few disadvantages (e.g., non-uniform gain profile, etc.), as listed in Table 7-10. Research goal for developing EDFAs is to overcome these disadvantages.

Table 7-10   Advantages and Disadvantages of EDFA.

Advantages	Disadvantages
• High gain at 1550 nm • Nearly ideal noise performance (low noise figure) • Low signal distortion and crosstalk • High efficiency and output power • Large optical bandwidth • No polarization dependence • Simple design, pump laser commercially available	• Non-uniform gain profile • No power gain at 1300 nm • Unpumped optical amplifier attenuates signal power • Impossible to have very short lengths

Existing WDM systems generally use Erbium-Doped Fiber Amplifiers (EDFAs), with a capacity limited to 8 or 16 optical channels. Newer systems are implemented with gain-flattened EDFAs using 35 nm bandwidth, and having 32 or 64 channels for both terrestrial and undersea applications. During terrestrial field trials, error free transmission has been achieved with 32 or 64 channels at speeds up to 10 Gbps, over distances greater than 500 km. Commercial systems carrying 80 optical channels with a total capacity of 400 Gbps are now available. Transoceanic transmission over distances of 7,000-to-9,000 km, having either 64 channels at 5 Gbps, or 25 channels at 10 Gbps have been implemented successfully. Longer distances, wider bandwidth, higher power, and greater capacity undersea systems are expected to be available in the near future.

In all the existing (deployed) WDM systems, EDFAs are operated in the "C-band" region (defined later in this chapter). However, optical amplifiers with higher gain and good S/N performance are believed to be capable of operating in the 1565 to 1615 nm ("L-band") region. Many experimental systems have been successfully demonstrated, practical systems will soon be commercially available.

Several other technologies have reported amplifiers achieving gain-bandwidths of 80 nm. These technologies are described in the following documents:

- "Changing fiber (glass) host", reported by H. Masuda et al., IEEE Photon Technology, Letter 10, pp. 516-517, 1998.

- "A parallel configuration", reported by T. Sakamoto et al., OAA98, TuB3

- "Combining a Distributed Raman Amplifier (DRA) with an Erbium-Doped Fluoride Fiber Amplifier (EDFFA)", reported by A. Srivastava et al., OFC98, post-deadline paper PD10, 1998, see Example 7-33

These techniques are being developed to improve the gain-bandwidth of an amplifier, in addition to maintaining high output optical power and/or gain (e.g., over 20 dB). Some of these devices have also reported error-free operation for WDM transmission. It is important to recognize that "imperfect gain-flatness" will produce a difference in the signal power level across the wavelength range, especially when several amplifiers are connected in cascade. This power difference leads to signal-to-noise degradation of the lowest power optical channels. Amplifiers with a total o 3-dB bandwidth of 80 nm, an output power of 25 dBm, and a noise figure of approximately 6 dB are considered to be acceptable.

Example 7-33 [The information shown in this example is taken from a report by S. Kawai, K. Suzuki and K. Aida, FC3-1, Technical Digest, Conference Edition, OFC/IOOC99]: Describe the characteristics of a hybrid optical amplifier used for wide bandwidth and long distance WDM transmission having highly gain flattened between 1530 nm and 1600 nm.

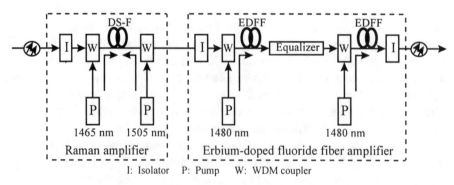

Figure 7-72   A Hybrid Optical Amplifier.

Figure 7-72 illustrates a hybrid amplifier that consists of a distributed Raman amplifier and an erbium-doped fiber amplifier. In the Raman amplifier portion, one isolator is applied at the input of the amplifier, Dispersion Shifted Fiber (DS-F) is used for internal connections, and two bidirectional Fabry-Perot pump laser diodes (operating at 1465 nm and 1505 nm, respectively) are used. The typical launched powers are 70 mw for the 1465 nm laser, and 200 mw for the 1505 nm laser. The Erbium-Doped Fluoride Fiber (EDFF) amplifier portion contains two backward-pumped 1480 nm Fabry-Perot laser diodes, EDF fibers for interconnections, and an intermediate gain equalizer. In this example, a transmission distance greater than 900 km with 15 dB amplifications is possible. This configuration uses fiber arranged in a circulating loop, and dispersion-shifted fiber deployed with a Raman amplifier having a length of 60 km. The typical signal speed is 2.5 Gbps ranging from 1535 nm to 1600 nm, with a signal power of −7 dBm.

Figure 7-73(A) shows the gross gain spectra for the Raman amplifier, the gross gain of the Erbium-Doped Fluoride Fiber (EDFF) amplifier, and the total gain of the combined hybrid amplifier. Figure 7-73(B) illustrates the net gain when using one hybrid amplifier, and the net gain of three amplifiers in cascade. Note that the resultant net gain has an excursion of only 1.5 dB over the entire 67 nm range from 1534 nm to 1601 nm. Figure 7-73(C) illustrates the comparison of the S/N characteristics of a discrete EDFA and the hybrid amplifier shown in Figure 7-72.

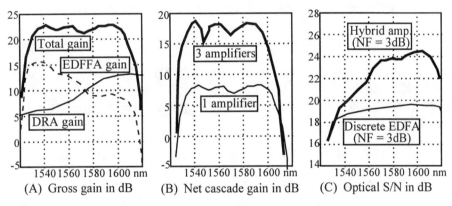

(A) Gross gain in dB   (B) Net cascade gain in dB   (C) Optical S/N in dB

Figure 7-71   Gross/Net Gains and S/N Comparison.

7.9.3 All-Optical Regenerators and Wavelength Converters

As communications networking moves toward all-optical configurations, the amplifiers used in the transmission path create problems resulting from the accumulated amplifier noise. Present system speeds/capacities still allow combinations of electronic regenerator and optical amplifiers, as shown in Figure 7-71(E). However, as system speeds continue to increase, the electronic regenerators/amplifiers will become a "bottleneck" in high-speed optical networks. "*All-optical techniques*" must be developed for signal regeneration to overcome this electronic bottleneck. One promising technique for implementing *all-optical regeneration* is based on Semiconductor Optical Amplifier (SOA) technology. Two versions of implementing all-optical regeneration using SOA are:

(1) **2R Regeneration** [**R**e-amplification and **R**egenerating (reshaping), see Figure 7-69] with full bit-rate transparency (see D. Chiaroni et al. ThD 1.4, ECOC96). Since "all-optical regenerators" are a new technology, a 2R regeneration has been considered to be a first generation technology of "all-optical regenerators". However, researchers have advanced into 3R regeneration because the 2R technique does not implement the third required function of a regenerator (i.e., re-timing, see Figure 7-69).

(2) **3R regeneration** [2R plus **R**e-timing, see Figure 7-69] (see D. Chiaroni et al. PD15, ECOC96). This full-function all-optical method will eventually be adopted for optical networking to overcome optical degradation caused by fiber dispersion, non-linearity, and optical amplifier noise.

A theoretically perfect regenerator would have a step function as its system transfer function [a linear system can always be represented by a transfer function, H(s); a perfect regenerator is the one with (Hs) = a step function]. Regenerative capabilities can be improved by cascading several devices, and the transfer function will approach a step function by using a cascaded structure. However, if the same regenerative capabilities can be achieved by using fewer cascaded devices, the result is a better regenerator. Several proposals for implementing all-optical 3R regenerative functions include:

(1) An SOA under cross-gain modulation.

(2) An SOA-based Mach-Zehnder Interferometric (MZI) or Michelson converter under cross-phase modulation [B. Mikkelsen, et al. "Electron Letter, vol. 33, No. 25, pp2137-2139, 1997; and, H. Spiekman, et al. "Photon Technology Letter, vol. 10, No. 8, pp1115-1117, 1998].

(3) Integrating SOA with a DFB laser.

Many experimental regenerators have been described in papers presented at technical conferences. For example, (1) cascading an SOA and a MZI, and (2) cascading an MZI and another MZI have been reported in TUJ-3, Technical Digest, Conference Edition, OFC/IOOC99, by B. Lavigne et al. Example 7-34 describes an integrated SOA/DFB laser regenerator.

Example 7-34: Describe the technologies for performing wavelength conversion and **all-optical regeneration** by applying an integrated Semiconductor Optical Amplifier (SOA) with a Distributed FeedBack (DFB) (SOA/DFB) laser (Reported by M. Stephens et al., TuJ2-1, Technical Digest, Conference Edition, OFC/IOOC99).

The integrated SOA/DFB device is fabricated using "InP/InGaAsP" techniques. The device incorporates a 500 μm SOA section [i.e., shown as amp in the upper-left portion of Figure 7-74(A)] and a 800 μm DFB laser section, with independent current biasing applied to both via a split contact [Figure 7-74(A)]. This device is designed to perform optical wavelength conversion. The optical injected signal has a 1560 nm wavelength, and the emitted optical signal has a wavelength of 1553.5 nm.

The regeneration of an optical signal is actually the result of "**the noise removal properties of the device**". Figure 7-74(B), from bit error rate viewpoint, shows that the "**all-optical wavelength conversion process can support regeneration**". The optical signal has a wavelength of 1560 nm before being converted to a signal with a wavelength of 1553.5 nm. This signal [shown by "bolded" line in Figure 7-74(B)] is transmitted over a distance of 93 km, the received signal has a degraded BER performance shown by the "dotted" line in Figure 7-74(B). This signal with a wavelength of 1553.5 nm is converted to an optical signal with a wavelength of 1548 nm. The BER performance [shown by a "thin" line in Figure 7-74(B)] is practically identical to the BER performance of the 1553.5 nm signal. "**The wavelength conversion supports regeneration of optical signal**".

In addition (in the same report of Example 7-34), simultaneous all-optical wavelength conversion and regeneration can be achieved by a compact, integrated device operating at 2.488 Gbps. Simulation has shown that this regenerative capability theoretically supports transmission over 10,000 km of dispersion shifted single mode fiber by cascading several of these devices. Actual implementation, using a 10 Gbps signal, has shown similar regeneration results.

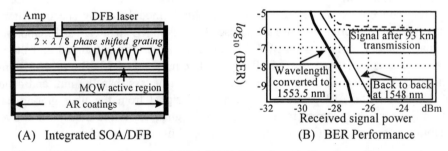

(A) Integrated SOA/DFB (B) BER Performance

Figure 7-74   An Integrated SOA/DFB Wavelength Converter/Regeneration.

Example 7-35: Describe the technique for using a SOA/fiber-grating hybrid wavelength converter to achieve error-free propagation of a 10 Gbps RZ signal over 20,000 km in a dispersion-managed system (Reported by P. Cho et al., ThQ2-1, Technical Digest, Conference Edition, OFC/IOOC99).

It is expected that RZ dispersion-managed high-speed transoceanic and long-haul terrestrial systems will evolve into all-optical networks. An all-optical RZ wavelength converter can be used for the following applications:

(1) When a faulty optical network is to be restored, a wavelength converter can reduce blocking during dynamic routing.

(2) A converter can connect optical links having different optimum wavelengths for their dispersion maps.

In addition to being compact and low-cost, an all-optical wavelength converter must have the following properties:

- Consuming low optical power
- Preserving data polarity
- Regenerating signals without degradation

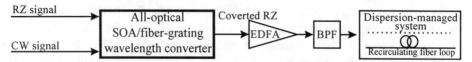

Figure 7-75   Schematic Diagram for An SOA/Fiber-grating Wavelength Converter.

The wavelength converter using an SOA-based Mach-Zehnder interferometric (MZI) or Michelson converter under cross-phase modulation can achieve the three properties listed above. This technique involves a complicated fabrication process, and requires an "all active" device design for high-speed operation.

In comparison, an all-optical SOA/fiber-grating hybrid wavelength converter has a much simpler design without requiring an interferometer, as shown in Figure 7-75. The fiber grating converts phase modulation into amplitude modulation by spectral filtering. This experimental converter launches a long pseudo-random RZ bit stream at 10 Gbps and a CW signal simultaneously into an SOA biased at 208 mA. The RZ signal and the CW signal have an average powers of –7.5 dBm (at 1545.6 nm) and –4.75 dBm (at 1551.8 nm), respectively. The converted signal has a spectral width of less than 0.2 nm, and a pulse width of approximately 17 ps. Since the converted signal from the fiber grating has a significantly narrower spectrum than the input RZ signal, error free transmission over a dispersion-managed system, (i.e., a recirculating looped fiber) can be achieved.

7.9.4 Wavelength Division Multiplexing

"Sharing" is an important aspect in all human activities, including communication. In the field of telecommunications, "sharing" provides an efficient way to utilize available system bandwidth. As technology advances, many "sharing" schemes have been developed, as shown in Figure 7-76:

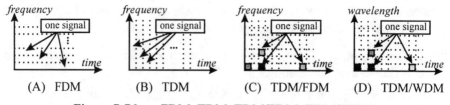

| (A) FDM | (B) TDM | (C) TDM/FDM | (D) TDM/WDM |

Figure 7-76   FDM, TDM, TDM/FDM, TDM/WDM.

- Frequency Division Multiplexing (FDM): a frequency "sharing" scheme.
- Time Division Multiplexing (TDM): a time "sharing" scheme.
- Time Division Multiplexing with Frequency Division Multiplexing (TDM/FDM)
- Time Division Multiplexing with Wavelength Division Multiplexing (TDM/WDM)

The FDM, TDM, and TDM/FDM technologies have been described in Chapters 6 and 8. Wavelength Division Multiplexing (WDM) will be discussed in this section. It should be noted that a WDM system is actually a system applying both time division multiplexing plus wavelength division multiplexing. For example, the 2-WDM FT-G optical fiber systems, which have been widely spread over the North American digital network, have a system speed of 3.4 Gbps. The system applies many stages of TDM for multiplexing every 672 voice or voice-equivalent channels into a DS3 signal, then 36 DS3 signals are TDM multiplexed into one 1.7 Gbps signal. Finally, two 1.7 Gbps signals are multiplexed into one 3.4 Gbps using 2-WDM technology.

As illustrated in Figure7-76(D), WDM can be viewed as a system that divides time into many intervals, and divides wavelength into many separate divisions. Thus, each block in the diagram occupies a "time range" and a "wavelength range", and is assigned to a signal, which can be a low-, medium- or high-speed signal (e.g., OC-3, OC-12, OC-48). Throughout this book, a WDM system refers to a "WDM plus TDM" system.

As the needs for larger bandwidth (higher-speed) transmission grow, and various all-optical technologies advance, WDM and Dense WDM (DWDM) systems will be deployed in global digital networks. WDM or DWDM is typically applied in the following scenarios:

(1) **For existing optical networks**: To increase the system speed, there are two alternatives that can be adopted. In this example, assume that the existing system is a four-node SONET ring (see Chapter 6, or ANSI T1.105.01) carrying OC-12 signals.

- Upgrading the system speed by applying TDM: To upgrade the system speed from OC-12 to the next higher rate OC-48, assuming the fibers will support the higher speed, still requires that all ring nodes be replaced with new higher speed OC-48 ADM equipment.

  But, if the distance between two nodes is too far to support the higher speed, even the fibers must be replaced. Therefore, the cost for upgrading may not favor using the TDM scheme. In addition, if TDM is implemented and only the capacity between two nodes (not the entire ring) needs to be upgraded, the entire ring must be upgraded.

- Upgrading the system speed by applying WDM: If the capacity between two nodes needs to be increased, using WDM requires only one new ring be added between these two nodes. The remaining nodes of the ring can continue to operate at the existing speed (Example 7-36 illustrates this technique).

(2) **For a new optical network**: For a new optical fiber network, WDM is the logical choice because technology advances constantly, and there is no restriction on the WDM speed hierarchy. For example, if OC-12 is adequate at the time of deployment, but in five years system capacity needs to be upgraded, TDM and WDM offer two different results:

- If the TDM scheme is adopted, the new speed will be OC-48, which is four times higher than OC-12.

- However, if the WDM is adopted, the system speed can be upgraded to 2 times OC-12 if 2-WDM is used, 4 times OC-12 if 4-WDM is used, 16 times OC-12 if 16-WDM is used, etc. depending on the needs.

Therefore, the WDM design is more flexible, and thus the cost of upgrading will be less expensive. Presently, there are some issues regarding system quality (for certain system needs) that may cause WDM to be rejected as the "best" approach. However, progress in developing all-optical network components will definitely position WDM technology as "the choice" for high-speed, long-haul transoceanic and terrestrial networks.

Example 7-36: Describe the method for upgrading a SONET ring speed (from OC-3 of 155.52 Mbps to 2 × 155.52 Mbps) by using WDM technology versus TDM technology referring to Figure 7-(Reported by O. Gerstel et al., ThE4-1, Technical Digest, Conference Edition, OFC/IOOC99).

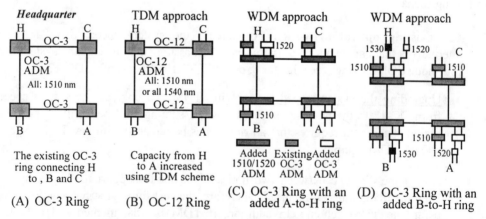

Figure 7-77   TDM versus WDM Implementations.

- The existing SONET OC-3 ring is assumed to operate at 1510 nm. The network node "H" is assumed to be the headquarters location, and "A", "B", and "C" are three branch nodes. Each node is equipped with an OC-3 ADM, as shown in Figure 7-78(A), according to the requirements of a SONET OC-3 ring.

- Assume the TDM scheme (instead of the WDM scheme) is adopted for upgrading. IF the traffic between "A" and "H" increases (e.g., to two times OC-3 speed), then all nodes must be upgraded to OC-12 ADM, because the next available SONET speed is OC-12. The fibers between any two adjacent nodes may need to be upgraded in some cases [Figure 7-77(B)]. Two issues need to be addressed: (1) the change (upgrade) is abrupt (not a smooth procedure) and (2) the cost may be unacceptable since nodes "B" and "C" do not actually "require" upgrading.

- If the WDM scheme is used for upgrading the capacity between nodes "H" and "A", an additional OC-3 ring operating at 1520 nm is added. The existing ring is retained, and each node, including "H", needs to be equipped with a 1510/1520 ADM (1520 nm is the second wavelength used in WDM). Nodes "A" and "H" require a new OC-3 ADM which are WDM compliant at 1520 nm. However, the existing OC-3 ADMs used at nodes "B" and "C" are retained. This arrangement is shown in Figure 7-77(C).

- Assume that the capacity between "H" and "B" needs to be increased. Now, an additional OC-3 ring operating at 1530 nm is required between "B" and "H". In addition, (1) at node "A" an optical add/drop multiplexer (OADM) is used to add/drop 1520 nm signal; (2) at node "B" an OADM is used to add/drop 1530 nm signal; (3) at node "H", one OADM is required to add/drop 1520 nm signal, and one OADM to add/drop 1530 nm signal, and (4) node "C" does not require an OADM, and its original OC-3 ADM is retained. This is illustrated in Figure 7-77(D).

- Eventually, if the capacity between "H" and "C" increases, a new OC-3 ring operating at 1540 nm can be added between "H" and "C". Both nodes "C" and "H" will require an additional OC-3 ADM, and an OADM (not shown in Figure 7-77).

It is clear that the WDM upgrading approach is implemented on an "as needed". Hence, the cost of upgrading can be spread over a longer period of time, and may be more acceptable to service providers. However, the TDM approach provides a much higher speed capability (from OC-3 to OC-12), which actually improves the overall system performance and increases the capacity for future growth.

7.9.4.1 Unidirectional WDM

WDM is a "wavelength sharing" system that divides the available spectral width into many smaller spectra distributed across several spectral channels. WDM technology has been deployed since the early 1990s.

The WDM technology can be classified into three eras. The (first generation) 1G-WDM is a 2-wavelength technology (one central wavelength at 1310 nm, and the other at 1550 nm). The (second generation) 2G-WDM is a 4-, 8-, 16-, or 32-wavelength technology, and all channels are located at the 1550 nm window. Both 1G- and 2G-WDM have been deployed in the field. The (third generation) 3G-WDM system is expected to be soon deployed widely in the field, and will allocate more than 32 wavelengths (e.g., 80 wavelengths). Figure 7-76 shows the concept of unidirectional n-channel WDM technology. The available fiber spectral width of one single fiber is divided into n segments that are occupied by n channels for one way transmission. The central (nominal) wavelengths for these channels are λ_1, λ_2, λ_3, ..., and λ_n. It can be simply stated that this WDM system operates at λ_1, λ_2, λ_3, ..., and λ_n. In both 2G- and 3G- WDM technologies, these wavelengths are located at the 1550 *nm* window. The WDM industry consider any system using 8 or more channels (wavelengths) to be a Dense WDM (DWDM) system, but this terminology does not have a specific definition.

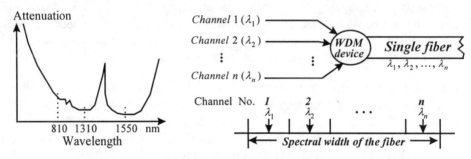

Figure 7-78   Unidirectional WDM.

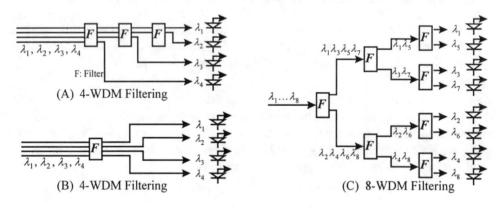

Figure 7-79   Various Filtering Architectures.

When implementing a DWDM system for high-speed applications, filtering is an important design consideration. Various filtering architectures are illustrated in Figure 7-79, and many filtering schemes have been deployed including:

- Fiber Fabry-Perot filters
- Acoustic-optic active tunable filters
- Semiconductor active tunable filters
- Multi-layer interference thin film filters

- UV-induced grating fiber based filters
- Planar Si/SiO$_2$ waveguide mux/demux technology
- Integrated InP based technology

7.9.4.2 Bidirectional WDM

WDM or DWDM technology have been primarily applied to unidirectional transmission. That is, when several wavelengths (channels) are multiplexed together for transmission over a single fiber, all channels originate at the same point and are terminated at the next node (where an add/drop function may be implemented).

However, the WDM/DWDM technology can be also applied to bidirectional transmission. In this arrangement, half of the available channels are assigned for transmission from a "West" location to an "East' location, and the other half of the channels are assigned for transmission from an "East" location to a "West" location (as shown in Figure 7-80). Assume that the optical fiber spectral width can handle 8 channels (λ_1, λ_2, λ_3, λ_4, ... λ_7, and λ_8). Four of these channels (λ_1, λ_3, λ_5 and λ_7) are used to carry signals from "West" to "East", and four (λ_2, λ_4, λ_6 and λ_8) are carrying signals from "East" to "West". Hence, the WDM devices shown in Figure 7-80 perform both optical multiplexing and demultiplexing functions.

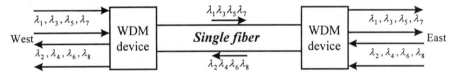

Figure 7-80   Bidirectional Wavelength Division Multiplexing.

7.9.4.3 All-wave Fibers and Channel Spacing for DWDM Systems

The demand for higher speed systems grows constantly and rapidly. In response, the number of channels in DWDM systems will continue to increase. Besides the technologies mentioned earlier (e.g., all-optical amplifiers with flattened gain over a broadband, optical filtering, optical add/drop multiplexing, optical cross-connect/switch/routing devices, and low dispersion fibers), increasing the number of channels in a DWDM system requires a larger wavelength window as shown in Figure 7-81.

A standard single mode (Ge-doped) fiber has a typical attenuation characteristic as shown in Figure 7-81(A). At approximately 1385 nm, the attenuation is extremely large due to hydroxide ions (OH⁻) existing in the fiber. For many years optical fibers used for long-haul transmission systems have been operating at approximately 810 nm, 1310 nm or 1550 nm. Generally speaking, low rate and short-haul optical fiber systems operate around 810 nm, low-medium rate systems operate around 1310 nm, and higher-rate systems operate around 1550 nm, that is known as the long wavelength technology.

With advances in technology, fiber characteristics have been improved drastically. Commercial single mode fibers behave very much as indicated in Figure 7-81(B). With the 1310 nm and 1550 nm windows widened considerably. However, the DWDM (1550 nm) window [shown in Figure 7-81(B)] is still bounded by the "OH⁻ hump". Therefore, the OH⁻ hump has become the dominating factor for DWDM systems. Researchers have searched for ways to reduce the size of this OH⁻ hump, or to eliminate it entirely. The first report of having eliminated this hump in a laboratory environment was 1995. For years leading fiber manufactures tried to make the process mass-producible, and now this type of single mode fiber is available commercially. One trade name is "*all-wave fibers*", as indicated in Figure 7-81(C). It is clear that the available DWDM window is much wider than current single mode fibers.

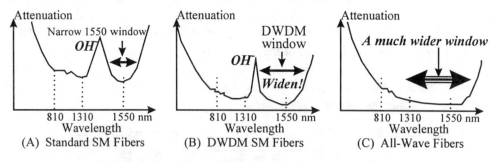

Figure 7-81   DWDM Window and All-Wave Fibers.

As DWDM systems are deployed in the field, the spacing and number of WDM channels must be standardized to insure compatibility. The 1550 nm window has been divided into three bands: the "C-band", the "L-band", and the S-band" wavelength. Existing commercial DWDM systems have either 8, 16, or 32 channels. Higher numbers have been actively studied/designed and may soon be available. Channel spacing is an important issue. Channels can be equally or unequally spaced. This issue needs to be resolved for global compatibility. Currently, ITU-T has proposed 8-channel, 16-channel, and 32-channel DWDM systems. Table 7-11 shows the channel frequencies for an 8-channel, 16-channel, and 32-channel DWDM systems. The frequency of 193.1 THz is called the reference frequency, f_{ref}.

Table 7-11   DWDM Channel Frequency Spacing

Channels	Channel frequency
8	$193.1 \pm (N \bullet 0.2)$ THz
16	$193.1 \pm (N \bullet 0.1)$ THz
32	$192.35 \pm (N \bullet 0.05)$ THz

N = an integer

Example 7-37: Calculate the channel frequencies for a 32-channel DWDM system if the wavelength for the highest frequency channel is set to be 1546.12 *nm*.

$$\lambda f = c = 2.997923 \times 10^8 \text{ m/s} \tag{7-78}$$

By substituting f = 193.90 THz (10^{12}) into Eq.(7-78), the wavelength can be calculated as follows:

$$\lambda = 2.997923 \times 10^8/(193.9 \times 10^{12}) = 1546.119 \text{ } nm$$

This is the frequency and wavelength of Channel *0* shown in Table 7-12.

Table 7-12   Frequency Assignment Proposal for a 32-Channel DWDM System.

Channel	Frequency (THz)	Wavelength (nm)	Channel	Frequency (THz)	Wavelength (nm)
0	193.90	1546.119	16	193.10	1552.524
1	193.85	1546.517	17	193.05	1552.926
2	193.80	1546.916	18	193.00	1553.328
3	193.75	1547.316	19	192.95	1553.731
4	193.70	1547.715	20	192.90	1554.134
5	193.65	1548.115	21	192.85	1554.537
6	193.60	1548.514	22	192.80	1554.940
7	193.55	1548.914	23	192.75	1555.343
8	193.50	1549.315	24	192.70	1555.747
9	193.45	1549.715	25	192.65	1556.150
10	193.40	1550.116	26	192.60	1556.554
11	193.35	1550.517	27	192.55	1556.959
12	193.30	1550.918	28	192.50	1557.363
13	193.25	1551.319	29	192.45	1557.768
14	193.20	1551.720	30	192.40	1558.173
15	193.15	1552.122	31	192.35	1558.578

7.9.5 Wavelength Add/Drop, Routing, Cross-Connect/Switching

An Optical Add/Drop Multiplexer (OADM) is a primary element required in a Wavelength Division Multiplexing (WDM) system. Figure 7-82(A) shows the block diagram of a WDM add/drop device, and Figure 7-82(B) is a WDM add/drop & cross-connect device.

The add/drop device consists of two main components: (1) a one by n (1:4 shown in Figure 7-82) wavelength filter (demultiplexer) that takes the WDM (λ_1, λ_2, λ_3, ..., λ_n) signal and filters the signal into n (wavelength) signals, with wavelengths of λ_1, λ_2, λ_3, ..., and λ_n (some wavelengths will be through traffic, and some are dropped traffic); and (2) a n by one (n:1 shown in Figure 7-82) multiplexer to combine n wavelengths (some are through traffic and some are added) to form a WDM signal.

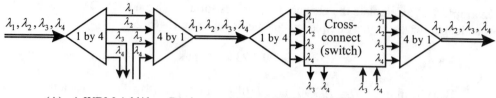

(A) A WDM Add/drop Device   (B) A WDM Add/drop & Cross-connect Device

Figure 7-82   A WDM Add/Drop Device.

Various "all-optical" ADM devices have been developed. The most commonly used techniques apply: (1) fiber Bragg gratings combined with a circulator, and (2) a Mach-Zehnder interferometer, or a directional coupler (based on grating-assisted coupling). That is, an "all-optical" ADM presently can be implemented with or without an interferometer. Figure 7-83(A) is the schematic diagram for an ADM using a interferometer. This scheme consists of three elements: (1) circulators for drop/add functioning, (2) a phase shifter, and (3) a fiber Bragg grating for all other wavelength operating functions. Figure 7-83(B) is the schematic diagram for ADM without an interferometer. This ADM contains the following components: (1) Mode Selective Couplers (MSCs), and (2) one tilted Fiber Bragg Grating (FBG). Between the input/output ports and MSC, and between drop/add port and MSC single mode fiber (SMF) is used for internal connections. Between MSC and FBG, Two-Mode Fiber (TMF) is used. Because the technique using the interferometer requires a precise length control in the interferometer. Therefore, the non-interferometer schemes, presently seem more desirable and promising. Another proposal for all-optical demultiplexing is to use the Nonlinear Optical Loop Mirror (NOLM) for 40 Gbps or higher rate data stream [the detail technical description of these "all optical" ADM is beyond the scope of this book; see Y. Liang et al., ThA3-1, Technical Digest, Conference Edition, OFC/IOOC99].

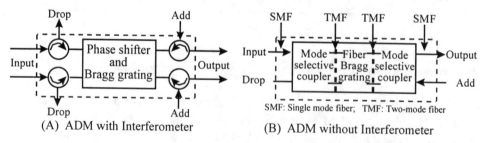

(A) ADM with Interferometer   (B) ADM without Interferometer

Figure 7-83   ADM Implementations with/without Interferometer.

A promising "all-optical" demultiplexing scheme at the receiver stage of high-speed Optical Time Division Multiplexing (OTDM) system is based on the four photon interaction in optical fibers [also known as Four-Wave Mixing (FWM)]. This is a nonlinear parametric process that has two advantages: (1) using the fiber approach (which is easier technology); and (2) using the non-interferometric scheme (for better stability; because the technique using the interferometer requires a precise length control in the interferometer). However, its major drawback is a strong dependence on the pump and signal state of polarization. To overcome this drawback, various schemes based on the polarization diversity principle have been proposed. The results of this experimental system, and its potential commercial value is expected to be announced in the near future.

The cross-connect function is conceptually illustrated in Figure 7-84 (the top graphs). For example, at the time instance $t = t_1$, the signal on input port 1 is required to be connected to output port 1, the signal on input port 2 is required to be connected to output port 2, the signal on input port 3 is required to be connected to output port 3, and the signal on input port 4 is required to be connected to output port 4. Later, at the time instance $t = t_2$, the signal on input port 1 is still connected to output port 1, the signal on input port 2 is required to be connected to output port 3, the signal on input port 3 is required to be connected to output port 4, and the signal on input port 4 is required to be connected to output port 2. Yet, at a even later time instance $t = t_3$, the signal on input port 1 is required to be connected to output port 4, the signal on input port 2 is still connected to output port 3, the signal on input port 3 is required to be connected to output port 2, and the signal on input port 4 is required to be connected to output port 1. The cross-connect function at various time instances is performed by a "Digital Cross-connect System (DCS)", that has been described in Chapter 6. From the "all-optical" network viewpoint. This type of DCS is referred to as an electronic DCS, that is a relatively low-speed electronic switch (router) with a fully "non-blocking" capability.

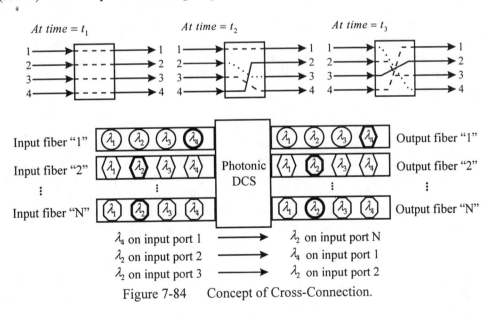

Figure 7-84   Concept of Cross-Connection.

To implement true "all-optical" networks, "all-optical" or photonic (instead of electronic) DCSs must be used. Figure 7-84 (the bottom graph) illustrates the function of a photonic DCS. For simplicity, assuming 4-WDM (λ_1, λ_2, λ_3, λ_4) is used on each fiber. To distinguish the same wavelength signal from different fibers, "circle", "hexagon" and "octagon" shapes are used. In this example, the following cross-connections are requested:

Wavelength λ_4 on input fiber No. 1 $\Rightarrow$ Wavelength λ_2 on output fiber No. N

Wavelength λ_2 on input fiber No. 2 $\Rightarrow$ Wavelength λ_4 on output fiber No. 1

Wavelength λ_2 on input fiber No. N $\Rightarrow$ Wavelength λ_2 on output fiber No. 2

The signal carried by λ_4 (on input fiber No. 1) is to be cross-connected to the "λ_2" position (on the output fiber No. N), a "wavelength conversion" is thereby required. The signal carried by λ_2 (on input fiber No. 2) is to be cross-connected to the "λ_4" position (on the output fiber No. N), a "wavelength conversion" is also required. However, the signal carried by λ_2 (on input fiber No. N) is to be cross-connected to the "λ_2" position (on the output fiber No. 2), a "wavelength conversion" is thereby not required.

7.9.6 Polarization Mode Dispersion (PMD), and Dispersion Compensation

The dispersion slope characteristic of transmission fiber presents a performance constraint in high-speed optical fiber systems. For example, a 16-WDM, 10.66 Gbps over 10,850 km system has been evaluated by KDD and TYCO [see M. Suzuki et al., PD17, Technical Digest, Conference Edition, OFC/IOOC98]. The results show that the edge channels have much worse transmission performance compared to center channels. The reason for this performance degradation is attributed to the dispersion slope of the transmission fiber. It is expected that dispersion slope degradation will become a serious problem as system capacity (i.e., the number of channels in DWDM systems) is increased. Another dispersion degradation that has received attention is "Polarization Mode Dispersion (PMD) on non-dispersion shifted fibers.

Many techniques have been used to overcome edge-channel and PMD transmission performance degradation: (1) using linearly chirped fiber gratings for chromatic dispersion compensation, (2) using chirp reduction schemes, (3) using nonlinearly-chirped Fiber Bragg Grating (FBG), (4) using polarization maintaining fiber with a polarization controller, (5) using electrical-domain compensation methods at the receiver [it is limited to the receiver], (6) using automatic PMD compensation schemes (requires monitoring the state of polarization of received optical signals), (7) using phase-conjugation by four-wave mixing in an SOA, (8) periodic dispersion slope compensation, etc.

Advances in dealing with fiber dispersion and nonlinearities that has supported the rapid increase in optical fiber system speeds. However, research indicates that PMD will be the next hurdles to overcome in achieving high-capacity, high performance optical fiber systems and networks. *"**The different transmission speeds of the two States-Of-Polarization (SOP) in the fiber**"* is generally the cause of PMD. Even though PMD is a statistically random quantity, it can be emulated and compensated by simply delaying one state-of-polarization with respect to the other. This scheme is often referred to as "the *Differential Group Delay* (DGD)" technique. The technical detailed description of FBG, PMD, SOA, SOP, and DGD is beyond the scope of this book (see "Optical Networks" by R. Ramaswami and K.N. Sivarjan).

Periodic dispersion slope compensation and reduced dispersion slope fiber is the most promising scheme for systems approaching speeds up to hundreds of Gbps. This scheme has been implemented by Taga, and is described in Example 7-38.

Example 7-38: Describe the Dispersion Slope Compensation (DSC) techniques [see H. Taga, TuD3-1, Technical Digest, Conference Edition, OFC/IOOC99] used "*to overcome the effect of the dispersion slope by compensating for the difference in accumulated dispersion on each channel periodically along the transmission path*".

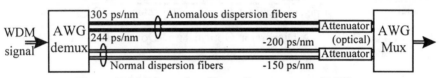

Figure 7-85 Dispersion Slope Compensator (DSC).

Figure 7-85 is a schematic diagram of the DSC technique developed by H. Taga. The major components of this DSC are: Arrayed Waveguide Gratings (AWGs), normal dispersion fibers, anomalous dispersion fibers, and optical attenuators. The input AWG demultiplexes the WDM signal into individual signals that are fed to fibers that are compensated for the difference of the accumulated dispersion. After being transported, these signals are multiplexed by the output AWG. Optical attenuators are used to adjust the insertion loss imbalance of Dispersion Compensating Fibers (DCFs). Taga's experiment using a 20-WDM, 10.66 Gbps over 9,000 km optical fiber link illustrated that periodic dispersion slope compensation results in uniform transmission performance over all channels and achieved a BER of 10^{-9}. This scheme is predicted to be effective for high-channel WDM applications with rates in hundreds of Gbps. In some cases, EDFAs are used instead of attenuators to compensate for the insertion loss of the DCFs. Taga's experiment also confirmed that use of a Reduced Dispersion Slope Fiber (RDSF), by applying Dispersion-Flattened Fiber (DFF) or a combination of different fibers, is an important Dispersion Slope Compensation technology for high speed systems.

Example 7-39: Describe the operation of a dispersion Optical Time Domain Reflectometer (OTDR) [see J. Gripp et al., TuS4-1, Technical Digest, Conference Edition, OFC/IOOC99].

Dispersion management has become the most important task in the design of high-speed optical fiber transmission systems and networks. Therefore, it is essential to perform a fast, and accurate measurement of fiber's chromatic dispersion as a function of distance (known as the "***dispersion map***"). J. Gripp's proposal, as shown in Figure 7-86, presents a efficient, reliable, and convenient way of obtaining a dispersion map. The proposed dispersion-OTDR requires only one end access to the fiber. Data acquisition takes place only in a few seconds. The spatial information is direct, unambiguous, and of high resolution. That is, the dispersion uncertainty is rather small.

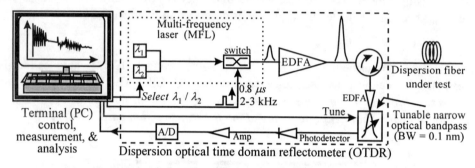

Figure 7-86   A Dispersion Optical Time-Domain Reflectometer.

7.9.7 Nonlinear Optical Effects

The study of nonlinear effects in optical fibers has grown rapidly since single mode fibers have been widely deployed in modern high-speed optical fiber systems. When an optical signal at a given wavelength is launched into a single mode fiber, the signal is transferred into a set of longer wavelength signals. This is called an "***optical nonlinear effect***". This nonlinear process is influenced by certain characteristic vibrations of the fiber material that induce changes in its refractive index. The refractive index change, which is caused by the applied optical field perturbing the atoms in the fiber material, induces various scattering processes. When the optical signal is weak (i.e., the optical signal intensity is low) scattering is a linear function of the applied field. This process is known as Rayleigh scattering. As the optical intensity increases, scattering becomes a nonlinear process. Examples of this phenomena are the Kerr effect, Raman scattering, and Brillouin scattering. Some applications of these nonlinear effects are Raman amplifiers, Brillouin amplifiers, soliton transmission, pulse compression, all-optical switching techniques, etc.

The Kerr effect is an intensity-dependent refractive index, which is caused by the "***phase shift***" in the fiber resulting optical intensity variations. Therefore, Kerr nonlinearities can alter the frequency spectrum of a pulse traveling through an optical fiber. This effect has been utilized to perform pulse compression.

When a molecule in the fiber absorbs a photon at one frequency, it will emit a photon at a shifted frequency. At the same time it makes a transition between vibrational states. In other words, the Raman effect not only causes a frequency shift of the incident light, but also gives rise to an optical gain at the shifted frequency (if the incident light intensity is high enough). A Raman amplifier is "self-phase-matching" between the pump and the input signal, and has a high-speed response characteristic.

A high-intensity optical field can also interact with the acoustic vibrations of a glass fiber, and as a result shift the frequency of the light by acoustic frequency (approximately 10^{10} Hz). This nonlinear phenomenon is called the Brillouin effect. The drawback of a Brillouin amplifiers is the narrow bandwidth (e.g., about 20 MHz).

7.9.8 Soliton Transmission

A special phenomenon occurs when the Kerr effect takes place in the negative-dispersion region of an optical fiber. For a high-intensity pulse, the Kerr effect can cause the pulse's leading edge to develop a "***red shift***". That is, a shift toward lower optical frequency or toward longer optical wavelengths. At the same time, the pulse's trailing edge will develop a "***blue shift***". That is, a shift toward higher optical frequency or toward shorter optical wavelengths.

Because of ***negative*** group-velocity dispersion, the "red shifted" light will travel through the fiber slower than the unshifted pulse center, and the "blue-shifted" light will travel through the fiber faster than the unshifted pulse center. If the pulse shape, pulse intensity, and pulse width are properly chosen, the Kerr nonlinear effect and the negative dispersion will cancel each other exactly. A "light pulse" of this type, which travels through an optical fiber without pulse-shape changing, is called a "***soliton***". Therefore, a soliton light pulse can propagate over extremely long distances without distortion.

The principle of soliton transmission is based on the exact compensation of dispersion and nonlinearities. A fixed pulse energy level throughout the entire transmission link must be maintained at a nearly constant level. Therefore, optical amplifiers must be applied along the entire transmission path to compensate for the fiber attenuation. In a traditional optical fiber link, when the light signal is distorted due to fiber attenuation and/or dispersion, regenerators are applied along the path. The signal is re-amplified, retimed, and regenerated to retain the incident pulse shape. However, in a modern optical network, the system speed is very high and DWDM is widely deployed. As a result, electronic regenerators suffer the following drawbacks:

- They are inherently bit-rate limited. If the link is upgraded to a higher speed, all the regenerators must be replaced. This is practically impossible or too expensive for undersea optical fiber systems.

- They are unidirectional. Each direction of transmission requires its own regenerators. Thus, system costs can be very high.

- For DWDM systems, regenerators are not compatible. That is, each regenerator processes only one wavelength. Thus, an *n*-channel DWDM system must be implemented by using one demultiplexer, *n* regenerators, and one multiplexer as shown in Figure 7-87.

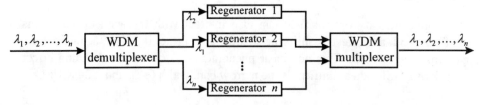

Figure 7-87   Regenerators for a DWDM System.

Because of these drawbacks, soliton transmission with optical amplifiers is considered more effective for high-speed WDM systems, and offers several advantages:

- The system can support very high signal speeds because optical amplifiers are bit-rate independent (unlike electronic regenerators, that are bit-rate limited).

- Theoretically optical amplifiers are bidirectional (i.e., one amplifier can serve both directions). However, to ensure system stability and to prevent noise accumulation, optical isolators are used with optical amplifiers. Since the isolators are unidirectional devices, optical amplifiers are also utilized as unidirectional devices.

- The system is WDM compatible. That is, it in not necessary to demultiplex a WDM signal to individual signals so that they can be regenerated to overcome attenuation and dispersion. Figure 7-88 illustrates the soliton for WDM applications.

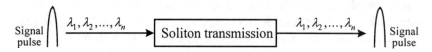

Figure 7-88   Soliton Transmission: without Pulse Distortion.

Review Questions II for Chapter 7:

(15) In a PDH environment, an optical link of 180 Mbps have been used to carry ____ μ-law DS3 signals, or _____ voice (or voice-equivalent) signal.

(16) In a PDH environment, a 1.7 Gbps FT-G optical fiber system can carry _____ voice signals. Likewise, a 565 Mbps optical fiber system can carry _____ voices.

(17) An ISM-2000 access multiplexer supports SDH STM-1 or STM-4. When it supports STM-4 for voice transport, the capacity is _____ voice signals.

(18) The inter-office optical fiber link used in the SONET environment is _____, which supports SONET _____ signal, and has a capacity of _____ voices. Its upgraded system is _____, which supports two _____.

(19) SDH inter-office optical fiber link supporting STM-16, and equivalent to FT-2000 is known as _____.

(20) Most undersea systems employ optical fiber technology. Several well-known systems are: _____, _____, _____, _____, _____, etc.

(21) For modern optical networking, the trends are to move towards all-optical networks. Two "break-through" technologies for all-optical network are _____, and _____.

(22) Erbium-doped fiber optical amplifiers have several applications: _____, _____ and _____.

(23) EDFA has several advantages: low _____, low _____, High _____, etc. But, it has several disadvantages: _____, _____ at 1300 nm, etc.

(24) Several WDM systems have been deployed recently, such as 2-WDM, 8-WDM, 16-WDM, etc., is a _____ WDM. A bidirectional WDM may soon be available in the near future.

(25) When implementing a DWDM system for high-speed applications, _____ is an important design consideration. Several schemes have been developed: _____, _____, _____, _____, etc.

(26) An "all-wave" fiber has been developed to eliminate the attenuation around 1400 nm and caused by _____ ions. This type of fiber will increase the _____ of channels for DWDM systems.

(27) To implement all-optical network, besides EDFA, DWDM, filters, the following are necessary technologies: _____, _____, _____, _____, etc.

Appendix 7-1

Error and Complementary Error Functions

In evaluating a system used for transporting a digital signal, the complementary error function is often used. A brief discussion of the error function and the complementary error function is provided in this appendix.

The definition of an error function is given as follows:

$$erf(a) = \frac{2}{\sqrt{\pi}} \int_0^a e^{-y^2} dy \qquad (7A-1)$$

The complementary error function is defined as follows:

$$erfc(a) = 1 - erf(a) \qquad (7A-2)$$

Eqs.(7A-1) and (7A-2) are the mathematical definitions for erf(a) and erfc(a). The meaning and the applications of these two functions are often derived from the identity given in Eq. (7A-3). Let x be a Gaussian random variable, $x \sim N(\mu, \sigma^2)$, with the probability density function of:

$$f(x) = \frac{1}{\sigma\sqrt{2\pi}} e^{-\frac{(x-\mu)^2}{2\sigma^2}} \qquad (7A-3)$$

In the Gaussian random variable expression, μ is the expected value of x, and σ^2 is the variance of x. (i.e., σ is the standard deviation of x). The meaning of the error function is given as follows:

$$erf\left(\frac{k}{\sqrt{2}}\right) = P[\mu - k\sigma \le x \le \mu + k\sigma] \qquad (7A-4)$$

The meaning of Eq.(7A-4) is: "the probability that the random variable x has a value between $\mu - k\sigma$ and $\mu + k\sigma$ is equal to the function value of the error function of 'a', erf(a) with $a = k/\sqrt{2}$".

For communication applications, especially digital communication applications, the approximation expressed in Eq.(7A-5) is important:

$$erfc(a) \approx \frac{e^{-a^2}}{a\sqrt{\pi}} \qquad if \quad a \gg 1 \qquad (7A-5)$$

The applications for Eq. (7A-5) can be classified into two cases. Two separate examples are provided to illustrate these applications. Assume the system performance of a digital optical fiber system, P_e is given by the following equation:

$$P_e = \frac{1}{2} \times erfc \left(0.354\sqrt{\frac{S}{N}}\right) \qquad \text{(7A-6)}$$

- **Case 1 applications:** Knowing S/N, compute P_e (Example 7A-1).
- **Case 2 applications:** Knowing P_e, compute S/N (Example 7A-2).

Example 7A-1: Compute the Probability of error (P_e) if an optical fiber system has a Signal-to-Noise ratio of (S/N) 19 dB.

This application is a straight forward computation. First, convert the signal-to-noise ratio from dB to a unit-less value by applying the anti-logarithm formula as follows:

$$\frac{S}{n} = 10^{\frac{19\,dB}{10}} = 79.43$$

Then substitute this value into Eq. (7A-6) to obtain $P_e = 0.5$ erfc (3.155). Finally, using the approximation given in Eq. (7A-5) obtain the following:

$$P_e = 0.5 \times erfc\ (3.155) \approx 0.5 \times \frac{e^{-3.155^2}}{3.155\sqrt{\pi}} = 0.5 \times 8.5 \times 10^{-6} = 4.25 \times 10^{-6}$$

Before proceeding to the case 2 application, the following iteration formula derived from Eq. (7A-5) must be explained.

$$a_{j+1}^2 = -\log_e \left[a_j\ \sqrt{\pi}\ erfc\ (a_{final})\right] \qquad j = 1, N \qquad \text{(7A-7)}$$

Instead of using mathematical tables, the iteration formula in Eq. (7A-7) can be applied for practical communication systems. The experience the author has had confirms that the above iteration procedure will converge quickly, and the final "a" value for erfc(a) can be obtained after approximately six iterations.

Example 7A-2: Assume that a probability of $P_e = 10^{-7}$ is required for the system described by Eq. (7A-6) [Note that this formula is used for the optical fiber systems, known as a thermal noise-limited system, deployed in the field]. Determine the minimum required signal-to-noise ratio in dB.

$$P_e = 10^{-7} = \frac{1}{2} \times erfc\,(0.354\sqrt{\frac{S}{N}}) \implies erfc(a) = 0.5 \times 10^{-7} \equiv erfc\,(a_{final})$$

$$with\ a = 0.354 \times \sqrt{\frac{S}{N}}$$

(7A-8)

By substituting erfc (a_{final}) into Eq. (7A-7), one can obtain the following:

$$a_{j+1}^2 = - \log_e\,[a_j\,\sqrt{\pi}\,erfc\,(a_{final})] = - \log_e\,[2.0 \times 8.86227 \times 10^{-8}] \qquad \text{(7A-9)}$$

where the initial value a_1 was chosen to be 2.0 (Note that any value larger than 1 can be used). From Eq. (7A-9), the next iterated value obtained is:

$$a_2 = 3.9428$$

Substituting this value into Eq. (7A-9) obtain the third iterated a_j (j = 3) value as:

$$a_3 = 3.8557$$

This procedure is continued to obtain several iterations as follows:

$$a_4 = 3.8587$$

$$a_5 = 3.8586$$

$$a_6 = 3.8586$$

It can be seen the value of "a" converges quickly to "3.8586". Finally, by substituting this value into Eq. (7A-8), the system signal-to-noise ratio can be calculated as:

$$\frac{S}{N} = 10 \log\,[3.8586 / 0.354]^2 = 20.7 \quad dB$$

Therefore, for an optical fiber system to achieve a probability of error 10^{-7}, the system must have a signal-to-noise ratio of 20.7 dB.

References

- **General Communication Topics:**

    1. D. Smith, "Digital Transmission Systems", Van Nostrand Reinhold, 1985
    2. Taub & Schilling, "Principles of Communication Systems", 2nd Ed., McGraw-Hill, NY, 1986
    3. Rainer Handel, Manfred N. Huber, and Stefan Schroder, "ATM Networks: Concepts, Protocols, Applications", Addison-Wesley Publishers Ltd. 1994
    4. Taub & Schilling, "Wireless Communications: Principles & Practice", 2nd Ed., McGraw-Hill, NY, 1996
    5. Chow, Ming-Chwan, "Understanding SOENT: Standards and Applications", 1st Ed., Andan Publisher, NJ, 1996
    6. Sexton, Mike, and Reid, Andy, "Broadband Networking: ATM, SDH, and SONET", Artech House, Inc. MA, 1997
    7. Chow, Ming-Chwan, "Understanding Wireless: Digital Mobile, Cellular and PCS", 1st Ed., Andan Publisher, NJ, 1998
    8. J. Bellamy, John, "Digital Telephony", Wiley & Sons
    9. B. Sklar, "Digital Communications", Prentice Hall
    10. SONET Rates and Formats, ANSI T1.105
    11. Fiber Optic Communications, J. Palais, Prentice Hall

- **ITU-T Recommendations:**

    12. E.163, "Numbering Plan for the Global Telephone Network"
    13. E.164, "Numbering Plan for the ISDN Era"
    14. G.701, "Vocabulary of Digital Transmission and Multiplexing, and Pulse Code Modulation (PCM)"
    15. G.702, "Digital Hierarchy Bit Rates"
    16. G.704, "Synchronous Frame Structure used at 1544, 6312, 2048, 8448 and 44 736 kbit.s"
    17. G.707, "Network Node Interface for the Synchronous Digital Hierarchy"
    18. G.726, "40, 32, 24, 16 kbit/s Adaptive Differential Pulse Code Modulation (ADPCM)"
    19. G.729, "Coding of Speech at 8 kbit/s using Conjugate-Structure Algebraic-Code-Excited Linear-Prediction"
    20. G.731, "Primary PCM Multiplex Equipment for Voice Frequencies"
    21. G.732, "Characteristics of Primary PCM Multiplex Equipment Operating at 2048 kbit/s"
    22. G.733, "Characteristics of Primary PCM Multiplex Equipment Operating at 1544 kbit/s"
    23. G.734, "Characteristics of Synchronous Digital Multiplex Equipment Operating at 1544 kbit/s"
    24. G.735, "Characteristics of Primary PCM Multiplex Equipment Operating at 2048 kbit/s and Offering Synchronous Digital Access at 384 kbit/s and/or 64 kbit/s"
    25. G.736, "Characteristics of a Synchronous Digital Multiplex Equipment Operating at 2048 kbit/s"
    26. G.741, "General Consideration on Second Order Multiplex Equipment"

27. G.747, "Second Order Digital Multiplex Equipment Operating at 6312 kbit/s and Multiplexing Three Tributaries at 2048 kbit/s"

28. G.751, "Digital Multiplex Equipment Operating at the Third Order Bit rate of 34 368 kbit/s and the Fourth Order Bit Rate of 139 264 kbit/s and Using Positive Justification"

29. G.753, "Third Order Digital Multiplex Equipment Operating at 34 368 kbit/s and Using Positive/Zero/Negative Justification"

30. G.754, "Fourth Order Digital Multiplex Equipment Operating at 139 264 kbit/s and Using Positive/Zero/Negative Justification"

31. G.755, "Digital Multiplex Equipment Operating at 139 264 kbit/s and Multiplexing Three Tributaries at 44 736 kbit/s"

32. G.775, "Loss Of Signal (LOS), Alarm Indication Signal (AIS) and Remote Defect Indication (RDI) Defect Detection and clearance Criteria for PDH Signals"

33. G.780, "Vocabulary of Terms for Synchronous Digital Hierarchy (SDH) Networks and Equipment"

34. G.783, "Characteristics of SDH Equipment Functional Blocks"

35. G.784, "SDH Management"

36. G.785, "Characteristics of a Flexible Multiplexer in a SDH Environment"

37. G.791, "General Consideration on Transmultiplexing Equipment"

38. G.797, "Characteristics of a Flexible Multiplexer in a PDH Environment"

39. G.802, "Internetworking between Networks Based on Different Digital Hierarchy and Speech Encoding Laws"

40. G.803, "Architecture of Transport Networks Based on the SDH"

41. G.804, "ATM Cell Mapping into Plesiochronous Digital Hierarchy (PDH)"

42. G.810, "Definitions and Terminology for Synchronous Networks"

43. G.811, "Timing Characteristics of Primary Reference Clocks"

44. G.812, "Timing Requirements of Slave Clocks Suitable for Use as Node Clocks in Synchronous"

45. G.842, "Interworking of SDH Network Protection Architectures"

46. G.861, "Principles and Guidelines for the Integration of Satellite and Radio Systems in SDH Transport Networks"

47. G.901, "General Consideration on Digital Sections and Digital Line Systems"

48. G.941, "Digital Line Systems Provided by FDM Transmission Bearers"

49. G.950, "General Considerations on Digital Line Systems"

50. G.951, "Digital Line Systems Based on the 1544 kbit/s Hierarchy on Symmetric Pair Cables"

51. G.952, "Digital Line Systems Based on the 2048 kbit/s Hierarchy on Symmetric Pair Cables"

52. G.953, "Digital Line Systems Based on the 1544 kbit/s Hierarchy on Coaxial Pair Cables"

53. G.954, "Digital Line Systems Based on the 2048 kbit/s Hierarchy on Coaxial Pair Cables"

54. G.955, "Digital Line Systems Based on the 1544 & 2048 kbit/s Hierarchy on Optical Fiber Cables"

55. G.957, "Optical Interfaces for Equipment and Systems Relating to the SDH

56. G.961, "Digital Transmission System on Metallic Local Lines for ISDN Basic Rate Access"

57. G.971, "General Features of Optical Fiber Submarine Cable Systems

58. G.972, "Definition of Term Relevant to Optical Fiber Submarine Cable Systems"

59. G.981, "PDH Optical Line Systems for the Local Network"

Answers for Review Questions

Chapter 1:

(1) amplitude (magnitude), frequency, phase, bandwidth/rate (2) True (3) True (4) single, double (5) 4, 64 (6) True (7) C, 1000 (8) common, channel, CCS7, SS7 (9) Canada & USA, Africa, Europe, Central/South America, South Pacific, (old) USSR, North Pacific, Far/Middle East, spare (10) 15 (11) two, 941, 1336 (12) busy, congestion, reorder (13) True (14) switching, signaling, network management (15) True (16) 5,160,960 (17) 3-dB, 3-dB, frequency response curve (18) PCM, IDLC, BRT, MODEM, multiplexers, ADMs, DCSs, regenerators, echo cancellers (19) concentration, grooming, test access, network restoration, centralized operation (20) 0.001% (21) speech encoding, channel encoding, bit-interleaving, modulation, multiplexing (22) line codes, B3ZS, B6ZS, B8ZS, scramblers, $g(x) = 1 + x^6 + x^7$ (23) control signaling, protocol conversion, billing (24) circuit, public data network, packet, public data network (25) basic rate, primary rate, BRI, PRI (26) Centrex

Chapter 2:

(1) regional, national (2) transmission, switching, signaling, microelectronics, opto-electronics, software, OAM&P (3) 2, 3, 4 A4 (4) cameras, monitors, codecs (5) True (6) PSTN, customer (7) Universal Personal Telecommunications (UPT) (8) local exchange, trunk exchange, gateway exchange (9) digit, analysis/translation, alerting, disconnecting (10) A, B, C, D (11) True (12) True (13) False (14) ¾, DECnet, Appletalk (15) Transmission Control Protocol (16) routers, gateways (17) website, World Wide Web, HyperText Markup Language (18) PC, PC, PC, PSTN, PSTN, PSTN, PBX, PSTN (19) True (20) IP address, dial-up, dedicated (21) contract, logon, dial-up (22) LAN hubs, routers (23) X.25, frame relay, ATM, data, narrowband, frame relay, data, ATM (24) efficiently, bandwidth, insensitive (25) 2^{12}, 2^{16} (26) virtual channel, unidirectional (27) Virtual Path, Virtual Channel (28) switched, dedicated (29) True (30) switched (31) STS-1 (OC-1), STS-3 (OC-3), STS-12 (OC-12), STS-48 (OC-48), STS-192 (OC-192) (32) STM-1 (STM-1O), STM-4 (STM-4O), STM-16 (STM-16O), STM-64 (ST-64O) (33) DS1, E1, DS1C, DS2 (34) DS1, E1, DS2, E3 (35) modularity, 12, 64 (36) path, line (multiplex section), section (regenerator section) (37) plesiochronous (38) True

Chapter 3:

(1) BRI-ISDN, cable modem, HDSL, VDSL, HFC (2) distribution, service area interface (3) improve transmission quality, provide pair gain (of wiring), relief duct congestion (4) True (5) OC-3, OC-12, STM-1,STM-4 (6) QAM, convolutional code (7) Integrated Service Digital Network (ISDN), ADSL, downlink, downlink, VDSL (8) thermal noise, reflection, echo, crosstalk, loading coils, bridge taps, RFI (9) splitterless, G.Lite (10) True (11) Echo canceler hybrid (12) 2-binary/1-Quaternary (2B1Q), Discrete Multi-Tone (DMT), Carrierless AM/PM (CAP) (13) Pre-qualification, Turn-up verification (14) low (15) LPF, HPF, HPF, LPF (16) STM, ATM, 68, DMT (17) True (18) 300, 30 (19) true (20) ATM (21) Synchronized Discrete Multi-Tone, Discrete Multi-Tone, Time Division Duplexing, ping-pong (22) access (23) True (24) speech encoder, channel encoder, encryptor, bit-interleaver (25) infrared (26) True (27) active, protection, working channel, protection channel

Chapter 4:

(1) True (2) subscriber, trunk (3) customer loop, inter-exchange circuit, special service, network management (4) True (5) directory number (DN) (6) True (7) R1, R2, 5, 6 (8) per circuit signaling (9) fast call setup time, high signal capacity, improved maintenance procedures, expanded customer services (10) True (11) quad (12) True (13) originating, destination point codes, signaling route ID (14) True (15) PDH, SONET, SD (16) PSTN, mobile, operations, administration and maintenance (17) user part (UP), message transfer part (MTP) (18) 64,56,physical,electrical,functional (19) flag insertion, delimitation, alignment, stuffing, error, control, failure, control (20) flag, flag, CRC-16, $g(x) = 1 + x^5 + x^{12} + x^{16}$ (21) routing label, originating point code (OPC), destination point code (DPC), signaling link selection (22) signaling connection control part, telephone user part, ISDN user part, data user part (23) message type, fixed, variable (24) link-by-link, end-to-end (25) international signaling point code (26) True

Chapter 5:

(1) multiple (2) $n(n - 1)/2$ (3) the request for service, the duration of the service (4) True (5) local, gateway (6) control, signaling, operations support (7) False (8) True (9) more economical, more flexible in design, less complicated (10) False (11) map (12) True (13) Station line interfaces, trunks/signaling interfaces (14) intra-location calls, incoming calls, outgoing calls, communication system management, office automation (15) customer's, local exchange's, requires (16) low, continuous, fixed, constant, greater (17) 38~40% (18) message, packet (19) X.25, frame relay, ATM (20) fixed, small (21) 2^{12}, 2^{16} (22) permanent virtual, switched virtual (23) administration, communication, switch, remote switch (24) scanning, routing, making, supervising, ringing, tone, tone, conference (25) administration (26) True

Chapter 6:

(1) North American, Japanese, ITU-T (2) digitized, 64 kbps (3) sampling, quantizing, coding, time division multiplexing, control signal insertion (4) 125, frame, 8000 frames (5) μ, A (6) DS1, DS1C, DS2, DS3, E1, E2, E3, E4, CEPT1, PS-1, European DS1 (7) Gray (8) modulation, multiplexing, ASK, FSK, PSK, QAM, TDM, WDM (9) 24, 672, 32, 30 (10) framing pattern, 100011011100, 001011, 4, 8, 12, 16, 20, 24 (11) AMI, B8ZS, B6ZS, B3ZS, HDB3, CMI, scrambler, $g(x) = 1 + x^6 + x^7$ (12) E4, 1920 (13) 30, 128 (14) 1, 2, 4 (15) 192, 1344, 2688, 5376, 129024, 24 (16) 64, 192, 4032, 122880, 30 (17) path, line, multiplex section, section, regenerator section (18) VT1.5, VT2, VT3, VT6, TU-11, TU-12, TU-2, TU-3 (19) Virtual Tributary Group , VTG, Tributary Unit Group-level 2, Tributary Unit Group-level 3, TUG-2, TUG-3 (20) path, line, multiplex section, section, regenerator section

Chapter 7:

(1) Optical fiber, coaxial cable, twisted-pair wire (2) infrared (3) total internal reflection (4) spectral width, rise time, central (nominal) wavelength, radiated power (5) True (6) 62.5, 8.5, dispersion (7) rise time, dispersion, attenuation, 820, 1310, 1550 (8) True (9) connector, coupling, splicing, fiber (10) rise time, responsivity, sensitivity (11) True (12) material, waveguide, inter-modal, inter-mode (13) PIN, APD (14) central limit theorem (15) 4, 2688 (16) 24192, 7680 (17) 7680 (18) FT-2000, OC-48, 32256, LCT-2000, OC-48 (19) SLM-16 (20) TAT-12, TAT-13, SL-2000, TPC-5, APCN (21) EDFA, DWDM (22) power amplifier, receiver pre-amplifier, in-amplifier (23) noise figure, signal distortion and cross-talk, efficiency and output power, non-uniform gain profile, no power gain (24) unidirectional (25) Fiber Fabry-Perot filters, Acoustic-optic active tunable filters, semiconductor active tunable filters, Multi-layer interference thin film filters, integrated InP based technology (26) hydroxide, channels (27) wavelength conversion, wavelength add/drop, wavelength routing, wavelength cross-connection

Abbreviations and Acronyms

A/D	Analog-to-Digital (conversion)
AAL	ATM Adaptation Layer
ABM	Accunet Bandwidth Management
AC	Address Control
AC	Administrative Computer
AC or ac	Alternate Current
AC	Access Code
ACC	Alarm Conversion Circuit
ACD	Automatic Call Distributor
ACP	Action Control Point
ACRU	Access Channel Receiver Unit
ACS	Address Complete Signal
ACTU	Access Channel Transmitter Unit
ACX	Asynchronous Cross-connect
ADE	Adaptive Design Engineering
ADM	Adapt Delta Modulation
ADM	Add-Drop Multiplexer
ADM	Adaptive DM (Delta Modulation)
ADPCM	Adapt Differential Pulse Code Modulation
ADSL	Asymmetric Digital Subscriber Line
AE	Application Entity
AIN	Advanced Intelligent Network
AIOD	Automatic Identified Outward Dialing
AIS	Alarm Indication Signal
AM	Administrative Module
AMA	Automatic Message Accounting
AMAR	Automatic Message Accounting Recording
AMASE	Automatic Message Accounting Standard Entries
AMI	Alternate Mark Inversion
ANI	Automatic Number Identification
ANM	Answer Message
ANSI	American National Standards Institute
AOC	Advise Of Charge (is under further study)
AOC	ASDL Overhead Control
AP	AM Processor
APCN	Asia Pacific Cable Network
APD	Avalanche Photo-Diode
APN	Action Point Number
APS	Attached Processor System
APS	Automatic Protection System (switching)
ARM	Access Resource Manager
ARP	Address Resolution Protocol
ASC T1	Accredited Standards Committee T1
ASC X3	Accredited Standards Committee X3
ASDS	Accunet Spectrum Digital Service
ASE	Application Service Elements
ASK	Amplitude Shift Keying

ASN	AT&T Switched Network
Async	Asynchronous
AT	Access Tandem
ATA	Asynchronous Terminal Adapter
ATM	Asynchronous Transfer Mode
ATU	Analog Trunk Unit
AWGN	Additive White Gaussian Noise
BASK	Binary (Basic) Amplitude Shift Keying
BER	Bit Error Rate (Ratio)
BFSK	Binary (Basic) Frequency Shift Keying
BHCA	Busy Hour Call Attempt
B-ICI	Broadband Inter-Carrier Interface
B-ISDN	Broadband-ISDN
BIB	Backward Indication Bit
BIP	Bit-Interleave Parity-check
BNZS	Bipolar N (N = 3, 6, 8) Zero Substitution (e.g., B3ZS, B8ZS)
BORSCHT	Battery, Over-voltage, Ringing, Supervision, Codec, Hybrid, and Testing
BPF	Band Pass Filter
bps or b/s	bits per second
BPSK	Binary (Basic) Phase Shift Keying
BR-ISDN	Basic Rate ISDN
BRI	Base Rate Interface
BRT	Business Remote Terminal
BS	Base Station
BSN	Backward Sequence Number
BSS	Base Station System
BSS	Broadband Switched Service
BSS	Broadband Switch System
BSTJ	Bell System Technical Journal
C.E./G.S.	Constellation Encoder/Gain Scaling
C/R	Command/Response
CAP	Carrier AM/PM
CAC	Connection (or Call) Admission Control
CAD	Computer Aided Design
CAI	Common Air Interface
CAM	Computer Aided Manufacture
CAM	Carrierless Amplitude Modulation
CAMA	Centralized Automatic Message Accounting
CAP	Computer Aided Topo-graphical
CAP	Carrierless Amplitude/Phase
CAT	Condition Access Table
CATU	Central Access and Transcoding Unit
CATV	Cable Television (TV)
CBOS	Computer Based Operation System
CBR	Constant Bit Rate
CC	Country Code
CCBS	Completion Call Busy Subscriber
CCC	Clear Channel Capability
CCIR	International Radio Consultative Committee
CCS	Common Channel Signaling
CCS7	Common Channel Signaling System

CCT	Continuity Check Transceiver
CCU	Call Control Unit
CD	Call Deflection
CD	Code Disk (with prestored information, data, voice and video)
CD	Channel Decoder
CDMA	Code Division Multiple Access
CDN	Cellular database Node
CDN	Cell Distribution Network
CDR	Customer Digital Receiver
CDS	Code Division Switch
CDX	Compact Digital eXchange
CE	Channel Encoder
CEI	Connection Endpoint Identifier
CELP	Code Excited Linear Predictor
CEPT	Conference of European Posts and Telecommunications
CF	Call Forwarding
CFB	Call Forward Busy
CFNR	Call Forward No Reply
CFU	Call Forward Unconditional
CI	Customer Installation
CI	Cable Interface
CIC	Circuit Identification Code
CK	Check Field
CLID	Calling Line Identification
CLP	Cell Loss priority
CM	Communications Module
cm	centimeters
CMCU	Communications Module Control Unit
CNI	Common Network Interface
CNI	Carrier Network Interface
CNO	Customer Network Option
CO	Central Office
CON	Connect
COT	COnTinuity
COT	Central Office Terminal
CP	Customer Premises
CP	Central Processor
CPE	Customer Premises Equipment
CPG	Call ProGress
CPN	Calling Party (Identification) Number
CRC	Cyclic Redundancy Check
CRC	Channel Control Code
CRED	Credit Card Calling
CREG	Concentrated Range Extension with Gain
CRSAB	Centerized Repair Service Answering Bureau
CRV	Call Reference Value
CS	Call Store
CS-ACELP	Conjugate-Structure Algebraic-Code-Excited Linear-Prediction
CSA	Carrier Serving area
CSL	Component Sub-Layer
CSM	Customer Service Multiplexer
CSPDN	Circuit-Switched Public Data Network
CSS	Computer Sub-System

CSU	Channel Service Unit
CT	Connection Type
CT	Call Transfer
CT	Customer Transceiver
CTRU	Central Transceiver Unit
CTU	Channel Transmitter Unit
CU	Customer Unit
CUG	Closed User Group
CW	Call Waiting
D/A	Digital-to-Analog (conversion)
DAS	Dual Attached Station
dB	Decibel
dBm	dB with respect to 1 mw
DBC	Digital Broadband Capability
DC or D.C.	Direct Current
DCB	Digital Channel Bank
DCF	Dispersion Compensating Fiber
DCIU	Data Communications Interface Unit
DCP	Digital Communications Protocol
DCS	Digital Cross-connection System
DCS	Digital Cellular Switch
DCTU	Directly Connected Test Unit
DCU	Digital Channel Unit
DDD	Direct Distance Dialing
DDI	Direct Dialing In
DE	Dielectric
DeE	De-encryptor
DeS	De-scrambler
DF-SMF	Dispersion-Flattened Single Mode Fiber
DFB	Distributed Feedback
DFI	Digital Facility Interface
DID	Direct Inward Dialing
DIF	Digital Interface Frame
DLC	Digital Loop Carrier
DLCI	Data Link Connection Identifier
DLTU	Digital Line and Trunk Unit
DM	Delta Modulation
DMT	Discrete Multitone
DN	Directory Number
DNS	Domain Name System
DOD	Direct Outward Dialing
DP	Dial Pulse
DPC	Destination Point Code
DPCM	Differential PCM (Pulse Code Modulation)
DPSK	Differential PSK (Phase Shift Keying)
DQDB	Distributed Queue Dual Bus
DS	Dispersion Shift
DS-F	Dispersion Shifted Fiber
DS-SMF	Dispersion-Shifted Single Mode Fiber
DS1	Digital Signal level 1
DSA	Digital Switched Access
DSC	Dispersion Slope Compensation

DSD	Dial Service Database
DSD	Direct Signaling Database
DSI	Digital Speech Interpolation
DSL	Digital Subscriber Line
DSLAM	DSL Access Module
DTE	Data Terminal Equipment
DTMF	Dual Tone Multi-Frequency
DTV	Digital TV
DUP	Data User Part
DVC	Desktop Video Conferencing
DWDM	Dense Wavelength Division Multiplexing
E/O	Electrical to Optical (conversion)
EA	Extended Address
EA	Express Amplifier
EC	Echo Canceler
ECH	Echo Canceler with Hybrid
ECMA	European Computer Manufactures Association
ECSA	Exchange Carrier Standards Association
EDFA	Erbium-Doped Fiber Amplifier
EDFF	Erbium Doped Fluoride Fiber
EDFFA	Erbium Doped Fluoride Fiber Amplifier
EIA	Electronics Industry Association
ELP	Excited Linear Predictor
EM	Error Monitoring
EN	Encryptor
EOC	Embedded Operation Channel
ESF	Extended Super Frame
ESPCS	Enhanced Private Switched Communications Services
ESS	Electronics Switching System
ETN	Electronic Tandem Network
ETSI	European Telecommunications Standards Institute
EXM	Exchange Multiplexer
FA	Frame Alignment
FAA	FAcility Accepted
FAR	FAcility Request
FBG	Fiber Bragg Grating
FCS	Frame Check Sequence
FDD	Frequency Division Duplexing
FDDI	Fiber Distributed Data Interface
FDHM	Full Duration Half Maximum
FDM	Frequency Division Multiplexing
FDMA	Frequency Division Multiple Access
FDX	Full-Duplex
FEC	Forward Error Correction (Control)
FEXT	Far-End Crosstalk
FFT	Fast Fourier Transform
FGA	Feature Group A
FGB	Feature Group B
FGC	Feature Group C
FGD	Feature Group D
FIB	Forward Indication Bit

FIFO	First-In-First-Out
FISU	Fill-In Signal Unit
FITL	Fiber In The Loop
FLAG	Fiber Link around the Globe
FM	Frequency Modulation
FNEXT	Foreign Near-End Crosstalk
FOT	Forward Transfer
FPOS	First Point Of Switching
FRJ	Facility ReJect
FSK	Frequency Shift Keying
FSN	Forward Sequence Number
ft	feet
FTTC	Fiber To The Curb
FTTH	Fiber To The Home
FTTN	Fiber To The Neighborhood
FTTO	Fiber To The Office
FTTS	Fiber To The Subdivision
FWM	Four-wave Mixing
FX	Foreign Exchange
GC	General Communication
GDSU	Global Digital Service Unit
GEO	Geostationary Orbit
GFC	Generic Flow Control
GOS	Grade of Service
GRIN	GRaded INdex
GRQ	General ReQuest
GSDN	Global SDN (Software Defined Network)
GSM	General Setup Message
GUI	Graphic User Interface
HDB3	High Density Bipolar 3
HDLC	High-level Data Link Control
HDSL	High-speed Digital Subscriber Line
HDT	Host Digital Terminal
HDTV	High Definition TV
HEC	Header Error Control
HFC	Hybrid Fiber Coax
HLR	Home Location Register
HMDF	Horizontal Main Distributing Frame
HPF	High Pass Filter
HSM	Host Switching Module
HTML	Hypertext Makeup Language
HTTP	Hyper Text Transfer Protocol
HTU	HDSL Terminal Unit
HTU-C	HDSL Terminal Unit at Center office
HTU-R	HDSL Terminal Unit at Remote
IAM	Initial Address Message
IB	Indicator Bit
ICC	International Carrier Company
ICMP	Internet Control Message Protocol
ICW	Intelligent Communications Workstation

IDDD	International Direct Distance Dial
IDFT	Inverse Discrete Fourier Transform
IDL	Integrated Digital Loop
IDLC	Integrated Digital Loop Carrier
IDN	Integrated Digital Network
IEC	International Electro-technical Commission
IEC or IXC	Inter-Exchange Carrier
IEEE	Institute of Electrical and Electronics Engineering
IETF	Internet Engineering Task Force
IFFT	Inverse Fast Fourier Transform
ILD-T	International Long Distance Translation
IM	Input Module
IM&M	Information Movement and Management
IN	Intelligent Network
INF	INFormation
INI	Inter Network Interface
INR	Information Request
IP	Internet Protocol
IrDA	Infra-Red Data Association
ISC	International Switching Center
ISDN	Integrated Services Digital Network
ISDN-UP	Integrated Services Digital Network User Part
ISI	Inter Symbol Interference
ISLU	Integrated Services Line Unit
ISO	International Organization for Standardization
ISP	Internet Service Provider
ISPC	International Signaling Point Code
ITE	International Transit Exchange
ITM	Intelligent Telecommunications Management
ITS	Intelligent Telephone Socket
ITU	International Telecommunications Union
ITU-R	ITU-Radio Communications Sector (Formally CCIR)
ITU-T	ITU-Telecommunications Standardization Sector (Formally CCITT)
IXC or IEC	Inter-Exchange Carrier
kbps or kb/s	kilo-bits per second
km	kilometer
KTS	Key Telephone System
KTU	Key Telephone Unit
L	Length
LA	Launch Amplifier
LADS	Local Area Data Set
LAMA	Local Automatic Message Accounting
LAN	Local Area Network
LAPB	Link Access Protocol Balanced
LAPD	Link Access Procedure on the D-channel
Laser	Light amplification by stimulated emission of radiation
LASS	Local Area Signaling Service
LATA	Local Access and Transport Area
LAU	Link Adapter Unit
LBO	Line Built Out (Lightwave Built Out)
LD	Laser Diode

LDSU	Local Digital Service Unit
LE/S	Line Extended with Splitter
LEC	Local Exchange Carrier
LEC	LAN Emulation Client
LEC	Local Exchange Central
LED	Light Emitting Diode
LEO	Low-Earth Orbit
LES	LAN Emulation Sever
LF	Low Frequency
LH	Line Hunting
LI	Length Indicator
LMDS	Local Multipoint Distribution System
LMOS	Local (loop) Maintenance Operating system
LMT	Local Maintenance Terminal
LOH or MSOH	Line (Multiplex Section) Overhead
LPF	Low Pass Filter
LRN	Local Reference Number
LSB	Least Signification Bit
LSSU	Link Status Signal Unit
LT	Local Tandem
LTE	Line Terminating Equipment
LTP	Line Trunk Peripheral
LU	Line Unit
LXE	Lightguide Express Entry
m	meter
MAAP	Maintenance and Administration Panel
MAC	Medium Access Control
MAS	Multiple Access System
Mbps or Mb/s	Mega-bits per second
MCC	Master Control Center
MCTU	Module Control and Timeslot interchange Unit
MCU	Multipoint Control Unit
MDF	Main Distribution Frame
MDSL	Moderate rate DSL
MF	Multi-Frequency
MF	Medium Frequency
MFC	Multifrequency Code
MFC	Multi-Frequency Control
MFJ	Modified Final Judgment
MFT	Multi-Frequency Transmitter
MLM	Multiple Longitudinal Mode
MLT	Mechanized Loop Testing
MMDS	Multichannel Microwave Distribution Service
MMF	Multi-Mode Fiber
MMP	Module Message Processor
MMRSM	Multi-Module Remote Switching Module
MMSU	Module Metallic Service Unit
MOS	Mean Opinion Score (Mean Open Score)
MPEG-2	Moving Picture Expert Group version 2
M-PSK or DPSK	M-ary PSK or Differential PSK (Phase Shift Keying)
MA	Maintenance and Adaptation
ME	Metallic

MIFL	Maximum Individual Fiber Loss
MLM	Multi-Longitudinal Mode
MS	Multiplex Section
MSB	Most Significant Bit
MSC	Mobile Switching Center
MSC	Mode Selective Coupler
MSCU	Message Switch Control Unit
MSDT	Multi Service Distance Terminal
MSISDN	Mobile Subscriber ISDN
MSPU	Message Switch Peripheral Unit
MSRN	Mobile Subscriber Roaming Number
MSU	Message Signal Unit
MTP	Message Transfer Part
MTS	Message Telecommunications Service
MTSC	Mobile Telephone Switching Center
MTSO	Mobile telephone Switching Office
MVL	Multiple Virtual Line
MZI	Mach-Zehnder Interferometric
N/S	Noise power/Signal power
NANP	North American Numbering Plan
NBS	National Bureau of Standards
NCC	Network Control Center
NCD	Network Call Denial
NCP	Network Control Point
NCT	Network Control and Timing
NDC	Network Destination Code
NE	Network Element
NEXT	Near-End Crosstalk
NGLN	Next Generation Lightwave Network
NI	Network Interface
NID	Network Interface Device
NIT	Network Information Table
NIU	Network Interface Unit
NNI	Network Node Interface
NOB or NR	Network Operator Byte
NOC	Network Operations Center
NOLM	Nonlinear Optical Loop Mirror
NPC	Network Parameter Control
NRZ	Non-to-Return-to-Zero (waveform)
NSC	Network Service Complex
NSN	National Significant Number
NSP	Network Services Part
NT	Network Terminal
NT	Network Timing
NTR	Network Timing Reference
O/E	Optical to Electrical (conversion)
OA&M	Operations, Administration and Maintenance
OADM	Optical Add-Drop Multiplexer
OAM&P	Operation, Administration, Maintenance and Provisioning
OC-N	Optical Carrier level N
OH	Overhead

OM	Output Module
OMC	Operation Maintenance Center
ONI	Operator Number Identification
ONU	Optical Network Unit
OPC	Originating Point Code
OPC	Operator Position Controller
OPS	Operator Position System
OPSM	Operator Position Switching Module
OPSM	Output Packet Switch Module
OS	Operation System
OSI	Open System Interface (Interconnection)
OSPS	Operator Service Position System
OTDR	Optical Time-Domain Reflectometer
PAM	Pass Along Message
PAM	Pulse Amplitude Modulation
PAT	Program Association Table
PBX	Private Branch Exchange
PC	Point Code
PC	Protection Channel
PCM	Pulse Code Modulation
PCR	Program Clock Reference
PCS	Personal Communications Service
PD	Payload Dependent
PDF	Probability Density Function
PDH	Plesiochronous Digital Hierarchy
PDN	Passive Distribution Network
PDU	Protocol Data Unit
PES	Program Element Stream
PH	Protocol Handler
PHRISLU	Protocol Handler and Remote Integrated Service Line Unit
PIDB	Peripheral Interface Data Bus
PIN	Personal Identification Number
PIN	Positive, Intrinsic & Negative
PLMN	Public Land Mobile Network
PLN	Private Line Network
PLP	Packet Level Procedure
PM	Performance Monitoring
PM	Phase Modulation
PMD	Polarization Mode Dispersion
PMT	Program Map Table
PNP	Private Numbering Plan
POH	Path Overhead
POP	Point Of Presence
POT	Point Of Termination
POTS	Plain Old Telephone Service
PPM	Periodic Pulse Metering
ppm	parts per million
PPMU	Periodic Pulse Metering Unit
PPP	Point-to-Point Protocol
PRI	Primary Rate Interface
PS	Program Store
PSD	Power Spectral Density

PSIU	Packet Switch Interface Unit
PSK	Phase Shift Keying
PSPDN	Packet-Switched Public Data Network
PSTN	Public Switched Telephone Network
PSU	Packet Switch Unit
PTE	Path Terminating Equipment
PTI	Payload Type Identifier
PTS	Public Telephone Service
PVC	Permanent Virtual Connection
QAM	Quadrature Amplitude Modulation
QoS	Quality of Service
QPSK	Quadrature Phase Shift Keying
RADSL	Rate-Adaptive DSL
RDI	Remote Defect Indication
RDSF	Reduced Dispersion Slope Fiber
REI	Remote Error Indication
REL	RELease
RES	RESume
REV	REVerse charging (is under further study)
RF	Radio Frequency
RFC	Request for Comment
RFI	Radio Frequency Interference
RL	Rodent Lightning
ROSC	Remote Operator Service Center
RPE-LTP	Regular Pulse Excited-Long Term Predictor
RPOA	Recognized Private Operating Agencies
RS	Regenerator Section
RS	Reed-Solomon (Error Control Code)
RSM	Remote Switching Module
RT	Remote Terminal
RTP	Real Time Protocol
Rv	Receiver
RZ	Return-to Zero (wavrform)
S&PRU	Sync and Paging Receiver Unit
S/N	Signal power/Noise power (Signal-to-Noise ratio)
SAC	Serving Area Configuration
SAM	Subsequent Address Message
SANC	Signaling Area Network Code
SAPI	Service Access Point Identifier
SAS	Switched Access Services
SAS	Single Attached Station
SBTU	Satellite Broadcast Transmitter Unit
SC	Synchronization Control
SCA	Selective Call Acceptance
SCC	Switching Control Center
SCCP	Signaling Connection Control Part
SCF	Selective Call Forwarding
SCM	Synchronous Carrier Module
SCN	Service Control Node
SCP	Service Control Point

SCR	Selective Call Rejection
SD	Speech Decoder
SDA	Selective Distinct Alerting
SDH	Synchronous Digital Hierarchy
SDMT	Synchronized Discrete Multitone
SDN	Software Defined Network
SDS	Switched Digital Service
SDSL	Single line DSL
SDV	Switched Digital Video
SE	Subscriber Equipment
SE	Speech Encoder
SE-CDMA	Spectrally Efficient CDMA
SEP	Signaling End Point
SF	Single Frequency
SF	Status Field
SG	Study Group
SHF	Super High Frequency
SI	Step Index (Fiber)
SIF	Signaling Information Field
SIO	Service Information Octet
SIO	Scientific and Industrial Organization
SLC	Subscriber Loop Carrier
SLM	Single Longitudinal Mode, Synchronous Line Multiplexer
SLS	Signaling Link Selection
SLTE	Submarine Line Terminating Equipment
SM	System Management
SM	Switch Module
SM	Service Module
SMDS	Switched Megabit Data Service
SMF	Single Mode Fiber
SMM	Service Management Module
SMP	Switching Module Processor
SMS	Service Management System
SMTP	Simple Mail Transfer Protocol
SN	Subscriber Number
SNEXT	Self Near-End Crosstalk
SNMP	Simple Network Management Protocol
SOA	Semiconductor Optical Amplifier
SOH	Section Overhead
SOH or RSOH	Section (Regenerator Section) Overhead
SONET	Synchronous Optical Network
SOP	States-of-Polarization
SP	Signaling Point
SPC	Stored Program Control
SPC	Signaling Point Code
SPM	Subscriber Pulse Metering
SPR	Signaling Point Relay
SQ	Stuffing Quat
SS/CDMA	Satellite Switched/CDMA
SS7	Signaling System No.7
SSBAM	Single Side Band AM (Amplitude Modulation)
SSM	Switched Service Module
SSP	Service Switching Point

SSR	Side-mode Suppression Ratio
STE	Section Terminating Equipment
STM	Synchronous Transfer Mode
STM-N	Synchronous Transport Module level N
STP	Signaling Transfer Point
STRU	Subscriber Transceiver Unit
STS	Synchronous Transport Signal
STS-N	Synchronous Transport Signal level N
SU	Subscriber Unit
SUS	SUSpend
SVC	Switched Virtual Connection
SVC	Switched Virtual Circuit
T/R	Transmitter/Receiver
TAB	Telephone Answering Bureau
TASI	Time Assignment Speech Interpolation
TC	Transaction Capabilities
TC	Transmission Convergence
TCAP	Transaction Capabilities Application Part
TCM	Trellis Coded Modulation
TCP	Transmission Control Protocol
TCP/IP	Transmission Control Protocol/Internet Protocol
TCTU	Traffic Channel Transmitter Unit
TDD	Time Division Duplexing
TDM	Time Division Multiplexing
TDMA	Time Division Multiple Access
TDR	Time Domain Reflectometer
TDS	Time Division Switching
TEI	Terminal Endpoint Identifier
TIE	Terminal Interface Equipment
TIU	Terminal Interface Unit
TLP	Transmission Level Point
TM	Timing Marker
TMF	Two Mode Fiber
TMS	Time Multiplexed Space (switch)
TMS	Time Multiplex Switch
TPD	Transmission Performance Designator
TR	Trail Tracing
TSI	Time Slot Interchange
TSL	Transaction Sub-Layer
TTL	Transistor-to-Transistor
TU	Trunk Unit
TU	Tributary Unit
TUP	Telephone User Part
TV	Television
Tx	Transmitter
UDP	User Datagram Protocol
UHF	Ultra High Frequency
UNI	User-to-Network Interface

UP/AP	User Part/Application Part
UPC	Usage Parameter Control
UPS	Uninterruptible Power Supply
UPT	Universal Personal Telecommunications
URP	Universal Receiver Protocol
USR	User-to-UseR
UUI	User-to-User Information
UUS	User-to-User Signaling
VBR	Variable Bit Rate
VC	Virtual Channel
VCC	Virtual Channel Connection
VCDX	Very Compact Digital eXchange
VCI	Virtual Channel Identifier
VCS	Virtual Circuit Switch
VDSL	Very high-speed Digital Subscriber Line
VF	Voice Frequency signal
VHF	Very High Frequency
VLF	Very Low Frequency
VLR	Visitor Location Register
VLSI	Very Large Scale Integrated
VMDF	Vertical Main Distributing Frame
VoIP	Voice over IP
VP	Virtual Path
VPC	Virtual Path Connection
VPI	Virtual Path Identifier
VPN	Virtual Private Network
VSELP	Vector Sum Excited Linear Predictor (ELP)
VT	Virtual Tributary
VTG	Virtual Tributary Group
WAN	Wide Area Network
WATS	Wide Area Telecommunications Service
WC	Working Channel
WDM	Wavelength Division Multiplexing
WGSM	Wireless Globe Switching Module
WIN	Wireless In-building Network
WLAN	Wireless Local Area Network
WLL	Wireless Local Loop
WLT	Wireless Line Transceiver
WP	Working Parties
WSM	Wireless Switching Module
WSS	Wireless Subscriber System
WWW	World Wide Web
ZBLAN	Zirconium, Barium, Lanthanum, Aluminum, and Sodium Fluorides
2B1Q	2-binary/1-quaternary
3PTY	Three ParTY

Index